Cities of Medieval Iran

Cities of Medieval Iran

Edited by

David Durand-Guédy, Roy P. Mottahedeh
and Jürgen Paul

BRILL

LEIDEN | BOSTON

Originally published as Volume 16, No. 1-2 (2018) of Brill's journal *Eurasian Studies*.

Cover illustration (from top to bottom): Ms. Köprülü 01589, Türkiye Yazma Eserler Kurumu Başkanlığı Süleymaniye Yazma Eser Kütüphanesi, fol. 116r.
Ruins of the citadel of Nishapur. Photo by David Durand-Guédy 2008.
Siege scene. Detail of an enamel painted plate, courtesy of the Freer Gallery of Art (item F1943.3).
The region of Isfahan with representation of altitude and visibility from the citadel of Kūh Buzī. ArcGIS produced map by Ali Shojaee Esfahani, 2018.

The Library of Congress Cataloging-in-Publication Data is available online at http://catalog.loc.gov
LC record available at http://lccn.loc.gov/2020019078

Typeface for the Latin, Greek, and Cyrillic scripts: "Brill". See and download: brill.com/brill-typeface.

ISBN 978-90-04-41960-5 (hardback)
ISBN 978-90-04-43433-2 (e-book)

This book is printed on acid-free paper and produced in a sustainable manner.

Contents

Preface

This volume is a reprint from *Eurasian Studies*, vol. 16, published in 2018. It is based upon the conference "Iranian Cities from the Arab Conquest to the Early Modern Period", held at Harvard University on 1 and 2 May 2015. In addition to a selection of the conference papers, the volume includes essays from scholars who were later invited to contribute.

The subject of the volume, medieval Iranian cities, has long been a favourite topic for historical research. Southwestern Iran, at the extremity of the Fertile Crescent, and Transoxiana at the northeastern end of the Iranian world, are home to some of the world's oldest cities. Sites occupied continuously since ancient times, such as Susa and Samarqand, have been the focus of more than a century of archaeological research and excavation. Many other cities such as Hamadan, Herat, Nishapur, Rayy and Tabriz have also been major centres of civilization since pre-Islamic times.

The time period considered here, from approximately 700 to 1500 CE, corresponds to the pre-modern phase of Islamic Iran. The eight centuries under discussion include such transformative events as the Arab conquest, the subsequent conversion to Islam, and the arrival of central Asian nomads during the eleventh to the fourteenth centuries, including the Mongol conquest of the thirteenth century. During this period Iranian cultures flourished, making their mark on world history. Iranian territories absorbed many peoples, first from the Arabian Peninsula and then from Central Asia, resulting in dramatic cultural changes. The continual revival of most cities following these migrations, including some of great destruction, demonstrates the vitality of urban life in the Iranian setting. The overall period encompasses both strong growth in the first centuries of Islam (accompanied by the writing of local histories), followed by centuries in which urban life was challenged and in some cases entered into decline.

The volume entitled *The Islamic City: A Colloquium*, edited by Albert Hourani and Samuel Stern, issued in 1970, dealt primarily with the Arab world – a reflection of the state of research at that time. It did include Jean Aubin's influential programmatic essay on Iran, reprinted here in English translation, with commentary. Since then, there has been continued increase in the publications dedicated to city-related matters in medieval Iran. In 1994, Masashi Haneda and Hisao Komatsu offered a synthesis of scholarship on medieval cities in present-day Iran and Afghanistan, in the volume *Islamic urban studies: history, review and perspectives*. Other useful volumes dedicated to cities have

 | DOI:10.1163/9789004434332_002

since been published, on the pre-modern Islamic world in general and on specific regions of that world (see the introductory review essay by Jürgen Paul).

This current volume deals with Iran as a whole. However, the aim is not to assemble a new general compendium, following Paul Schwarz's famous *Iran im Mittelalter* (1896-1936), or the more recent compilation of articles from the *Encyclopaedia of Islam*, gathered by C.E. Bosworth (*Historic cities of the Islamic world*) with some updated bibliography. A glance at our Table of Contents reveals that many regions are missing (Khuzistan, the Caspian region, the Caucasus, and the Persian Gulf). The cities under examination are primarily located on the Iranian plateau and in Central Asia. Not all the articles focus on specific cities: some deal with transverse problems, such as the role of irrigation in urban design or the transition from Sasanian to Islamic cities. The essays take very different approaches, and in doing so suggest still further ways in which cities can be studied. The authors clearly view Iranian cities not only as settings where history happened, but as phenomena to be understood in their own right due to their continuity across the centuries.

Many contributions consider geographical factors (and geographical writings in the original sources), bringing the relationship between city and hinterland into focus. The water supply of city and its hinterland is a central subject for Rocco Rante and for Etienne de la Vaissière in his discussion of Balkh. It also appears as an important issue in the discussions of Jamsheed Choksy on Yazd and of Roy Mottahedeh and Mehrdad Amanat on Kashan. Richard Bulliet and David Durand-Guédy, in examining the respective cases of Nishapur and Isfahan, explore the significance of the geographic setting in the positioning and history of major cities. The contributions by Ali Shojaee Esfahani and Denise Aigle focus more on features created for specific purposes in the city and its hinterland. Ali Shojaee Esfahani demonstrates how defensive works can extend far beyond the city walls, involving development of networks of fortresses and fortifications in the surrounding landscape, as at Isfahan. Denise Aigle demonstrates how the sacred landscape of Shiraz extends to the hinterland, creating a network of places interwoven with those inside the city.

The upheavals of the Arab conquest, the coming of the Turks in the eleventh century, and the Mongol invasion in the thirteenth century feature in many of the essays. Donald Whitcomb focuses on the transition from the Sasanian to the early Islamic period. He compares the specificities of the Iranian circumstances to those of the Arab Near East and discusses the events and motivations behind the changes in the shape of the city from Antiquity to the Islamic period. Etienne de la Vaissière shows that as early as the third-fourth/ninth-tenth century Muslim authors misunderstood the remnants of the pre-Islamic past and that, in the case of Balkh, there were remarkable gaps in the people's understanding of their own city's history. Jamsheed Choksy shows how in

Yazd, the pre-Islamic past was present not only in architectural remains and irrigation networks but also lived on in the religious sphere. Daniel Zakrzewski's essay on Tabriz and David Durand-Guédy's on Isfahan both bridge the Mongol period and concur that the Mongol invasion did not create a total discontinuity in the urban social-political order. In some respects, the Saljuq period in the eleventh century and Timurid period in the fourteenth century may have had more important effects.

In recent decades, a major subject of research has been the issue of cooperation and conflict between city notables and the representatives of regional or imperial powers. The current volume adds new case studies in this area. The essays by Jürgen Paul of Saljuq and post-Saljuq Balkh, by David Durand-Guédy of Isfahan, and by Daniel Zakrzewski of Tabriz all take long-term perspectives, showing that there was remarkable continuity in the patterns of what Albert Hourani has called the "politics of notables". Roy Mottahedeh and Mehrdad Amanat also stress the continuities bridging the pre and post-Mongol periods in the history of Kashan.

Many contributions take a long-term perspective, with the result that some go far beyond the nominal time frame for this volume. Rocco Rante offers a characterization of general features of Iranian cities, beginning with their foundation some two millennia ago. Similarly, Jamsheed Choksy uses the oldest written Iranian and Greek sources known for the history of Yazd, and Denise Aigle deals with the history of Shiraz and its sacred landscapes over much of the pre-Safavid period.

Many contributions bring together written sources and evidence from material remains, an approach necessitating close cooperation between the disciplines of history (including the various branches of art history, numismatics, epigraphy and others), archaeology, and geography. However, some contributions have taken still more innovative methodological approaches. Rocco Rante introduces geomorphology, examining how humans have changed the landscape in permanent fashion and how human activities can depend on the prevailing geographical contexts of the different periods. Ali Shojaee Esfahani applies GIS software to reveal and gain understanding of the motivations behind construction of different fortifications at their specific locations. Roy Mottahedeh's essay on dictionaries draws on lexicography as an instrument for historical study: an altogether new approach as far as the history of Iranian cities is concerned. Finally, Sarah Savant and Majid Mahdi show how manuscript heritage can be used as a source for the social history of a given city, in this case Timurid Shiraz.

The organisation of the conference and the publication of this volume have been made possible by the generous contribution of an Iranian-American businessman, whose friendship with a faculty member of Harvard University led to his interest in supporting a project on Iranian history sponsored by the university's Center for Middle Eastern Studies.

It is our pleasure to recognise the strong support of the staff of the Center for Middle Eastern Studies of Harvard University, in particular Lauren Montague and Elizabeth Flanagan, and to thank the Director, William Granara, for the general support and funding received from the Center. We also wish to thank Neal Putt, our copy editor, for his painstaking and meticulous work. In addition, we extend our warm appreciation to Michele Bernardini and the team of *Eurasian Studies* for their outstanding technical support.

David Durand-Guédy (Tehran)
Roy P. Mottahedeh (Cambridge, USA)
Jürgen Paul (Hamburg)
4 July 2018

Note on Transliteration

Proper names are transliterated as if written Arabic (the rule of elision applies). Words, sentences, titles of book and verses from Persian texts are transliterated according to the rules for *fārsī*. Words of Turkish or Mongol origin are written using a more elaborate vowel system (hence Hülegü). Well-known Iranian place names are not transliterated (hence, Khurasan, Bukhara, Istakhr, Herat), nor the names of dynasties (hence, Timurids).

Note on This Edition

We extend our thanks to Brill Publishers who suggested to reprint this work as a book. We have changed only a small number of typos and misprints. Some maps have been enhanced for clarity. The pagination has not changed, except for marginal text overruns on 28 pages. Therefore, the journal edition and the present book can be used and quoted without problems. The only major change is that this edition has indexes (prepared by Laura Emunds).

Cities in Medieval Iran: a Review of Recent Publications

Jürgen Paul
Martin-Luther-Universität Halle-Wittenberg

Two monumental compilations emerging at the beginning of the current century marked the then state of the art on cities in the Muslim world.[1] Wheatley[2] was interested in the hierarchy of towns and cities in the Islamic Empire, and so based his study on medieval (ninth-tenth centuries CE) 'Arab geographers'. Here we learn much about the ways scholars and administrators of the Abbasid empire perceived and structured the larger living space. In individual chapters dealing with the regions, Wheatley discusses aspects of urban life: market and service centres, the foci of transport, industrial and craft centres, religious centres, fortified settlements. Many sections supplement or even substitute the information contained in the entries of the Encyclopaedia of Islam, Second Edition.[3] The author also deals with certain unique features, among these the coastal fortifications along the Persian Gulf littoral, in the chapter on Fars (p. 150).

The other monumental compilation, Wirth's *summa*, concentrates more on contemporary cities, but also deals with the classical Islamic era. The volume again covers the entire Muslim world (excluding peripheral regions such as India and sub-Saharan Africa) and is written from a geographical perspective.[4] Wirth's leading question has by this time become almost a tradition in itself: whether we can meaningfully speak of something called the 'Islamic city'.[5] Briefly put, he concludes that the term is misleading: the characterizing features of these cities, apart from the central mosque, were in place before the

1 This review focuses on publications dealing with Iranian cities prior to 1500 CE, which is the subject era for the entire current volume. Regretfully, it deals only with publications in Western European languages (English, French, German). Persian and Central Asian languages are omitted for practical reasons. I have not found relevant recent publications in Russian.

2 Wheatley, Paul, *The Places Where Men Pray Together: Cities in Islamic Lands, Seventh Through the Tenth Centuries* (Chicago: Chicago University Press, 2001).

3 For example, sections on: Nishapur: pp. 305-8; Balkh pp. 308-9; Zaranj in Sīstān pp. 310-2.

4 Wirth, Eugen, *Die orientalische Stadt im islamischen Vorderasien und Nordafrika*. 2 vols. (Mainz: Philipp von Zabern, 2000).

5 See my introductory remarks to the translation of Jean Aubin's article, this volume.

 | DOI:10.1163/9789004434332_003

coming of Islam. Wirth therefore proposes the alternative, more encompassing term of "Oriental city". However, his proposal has not gained broad acceptance, and indeed more recent research has bypassed any such questions of macro-sociological classification or analysis. The focus of studies has instead been on individual cities, and has been accompanied by the development of new lines of research, particularly those involving cooperation between archaeology and history.

No major summarising publications on cities in Muslim or medieval Iran have emerged since the works by Wheatley and Wirth, however Gaube did provide a synthesising article, in 2008.[6] The author inquires into the general layout of the city, comparing round versus rectangular grid layouts, thoroughfares and open spaces versus irregular structures, and dead ends and houses with inner courtyards. Isfahan and Herat receive the most attention, and Gaube concludes that Iranian cities were frequently planned and founded with a circular perimeter around a rectangular grid. As a possible origin of such regular rectangular planning, he examines the fortified settlements known as *qalʿa* or *kalāt*, widespread in pre-Islamic Iran and still an important feature in the Medieval and Early Modern eras.

In recent years the archaeological sector has been quite productive of publications on Iranian cities, while we have seen somewhat less from historians. As noted, there has also been a trend towards the combination of written and archaeological sources in scholarship, as well as more frequent attempts to broaden the perspective beyond the city alone, taking account of the hinterland and oasis.

There have been only a limited number of monographs on individual cities in medieval Iran, with the exception of Herat (in particular Timurid Herat),[7] and only four books on specific cities have appeared: David Durand-Guédy's study on pre-Mongol Isfahan;[8] John Limbert on Shiraz in the age of Ḥāfiẓ (fourteenth century);[9] more recently, Arezou Azad on Balkh, likewise in the pre-Mongol

6 Gaube, Heinz, "Iranian cities", in Jayussi, Salma K., Holod, Renata, Petruccioli, Attilio, and André Raymond (eds.), *The City in the Islamic World* (Leiden: Brill, 2008): pp. 159-80. In this article, Gaube takes up much of his argument from his *Iranian Cities* (New York: New York University Press, 1979).

7 References are provided in the Introduction to Jean Aubin, this volume.

8 Durand-Guédy, David, *Iranian Elites and Turkish Rulers: a History of Iṣfahān in the Saljūq Period* (London: Routledge, 2010).

9 Limbert, John, *Shiraz in the Age of Hafez: The Glory of a Medieval Persian City* (Seattle-London: Washington University Press, 2004).

period,[10] and finally, Drechsler's study of the history of Qum. Drechsler's work is in fact not so recent but deserves attention. Although published in 1999, it has rarely been mentioned.[11] Also in this group of 'single city' volumes, we have the inter-disciplinary endeavour of Gangler et al., on Bukhara.[12]

Durand-Guédy's book, appearing in 2010, amounted to a benchmark in many respects. The author makes use of all available written sources, which in the case of Isfahan means chronicles and historiography, local and general biographical dictionaries, geography, poetry, as well as the compilations known as *inšāʾ*: copies and drafts of official correspondence and documents. Taking a radically local perspective, he sets out to write a 'non-dynastic' history, putting local actors in their proper place as fully active participants. The local context is not a mere setting for events. Durand-Guédy's stresses the importance of urban notables, and in many contexts of urban militiamen. We see that Isfahani militiamen, led by a faction of urban notables, contributed to the victory over the Ismāʿīlīs, dislodged from the fortress of Šāhdiz at the opening of the twelfth century. We learn that it was again the support of the notables that made the second career of Jalāl al-Dīn Mangburnī possible, during the first Mongol invasions of the 1220s. The actors at the centre of this analysis are urban, with the rulers considered on the margins. The author provides maps of the city, and of the network of trade routes between Isfahan and regions near and far, as well as genealogical tables for the leading houses of urban notables, related to those of the regional and imperial dynasties. In taking these approaches, the volume qualifies as one of the very first truly urban histories of Iranian cities.

Some years earlier, Limbert had focused on fourteenth century Shiraz, also in quite holistic fashion. City notables are a major topic. A very useful chart of *qāḍīs*, from the Būyids to the Mongols, shows the potentials of this kind of presentation, even with the presence of lacunae. The most important families

10 Azad, Arezou, *Sacred Landscape in Medieval Afghanistan: Revisiting the Faḍāʾil-i Balkh* (Oxford: Oxford University Press, 2013). The groundwork for all research on the history of Balkh and its oasis had been laid previously, by Robert McChesney: *Waqf in Central Asia: Four Hundred Years in the History of a Muslim Shrine, 1480-1889* (Princeton: Princeton University Press, 1991).

11 Drechsler, Andreas, *Die Geschichte der Stadt Qom im Mittelalter (650-1350): politische und wirtschaftliche Aspekte* (Berlin: Klaus Schwarz, 1999).

12 Gangler, Anette; Gaube, Heinz, and Attilio Petruccioli (eds.), *Bukhara, the Eastern Dome of Islam: Urban Development, Urban Space, Architecture and Population* (Stuttgart: Axel Menges, 2004).

are listed and traced, for at least several generations.[13] The treatment of the city quarter as basic building block of urban society, including the aspect of the "headman of the quarter", is a very welcome consideration. We begin to understand the essentiality of these figures in operating the bazaar and the entire neighbourhood, both of which are essential structures of urban social life. The headmen also seem involved in organizing the city militia, and like Durand-Guédy, Limbert stresses the importance of these urban fighting forces. The sacred landscape, composed of many holy places and structures within and around the city, again relates to the city quarter, but the author has been unable to provide details given the scarcity of information in the sources. Limbert concludes (p. 96): "The sources draw a picture of fourteenth-century Shiraz that is not one of orderly hierarchies, pyramids and webs. Rather, it is a picture of overlapping and undefined jurisdictions that changed according to the personalities of the holders of various offices". This is in fact a very similar picture to what emerges from Durand-Guédy's study of pre-Mongol Isfahan.

Azad's intention, with her 2013 volume, was primarily to examine a specific source: the *Fażā'il-i Balkh*. She draws on this to describe the creation of a Muslim sacred landscape, a process in which the city and the oasis were Islamised through a 'new narrative', which involved the reinterpretation of pre-Islamic Buddhist monuments.[14] Azad focuses on the urban notables, in this case Hanafi scholars and judges, providing a list of local *qāḍī*s and the genealogical tables of two scholarly lineages. She also projects the medieval monuments and the contours of Balkh's successive stages onto a modern map (plate 12). In the short time since publication, this work has already stimulated new research into the city and oasis of Balkh.[15]

Drechsler, writing in 1999, also based his study of medieval Qum on one main source, but the *Ta'rīḫ-i Qum* differs from the *Fażā'il-i Balḫ*, in offering a wealth of political and economic data compared to what is found in typical local historiography. Drechsler's treatment of the Abbasid period thus abounds in detail, while the coverage from the mid-tenth century onward

13 Limbert mentions a family prominent in the Būyid period, which he calls the *mard-āsā* (p. 80). He reports being unable to trace it. The name cited would point to an Iranian origin, however the family is in fact called the Mirdāsiyān in other sources, and its origins are Arab. This is indeed one of the most important families of the ninth century. See Paul, Jürgen, "Who were the '*mulūk Fārs*'?", in Stefan Heidemann, Hannah-Lena Hagemann (eds.), *Transregional and Regional Elites: Connecting the Early Islamic Empire*, (Berlin: de Gruyter, in press, 2018).

14 This is also seen in the newer historiography of other Iranian cities: e.g. below, regarding Ghazni.

15 See de la Vaissière and Paul in this volume; and articles by de la Vaissière and Wordsworth as discussed below.

suffers in comparison. The *Taʾrīḫ-i Qum* is revealed as a unique but challenging source for study of on-the-ground taxation processes. Taking this, Drechsler completed a remarkable work in disentangling the frequently arcane text. He very clearly shows how the local notables and landowners cooperated with central authorities, and from their side, how the authorities depended on this cooperation to bring in revenues. From this examination, we learn that even at such early times, the apparatus of the central administration necessarily operated in co-function with local social structures. As with later works by other authors, Drechsler compiles useful lists: governors, tax agents, local leaders, *qāḍī*s, headmen of sayyid families.

Bukhara – the Eastern Dome is by three authors: the historian and archaeologist Heinz Gaube; Anette Gangler, a specialist in urban planning; and Attilio Petruccioli, architect.[16] Gaube was the author for the chapters dealing with the pre-Mongol era. Other chapters, credited to Gangler and Petruccioli, deal with the city from the sixteenth century onward, beyond the period of interest for our current volume. Gaube provides a summary of the years of Soviet archaeological research. Again, his main interest is the urban layout, with very much the same results as an earlier article and book, published by Gaube as part of an edited volume on the theme of the 'Islamic City'. In his earlier 1979 and 2008 work, the author had identified an "eastern Iranian" type of city: with a square plan, rectangular grid, one gate in each of the four walls (although at Bukhara the west wall alone had four gates), and a citadel, which might be part of the defensive wall or stand at some distance from the city proper (as at Bukhara).[17] The research methods for the Bukhara volume were remarkable. For example, the interdisciplinary team developed a map based on elevations recorded throughout the city. This revealed two mounds standing in the oldest parts of the city,[18] and on this basis, Gaube advances the argument that Bukhara could have grown out of a twin settlement.[19]

All these books have chapters introducing the city and its landscape context, and discuss the needs for irrigation, and systems used to divert, store, and distribute water. They describe the built environment, including the city walls. Durand-Guédy and Azad provide maps; Drechsler offers a sketch showing the region of Qum; Limbert provides no maps.[20] But the main focus of most of

16 Gangler, Gaube, Petruccioli, *Bukhara*.

17 See note 6 for references to Gaube's publications, both entitled "Iranian Cities".

18 Gangler et al., *Bukhara*: fig. 4.03 on page 41, the map: fig. 4.02 on page 38.

19 Compare this to Aubin's proposition, this volume, where he talks about the possibility two have two "kernels" at the origin of an urban settlement.

20 Compensating for Limbert, Aigle does provide a map of Shiraz, in her contribution to this volume.

these works is on the social history of the city. Given the fact that the sources rarely speak of the middle and lower social strata, instead concentrating on the notables, it is no surprise that the research then tends to concentrate on this group of people. However in some cases the biographical dictionaries do offer a wealth of additional prosopographic information, which historians could put to good use.[21]

As noted, in recent decades, scholarship seems to have made more progress in the archaeology of Iranian cities than in their history. Substantial attention has been devoted to cities now outside the borders of Iran, such as Balkh, Ghazni and Marw, Bukhara and Samarqand, but also to cities on the plateau: Rayy and Nishapur. The paragraphs that follow will discuss the monographs by Georgina Herrmann on Marw,[22] and by Rocco Rante on Rayy.[23] *Eurasian Studies* has published a dossier on Ghazni with four contributions,[24] and mention will be made of a much older monograph on Gurgān.[25] Bukhara is the subject of a number of recent projects, but as yet there has been no news of a major publication.[26] There has been much archaeological work at Samarqand,

21 Note that the *Ta'rīḫ-i Qum* and *Fażā'il-i Balḫ* arrive to us as 'atypical' biographical dictionaries. The purely biographical part of the former work is completely lost. The *Fażā'il-i Balḫ* has only 73 entries, which are heavily oriented towards a pro-Ḥanafī hagiographical style. The extant biographical dictionaries for Isfahan end too early to serve in the study of the Seljuq period. In spite of this, the authors discussed here have been able to prepare worthy monographs on the city. This should then also be possible in other cases, even with the lack of city-specific sources.

22 Herrmann, Georgina, *Monuments of Merv: Traditional buildings of the Karakum* (London: The Society of Antiquaries of London, 1999).

23 Rante, Rocco, *Rayy: from Its Origins to the Mongol Invasion, an Archaeological and Historiographical Study* (Leiden: Brill, 2015).

24 *ES*, XIII (2015), with contributions: Allegranzi, Viola, Agnese Fusaro, Valentina Laviola, Martina Massullo, "Islamic Ghazni: New Data and Research Perspectives, an Introduction": pp. 1-4; Fusaro, Agnese, "From Ceramics to History: Pottery Contribution to the History of Ghazni": pp. 5-22; Allegranzi, Viola, "The Use of Persian in Monumental Epigraphy from Ghazni (Eleventh-Twelfth Centuries): pp. 23-41; Laviola, Valentina, "*Ziyārāt* of Ghazni: Three Case Studies about Marble Re-Employments": pp. 42-53; Massullo, Martina, "Marble Tombs from Ghazni (Fifteenth-Eighteenth Centuries): Morphological and Epigraphic Remarks": pp. 54-67.

25 Kiani, M. Yūssuf, *The Islamic city of Gurgan* (Berlin: Reimer, 1984 [Archäologische Mitteilungen aus Iran: Ergänzungsband, 11]). Now in print for almost 40 years, this book has gone almost unnoticed. The intention here is to draw attention to a work of value.

26 Projects under the direction of the Institute for the Study of the Ancient World (ISAW), New York University, and Musée du Louvre, Paris (project title: Mission Archéologique Franco-Ouzbèke dans l'Oasis de Boukhara (MAFOUB). The ISAW website www.isaw.nyu.edu/research/bukhara-project/results-resources-news has seen little activity since mid-2015. For MAFOUB see www.mafoub.com (last consulted June 2018).

but once again a major synthesis is still lacking.[27] A number of articles address the archaeology of individual sites: Marw,[28] Balkh,[29] Transoxiana,[30] Nishapur,[31] Rayy,[32] and Herat.[33]

The publications originating from archaeological research generally treat the city in its geographical context, typically concentrating on the oasis. They address some of the same questions as historians raise: identifying toponyms

---

27 The publication most relevant to the current review is Stride, Sebastian, Rondelli, Bernardo, and Simone Mantellini, "Canals versus Horses: Political Power in the Oasis of Samarkand", *World Archaeology*, XLI/1 (2009): pp. 73-87, discussed in the following paragraphs. Here, the city itself is seen in a marginal role. Archaeological teams operating at Samarqand, particularly the French-Uzbek teams, have issued many publications dealing with a very wide range of specialized issues: too numerous and far from the focus of this review to list. See Karev, "From Tents to City": note 30.

28 Williams, Tim, "The city of Sultan Kala, Merv, Turkmenistan. Communities, neighbourhoods and urban planning from the eighth to the thirteenth century", in Bennison, Amira K., Gascoigne, Alison L. (eds.), *Cities in the Pre-Modern Islamic World: The Urban Impact of Religion, State and Society* (London: Routledge, 2007): pp. 42-62; Wordsworth, Paul, "Sustaining Travel. The Economy of Medieval Stopping Places Across the Karakum Desert, Turkmenistan", in McPhillips, Stephen, Wordsworth, Paul D. (eds), *Landscapes of the Islamic World: Archaeology, History, and Ethnography* (Philadelphia: Penn University Press, 2018): pp. 219-36; Wordsworth, Paul, "Merv on Khorasanian trade routes from the 10th-13th centuries", in Rante, Rocco (ed.), *Greater Khorasan: History, Geography, Archaeology and Material Culture* (Berlin: de Gruyter, 2015): pp. 51-62; de la Vaissière, Étienne, "The 'Abbāsid Revolution in Marw: New Data", *Der Islam*, XCV/1 (2018): pp. 110-46; Simpson, St. John, "Merv, an archaeological case-study from the northeastern frontier of the Sasanian Empire", *Journal of Ancient History*, II/2 (2014): pp. 116-43.

29 de la Vaissière, Étienne, "Early Medieval Central Asian Population Estimates", *JESHO*, LX/6 (2017): pp. 788-817; Wordsworth, Paul, "The hydrological networks of the Balkh oasis after the arrival of Islam: a landscape archaeological perspective", *Afghanistan*, I/1 (2018): pp. 182-208. Thanks to Étienne de la Vaissière and Paul Wordsworth for sharing their papers before publication.

30 Karev, Yuri, "From Tents to City. The Royal Court of the Western Qarakhanids Between Bukhara and Samarqand", in Durand-Guédy, David (ed.), *Turko-Mongol Rulers, Cities and City-Life* (Leiden: Brill, 2013): pp. 99-147.

31 Laleh, Haeedeh, Mokarramifar, Abolfazl, and Zahra Lorzadeh, "Le paysage urbain de Nishapur", in Rante (ed.), *Greater Khorasan*: pp. 115-23.

32 Rante, Rocco, Di Pasquale, Carmen, "The Urbanisation of Rayy in the Seljūq Period", in Ansari, Hassan, Schmidtke, Sabine, and Deborah G. Tor (eds.), *The Religious and Intellectual History of Rayy from 900 through the Seljūq Period*, special issue of *Der Islam*, XCIII/2 (2016): pp. 413-32; Tor, Deborah, "Rayy and the Religious History of the Seljūq Period", ibid.: pp. 374-402; Ansari, Hassan, "Sunnism in Rayy during the Seljuq Period: Sources and Observations", ibid.: pp. 460-71. These two contributions are noted because of their material on factional fighting in Rayy, climaxing in the twelfth century; the others are deeply ensconced in intellectual history and contribute little to urban history.

33 Herat: Franke, Ute, "Ancient Herat Revisited. New Data from Recent Archaeological Fieldwork", in Rante (ed.), *Greater Khorasan*: pp. 63-88.

and typologies of buildings, deciphering the landscape, and making its features legible for those scholars who would otherwise work from texts alone. However they leave still more questions unanswered, in particular about the agents and interests responsible for the features identified through excavation and survey. If answers are at all possible, they must be elaborated through the shared effort of historians and archaeologists, as well as still other disciplines.

Given the approach of this archaeological research, the following remarks deal with works addressing not only the city, but also the oasis and city. There has still been relatively little research addressing the relations between the sedentary world, of city and oasis, and the broader landscape, which would frequently be the nomadic habitat, of steppe, desert and mountains. Works on the management and maintenance of trade routes are an exception.

Some publications originating from archaeology also have chapters on history, of varying merit, but rarely qualifying as true urban history. Instead they tend to relate what happened to the city, mostly in terms of who ruled and for how long. This is a structure shared with many encyclopaedia entries and handbooks.[34] Herrmann's monograph on Marw offers two such chapters, one by herself[35] and one by Hugh Kennedy.[36] Kennedy summarises the scarce material on the topography of Marw from the available written sources, and attempts to locate some of the "monuments of Merv", as indicated in the monograph title. He discusses the transfer of the urban centre from the pre-Islamic city (Gyaur Kala)[37] to the newly built western suburb (locally known as 'Sultan Kala').[38] We are provided a more detailed report for the earlier periods, where the geographers can be used as sources, compared to very cursory treatment for the periods beginning around 1000 CE. Here, we return to the characteristic recitation of who ruled over Marw, for which periods of time.[39] The opening chapter of the monograph, by Herrmann, attributes the destruction of

34 See my remarks in the introduction to Aubin.

35 Chapter "Central Asia and Marw": pp. 3-24.

36 Kennedy, Hugh, "Medieval Merv: An Historical Overview", in Herrmann, *Monuments of Merv*: pp. 25-44.

37 For pre-Islamic Marw, see Simpson, "Merv, an archaeological case-study".

38 Much clearer in Williams, "The city of Sultan Kala".

39 The possibilities of written sources for Seljuq and post-Seljuq Marw have not been exhausted. For example, to the best of my knowledge, Sam'ānī's *Ansāb* has not been tested; Sam'ānī was born and spent most of his life in the Marw area, and lists many toponyms related to the oasis. Neither archaeologist nor historian can complete a full inquiry in the scope of a brief chapter. Abū Sa'd 'Abd al-Karīm al-Sam'ānī, *al-Ansāb*, ed. Yamānī (Hyderabad (D.): Dār al-ma'ārif al-'uthmāniyya, 1962-82). The existing historiography of the Seljuq and post-Seljuq periods has again not been fully applied to the city history of Marw.

Sultan Kala solely to the Mongol conquest(s) in 1220. However, Kennedy points to the earlier destruction wrought by the Khwarazmians (under Atsïz, in 1142) and the Ghuzz (1153-6). This is an important difference, given that the correct dating of disastrous large-scale destructive events would be a central concern for productive archaeological research.

Given the subtitle of Rante's volume on Rayy, "an Archaeological and Historiographical Study", the introduction is particularly disappointing.[40] The author embarks on an external perspective, more concerned with dynasties than the city and its history, which is maintained throughout. An article co-authored with Carmen Di Pasquale, published slightly later, compensates for some omissions. The article speaks of factional strife, and above all works towards the identification of city quarters. In this way we finally obtain a map of Rayy, provided in the article rather than the book.[41] In the numismatic appendix to Rante's "The Governors of Rayy" (pp. 149-55) the dynastic bias is very clear. For example, the author applies a single-level term for all those who ruled over Rayy, resulting in the oddity that Sanjar b. Malik-Šāh, the Seljuq sultan, like all the rest in the list, appears as "governor" (AH 511-49/1117-54). In fact the matters of rule and governance were highly complicated, with the caliph and up to three other persons named on the coinage of Rayy. On coins dated 525/1130-1, we have Sanjar, then the western Seljuqid ruler Maḥmūd, and then the regional man, also a Seljuqid prince, Sulaymān. Besides this Sulaymān, the narrative sources also give the names of the 'real' governor, a slave general in Sanjar's service, and his deputy, the one truly in charge. Such situations are by no means exceptional. If the intention is to understand city history, particularly the local power structures, we must seek out and pursue the implications of such information.[42]

I refer to the manner of dealing with city history seen in these publications as the 'dynastic' approach. We can observe this approach not only in other archaeological monographs,[43] but also in historiography proper.[44] Indeed

40 Rante, *Rayy: from Its Origins to the Mongol Invasion, an Archaeological and Historiographical Study*. The subtitle of the book is very ambitious, both in terms of time frame and interdisciplinarity.

41 Rante, Di Pasquale, "The Urbanisation of Rayy".

42 I address the question of the power structure of Rayy under the rule of Sanjar, in *Lokale und imperiale Herrschaft im Iran des 12. Jahrhunderts. Herrschaftspraxis und Konzepte* (Wiesbaden: Reichert, 2016): pp. 193-200. Information about coins is from Miles, George, *The numismatic history of Rayy* (New York: American Numismatic Society, 1938): pp. 213-4. This has been the source for both myself and Rante.

43 For instance, in Kiani, *Gurgān*.

44 See the example of Fragner, Bert, *Geschichte der Stadt Hamadān und ihrer Umgebung in den ersten sechs Jahrhunderten nach der Hiǧra. Von der Eroberung durch die Araber bis*

the fact that archaeologists cannot draw on good city histories is symptomatic of the state of the art in historical research. Exceptional cases would be the histories of Isfahan and Qum in the pre-Mongol period, of Balkh under certain aspects, Shiraz for the fourteenth century, and Herat, above all for the Timurid century. Historians too, when called to collaborate in a coauthored monograph, must typically start from scratch, due to the continuing lack of a thorough basis of historical research on Iranian cities. This observation serves as a reminder of how much work still must be done.

It is still rare that written sources and material culture, meaning history and archaeology, have contributed to a balanced research agenda. A positive example[45] can be found in Yuri Karev's article on the way the Qarakhanid rulers of Samarqand and Bukhara proceeded to move their residences closer to the city, ultimately entering the urban centre to construct palaces. From the archaeological point of view, he records the parallel advancement from light and relatively ephemeral buildings to constructions with thick walls in fired and mud brick. Karev also describes a kind of "royal landscape" around the Transoxanian metropolises of Bukhara and Samarqand, identifying more or less exact locations of the residences. His approach could inform research into other royal or 'noble' landscapes around cities such as Marw, but also Hamadan[46] and Tabriz.[47]

The advancements in knowledge provided by archaeology serve to 'ground' historical research. It is revelatory to know where specific buildings were located, the relative strengths of defensive walls, the structural types meant by the terms in the original sources. The combination of archaeology and history serves above all to understand the combination of city and hinterland: the irrigation network, trade route systems, the villages providing food and commodities. Studies at the oasis level seem to be a promising line of combined archaeological-historical research, and a number of publications take this approach. The following paragraphs note some innovative works, offering very encouraging results.

---

zum Untergang der ʿIrāq-Selčuken (Wien: Notring, 1972). This is a particularly interesting case, since Fragner later used altogether different approaches, no longer limited to a dynastic outlook.

45 See below, concerning Étienne de la Vaissière, for further examples. Note that in both cases, the 'interdisciplinary' approach has been achieved by one author working from both historical and archaeological perspectives.

46 See David Durand-Guédy's discussion of the term *kūšk* in Durand-Guédy, David, "Where did the Saljūqs live?", *StIr*, XL (2011): pp. 234, 239.

47 See the contribution by Daniel Zakrzewski in this volume.

The distribution of villages in the landscape can yield important insights concerning the stability of life in the city. Information on the number and size of the villages can corroborate that on declines or crises in urban life, and the expansion and dwindling of the city itself.[48] Researching the rural landscape of the city requires that the villages first be identified. A second step would be to study the development of settlement patterns. Paul Wordsworth therefore investigates the hydrology of the Balkh oasis, to then better understand "the connectedness of the city of Balkh as part of a network of medium-sized towns, and the relationship between changing settlement patterns in the delta and water distribution". Wordsworth's analysis includes sketching out the history of the irrigation system of Balkh from the tenth through the seventeenth century.[49]

The importance of water to Iranian cities is evident. In studying the history of a city and hinterland, or even a simple desert stopping-place, we must inevitably include the matters of how and from which sources (river, subterranean channel *qanāt*, wells, sources, rainfall, water harvesting) these derived their water.[50] Archaeology is indispensable in this matter, and there have been some truly innovative studies from this field. The most interesting results are perhaps those obtained by Paul Wordsworth and team, concerning the Karakum desert between Marw and Amul on the Amu-Darya. Given that Marw-Amul was part of a major trunk road within the transcontinental trade network, the research question was how the stopping places were maintained so that this part of the caravan route would function over the longest possible annual season. The solutions included both water storage and harvesting. Water was obtained from relatively level localities of nearly impermeable heavy clay, known as *takyr*, and conducted via ceramic pipes to storage facilities near the stopping place.[51] Supplies were sufficient to support small agriculture around some stopping places, with development of settlements. One of these was Dandānaqān, where the Seljuqs won their decisive victory over the Ghaznavids in 1040.[52]

Archaeological research offers much needed insight concerning the problem of distributing water within the city. Arab geographers describe the administration of water distribution within the oasis of Marw,[53] however much

48 Laleh and others, "Le paysage urbain", on Nishapur.

49 Wordsworth, "Hydrological networks": p. 184.

50 See Rante, in this volume.

51 Wordsworth, "Sustaining travel". For an outline of the research design: Williams, Tim and Paul Wordsworth, "Merv to the Oxus: a desert survey of routes and surviving archaeology", *Archaeology International*, XII (2011): pp. 27-30.

52 Wordsworth, "Merv on Khorasanian Trade Routes".

53 Bosworth, C. E. (Yakubovskii), s.v. "Marw al-Shāhidjān", *EI* 2, VI (1991): pp. 618-9.

detail is lacking, including on technical aspects. Archaeology can fill some of the gaps. From this field, we learn that in cities such as Marw and Gurgān, the main canal feeding the city with was built along the line of maximum elevation, enabling gravitational flow into secondary channels and city quarters. There, the water could be stored in open or covered tanks (*sardāba, āb-anbār*).[54] Other cities depended on pond-like open reservoirs (*ḥawḍ*).[55]

One of the questions which immediately arises when confronted with such results is whether the infrastructure of trade routes was planned at a higher level. For some stopping places we have the names of royal or noble patrons, but that does not say much about the management, operation and maintenance of these locations. Did the development and/or operation of caravan routes depend on a centralised state institution? What happened once a ruler, local or imperial, had a line of stopping-places built? Can we identify who was specifically responsible for the governance or administration? Who manned and managed the individual places? As well as for trade routes, very similar questions arise concerning the 'long walls' that surrounded a number of eastern oases, and of course concerning the entire matter of irrigation technology and management. Did such infrastructure presuppose a centralised state institution? The answer is that in fact such systems did not necessarily require the structures of "Oriental despotism": a city state would be sufficient to develop and control an oasis, so the question becomes that of the balance between state (or royal) power and local initiative. Stride, Rondelli and Mantellini deal with this balance in examining the irrigation systems of the middle Zerafshan valley, for which Samarqand was the urban centre, and in particular the renowned Dargom canal, branching to the left of the main course of the river at a point many kilometres above the city. The authors inquire into possible preceding irrigation technologies, the chronology of canal construction, and the overriding question of whether the development of the Dargom implied a strong state based at Samarqand or elsewhere. They convincingly argue that this did not have to be the case, and that even a major irrigation system could originate piecemeal from much smaller earlier systems of obtaining water from creeks and even non-perennial water courses descending the mountains. All these appear to have been gradually integrated into the overall Dargom system.[56]

---

54 For Gurgān: Kiani, *Gurgān*. For Marw: Williams, "The city of Sultan Kala". For the *āb-anbār*, see the eponymous article in *EIr*, by Renate Holod: www.iranicaonline.org/articles/ab-anbar-i-history (last consulted November 2017).

55 Bukhara: Gaube in Gangler et al., *Bukhara*.

56 Stride, Rondelli, Mantellini, "Canals versus Horses".

Who then controlled and operated these irrigation networks? We could surmise that the answers would be different depending on the irrigation technology in use in a given region. Marw, Balkh and Bukhara, Samarqand and Herat were almost entirely or completely irrigated by water from rivers, whereas Nishapur and many other places along the rim of the central Iranian plateau depended to varying but always important extent on subterranean channels (*qanāt*). A necessary precondition to answering our questions seems therefore to know which technologies were used in the different places.[57]

Further questions where archaeological research has recently resulted in important contributions to social history are those concerning the agricultural production of the oases, and therefore the capacities for feeding and supporting a population, given pre-modern technologies. Étienne de la Vaissière has addressed this complex of topics in two publications.[58] These works are again a demonstration of the effective combination of historical and archaeological evidence, both of which de la Vaissière puts to good use. The author draws on source material from different, independent sources: Chinese and Muslim, historiographical, ethnographical, and administrative. The information adduced from these sources concurs and leads to the conclusion that the 'feeding power' of agriculture would not have exceeded 4 to 4.5 adults per hectare. Identifying a ratio of fallow to cultivated land and garden, and the availability of water, he then undertakes to define the total carrying capacity of a number of oases in Xinjiang, former Soviet Central Asia, and Afghanistan. In the case of Marw, de la Vaissière addresses the specific question of how many Arab soldiers (unproductive males, possibly with families) the oasis could support.

De la Vaissière's procedure will not work in cases such as Isfahan where we cannot identify the amounts of irrigated arable land. Still, our previous population estimates for other cities will have to be checked against such results. To give an example, the oasis of Balkh, according to de la Vaissière's estimates, could have supported a peak population (including the city) of 300,000 to 400,000 people, or even slightly more. What does that mean for population estimates previously established for cities and oases such as Nishapur, Isfahan or Rayy? Isfahan has been put at between 100,000 and 200,000 inhabitants in the 11th century; how do these figures relate to the ones obtained by de la Vaissière?[59] A corollary question would be that of what shares of relative

57 See Rante, in this volume.

58 De la Vaissière, "Population Estimates" and "'Abbāsid Revolution in Marw".

59 Calculations have been made for a number of cities, including Isfahan and Nishapur, using a variety of methods. Nishapur: Bulliet, Richard, "Medieval Nishapur: A Topographical and Demographic Reconstruction", *StIr*, v/1 (1976): pp. 67-89; Isfahan: Durand-Guédy, *Iranian Elites*: pp. 25-6; western Iran: Aubin, Jean, "Chiffres de population urbaine en Iran

populations dwelt in the city versus the villages and towns. For the moment I leave it to others to grapple with this extremely difficult question, hoping we might receive answers that are at least partially informed.

Apart from the relations involving hydrology, water provision and food production, archaeologists have stressed the importance of other links between city and hinterland. Indeed, there are several 'landscapes' uniting city with oasis, and reaching as far as the open steppe or desert. Articles in the current volume deal with the 'sacred' and 'fortified' landscapes.[60] Limbert has described the sacred landscape of Shiraz,[61] and this is the central theme of Azad's volume on Balkh. In Ghazni, sacralisation of the landscape involved reuse of marble slabs from secular buildings, sometimes with secular inscriptions.[62] Ali Shojaee, in his article on Isfahan in this volume, is the first to deal with fortified landscapes.

As noted above, Karev has introduced the concept of a 'royal landscape', of regal residences, gardens, and preserves, within and beyond the city.[63] Given the observation of potentially fortified noble residences at many oases, historians should discard the ready assumption that nobles resided in cities, and could instead conceive of a 'noble landscape'.[64] Even administrative buildings (*dār al-imāra*), where present, could be located outside the city proper, particularly in the early Islamic period.[65] Palaces and administrative buildings could also be reused in very profane ways in periods when royal power was absent from the city, as has been shown for Ghazni.[66] This process seems to begin in the later twelfth century, between the 1173 Ghurid conquest, at the expense of Ghaznavids, and the Mongol invasion.

It could be useful for the archaeological and historical disciplines to conceive of still other 'landscapes', such as those concerning the forms of economic

---

occidental autour de 1500", *Moyen Orient & Océan Indien*, IV (1976-7): pp. 37-54. David Durand-Guédy offers a detailed and methodologically informed discussion of population estimates for Isfahan at various moments of the city's history, see his contribution to this volume.

60 See Aigle and Shojaee, both in this volume.

61 Limbert, *Shiraz*.

62 Laviola, "*Ziyārāt* of Ghazni".

63 Karev, "From Tents to City".

64 Marw is a comparatively well studied example, but even in this case, much remains to be done. Aubin already noted that nobles in Iran did not always reside in cities but often preferred rural locations. See p. 38 in this volume.

65 Herat: Franke, "Ancient Herat"; and Kennedy, Hugh, "From Shahristān to Medina", *StIs*, CII-CIII (2006): pp. 5-34.

66 Fusaro, "From Ceramics to History".

production. In the case of Nishapur, for example, we can observe a distribution of water-driven mills, all around the city, probably along water courses descending from the mountains, but also far into the plain where streams are no longer readily discerned.[67]

Archaeologists, and historians,[68] long ago introduced questions concerning the degree and kinds of 'higher level' urban planning, similar to those now raised regarding irrigation networks. Besides the matter of the overall city plan (round with regular rectangular grid, other forms, no plan at all), we must also ask to what extent water provisioning and drainage relied on local initiative or on 'master planning'. Not only agriculture, but also the daily functions of city life depend on reliable supply. In some cities, the main canal runs along the line of highest elevation, permitting gravity feed of secondary canals branching off into urban districts. In Marw, the main canal was fitted with fired brick and constructed in regular form, over a base of 1.7 metres width, rising 2.7 and probably widening towards the surface. Secondary canals were also constructed in fired brick. Water was stored in district reservoirs,[69] and edifices requiring high consumption, such as baths, appear strategically situated in relation to the main canal.[70] As with the development of rural irrigation systems, we are faced with the question: to what extent do such systems arise from local investment and expertise, and to what extent from a higher planning power?

At Gurgān, there seems to have been a combination of roads with water conducts. Archaeologists have identified a road of width between 10 and not less than five metres, running straight for around two kilometres. This was coated twice, in the early Islamic and Seljuq periods. In later times the length of the road was bisected by a central watercourse, using the coating as the bottom surface.[71] To what extent did such urban infrastructure developments result from 'master planning'?[72]

In all, there has been remarkable progress in the study of Iranian cities since the beginning of this century, in particular when archaeologists and historians have worked closely together. The scarcity of written sources which so frequently appears as an insurmountable obstacle can be balanced in some measure by archaeological findings. The interdisciplinary cooperation allows

67 See fig. 1, p. 257, in Laleh and others, "Le paysage urbain". The observations of these mills are not explained in the text.

68 Gaube in his various publications.

69 These have also been identified in many other cities.

70 Williams, "The city of Sultan Kala": p. 55.

71 Kiani, *Gurgān*: chapter 2, "Town planning".

72 For a discussion, see Whitcomb in this volume.

to much better address questions of fundamental importance such as water supply, fortification, and the relationship between the city and its hinterland.

But the survey of recent research also shows how much remains to be done. The present volume cannot aspire to fill the gaps, but hopefully, it will contribute to develop new questions.

Jean Aubin's Article "Elements for the Study of Urban Agglomerations in Medieval Iran" in Context

Translation and Introduction by Jürgen Paul

Introduction

Aubin's text, published here in English translation, was written for a conference on "The Islamic City" held at Oxford in 1965, and first issued as part of the conference papers in 1970.[1] Why translate and republish such an old text today? Evidently because we hold it still relevant. Moreover, we believe that it hasn't had the impact it deserves, and that in spite of all progress achieved over the intervening half century, many of the questions it raises still seek full answers.

Jean Aubin (1927-1998) was one of the most important of the European historians of Iran active in the second half of the twentieth century. He published in a wide variety of fields, of which urban history was only one. He was a leading specialist in the Mongol and post-Mongol periods, up to and including the Safavids. He also pursued a lively interest in the Indian Ocean territories, the cities on the Persian Gulf and the Portuguese trade in this area, and then embarked on a second career in the history of early modern Portugal. Twenty years after his untimely death, some selected papers have been published in reprint.[2]

As mentioned, the article republished here has not had the impact it deserves. The 1965 conference proceedings (issued 1970) received a number of scholarly reviews, but these tended to focus on Hourani's "Introduction" and

1 Hourani, Albert, and Stern, Samuel Miklos (eds.), *The Islamic City. A Colloquium* (Oxford: Cassirer, 1970). Jean Aubin's article: "Éléments pour l'étude des agglomérations urbaines dans l'Iran médiéval": pp. 65-75.

2 Aigle, Denise (ed.), *Jean Aubin. Etudes sur l'Iran médiéval. Géographie historique et société*, (Paris: Association pour l'Avancement des Etudes Iraniennes, 2018) (*Studia Iranica*: Cahier 60). For a detailed appraisal of Aubin's place in the writing of Iranian history, see Denise Aigle's introduction to this volume: "L'œuvre de Jean Aubin (1927-1998) et l'histoire globale": pp. 11-24. Thanks to Denise Aigle for having made this volume available to me before publication. The article translated here is also reprinted in the reprint volume, text 1: pp. 29-38.

 | DOI:10.1163/9789004434332_004

several other papers, and mentioned Aubin's piece only in passing.[3] Subsequent authors have rarely quoted the work, and still more rarely cited it as a particularly effective influence on studies of pre-modern Iranian cities.[4]

One of the reasons for this may be Aubin's style in the French language. His written texts are elegant but difficult, with many subordinations. In this particular article the author alludes to earlier research without providing direct quotes, sometimes giving the impression that he is speaking only to the initiated. Still, his explanations are perfectly clear, and we have made efforts to maintain his style in the English translation.

Another point is that Aubin was viewed as somewhat of an outsider to the debate on the Islamic city. In French academia in particular, this field was seen as the preserve of specialists of the Arab world of North Africa and Syria, and Iran was often viewed as a matter of very different study.[5]

Last but not least, Aubin was not followed by a group of disciples who would immediately continue his agenda. Nevertheless his substantial influence continued to grow over time, and his body of work is now recognised as exceptionally significant. For myself, Aubin's writings were a revelation. I recall one of my teachers speaking admiringly of his articles as so many miniatures: a

3 Reviews by Carré, Olivier, *Revue française de science politique*, XXI/5 (1971): pp. 1111-2; Cahen, Claude, *JESHO*, XIV/1 (1971): pp. 78-80; Holt, Peter M., *JRAS*, CIV/2 (1972): pp. 147-8; Bulliet, Richard, *Speculum*, XLVI/2 (1971): p. 385; Le Tourneau, Roger, *REMM*, VIII/1 (1970): pp. 189-92. In none of these reviews does Aubin's article have any special place; in most cases, he is mentioned only in passing, and sometimes *mal à propos*. – The list of reviews does not claim to be exhaustive.

4 One of the rare explicit references I have identified is Aubin's disciple, Haneda, Masashi, "The pastoral city and the mausoleum city. Nomadic rule and city construction in the Eastern islamic world", Tsugitaka, Sato, (ed.), *Islamic Urbanism in Human History. Political Power and Social Networks*, (London: Kegan Paul International, 1997): pp. 142-70; Haneda reproduces and, in places, enlarges Aubin's concept of the *association ville-pâturage* as propounded in this article (p. 34). Haneda also gives full credit to Aubin in his review of urban studies in Iran, "Iran", Haneda, Masashi and Miura, Toru (eds.), *Islamic Urban Studies. Historical Review and Perspectives*, (London and New York: Kegan Paul International, 1994): pp. 235-79, in particular p. 238 where Haneda says that Aubin's article "was influential and became the starting point for later studies of the history of pre-modern Iranian cities", without naming these studies, however. – In the larger field, note must be made of Ira Lapidus who quotes Aubin's paper as an important stimulation: Lapidus, Ira M., "Muslim Cities and Islamic Societies", in Lapidus, Ira (ed.), *Middle Eastern Cities* (Berkeley & Los Angeles: California University Press 1969): pp. 47-79; 48 note 1, with special reference to Aubin's way of seeing the city and the countryside as a unit comprised in a *pays*, and p. 68 with the oasis of Bukhara as a very appropriate example. Lapidus uses the French term.

5 Olivier Carré, in his review, (see note 3) puts Aubin on the same level as the article on Chinese cities in the volume, and he says that these two "permit a comparison between the Arab Near East and Muslim Asia" (my translation, JP).

well-chosen metaphor, when we think of his artful prose, careful treatment of details, and the scope of some of his theories, rarely made explicit. Indeed, we must recall these theories at all times as we proceed with the reading of Aubin's works.

In considering this current work by Aubin we must first note a few elements concerning the general state of research in the 1960s, which include some questions that still remain relevant today.

In the first part of the present essay, Aubin refers to the debate on the Islamic city. The genealogy of the question has been retraced many times, and there is no need to repeat it here.[6] The debate was composed of several threads. French scholars working on North African cities had published detailed studies with a wealth of empirical material, while a second group of French scholars worked in similar manner on Syrian cities. The concern was that such scholarship generalised about the 'Islamic city' on the basis of sources that concerned either North Africa or Syria, or at best the two areas together.[7]

The core text for the question in the 1960s without doubt was G. von Grunebaum's *Die islamische Stadt*, published simultaneously in English as *The Structure of the Muslim Town*.[8] The main thrust of Grunebaum's argument is

---

6 Beginning with Hourani, Albert, "Introduction: The Islamic City in the Light of Recent Research", Hourani, Stern (eds.), *The Islamic City*: pp. 9-24. Later prominent texts include: Abu-Lughod, Janet, "The Islamic City – Historic Myth, Islamic Essence, and Contemporary Relevance", *IJMES* 19.2 (1987): pp. 155-176; Haneda, Masashi, "Introduction", Haneda/Miura (eds.), *Islamic Urban Studies*: pp. 1-10; Garcin, Jean-Claude, "Problématiques urbaines", Garcin, Jean-Claude et al. (eds.), *États, sociétés et cultures du monde musulman médiéval* tome 3, Problèmes et perspectives de recherche, (Paris: PUF, 2000): pp. 93-109; one of the most interesting contributions is Neglia, Giulia Annalinda, "Some historiographical notes on the Islamic City with particular reference to the visual representation of the built city", Jayyusi, Salma et al. (eds.), *The City in the Islamic World*, I, (Leiden: Brill, 2008): pp. 3-46. The narrative has been challenged recently in an attempt to reassess the earlier works by the "French urbanist" school: Aldous, Gregory, "The Islamic City Critique: Revising the Narrative", *JESHO* 56.3 (2013): pp. 471-493. The evolution of the debate after the publication of Abu Lughod's 1987 article until roughly the beginning of the 21st century is retraced in Bennison, Amira K., "Introduction", Bennison, Amira K. and Gascoigne, Alison L., *Cities in the Pre-Modern Islamic World. The urban impact of religion, state and society*, (London: Routledge, 2007): pp. 1-12.

7 Neglia, "Some historiographical notes": p. 4. The influence of Max Weber has been stated by Eickelman as well: Eickelman, Dale, "Is There an Islamic City?", *IJMES* 5.3 (1974): pp. 274-94. For the Weberian style of Grunebaum's argument, Haneda, "Introduction": pp. 4-5.

8 v. Grunebaum, Gustave, "Die islamische Stadt", *Saeculum* 6 (1955): pp. 138-53; "The structure of the Muslim town", v. Grunebaum, Gustave, *Islam: essays in the nature and growth of a cultural tradition*, (Menasha, Wisconsin: The American Anthropological Association, 1955); second edition (London: Routledge & Kegan Paul, 1961), reprinted in: *Islam and Medieval Hellenism: Social and Cultural Perspectives* (London: Variorum Reprints, 1976), text IV, with the original pagination: pp. 141-58. Aldous, "The Islamic City Critique", ignores the German

a comparison between the Muslim town in Syria and North Africa on the one hand, and the Hellenistic or Roman town in the same areas on the other. He is deeply convinced that Islam as a religious system is and must be behind the outward form of the city and also the organizational patterns of its inhabitants; therefore, all variation is secondary, and in fact the object under study begins to assume an uncanny resemblance to a Weberian ideal type. We should note that Weber's ideal types were not meant to be empirical constructs, and indeed Weber had not written much on cities in the Islamic world, having at most examined the case of Mecca. This does not prevent Weber from making generalisations and strong assertions about the 'Islamic city' as a phenomenon to be observed in all parts of the Muslim world: the common fallacy being that if we know one Islamic city, we know them all.[9]

It would not serve to dwell on whether von Grunebaum in particular had read Weber, or to what extent he was influenced. The fact is that in many Western countries, the concept of an Islam as cultural model that formed the regions under Muslim rule, so that they became a unified civilization, was a constituent and widely shared feature of studies in Islam and Islamic civilizations.[10]

Albert Hourani, in his introduction to the volume, points out that the concept of an 'Islamic city' does not take historical change into account, and he warns against assuming the most recent pre-modern stage (the nineteenth century for most of the Muslim world) as the "traditional" form of cities in the region under study.[11] Aubin takes a much stronger, even extraordinary position for the time, going so far as to assert that the entire "Islamic city" approach is irrelevant. He cites two reasons for this: first, the debate concerned cities of the Arab world alone; second, the fact that Muslims live in a city does not make their city an 'Islamic' one. Moreover, Aubin argues that such an approach blocks the avenues to further research, and that we should instead deal with the geographic, climatic, economic, political, and social factors which lead to the foundation of a city, bear on its development, and ultimately its decline. A

---

publication and the 1955 English edition. For criticism of von Grunebaum's position in the later 1960s, see Kisaichi, Masatoshi, "Maghrib", Haneda/Miura, *Islamic Urban Studies*: pp. 11-82, in particular 34-7.

9 Paul, Jürgen, "Max Weber und die 'Islamische Stadt'", Lehmann, Hartmut, and Ouédraogo, Jean Martin (eds.), *Max Webers Religionssoziologie in interkultureller Perspektive* (Göttingen: Vandenhoeck & Ruprecht, 2003): pp. 109-37.

10 Becker, Carl-Heinrich, "Der Islam als Problem", *Der Islam*, 1 (1910): pp. 1-21; Becker founds his concept of *Islamwissenschaft* on the "unified culture of Islam" (*Einheitszivilisation des Islam*).

11 Hourani, "Introduction": p. 10.

central argument, repeated throughout Aubin's work, concerns the erroneous stipulation that 'Islam is an urban religion'. Instead the city and its hinterland form a unit, and in Iran in particular, such a unit is also present in the language.[12]

Aubin was very much ahead of his time in this refutation of the contemporary mainstream position. What he called for was instead true effort in historical research, based on sources, taking differences of time and space into account, and putting Islam – or religion in general – in context. Aubin sketched out a very ambitious research agenda, in an era when he himself acknowledged that the preconditions for such work were not even remotely established. Given this, it is no surprise that readers and reviewers failed to recognise the scope of his contribution, offered here in reissue. Even Abu Lughod, who appreciated the 1970 volume as "the beginning of some reasonable questioning of past work",[13] does not make explicit reference to Aubin's considerations. However, as she turns to outlining a research agenda, what we read is: "A modest list of the features that created the traditional Islamic city would include: a terrain/climate; a technology of production, distribution and transportation; a system of social organization; and a legal/political system which, in Islamic places and times, could vary considerably".[14] Lughod is not quoting from Aubin, but her thoughts are now very much in line with what he called for.

As Aubin himself notes, the bibliography for the study of the cities of medieval Iran was at the time very meagre. In his introduction, Hourani comments that there is "virtually nothing on Iran".[15] The situation has changed greatly since the 1960s, and it would now take many pages to even present an outline of the state of the art.[16] The pursuit of Islamic archaeology for Iranian cities can no longer be called the 'poor cousin' of Islamic studies. There are now many monographs in historical research for individual cities of Iran. In

12 Cahen alone, out of all the reviewers quoted note 3, sees this point clearly. Regarding the "Islamic" character of cities in the Near and Middle East, Cahen refuses to follow as well: "religions, in their practical application, are what the faithful make out of them" (my translation, JP).

13 Abu Lughod, "The Islamic City": p. 160.

14 Abu Lughod, "The Islamic City": p. 162.

15 Hourani, "Introduction": p. 11.

16 A short bibliography, listing works in English, French, German, and Spanish until the mid-1990s, in Garcin, Jean-Claude, *États, sociétés et cultures du monde musulman médiéval (Xe-XVe siècle)*, tome 2, *Sociétés et cultures* (Paris: PUF, 2000): pp. xxxv-xliv, on Iranian cities in particular: pp. xliii-xliv. See also the bibliography in Haneda "Iran": pp. 270-79; Haneda also includes works in Persian. For a general bibliography, see Bonine, Michael E., *The Middle Eastern City and Islamic Urbanism: an annotated bibliography* (Bonn: Dümmler, 1994). – I am not aware of a more recent bibliographical compilation.

this regard, Aubin had singled out Richard Bulliet's *Patricians of Nishapur*[17] as pioneering in the art of "get[ting] better results out of a uniform and hardly forthcoming documentation", in the "rather disappointing genre of biographical dictionaries of theologians". But even for cities for which such a source never existed or has not been handed down, such as Herat, the state of knowledge has advanced significantly.[18] The two Russian publications by Belenitskij and Masson, which Aubin quotes as examples of the potential results from topographic studies, have long been superseded by works such as those of Lisa Golombek, Heinz Gaube, Terry Allen, Maria Eva Subtelny and Maria Szuppe.[19] Not all of these works are strictly city histories or urban studies, but they all address essential questions, of the kind that Aubin raises in our current article.

It is important to note several issues in Aubin's research agenda that still remain 'open', in spite of much progress over the past half-century. The first is the great question of 'locality'. How did Iranians conceive of their city in the midst of its hinterland, including over the different periods? In fact we have not yet paid sufficient attention to the close links between urban and the rural spheres, perhaps because of the often repeated idea that 'Islam is an urban religion'.[20] Elaborating on Aubin's original thought, we can see that this larger area would contain rewarding fields of focus, of research into such essential matters as irrigation, food and fuel supplies, or the defence networks and surrounding 'sacred landscapes', or the linking networks between city and hinterland.[21]

---

17 Bulliet, Richard, *The Patricians of Nishapur. A study in medieval Islamic social history* (Cambridge (Mass.): Harvard University Press, 1972).

18 See the excellent contributions by Maria Szuppe in the *EIr*, "Herat iii: History, medieval period", and "Herat iv: Topography and urbanism", XII (2003): pp. 206-17 (bibliography at the end of part iv). In addition, for the later history of the city, Noelle-Karimi, Christine, *The Pearl in its Midst. Herat and the Mapping of Khurāsān (15th-19th Centuries*, (Vienna: Verlag der Österreichischen Akademie der Wissenschaften, 2014). For ancient Herat and archaeology, Franke, Ute, "Ancient Herat Revisited. New Data from Recent Archeological Fieldwork", Rante, Rocco (ed.), *Greater Khorasan. History, Geography, Archeology and Material Culture* (Berlin: de Gruyter, 2015): pp. 63-88.

19 Golombek, Lisa, *The Timurid Shrine at Gāzur Gāh* (Toronto: Royal Ontario Museum, 1969); Gaube, Heinz, "Herat: An Indo-Iranian City?", in Gaube, Heinz, *Iranian Cities*, (New York: New York University Press, 1979): pp. 31-63; Allen, Terry, *Timurid Herat*, (Wiesbaden: Reichert, 1983) (Beihefte zum TAVO, B 56); Szuppe, Maria, *Entre Timourides, Uzbeks et Safavides: questions d'histoire politique et sociale de Hérat dans la première moitié du XVIe siècle* (Paris: Association pour l'Avancement des Etudes Iraniennes, 1992) (Cahiers de Studia Iranica: 12); Subtelny, Maria Eva, *Timurids in Transition* (Leiden: Brill, 2007). The list could easily be extended.

20 This was a standard sentence in the 'Islamic city' debate. – Bulliet's questions in his contribution to this volume deal with this precise *problématique*.

21 This is shown in the contributions by Shojaee Esfahani and Aigle in this volume.

A second area of unanswered questioning concerns the subdivisions of the city, or the city with rural hinterland. We cannot expect to always penetrate to the levels of residential quarters or the streets of the bazaar, but in some cases even this level of detail is possible. The third and most important issue raised by Aubin is the rigorous necessity of adopting a local perspective when writing urban history. We continue to be tempted by the perspective of the historiographic sources, by that of princes and courts, and so we continue to have histories where the city is relegated to a mere backdrop for events, without any agency. Often we do not even meet the major urban players.[22] Here, David Durand-Guédy's study on pre-Mongol Isfahan has given an example of what can be achieved[23]: local urban players come out in the open, with their interests, their strengths and weakness, and we see that urban factional strife can sometimes be interrelated with the larger events.

Let us hope that more studies will pursue the directions so intelligently illustrated by Jean Aubin.

Translation

ELEMENTS FOR THE STUDY OF URBAN AGGLOMERATIONS IN MEDIEVAL IRAN

I. A GLIMPSE OF THE STATE OF THE ART

As yet, the cities history of Islamic Iran has not been considered as a distinct field. Apart from the notes written by Minorsky for the first edition of the *Encyclopedia of Islam*, in part on the basis of compilations of Qajar scholarship,

22 This is true of many entries in encyclopaedias. Aubin was correct in observing that the *Encyclopedia of Islam*, Second Edition, had not really achieved progress. This is partly explained by the limits of format: it is impossible to write a true city history in a thousand words. The articles in the *Encyclopedia of Islam-Three* will thus only exceptionally offer new perspectives. The *Encyclopedia Iranica* is another matter, since the format allows multi-part articles covering many pages, see note 19 above on Herat. Zakrzewski's article in the present volume is a good example of what can be achieved when this change of perspective truly takes place.

23 Durand-Guédy, David, *Iranian Elites and Turkish Rulers. A history of Iṣfahān in the Saljūq period* (London: Routledge, 2010). David Durand-Guédy has summarised some of his results in his "What does the history of Isfahan tell us about Iranian society during the Saljuq period?", Edmund Herzig and Sarah Stewart (eds.), *The Idea of Iran, VI: The Age of the Seljuqs* (London and New York: Tauris, 2014): pp. 58-73. See also his contribution to this volume.

and which describe the evolution of a given place in linear fashion through its external events, the bibliography on the subject would be quite meagre if we were to discard the titles with insignificant or dated content, and the works of general history where the relevant digressions relate more to social history than to the urban aspect.[24] The publication of the second edition of the *Encyclopedia of Islam* has not been the hoped-for occasion of an updating (most of the entries on Iranian geography seem to imitate Malte-Brun's *Géographie*).[25] Without doubt, the most positive contributions over recent years have been some monographs by non-orientalist geographers on present-day cities.

The difficulties are truly great. The indifference not only of public authorities, but also of cultivated circles, towards the disruption of sites and disappearance of ancient remains brought about by the current growth of Iranian cities, is not without precedent or parallel. However in Iran, when compared to elsewhere, these irrecoverable destructions are less compensated by the activities of those curious about the past. Although there has been a recovery of interest in local history over the past dozen or so years, witnessed by the appearance of books and brochures, the amateur scholars' lack of formal education too often makes their efforts unfruitful. They could render enormous services in recording architectural and epigraphic remains, in collecting toponyms at risk of disappearance, and traditions and memories – all services that they fail to offer in stringing together quotations from well-known published texts. Poor parent of Islamic studies, Islamic archaeology has not, in Iranian lands, been oriented towards the study of urban history, except in the regions under the Soviet Union.[26] While notably more active, pre-Islamic archaeology in Mesopotamia and Central Asia has above all explored marginal areas: the results achieved cannot be extended *a priori* to the Iranian world as a whole. In spite of the excavations, urban life in Sasanian Iran remains in obscurity – as reflected in later Arabic texts – and this constitutes a considerable hindrance in taking up the study of the urban reality in the Islamic period: we address the matter at a moment of evolution for which the preceding stages have escaped us.

---

24 Aubin uses variously *"fait urbain", "phénomène urbain"*, which we have translated here as "urban aspect" (at times "urban reality") and "urban phenomenon".

25 Conrad Malte-Brun (1775-1826), author/compiler of *Précis de la géographie universelle*, first published in Paris in eight volumes 1810-29, re-edited as *Géographie complète et universelle*, seven volumes, Paris 1856. – All notes are translator's notes; in the French original, there are no notes at all. Notes are restricted to a minimum: to give references to works which Aubin quoted but did not identify, and to some works by Aubin himself.

26 Recall that Aubin is speaking in the 1960s.

In contrast to these unfavourable conditions, there is the quantity and value of the Islamic textual sources, enriched over recent centuries by voluminous documentation in European languages. Alongside dynastic chronicles, legal treatises and theological controversies, in Arabic and Persian – itself a literature quite rich in material on urban history – the histories of cities are of particular value. The medieval Orient, in contrast to the Occident, possesses a rich urban historiography. Within this, the disappointing genre of biographical dictionaries dominates at the outset, but in Iran the material continues with a series of descriptive and narrative works, first in Arabic, later translated into or written directly in Persian.

Under the present circumstances, it is this written documentation, rich in both quantity and quality, that represents the almost exclusive basis of possibilities for understanding urban history. The classical auxiliary sciences can only rarely be used, and the nature of the texts offers no greater advantage in applying the newly developed auxiliary techniques. Lacking in statistical, demographic and economic content, it is not until the 16th century that these materials are supplemented by Persian or European archival collections, currently still inaccessible or unpublished; there remains little likelihood of obtaining satisfactory series of figures prior to the second half of the 19th century. In addition, the rudimentary level of Iranian cartography[27] prohibits the use of topographic data at the detailed level, and at the broader scale, the measure of density of points of agglomeration in terms of the real surface areas of cultivated lands and habitat. The insufficiency of studies in human geography condemns historical geography to conduct its work in the abstract.

It would be natural to think of addressing the gap in research on the urban history of medieval Iran by taking the works on Islamic cities as a model, meaning the ones in practice on the Arabised world. I don't believe that the study of the urban phenomenon has much to win if it is from the outset undertaken in the light of what might be learned by comparing with the "Islamic city". This isn't to deny that, if there are cities in Iran, they are Muslim. Just like Italian or Flemish cities are Christian: a patently secondary characteristic (apart from cases of pilgrimage cities), which no historian could accept as an explanation of the urban phenomenon. Certain characteristics that distinguish the city as it formed in Western Europe, and the eastern city as Islam knows it, could pertain to the religious aspect; but essentially, this contrast deals with nothing more than the comparative type of the western city, in the strict sense, and the eastern city in the broad sense. Our research perspectives would be blocked by simply assuming that one could define urban life in Islam by saying that the

27 Once again, referring to cartography available in the 1960s.

city is Islamic. As if there were only one Islam, as if the theoretical prototype of the ideal city were identical with all the diversified forms of the eastern city, which it has taken and still takes.

In the Iranian world alone, and in the period by convention named "medieval", the geographic settings, historical circumstances, and functional associations bring about the emergence of different types of cities. Agricultural market, artisanal centre, trading place, strategic position, site of pilgrimage, intellectual nucleus[28] – all functions that, alone or in combination, shape the form and destiny of a city. Not only these. The ports of the maritime coast are different from those of the desert border; cities of warm lands from those of cold lands,[29] those of the central plateau from those of the Caspian region or of Central Asia. The personality of cities depends on much more than the characteristic qualities discerned in their inhabitants by the medieval authors. The process of peopling, customs, and way of life mean that a city of Fars is unlike a city of Khurasan or Azerbaijan, in both aspect and atmosphere.

Rather than treating the Iranian city as an example or a variant of the "Islamic city", as a substantiating aim, it seems more urgent to search for the means to obtain the best results from a uniform and scarcely coherent documentation, and how to progress from descriptive inventory to dynamic classification and formulation of complex problems.

The following notes will contribute to establishing a programme of inquiry[30] into the historic evolution of Iranian cities. Only some of the questions pertaining to the matter will be mentioned, chosen from among those solvable even with the limited methodological tools presently at hand, and which can serve in advancing the research methodology.

II. THE CONTEXT FOR THE URBAN PHENOMENON

There has not been due attention to the concept of space, as implied in Iranian geographic terms. Considering only those immediately intelligible, they reveal a vision and an effort of organisation of the natural setting on the part of man. One is immediately struck, for example, by the ambivalence of the name indicating both a district and its administrative centre. Sometimes the administrative centre takes the name of its territory. Thus Kirman, a city which

28 Among others, the original terms here are "*place marchande*", and "*foyer intellectuel*".

29 Here the author evidently conceives of *garmsīrāt* "warm regions" and *sardsīrāt* "cold regions": an essential difference in Iranian landscapes.

30 "contribueront à faciliter l'établissement d'un questionnaire d'enquête"; literally: "contribute to easing the development of a questionnaire of inquiry/investigation"; perhaps "a research agenda".

earlier was called Bardasīr, was named *šahr-i Kirmān*, "the city (administrative centre) of Kirmān", and later simply Kirmān. The form *šahr-i Ray*, "the city (of the district) of Ray" is still in use for a small locality of the south suburbs of Tehran. Sometimes, what would seem to be a city name is not: unless it is preceded by the term necessarily specifying the administrative division, in fact the name of the administrative centre designates the territory under its jurisdiction. "Shiraz" frequently means not the city of Shiraz, but "the (territory under) Shiraz" – the province of Fars; likewise, in the context of an account of Fars, the word *šahr*, "the city", in absolute terms, denotes Shiraz. Medieval texts offer many examples of this double usage, which is general (cf. *šahr-i Ṭūs*, *šahr-i Sīstān*, etc.; but "*Ṭūs*" "the Ṭūs", "*Isfahān*" "the Isfahan" etc.).[31] The semantic evolution of the words meaning "city" and "district", *šahr* and *šahristān*, shows that the ambiguity at the level of proper nouns also existed at the level of common nouns.

This tendency in Persian to not distinguish between the territory, limited or vast, and its dominant locality, prompts us to situate the study of the urban aspect within the framework of the historic territorial divisions. The spatial unit to be observed is not that delineated by the city walls or their surrounding belt of suburbs: the agglomeration cannot be dissociated from its territory. The implantation of non-native nomadic groups has disrupted the ancient historical-geographic structure, in this way favouring the centralizing policies of the modern dynasties; so that, together with the degradation of economic mechanisms, it is only by means of a painstaking labour of reconstruction (leading well into the night of the pre-Islamic ages), that we can we recognise the fundamental importance of those "cantons" (*rustāq, rab*ʿ, *nāḥiyat*) or "districts"[32] – which were living cells in the social organisation of the Iranian lands, among which many have already lost their ancient names.

Each of these "districts" groups a number of villages; as administrative centre it has a small agglomeration of embryonic urban features – seat of the local administration, of intellectual life, of a specialised artisanship, of a permanent market. This small kernel, or momentarily sometimes even two kernels (cf. Nūqān and Ṭābarān in Ṭūs district) contained within every territorial cell, represents an urban reality of unequal future prospects. Within all the counties forming a district or province, only a few will exercise an attraction over the

31 Similarly to certain English uses, the definite article in Arabic, rendered here in Aubin's original French, "*le Ṭūs*" etc., refers to the district or territory.

32 Here, the author's original terms are "*cantons*" and "*pays*". We take the latter, in Aubin's thinking, as derived from Latin *pagus*, an administrative unit roughly corresponding to the English "county". This meaning endured as late as the twentieth century, in the familiar expression *'on est pays'*, in referring to a common place of birth.

surrounding counties, and only one will advance to the rank of regional metropolis. Do we then have to grant the title of town or city to all these points of demographic concentration, of which some will remain large villages, without succeeding in becoming small cities? An objective description will see them simply as agglomerations, situated within a rural landscape, surrounded by their gardens and fields. What are the conditions producing this phenomenon of agglomeration, and under what conditions will it disappear? This is the first matter in studying the urban aspect. Only subsequently can one deal with the organisation of the human communities.

III. FACTORS AND TYPES OF AGGLOMERATION

How then, given the rural environment of the Iranian civilisation, do cities originate and die? The evolutionary curve of cities during the Islamic period shows which factors are favourable to urban life, and how they transform it.

a. Geographical factors. In Iran, large cities are situated along the rim of a high relief, which provides for the water needs of the agglomeration and its crop lands. The most evident example is Ray/Tehran, at the feet of the Alborz range. Or again Tabriz and Marāġa, on the perimeter of Mount Sahand; Nishapur and Ṭūs/Mashhad, on either side of the Kūh-i Bīnālūd. But Isfahan also is a foothills city, as well as Yazd, erected at a distance from the mountainous massif and linked to it through series of subterranean channels. Only in exceptional cases does the mountain hydrographic network feed a powerful water course; where this is the case, the city can originate very far from the summits that ensure its life: an extreme example would be the oasis of Marw, nourished by the long offshoot of a river descended from the Hindukush.

No metropolis is without vast adjoining stretches of agricultural and garden lands, capable of producing the majority of foodstuffs near the places of consumption: we see Rayy, between the plains of Šahryār and Warāmīn; Isfahan in the lower Zanda-rūd valley; Istakhr on the edges of the Kurbāl plain, which later became the breadbasket of Shiraz. Whatever might be the support from caravan provisioning (cf. Nishapur, receiving the products of Ustuvā), there is a correlation between the size of the cities and the extension of the neighbouring arable lands. The great trading ports of the Persian Gulf are no exception. It's true that the coastal plain of Sīrāf is very limited,[33] and that Ormuz

33 Jean Aubin, "La ruine de Sīrāf et les routes du Golfe Persique aux XIe et XIIe siècles", *Cahiers de civilisation médiévale* 2.3 (1959): pp. 295-301; reprint volume text 4.

is completely sterile.[34] Nevertheless both cities, which received their non-perishable goods (rice, dates) from the exterior, by boat, had fresh foodstuffs brought from nearby – one from its hinterland, the other one from the island of Qišm.

b. The rise of trading cities within unfavourable natural environments is due to another factor of urban development: the economic factor. Every expanding city is unfailingly grafted on an active commercial artery. Once the flow of caravans diminishes, these cities die. It is for this reason that in southern Iran we find the constellations of abandoned cities. Every great traffic axis thus entails a chain of such dead centres. At the investigative level, the geography of routes is a good instrument for the revelation of economic history.

c. The political factor intervenes either as an expression of strategic opportunity or in terms of pressures from interested parties. The expansion of principalities is first determined by the framework of road and urban fabric. When the agglomeration is not found on the path of a heavily travelled route, its notables exert efforts to divert its course so that it passes by their city (cf. Faryūmad, Ḫunj).[35] There is then a correlation between the rise of a city and the accession of its aristocracy to positions in the central government.

In the decline of urban life, the same factors come into inverse play. There is an interaction, often rapid, between natural disruptions and those provoked by geographical imbalances, economic stagnation, and political disorder. We observe this clearly during the period of expanding pastoralism, which transformed urban life. Since the lands travelled by the herders were situated at a distance from the traditional network of routes and towns,[36] many new settlements then arose during the Mongol period in places that were geographically less favourable. Such settlements promptly declined, because apart from the ephemeral will of the princes, soon attracted by the traditional sites, nothing could ensure their survival: neither the permanence of pastoral routes, nor the fact of an artisanal or agricultural vocation. The only one to endure was the most celebrated of these creations, Sulṭāniyya, which is the primary example of the city-pasture type of association, already well known in Central Asia, but still new in Iran, where cities instead belonged to the city-oasis type of association. Although it soon declined after the death of its founder, Sulṭāniyya would not completely disappear until the 17th century; its survival is explained by

34 Aubin's first study on Ormuz was "Les princes d'Ormuz du XIIIe au XVe siècle", *JA* (1953): pp. 77-138; he returned to this city and its hinterland in numerous later works.

35 Aubin went into more detail in "La survie de Šīlāu et la route de Ḫunj-ō-Fāl", *Iran*, VII (1969): pp. 21-37; reprint volume, text 5.

36 Jean Aubin, "Réseau pastoral et réseau caravanier: les grand'routes du Khorassan à l'époque mongole", *Le monde iranien et l'Islam* 1 (1971): pp. 105-30; reprint volume, text 6.

its triple role as the marketplace for a region of tribal population, its strategic position, and as a staging point on the great commercial East-West route.

As they become acculturated, the nomadic aristocracies permit the old cities to develop the embellishments of public buildings; next to them they build their own settlements (Šanb-i Ġāzān) and ceremonial residences, surrounded by vast grounds where they continued to live the camp style of life (Samarqand, Herat). The substitution of the pastoral regime for the traditional one nevertheless leads to a palpable erosion of urban life, sometimes sudden, in other cases gradual, with the Türkmen invasions of the 15th century accelerating the collapse. Apart from a few regional metropolises (Herat, Shiraz, Tabriz), most cities wane to the level of country towns, the most fortunate surviving as markets serving pasture lands (Ṭūs, Warāmīn) or as stopping places. The urban collectivities were regarded by the nomadic powers as dangerous groups (reprisals, collective massacres), as reserves of labour (deportation) and of riches (tolls, holding for ransom, pillage). The accommodations allowed to cities that continue an industrial activity (Kashan or Yazd) reveal clearly that the attitude of the nomadic aristocracy concerning the urban aspect was anything but blind or inconsequential. The insertion of the pastoral economic system, after the failure of attempts at founding new cities, was accompanied by development of the *ordōbāzārs*: tent cities of traders, shifting location along with the royal or military camps (*ordo*), which duplicated and often competed with the urban bazaars.[37]

Paradoxically, the Safavid and post-Safavid period, for which the densest documentation is available, is also the most neglected. What were the effects of the halt to the great waves of invasion, of the Iranian colonial expansion towards India, of the European colonial expansion? Following the economic depressions of the 15th and 18th centuries, what was the rhythm of restored urban life, still flourishing sufficiently to stimulate the admiration of European travellers? When and how did the gap develop between European and Oriental urbanism? What was the part of religious minorities (Armenians, Zoroastrians, Banians) in urban prosperity? How did the town-country relationship evolve in the Safavid period? And what was the true hierarchy of the urban classes? All these are crucial questions for which we currently have no answers.

37 The last paragraph was translated by David Durand-Guédy, see "Introduction", in David Durand-Guédy (ed.), *Turko-Mongol Rulers, Cities and City Life* (Leiden: Brill, 2013): p. 3.

IV. URBAN TOPOGRAPHY

Separate study is required for the principal elements of the delimited, built-up space, which contains, subdivides and brings together the civic population: the defensive works, residential quarters, and public spaces.

a. It is well-known that in Iranian cities of the high Middle Ages, a suburb (*rabaḍ*, Persian *bīrūn* "the exterior") flanked the old city (*madīna*/*šahristān*), within which stood the citadel (*kuhandiz*), most frequently at its centre. In general, the *šahristān* perimeter continued the pre-Islamic form, with some modifications. Whether circular or square, the ancient perimeter wall is pierced by four gates, whose positioning is governed by astronomic calculations. This is the case at Samarqand, Herat, Nishapur, Bardasīr, Jīruft, Jayy, Hamadan, Jūr, Dārābjird. Other cities, since prior to Islam, seem to have had a greater number of gates, such as at Balkh; some smaller cities would have had only two or three gates. A classification is necessary, as well as the search for an explanation. Islamisation leads to change: if need be, a fifth gate is pierced, where the comfort of the inhabitants demands one; this is the "new gate" (*bāb al-jadīd*), at Zaranj, at Isfahan, etc. The age of the cosmogonic meaning of the city is over.

Beginning in the troubled years of the 11th century, the urban agglomerations surrounded themselves with ramparts. There are many imprecisions in the texts, with some indicating cities as open where others say they are closed. It is difficult to distinguish whether these are reporting on the old walls of the *šahristān* or on the defences erected outside these walls, around the gradually urbanised area, until then insufficiently protected: the once suburban zone, now become the new heart of the agglomeration. The confusion in the sources between *šahrband* enclosures and *kūčaband* enclosures is sometimes at the heart of these apparent contradictions.[38] Intended to protect against external menaces, the ensemble of the city walls, supported by the defensive works of the city quarters and the citadel (which in some cities is no longer central, but to one side), also serves to isolate and divide the turbulent and fluid mass of citizens. Finally, the walls represent a tariff barrier.[39] There is therefore no contradiction if a source speaks of a defensive rampart where according to other sources there is already a wall, which may have been only a modest means of

38 Whereas *šahrband* denotes the wall surrounding the city in its entirety, *kūčaband* walls are erected around city quarters.

39 Original: "*barrière fiscale*"; differently, in the next sentence: "*mur d'octroi*".

delimitation. (Cf. on the other hand, the maintenance of a tax barrier for entry in cities where Tamerlane had the fortifications torn down.)[40]

b. The Iranian city is in fact an agglomeration of autonomous quarters, or ones in competition. We should therefore attempt to trace the events of each quarter, and the relationships among the quarters. At origin, according to local traditions, the city would at times have been formed by the fusion of a number of villages. In other cases, divisions are perpetuated by the persistence of heterogeneous communities. It would be important to know if there were changes in the boundaries, names or number of quarters. Some consist of a single street (*kūy, kūča, maḥalla* are at times synonymous). The study of streets, and the positioning of buildings, are necessities in the social and economic definition of the quarter and the urban structure in which it is inserted. (The works by Belenitskii and Masson, on Herat, have shown how much can be extracted from the topographic data in the sources.)[41] The study of quarters should be complemented by inquiries into the semi-urbanised fringe extending from town to countryside: the external caravanserais, the "garden quarter" (*kūča-bāġāt*), the race grounds (formerly *asfrīs*, witnessed in the names of gates; later *maydān*, locally *lard*), etc.

c. The influence of the mosque on life in the Islamic city has been stressed. But this has been analysed, or affirmed, only over the short term. Greater nuance is required. Permanent element on the urban scene, favoured object of monumental ostentation, the mosque is subject to competition from countless mausoleums as a place of devotion, and above all from the *ḫānaqāh*. Over the centuries, it tends to become an annex within the structural complex that develops around it, and which includes, as well as the *madrasa* and the *ḫānaqāh*, a number of other public spaces: baths, dispensaries, hostels, bazaars, which serve as the living centres of the community life. Around these profane places gravitate all the activities that signify the existence of a city. In this regard, there is also an unavoidable yet always unspoken question: what could be more urban than the commercialisation of debauchery? The infamous quarters, the *ḫarābāt* so often mentioned in poetic works, the rendezvous of all outcasts, have had an insufficiently examined influence on psychological behaviour and on social life.

40 Jean Aubin, "Comment Tamerlan prenait les villes", *StIs*, XIX (1963): pp. 83-122.

41 The author probably intends the studies devoted to Timurid Herat: Belenitskij, A.M., "Istoričeskaja topografija Gerata XV v.", in *Committed Navoi* (Moscow-Leningrad: Institut Vostokovedeniia Akademii Nauk SSSR 1946,): pp. 175-202; M.E. Masson, "K istoričeskoj topografii Gerata XV v.", in *Velikij uzbekskij poėt* (Tashkent: Izdatel'stvo Akademii Nauk UzSSR 1948): pp. 120-145. Jean Aubin read Russian with ease (personal communication, Denise Aigle.).

V. SOCIAL HISTORY OF THE CITIES

Urban history is at the heart of Iranian social history. But it is far from covering it entirely. Is it justified that urban history should hold this privileged place, awarded by our sources, in comparison to the history of the rural or nomadic world? Has the city always been the source, and then the refuge, of social dynamism? Such a question is so clearly general that it requires prefacing by examination of individual points, touching less on the sociology of groups of not particularly urban character, and more on explaining the mechanisms of the agglomeration phenomenon.

a. The ties between an agglomeration and its rural surroundings are tight. C.E. Bosworth has identified them for Nishapur.[42] Nevertheless, he attributes an observed dislike of rural life to an Islamic prejudice. Would we be better advised to view this as simply the contemporary attitude of the city-dweller towards the *rūstā'ī*, the "bumpkin"?[43] Given that this is Iran, the conception of "Islam, religion of city-dwellers" requires qualification. The ideal type of Iranian man, in the classic period, is that of the *dihqān*, the rural landowner; and later, shaykhs and dervishes, who raised their *ḫānaqāh* in the rural towns, at a distance from the big cities. Research is necessary into the level of concentration of artisanship in the urban centre; into whether there are particular relations between a quarter and rural district; whether the recruitment of popular urban troops[44] always includes a component of rural *'ayyār*; whether seasonal migrations of labour are the rule; into what is the focus of attraction of the particular city; what is the exterior contribution to the recovery of population after wars, famines, epidemics. Thus, the history of our village societies, barely reflected in our sources, will become clear, through the understanding of the relationships between city and country.

b. Among popular troubles, for which methodical, focused inquiry is indispensable, there should be a distinction between the endemic factional fights, and those which in certain periods take on the character of class struggles.

c. The ruling classes would invest some of their revenues, of varying provenance (rural landholdings, leasehold farming, grain monopolies, participation in trading enterprises)in purchase of urban property holdings. How is urban

---

42 Clifford Edmund Bosworth, *The Ghaznavids. Their Empire in Afghanistan and Eastern Iran, 990-1040* (Edinburgh: Edinburgh University Press, 1963): p. 152.

43 For Aubin's *"rustaud"*, retaining his original use of quote marks.

44 Original *"recrutement des bandes populaires urbaines"*: the sense could be that of a militia, also squad, gang.

property ownership distributed among the families of notables? What is their hold over public spaces, over the city's surroundings, over the routes that provide its access? In what proportion are these holdings the establishments of social interest? Does the military aristocrat have an urban basis. In the absence of notary acts and cadastres, the *vaqf-nāme* documents can give precise information on these matters, compensation for the more general indications of the chronicles.

d. The body politic is not structured in terms of institutions (there are no estates of the realm).[45] The classes are not organised as such (in this sense there is no formal nobility). There is no legal link between the organised social groups (sayyids, merchants, artisans) and power. Therefore one cannot expect that the city would have a true administrative statute, with city forces, consular institutions, etc. A notable who might be chosen among others by the prince, for his wealth, his skills in relations, his clientele, will govern as intermediary for the central authority, who often controls this same notable through its delegated officers. The attempts at self-government, arising in times of peril or crisis, never endure. Even in cases where the geographic and economic situation seems favourable to the transformation of an oligarchy into a governmental body, no organ of municipal life develops. Ormuz, for instance, city state and merchant city, is governed like all the other seigneuries. One should not conclude that cities were deprived of all political existence. These negative observations leave the problem of urban administration completely intact.

If we were to take the word "city" in the meaning it has come to acquire in western history, one would be led, as the end of a series of false analogies, to conclude that there are no cities in the East, because there we do not find the features that make the cities of the West. But it becomes clear that, far from being the norm, what happens in the western part of Europe has represented the exception. Moreover, the recent inquiries into the "pre-urban" agglomerations of Europe open towards the comparatism of perspectives, already familiar to the historian of eastern cities, who studies agglomerations that have remained at the functional stage.

45 Original: "*(il n'y a pas de régime des ordres)*".

Iranian Cities: Settlements and Water Management from Antiquity to the Islamic Period

Rocco Rante
Musée du Louvre
rocco.rante@louvre.fr

Abstract

This article attempts a long-term perspective on cities and water from Late Antiquity to the early Islamic centuries (until ca. 1000 CE). It focuses on the question of how cities and their agricultural hinterland were supplied with water. The topography of the site, its geomorphological features, are shown to influence both the setup and subsequent history of the cities. The article uses two sets of examples, one chosen from the Iranian plateau where *qanāt* irrigation predominates, and the other one from Persianate Central Asia (Transoxiana), where water is derived from larger and medium-sized rivers. The type of irrigation influences the ways in which the city grows, and more generally, the layout of the city is also determined by the water supply. Cities tend to grow towards the source of water, and it can also be observed that in many cases, the political and administrative centre is located where the best water is available. One of the major questions is whether imperial will was behind the construction of irrigation systems or whether local players such as landlords were the decisive factor.

The article combines archaeological research and the study of textual sources but is mostly based on recent archaeological fieldwork.

Keywords

water – city – landscape – Central Asia – Iran

1 Introduction

What are the links between water management, landscape and urban patterns in medieval Iran? The foundation and development of the earliest cities required continuous access to water sources, and management of supply and drainage throughout the urban context.

 | DOI:10.1163/9789004434332_005

It is well known most of the agriculture in Iran depended on irrigation. Only irrigated agriculture permitted larger settlements to come into being. Through irrigation networks and the founding of settlements, the Iranian landscape was transformed. This paper is about these modifications. It asks what determined a particular modification and what impact it would have had on the urban morphology. Were other options available to achieve the same goal, and if so, why was a specific modification chosen over the others? Were solutions always available to adapt the landscape and water resources to human needs? If not, what were the alternatives?

The Iranian plateau and Central Asia offer a vast array of examples which assist in answering these questions. This paper discusses a selection of settlements and cities from the two regions, examining them in terms of their geomorphological and topographical context, and where possible in terms of their political and social conditions. First, the cities are introduced through the consideration of their geomorphological context and main natural features.[1] The paper then provides a general framework of settlement considerations relevant to the pre-Islamic period, followed by case studies developing observations on Iranian cities during the Islamic period. These examples serve to show certain continuities and transformations that have characterised the evolution of urban centres.

There is relatively little scientific literature focusing specifically on the relations between water and human settlement in the Iranian world, including in regards to cities.[2] However, the materials available seems sufficient for the

---

1 I am very grateful to Professor Eric Fouache for his suggestions and support in providing detailed geomorphological data on the regions and cities under investigation.

2 On relationships between first settlements and water management, the most important scientific works are: Wittfogel, Karl August, *Oriental Despotism. A Comparative Study of Total Power* (New Haven: Yale University Press, 1957), (classical but now contested theory on the political implication of irrigation systems); Adams, Robert McC., *Land behind Baghdad. A History of Settlement on the Diyala Plains* (Chicago: University of Chicago Press, 1965), Adams, Robert McC., "Agriculture and Urban Life in Early Southwest Iran", in Caldwell, J.R. (ed.), *New Roads to Yesterday: Essays in Archaeology* (London: Thames and Hudson, 1966): pp. 436-65 (general studies on irrigation and agricultural development in ancient Mesopotamia); Lisitsyna, Gorislava N., *Orošaemoe zemledelie ėpokhi ėneolita na juge Turkmenii* (Moscow: Nauka, 1965) and Ead., "The Earliest Irrigation in Turkmenia", *Antiquity*, XLVIII (1969): pp. 279-88 (general study of irrigation systems in Turkmenistan); Spooner, B., "City and River in Iran: Urbanization and Irrigation of the Iranian Plateau", *IrSt*, VII (1974): pp. 681-713 (illustrating the importance of the irrigation system in some major Iranian cities); Andrianov, Boris V. et al., "Rol' irrigatsii v sotsial'noj istorii drevnej srednej Azii", in Masson, Vadim (ed.), *Drevnie tsivilizatsii Vostoka* (Tashkent: Fan, 1986): pp. 34-44 (highlighting the social consequence of irrigation in Central Asia); Prickett, Martha, "Settlement during the Early Periods", in Beale,

intended investigations, which link water, landscape and urban patterns in historic and modern cities. The paper will present the results of geochronological studies and of archaeological and historical research concerning the times from the first centuries CE to the early Islamic period.

The observation of two major geochronological phenomena occurring in the ancient era give us the tools to identify and understand the same phenomena in later periods. Both are transformative changes that developed over long periods of time, and which affected settlement and urbanisation. The first change is the gradual aridification observed between 2500 and the 1900 BCE.[3] The second change is the decline of precipitation around 1000 BCE, observed particularly in south-eastern Iran,[4] but which also affected other areas. These climatic changes were responsible for two concurrent events: a decline of settlement, and the transformation of the landscape, undertaken by people for the achievement of sustainable water supplies. The latter change necessarily

T.W., (ed.), *Excavations at Tepe Yahya, Iran 1967-1975. The early periods*, (Cambridge/Mass.: Harvard University Press, 1986): pp. 215-46 (on early settlements in Iran); Askarov, Akhmed Ali et al., *Rannaja gorodskaja kul'tura ėpokhi bronzy juga Srednej Azii* (Samarqand: Institute of Archaeology of Samarqand, 1993) (on bronze age cultures in Central Asia); Allan, Tony, *The Middle East Water Question: Hydropolitics and the Global Economy*, London: Tauris, 2001) (on social and political aspects of water management in the Middle East); Francfort, Henri-Paul and Lecomte, O., "Irrigation et société en Asie Centrale des origines à l'époque achéménide", *Annales. Histoire et Sciences Sociales*, LVII/3 (2002): pp. 625-63 (exhaustive survey of irrigation systems in Central Asia, with bibliography); Gillmore, Gavin K., Coningham, R. A. E., Fazeli, H., Young, R. L., Magshoudi, M., Batt, C. M., and G. Rushworth, "Irrigation on the Tehran Plain, Iran: Tepe Pardis – the Site of a Possible Neolithic Irrigation Feature?", *Catena*, LXXVIII (2009): pp. 285-300 (case-study of irrigation on the Tehran Plain and Tepe Pardis); Wilkinson, Tony J., *Archaeological Landscapes of the Near East* (Tucson Arizona: The University of Arizona Press, 2003) (on archaeology of landscapes in the ancient Near East). For the modern era, see Beaumont, Peter, s.v. "Āb iii. The Hydrology and Water Resources of the Iranian Plateau", in *EIr*, I (1982): pp. 28-39 (dated, but thorough). For Central Asia: Lioubimtseva, Elena, "Climate and Environmental Change in Arid Central Asia: Impacts, Vulnerability, and Adaptations", *Journal of Arid Environments*, LXXIII (2009): pp. 963-77, focuses on the twentieth century but provides a useful examination of human vulnerabilities during climate change. It has to be noted that climatic literature on the proto-historical periods and Bronze-Iron Ages is more thorough than for later periods, see for example Fouache, E. Francfort, H.-P., Cosandey, C., and Ch. Adle, "La crise d'aridité climatique de la fin du 3ème millénaire av. J.-C., à la lumière des contextes géomorphologiques de 3 sites d'Iran Oriental (Bam, Tepe Damghani, Jiroft)", in Rante, R. (ed.), *Greater Khorasan. History, Geography, Archaeology and Material Culture* (Berlin: De Gruyter, 2015): pp. 27-37, where three cases in the Iranian world are studied, demonstrating the slow but inexorable climate change in this region.

3 Fouache *et al.* "La crise": p. 34.

4 Wilkinson, Tony J. et al., "From Human Niche Construction to Imperial Power: Long-Term Trends in Ancient Iranian Water Systems", *Water History*, IV/2 (2012): p. 168.

involved more complex management of the available water resources. This included the expansion of irrigation, beginning in the Achaemenid period and gradually extending throughout Central Asia and the Middle East.[5] Beginning in the sixth century BCE, events continually transformed the landscapes, and urban patterns were continually adapted.

The current discussion of the relationship between city and water in the Iranian world during the ancient and medieval serves as an initial summary of the different human approaches to the problems of settlement and urbanisation.

2 Geomorphology and Water Resources

The areas covered by the investigation are the Iranian Plateau and the adjacent steppes of Central Asia (Fig. 1), falling within the modern states of Iran, Turkmenistan, Uzbekistan and the northwestern part of Afghanistan.[6] The paper develops the thesis that the topographical and hydrological differences between the fringes of the Iranian plateau and the central Asian plains lead to two distinct forms of irrigation, which in turn affect the agricultural potential and urban structures of the two regions. These differences are demonstrated in two sections of the paper, each examining two cases: Nishapur and Rayy for the cities around the Iranian plateau; Marw and Paykand for the central Asian steppe regions. In a subsequent section, several further cases are briefly presented for comparative discussion of how urban development depended on the given topographical and hydrological factors, and how this development in turn shaped the landscape.

The central part of the Iranian plateau is characterised by vast deserts. High mountain ranges rise at the fringes: the Zagros to the west and south, the Alburz to the north, the Hindu-Kush to the east. The most important urban centres are located at the foot of mountains. Hamadan, Isfahan and Iṣṭaḫr, for example, are inserted along the eastern slope of the Zagros; Rayy, Damghan and Nishapur are situated to the south of the Alburz, and Herat and Balkh

5 Briant, Pierre (ed.), *Irrigation et drainage dans l'antiquité, qanats et canalisations souterraines en Iran, en Egypte et en Grèce* (Paris: Thotm éditions, 2001): pp. 12-4; Magee, Peter, "The Chronology and Environmental Background of Iron Age Settlement in Southeastern Iran and the Question of the Origin of the qanat Irrigation System", *Iranica Antiqua*, XL (2005): pp. 217-31; Wilkinson, "From Human Niche Construction": pp. 167-8.

6 For a presentation of the lands of Iran, see De Planhol, Xavier, s.v. "Iran i: Lands of Iran", in *EIr*, XIII (2004): pp. 204-12.

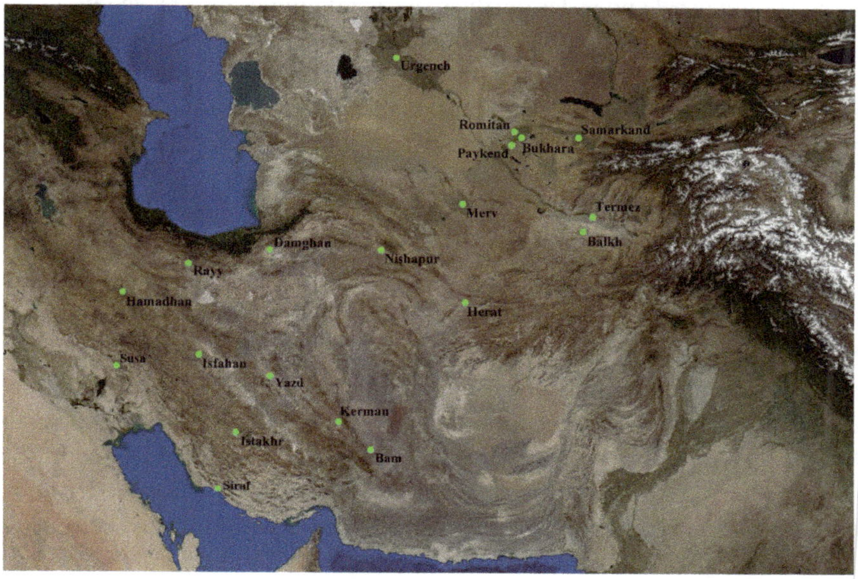

FIGURE 1 Map of the Iranian world
SOURCE: ESRI IMAGE 2016 ADAPTED BY THE AUTHOR

near the northwestern Hindu Kush.[7] The main roads between the western and eastern parts of the Iranian Plateau pass along the south slopes of the Alburz range, or through the southern part of the central deserts, making communication difficult.

Many Iranian cities obtained their water through *qanāt* (var.: *kāriz*): underground irrigation canals of up to several kilometres in length, conveying water from aquifers in the mountains to the alluvial plain.[8]

Beginning in the Iron Age, and especially with the onset of the Achaemenid period (sixth-fourth century BCE), the *qanāt* system was the basis for development of urbanisation over Iranian territories extending from the Caucasus

7 See the basic settlement pattern proposed by Christensen, Peter, *The decline of Iranshahr. Irrigation and Environments in the History of the Middle East 500 BC to AD 1500* (Copenhagen: Museum Tusculanum Press, 1993): p. 121, fig. 6.

8 The literature on *qanāt* is vast. For recent research with bibliography on the origin of the *qanāt* see Briant, *Irrigation*. Adle, Chahryar, "Qanāts of Bam: an archaeological perspective", in *Qanāts of Bam* (Tehran: UNESCO, 2006): pp. 33-85 reports on recent multidisciplinary research on the *qanāt* in the region of Bam. Bulliet, Richard, *Cotton, Climate and Camels in Early Islamic Iran*, (New York: Columbia University Press, 2009) provides a highly useful analysis on the interrelationships of *qanāt* and society during the Islamic period.

In the remainder of this paper, '*qanāt*' refers only to underground canals, distinct from surface channels (the latter also 'channels' or simply 'canals').

to the Hindukush,[9] and as far north as the Karakum desert. The geographical constraints led to intense urbanisation within oases.

The *qanāt* technique was little used in Central Asia, given the nature of the terrain. Mountainous aquifers are unavailable, and in their place, medium to large rivers offer comparatively rich water resources. In the Bronze Age, the foundation of urban centres was enabled by construction of surface channels deriving water from the riverine channels, thus leading to permanent modification of the landscape.[10] Cities were developed along the water channels within oases. The waterworks required continuous maintenance, and the urban context was organised in relation to both natural features and human factors. Through the Bronze and Iron Ages, and again especially with the onset of the Achaemenid period, cities assumed a recognisable water-based structure and morphology, which continued unchanged up to Islamic conquest. In the central Asian steppes, the management of oases permitted development of important urban centres, such as Marw, Bukhara, Samarqand, Urgench (modern Konye Urgench). The cities were connected by a network of precise caravan routes, with tanks and reservoirs (*sardāb*) fed by elaborate water harvesting techniques.[11]

By the time of the Islamic conquest, the urban patterns of both the Iranian Plateau and adjacent Central Asia were generally established.

3 Cities on the Fringes of the Iranian Plateau

3.1 *Case Study 1: Rayy*

Most of the urban nuclei of mountainous western Iran have existed for millennia.[12] A particular concentration was found on the foothills between the

9 De Planhol, Xavier, s.v. "Cities: Geographical Introduction", in *EIr*, V (1991): pp. 603-7, in particular pp. 603-5; see also Boucharlat, Rémy, "Les galeries de captage dans la péninsule d'Oman au premier millénaire avant J.-C.: questions sur leurs relations avec les galeries du plateau iranien", in Briant, P. (ed.), *Irrigation et drainage dans l'Antiquité, qanāts et canalisations souterraines en Iran, en Egypte et en Grèce* (Paris: Thotm éditions, 2001): pp. 157-83; here pp. 178-80.

10 Francfort and Lecomte, "Irrigation": p. 646.

11 Concerning the water resources along the caravan routes, see the informative article: Wordsworth, Paul, "Merv on Khorasanian trade routes from the tenth-thirteenth centuries", in Rante (ed.), *Greater Khorasan*: pp. 51-62, focusing on the stretch of caravan routes linking the city of Marw with the Oxus valley.

12 Another example for early settlement in this region is Kashan: see the contribution by Mottahedeh and Amanat in this volume. Kashan has hydrological conditions similar to those of Rayy.

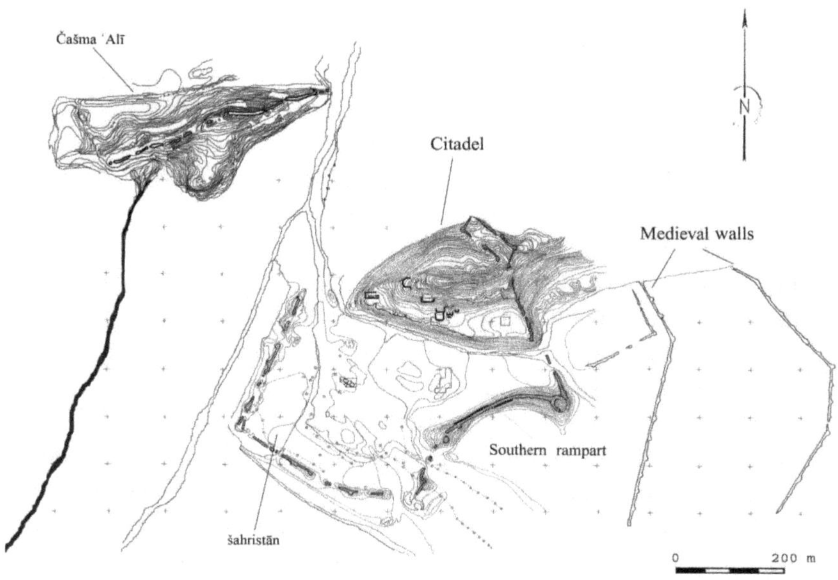

FIGURE 2 Topographic map of Rayy
AFTER RANTE 2015

Zagros and Alburz mountain ranges. Among these settlements was Rayy, situated on the lower southern foot slope of the Alburz. Beyond this extends the desert plateau, occupied by the western branch of the Dašt-i Kavīr.

The first known occupation of Rayy, concentrated on the spring of Čašma ʿAlī (Fig. 2), dates to the late Neolithic. Recent excavations reveal that the foundations of the two main components of Iranian cities, the citadel (*arg*) and the *šahristān* (the inner city), date to the Iron Age I and II (fourteenth-eighth centuries BCE). There is as yet little data concerning Iron Age III[13] (including the Median period, seventh-sixth centuries BCE).[14] Later Parthian (second century BCE-second century CE) and Sasanian (third-seventh century CE) occupations have been identified in the citadel and *šahristān*. The urban area of Rayy expanded south from the original nucleus during the Sasanian period, with construction of prosperous residences and the Fire Temple of Tepe Mīl

13 The Median period was likely important at Rayy: the fact that we have limited data seems due to a lack of archaeological investigation and study of material culture.

14 Rante, Rocco, *Rayy: from its Origins to the Mongol Invasion. An Archaeological and Historiographical Study* (Leiden: Brill, 2015): p. 11.

demonstrating the city's political and cultural importance at this time.[15] The suburbs, protected by walls, very likely developed during the Abbasid period.

Rayy is situated on an interfluvial plain between two rivers (Persian: *rūd*). Apart from these rivers and a stream descending from Časma ʿAlī, there are very few watercourses. The larger of the two rivers is the Karaj, to the west, passing through the vast and fertile plain of Šāhryār, an area occupied very early, which now includes the modern city of Karaj. To the east, the waters of the Jājirūd supply the plain of Warāmīn.[16]

By the time of the founding of the fortified city of Rayy, with its citadel and *šahristān*, Rayy was probably fed by water from the Jājirūd, which was channelled through smaller streams and ponds. The first *qanāt*s were very probably constructed after the city was founded. These would have significantly increased the water supplies and therefore the potential for demographic and economic growth. The development of the *qanāt*s seems to have permitted population growth within the morphological area of the original city location. It was only during the Parthian era that the city expanded westwards into what is now called Ḥusaynābād. During the Sasanian period, the *qanāt* network was expanded to the south, allowing Rayy to colonise lands nearer the desert and further from the mountains. No traces of pre-Sasanian occupations have been found in these southern lands, suggesting that they were first irrigated precisely at this time. The construction of the large Fire Temple of Tepe Mīl and the aristocratic residences of Čāl Ṭarḫān, Niẓāmābād and ʿIšqābād, seems to confirm the Sasanian implementation of important hydraulic programs for improved irrigation in the areas south of Rayy.

The above chronology with its hypotheses concerning irrigation works is based on geomorphological research conducted by the author, in collaboration with Eric Fouache. The theses presented require further investigation. One research question arises from the possibility that the new territorial arrangements south of the ancient nucleus of Rayy brought water not only from the older, probably insufficient system supplied from the Jājirūd basin, but also via a system fed from the northwest, derived from the more substantial waters of the Karaj basin. The 1947 topographic plans in fact show numerous *qanāt*s and surface channels, or 'canals', bringing water to southern Rayy from the villages of Kan and Dāniš (see Fig. 3). The proposed chronological sequence implies that a vast hydraulic network was constructed during the Sasanian

15 See Rante, *Rayy*: fig. 102.

16 For these two rivers see Valles, Vincent, Gholami, M. and R. Lambert, "Chimie des eaux et alimentation du bassin versant du Djajerud (Iran)", *Hydrologie Continentale*, V/1 (1990): pp. 61-9, and Hourcade, Bernard, s.v. "Karaj River", in *EIr* XV (2010): pp. 543-4 and Hourcade, Bernard, s.v. "Jājrud River", in *EIr*, XIV (2008): pp. 399-400.

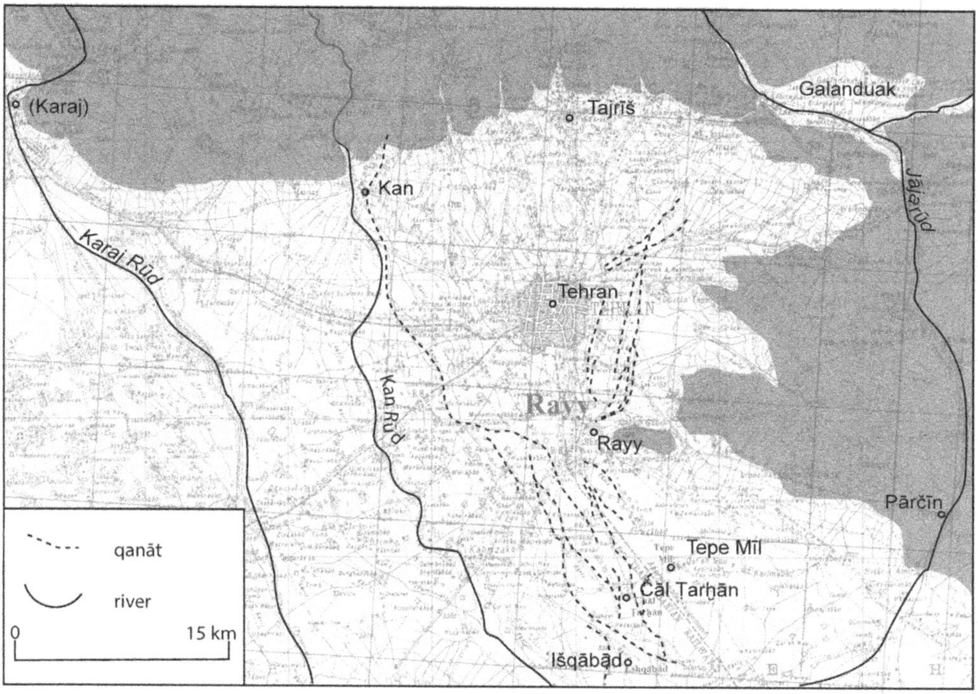

FIGURE 3 Geotopographic map of the plain of Rayy
DRAWING: D. DURAND-GUÉDY AND R. RANTE BASED ON UNITED STATES ARMY MAP SERVICE K501, 1947.

period. The question then arises who was behind the construction works: were they designed and implemented by the empire or the urban elite? Whoever the main actors, the aim was evidently agricultural and socio-political expansion towards the barren lands south of the original centre. Apart from the main agents, we must also inquire into the urban organisation that would have developed and maintained the waterworks. Given that Rayy was an important military outpost and major centre on the caravan route to Central Asia, it is not difficult to imagine that imperial would have lain behind this large-scale transformation of the landscape.

The archaeological and historical evidence is as yet insufficient for firm conclusions on these matters, however it seems that in the first centuries of the Muslim period the inhabitants of Rayy continued to rely on the system inherited from the Sasanian period. During the late eighth and the ninth century, the townspeople continued development of the newly irrigated lands to the east and south, especially those located immediately south of the ancient city nucleus, in which the Arab communities had settled in the seventh century. Here, according to the account of the tenth century geographer Ibn

Ḥawqal, the city advanced a well-organised network of both *qanāt*s and canals. The two most important canals were the Surqānā (var.: Suraqanī, Surīnī) and Gīlānī.[17] Building on the useful studies of Ḥusayn Karīmān, through field work and archival research, I suggest that the Surīnī canal would correspond to the current Čašma ʿAlī.[18] Fed by springs emerging from bedrock, the river flows south-southwest through the Rūḏa quarter. Concerning the Gīlānī canal, a nineteenth-century CE map drawn by the architect Pascal Coste shows this originating in the "Ṭabaristān". The canal entered the city from the north, probably passing through Bāb Dūlāb, the *šahristān* and the Sārabān quarter.[19] A third canal appears on Coste's map but it has not yet been possible to confirm its existence or dating through archaeological surveys or historical sources.

Although precise dating is impossible due to the scale of modern urbanisation, it is clear that the network of waterworks in Rayy has been managed for centuries. The fact that sectors of the irrigation network were present in the oldest parts of the inner city testifies to its existence since at least the founding of the fortified city, in the Iron Age. I also suggest that the later urban reorganisation undertaken by the Abbasid Caliph al-Mahdī, in which areas south of the ancient nucleus were made habitable, required both an additional water source and expansion of the network of channels.[20] Was the unconfirmed (or now hidden) third canal, which apparently flowed through these same southern areas, a part of this reorganisation?

Given the Rayy's strategic location, the restructuring of the water sources and related distribution networks was probably due to imperial will. The analysis of pottery recoveries suggests that Islamic policy in managing the southernmost lands previously occupied by Sasanian nobles and Zoroastrian priests was to retain and reuse the buildings, including the Fire Temple, at least until the fourth/tenth century. Ultimately the buildings were abandoned, but not purposefully destroyed, while management of the agricultural irrigation system has continued up to the present day.

---

17 See Ibn Ḥawqal, *K. Ṣūrat al-arḍ*, ed. de Goeje (2nd edition revised by J.H. Kramers, Leiden: Brill, 1967). Trans. J.H Kramers and G. Wiet, *Configuration de la terre*, 2 vols. (Paris and Beirut: Maisonneuve et Larose, 1964): p. 378, trans.: II, p. 368; Ibn Rusta, *K. al-Aʿlāq al-nafīsa*, ed. de Goeje (2nd edition, Leiden: Brill, 1967). Trans. G. Wiet, *Les atours précieux*, (Cairo: Société de géographie d'Egypte, 1955): p. 186, trans.: pp. 194-5. Le Strange, Guy, *The Lands of the Eastern Caliphate* (3rd edition, London: Frank Cass, 1966): p. 215.

18 Karīmān, Ḥusayn, *Rayy-i bāstān*, 2 vols. (Tehran, Anjuman-i Āṯār-i millī, 1345-1349/1966-70): pp. 130-2; Rante, *Rayy*: p. 134, fig. 6.

19 Karīmān, *Rayy*: pp. 143-7; Rante, *Rayy*: p. 36.

20 Bulliet, *Cotton*: pp. 14-6, discusses management and ownership of the wastelands in accordance with Muslim law.

3.2 *Case Study II: Nishapur*

Further to the east, in Khurasan, Nishapur presents an emblematic case of the ongoing dynamic between settlement and water management in the Iranian territories during the early Islamic period.[21]

The city is situated on the northern edge of the Iranian Plateau. Recent archaeological excavations have dated its foundation to the final years of the reign of the Sasanian king Šapūr II (309-379). The original fortified nucleus consisted of an approximately square citadel and a rectangular lower city, or *šahristān* (Fig. 4). The location is in an interfluvial zone, rich in water resources. To the northwest is the Mīrābād Rūd, to the northeast the Bužān Rūd (Fig. 5),[22] and beyond this are the foothills of the Bīnālūd range, rising 2000 metres above the plain. Rain and snowfall from the mountains irrigate the farmland below.[23] To the southwest are the brackish and marshy courses of the Šūra Rūd.

The Nishapur area is irrigated by both surface water and *qanāts*. The Sasanian city was supplied with water by *qanāts*, fed from the water table of the Bužān channel, to the north-east. Some of the *qanāts* surfaced before entering the city; others passed under the citadel and the *šahristān*, emerging in the agricultural fields south of the original nucleus.

Islamic communities soon settled in the ancient nucleus and founded nearby villages.[24] With the arrival of the Tahirids in the third/ninth century, other outlying areas were urbanised.[25] The dynasty established itself in Šādyāḫ, a little over two kilometres west of the ancient city. According to written sources, this new location first served to quarter the army,[26] however it rapidly assumed the features of an urban political centre. The most recent Iranian excavations demonstrate the continued existence of a strong rampart

21 For a discussion of why Nishapur evolved the way it did, see Richard Bulliet's contribution in this volume.

22 In this article I provide an excerpt from the topographic USSR Military Map of 1989 (1:50,000), specifically because this excludes *qanāt* developed after the founding of Šādyāḫ. More detailed Iranian mapping (1:25,000) is also available, but this shows all *qanāts*, including those constructed later.

23 Fouache, E. Cosandey, Cl., Wormser, P., Kervran, M., and R. A. Kh. Labbaf, "The River of Nishapur", *StIr*, XL/1 (2011): pp. 99-119, here p. 102.

24 Bulliet, *Cotton*: p. 41. Bulliet notes 68 villages founded around the ancient walled city of Nishapur, but a previous publication indicates 65; Bulliet, Richard, "Medieval Nīshāpūr: A Topographic and Demographic Reconstruction", *StIr*, V (1976): pp. 67-89, here p. 73.

25 Rante, Rocco and Collinet, A., *Nīshāpūr Revisited. Stratigraphy and Ceramics of the Qohandez* (Oxford: Oxbow Books, 2013): pp. 53-5.

26 Le Strange, *Lands*: p. 385.

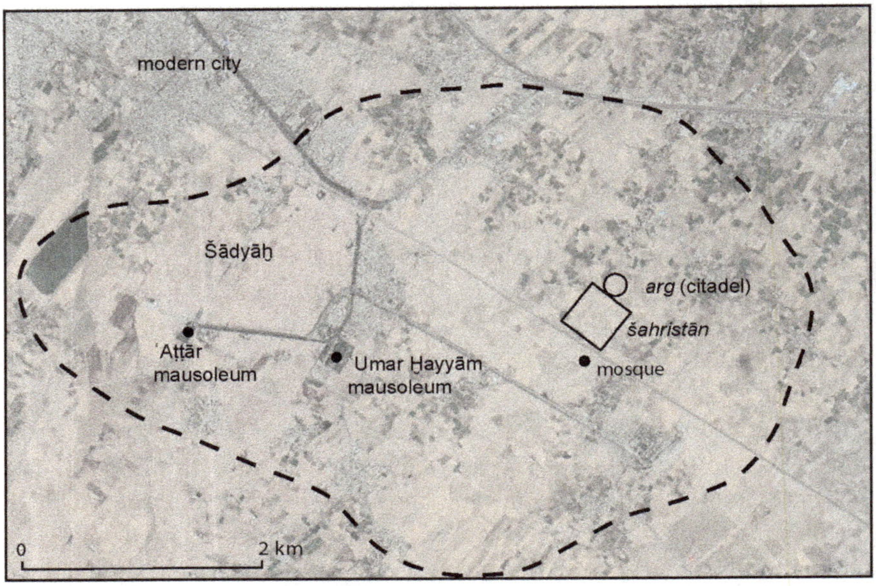

FIGURE 4 City limits of Medieval Nishapur
SOURCE: GOOGLE EARTH 2012, BULLIET, "MEDIEVAL NĪSHĀPŪR", PL. 1, RECTIFIED BY THE AUTHOR

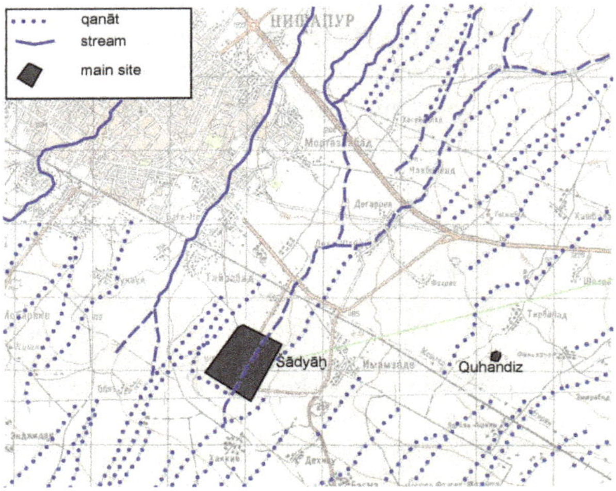

FIGURE 5 a. Water stream and *qanāts* on the site of Nishapur.
DRAWING D. DURAND-GUÉDY BASED ON RUSSIAN MILITARY MAP, 1989

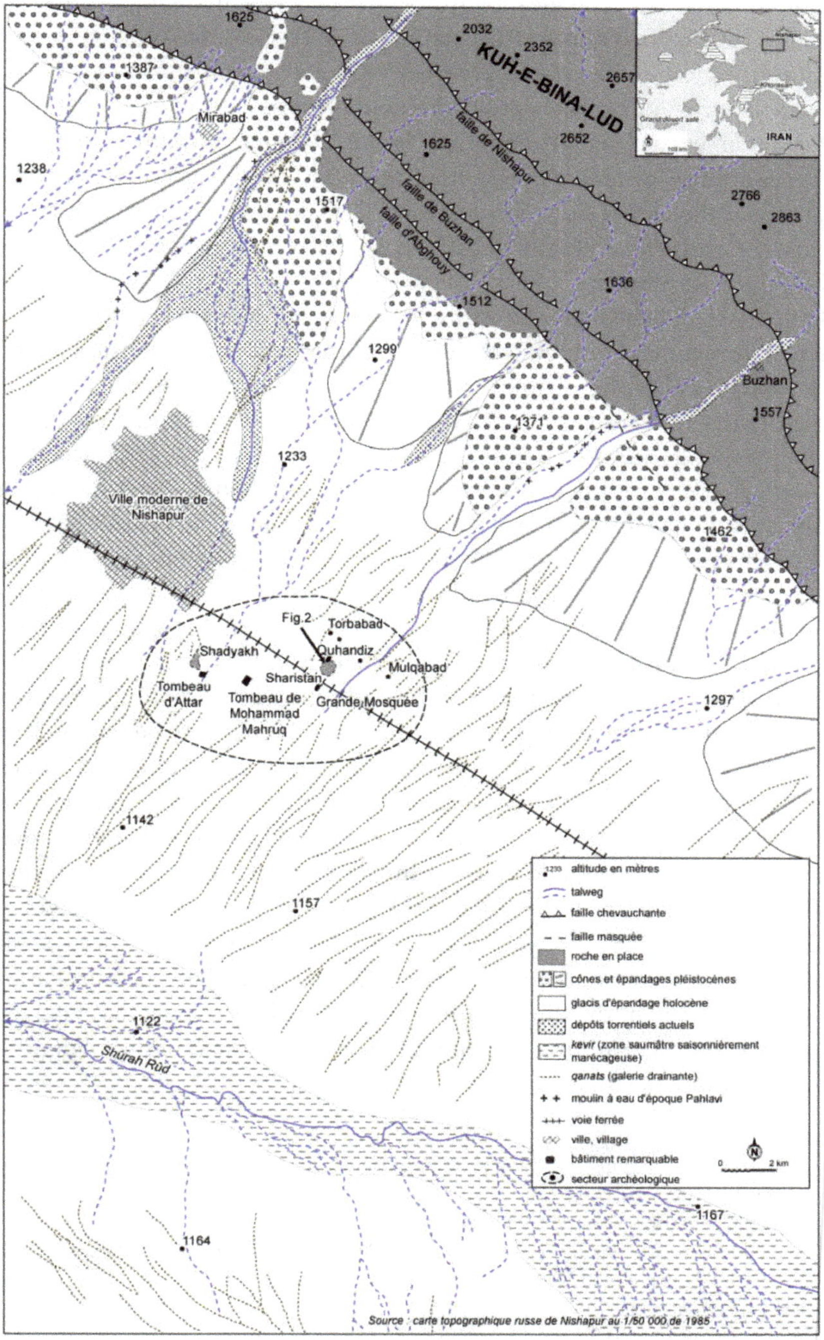

FIGURE 5 b. Geomorphological map of the Nishapur area
DRAWING: FOUACHE ET AL., "RIVER OF NISHAPUR", FIG. 1

around the nucleus.[27] The locality of the new Tahirid settlement was well irrigated by the Mīrābād Rūd and its channels. Nevertheless, according to recent geomorphological and archaeological fieldwork,[28] it seems that a new *qanāt* network was dug during this period, oriented towards Šādyāḫ and other later settlement areas, such as Tepe Madrasa, Sabz Pūšān and Vineyard Tepe.[29] This new network was clearly supplied from the same north-eastern water table that had fed the earlier Sasanian network. Thus, already in the fourth/tenth century, the entire area identified by Richard Bulliet as ancient Nishapur was well supplied with water.

Bulliet has published an interesting reflection on toponyms, using them as evidence to relate the founding of villages in Iran with the expansion of the *qanāt* networks after the Islamic conquest.[30] The aim of this current article is to identify the water networks and their various uses through the centuries, however it should be noted that in the oldest area of Nishapur there seems to a lack of general correspondence between *qanāt* names and village names. In particular, Bulliet identifies only three associations of *qanāt* and village name.[31] Perhaps this relative lack of unity between *qanāt* and village name would be justified by the fact that “digging *qanāt* close to an established city had less to do with agriculture than with bringing water to a rapidly growing urban population.”[32] Bulliet also supposes that “the cotton fields were presumably established farther from town and involved Iranian converts to Islam more than Arab settlers”. These arguments, although enticing, may be wrong. It is true that there are few correspondences between village and *qanāt* names around the old walled city of Nishapur (citadel and *šahristān*). However we now have geomorphological evidence of a large-scale system of *qanāt*s drawing from the same water table that had supplied the old city, to the northeast (the Bīnālūd piedmont), but supplying water to the area further west, where the Islamic settlement of Šādyāḫ was built. Given that this large water system clearly supplied the new urban centres (e.g. Šādyāḫ, Tepe Madrasa, Sabz Pūšān and Vineyard Tepe) of the third/ninth century, it must have been intended for human as well as agricultural use, and can likewise be dated to the beginning of the third/ninth century.

---

27 On recent excavations see Labbāf, Khānikī Rajabʿalī, *Reports of excavation* (Tehran: ICHHTO-ICAR, 2004).

28 Personal communication of Labbāf Khanikī.

29 See Wilkinson, Charles K., *Nīshāpūr, Some Early Islamic Buildings and Their Decoration* (New York: Metropolitan Museum of Arts, 1986).

30 Bulliet, *Cotton*: p. 20.

31 Ibid.: p. 41.

32 Ibid.: p. 41.

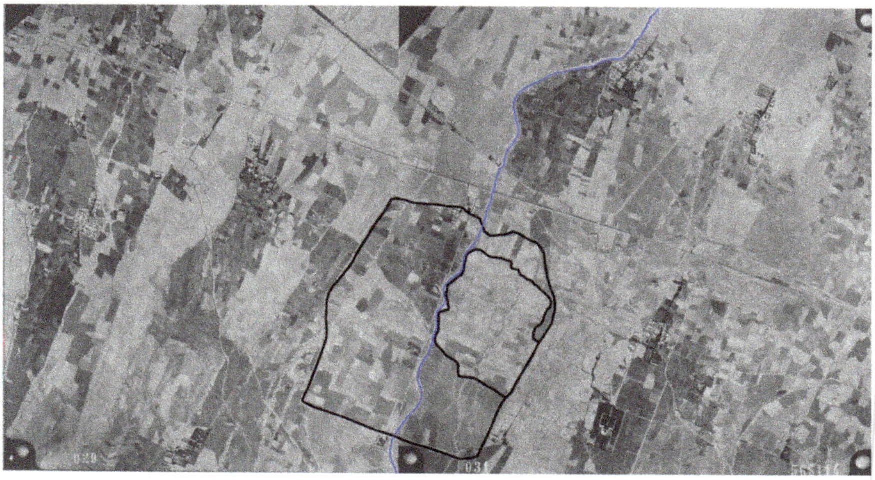

FIGURE 6 Aerial photo of the Nishapur area and limits of Šādyāḫ
AFTER *SĀZMĀN-I NAQŠA-BARDĀRĪ-YI ĪRĀN*, AROUND 1950, RECTIFIED BY THE AUTHOR

All segments of the population (and not only the Islamic communities, as postulated by Bulliet) would have been involved in this new large-scale public works project, under the aegis of the Tahirids. Together they would have constructed a vast water network to meet community needs, which could explain the absence of connections between village toponyms and *qanāt*s. The water network irrigating a territory would be thus fed by underground channels coming from the foothills.

Recent geomorphological research and examination of air photos shows that the Islamic settlement at Šādyāḫ city consisted of a quadrilateral fortified area, of approximately 150 hectares, crossed at the centre by a branch of the Mīrābād River. In the northeast of the fortified area is a zone that is slightly higher, probably due to both natural topography and subsequent agricultural development. This elevated zone is bordered on the east by the central channel, which is slightly higher than the entire fortified area, a position that allows it to function in a dual capacity. Elevated, the channel could supply the city with water using gravity flow, but also drain away excess water, down the slopes from the city centre to the exterior. (In the vast Iranian world, where mud brick was the basic structural material, cities needed not only an adequate water supply but also an effective drainage system, so that excess water in periods of heavy rainfall or flood could be removed before structures would be undermined). Lesser canals would have been employed to carry water from the main canal towards the borders of the fortified quadrangle. The slightly elevated central canal provides the necessary slope for both supply and drainage of excess water.

As concerns the topography of the Šādyāḫ area, its relatively minor elevation above the surrounding lands can be explained by a series of events and factors. Šādyāḫ was rapidly urbanised between the second/eighth and fourth/tenth centuries, and recent excavations have shown that it was never as densely occupied as the old city centre.[33] These factors would lessen the accumulation of topographic relief. On the contrary, Nishapur suffered structural destruction and accumulations of debris several times, through the sixth/twelfth centuries and the internal strife caused by the Turkmen takeover, and also from natural disasters. (After the 1405 earthquake the inhabitants moved to a new site to the west.)[34] Finally, the presence of the central channel did not allow for the accumulation of layers and the resulting formation of a high domed *tepe*: not being equipped with hydraulic pumps, the city managers could not continually raise the inner canal to higher levels.

4 Central Asian Cities: Steppe Zone

4.1 *Case Study III: Marw*

The best-known and most comprehensive Central Asian urban site is Marw.[35] The pre-Islamic city dates from Achaemenid times. It was surrounded by a city wall of essentially oval shape ('Arg Qalʿa ', vernacular 'Erk Kala') with a central elevated mound/citadel, and some distance south of this was a later Seleucid walled city (second half of fourth-second century BCE) erected on a square plan (Gyaur Kala).[36] The Parthians and the Sasanians did not significantly alter the resulting urban plan, although they continued the development of Marw as a major military, political, administrative and trade centre.

The sources report that Abū Muslim, leader of the Abbasid revolution, commissioned a mosque and a governor's palace outside the old city, alongside the Mājān channel, a locality which seems to have been inhabited since prior

33 Rante and Collinet, *Nīshāpūr Revisited*: pp. 53-5, 136-55.

34 Melville, Charles, "Earthquakes in the History of Nishapur", *Iran*, XVIII (1980): pp. 103-20, here pp. 113-14.

35 For a global study on Marw, see Herrmann, Georgina (ed.), *Monuments of Merv: Traditional Buildings of the Karakum* (London: The Society of Antiquaries of London, 1999).

36 Herrmann, Georgina, Masson, V. M., Kurbansakhatov, K. et al., "The International Merv Project. Preliminary Report on the First Season (1992)", *Iran*, XXXI (1993): pp. 39-62, 44-8.

to the Islamic conquest.[37] The sources also refer to a new settlement named Marw al-Šāhijān ('Royal Merv'), constructed to the west of the old town, on the west bank of the Razīq canal (Fig. 7), which probably remained without walls until the late fifth/eleventh century.[38] Important public buildings and housing are located along the Mājān canal, which flowed from the south to the north through the centre of Marw al-Šāhijān and provided the focus for this new urbanisation.[39]

The Islamic city of Marw received its water from four main channels arriving from the Murġāb river, of which one was the Mājān.[40] The furthest east was Asādī al-Ḫūrāsānī channel, east of Gyaur Kala. The Razīq canal, the best known, flowed between Gyaur Kala and Marw al-Šāhijān. As Tim Williams affirmed, the Razīq canal supplied water mostly to the eastern areas, including Gyaur Kala, more so than to Marw al-Šāhijān in the west, given the higher elevation of the latter and the general slope from west to east. The Mājān canal, at the center of Marw al-Šāhijān, was the main source of supply for the urban centre. The Ḫurmūzfarra canal, west of Marw al-Šāhijān, provided water for the lands immediately west of the city and flowed northwards to the town of Ḫurmūzfarra, a day's march away.

The Razīq canal was probably an early canalisation of the Murġāb River, possibly supplying the city since its founding. The Mājān canal is not as old. Although it existed prior to the reorganisation program begun by Abū Muslim in the second/eighth century, the date of construction remains unclear. Tim Williams considers it the main water source of the medieval city.[41] From the onset of the Abbasid period (751-1258), the surrounding area developed as the main centre of Marw al-Šāhijān, replacing the gradually abandoned Achaemenid and Seleucid centre.[42] Along its central path were clusters of important civic buildings, including mosques, baths and markets.

---

37 Williams, Tim, "The City of Sultan Kala, Merv, Turkmenistan. Communities, Neighbourhoods and Urban Planning from the Eight to the Thirteenth Century", in Bennison, A.K. and Gascoigne, A.L. (eds.), *Cities in the Pre-Modern Islamic World. The Urban impact of religion, state and society*, (London and New York: Routledge, 2007): p. 42-62, 45-7.

38 Ibid.: p. 54.

39 Ibid.: pp. 53-7.

40 The water system of Marw is described in Hermann, Masson and Kurbansakhatov, "International Marw Project": pp. 40-1, 48; Williams, "The City of Sultan Kala": pp. 44-52. I am very grateful to Professor Tim Williams for providing additional details.

41 Personal communication.

42 Kennedy, Hugh, "Islamic Geographies", in Herrmann, (ed.), *Monuments of Merv*: pp. 121-6, 30.

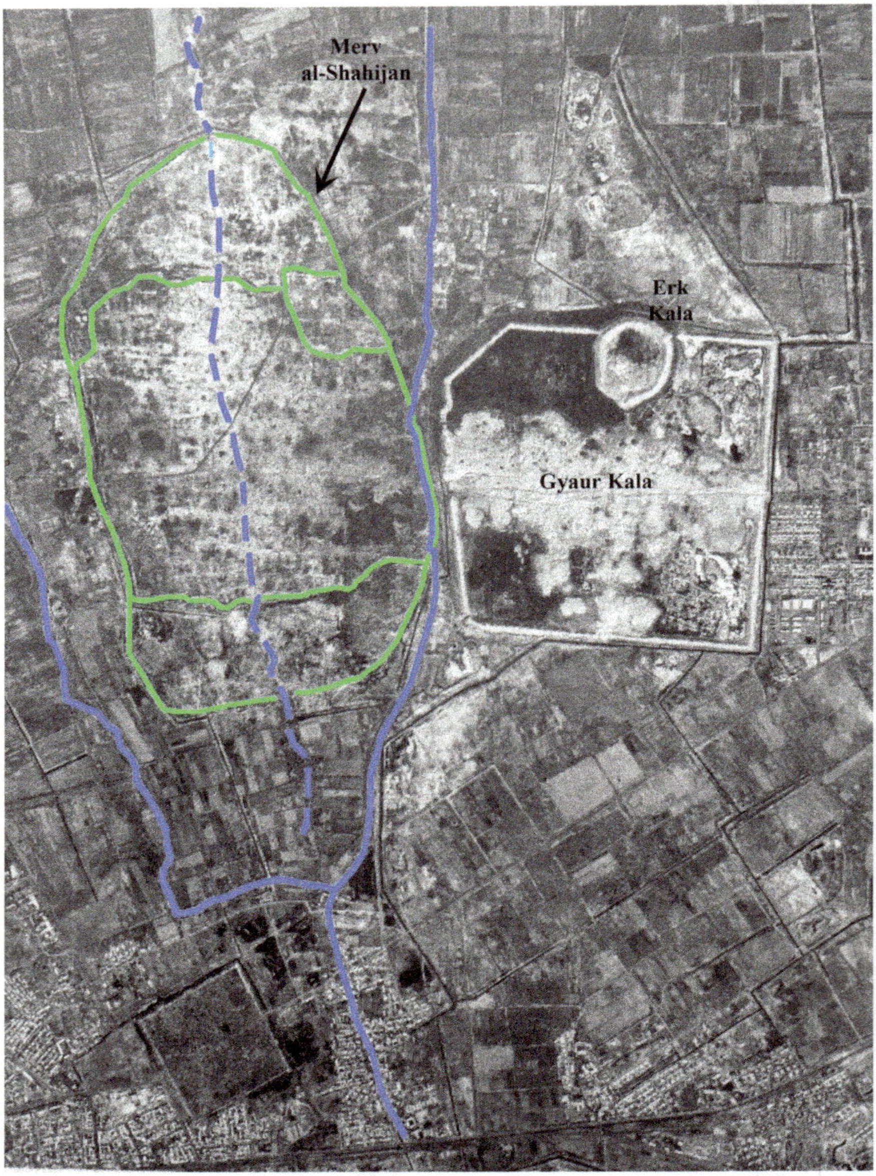

FIGURE 7 Aerial photo of Marw al-Šāhijān
AFTER WILLIAMS, "SULTAN KALA"

Archaeological surveys and excavations confirm the presence of buildings with high water consumption along the Mājān, including the public baths and mosques. These would have required an important network of canalisation for elimination of excess water. Although we as yet have no archaeological evidence, Marw al-Šāhijān must therefore have had an urban morphology adapted for drainage of this water. It would have been the higher position of the central Mājān canal that allowed canalisation of water away from the city on both eastern and western sides, operating by gravity alone.

In terms of occupation dynamics and political realities, Marw and Nishapur present strikingly similar innovations. In both cases, a new Islamic-period political power decided to found a new settlement at some distance from the old one. Marw al-Šāhijān was the earlier of the two cases. Here, water management was based on the redevelopment of the pre-existing Mājān canal, formerly external to the city, as the main water resource for the new urban centre. The development program may also have involved reuse of the external Razīq canal, further to the east, as a drainage canal. A very similar settlement dynamic operated at Nishapur/Šādyāḫ, where the new urban area was developed around a main central canal, and the *qanāt* system was extended to increase the water supplies for both the city and the lower areas to the east.

The main difference between the two cases was the use of *qanāt*, at the Iranian plateau site of Nishapur/Šādyāḫ. However, both here and at Marw al-Šāhijān, the new urban organisation was conceived around a surface water canal. This organisation had long-term topographical consequences: in both cases, these systems precluded the newly founded centres from ever developing sufficient elevations to constitute high-domed *tepe*.

4.2 *Case Study IV: Paykand*

Further to the north, in Transoxiana, the site of the city of Paykand was occupied as early as the third century BCE. Settlement was initially between the end of the Bukhara oasis and the Qaraqöl micro-oasis, on a plain east of the Zarafšān River. Recent geomorphological research has shown that this area was supplied with water from the Zarafšān, beginning around the fourth century BCE. This research also determined the paleo-path of the Zarafšān as flowing west, since at least the late Neolithic-early Bronze Age.[43] The territory between Farāb (or Firabr), on the Oxus, and the Bukhara oasis, was therefore desert land

43 A detailed geomorphological study of the Bukhara oasis can be found in Fouache, *et al.*, "The Role of Catastrophic Floods Generated by Collapse of Natural Dams Since the Neolithic in the Oases of Bukhara and Qaraqöl: Preliminary Results", *International Journal of Geohazards and Environment*, III/2 (2016): pp. 150-65.

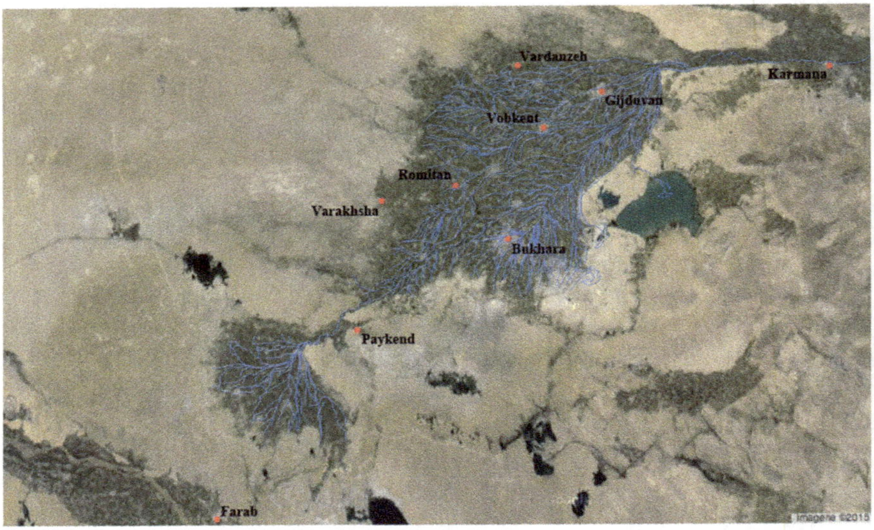

FIGURE 8 Map of the Bukhara and Qaraqöl oases
SOURCE: ESRI IMAGE AND GIS 2015

with marshy zones, where the water table of the Kaška Darya basin emerged at surface. The geomorphological formation of the Qaraqöl micro-oasis dates to around the fourth century BCE or later. The swampy area must have created serious difficulties for caravans attempting to reach the Bukhara oasis and points located further east (Fig. 8). It is around the time of the appearance of this micro-oasis that the first occupation of Paykand is attested, while recent excavations suggest that the strongly fortified citadel and *šahristān*-1 were constructed during the first centuries CE (Fig. 9).[44]

Settlement was initially situated south of a topographical depression, which benefited from periodic flooding of the Zarafšān. This made the locality ideal for agriculture and habitation. North of this micro-oasis, a thalweg (line of lowest elevation within a valley), with a profile in equilibrium with the Kaška Darya paleochannel, seems to have created a flow into the plain south of Paykand until at least the Bronze Age. The arrival of the Zarafšān in the fourth century BCE closed this Bronze Age thalweg, constantly raising its bed and banks until the upstream sector transformed into a basin, or topographical depression. Until the 1960s, when Soviet authorities widened the north-south canal flowing to Paykand, the only means of transporting water from Zarafšān was

44 Like other sites (e.g. Pendjikent), the urban plan of Paykand expanded in its lower city, or *šahristān*, described as *šahristān*-1 and *šahristān*-2.

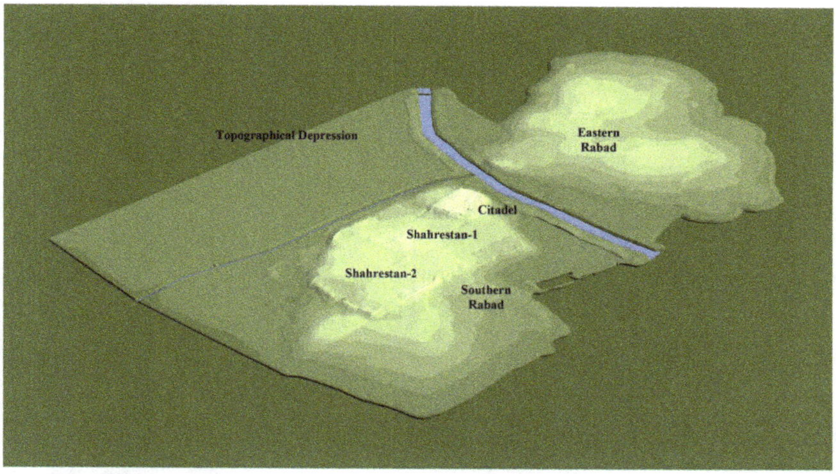

FIGURE 9 3D map of Paykand
AUTHOR: DANILO ROSATI, RILIEVI SRL

by a canal arriving from its left bank (today observed at Yangimazar), circumventing the basin from the north, along a route still visible, and then flowing southwards to the fortified city and beyond.[45] Paykand has significant natural slope towards the north and south. In general, the alignment of the longer side of a city along a channel, as frequently seen in the Middle East, ensures more rapid elimination of excess water.[46]

In Islamic Paykand, the urban policy was to maintain previous water resources and reuse previously occupied space, rather than construct new channels. Archaeological research shows that the organisation of the inner city retained the features of the pre-Islamic occupation, contrary to the experience of Marw and Nishapur.[47] The unusually high degree of reuse inside the city

45 Data brought to light by Eric Fouache and Lucie Cez during the 2016 archaeological mission organised in the Bukhara Oasis under my direction.

46 Concerning the Middle East, see Margueron, Jean-Claude, *Cités invisibles* (Paris: Geuthner, 2013): pp. 384-6.

47 For previous excavations, see Mukhamedzhanov, A.R., Semenov, G.L., Mirzaakhmedov, D.K. and Š.T. Adilov, *Gorodišče Pajkend* (Taškent: Fan, 1988), Semenov, Grigorij L. (1989), "Sogdijskij gorod v rannem srednevekov'e: formirovanie plana", in Smirnova, G.I. (ed.), *Itogi rabot arkheologičeskikh ėkspeditsij Gosudarstvennogo Ėrmitazha* (Leningrad: Gosudarstvennyj Ėrmitazh, 1989): pp. 128-40; Semenov, G.L. and Mirzaakhmedov, D.K., *Raskopki v Pajkende v 2000 godu. Materialy bukharskoj ėkspeditsii* (St Petersburg, 2004); Torgoev, A.I. and Mirzaakhmedov, D.K., *Raskopki v Pajkende v 2006 godu. Materialy bukharskoj ėkspeditsii* (St Petersburg, 2009); Semenov, G.L., "Dwelling Houses of Bukhara in the Early Middle Ages", in Compareti, M., Raffetta, P., and G. Scarcia (eds.), *Ērān ud Anērān.*

FIGURE 10 Real section of ditch-streets of Paykand
PHOTO: AUTHOR

is demonstrated by the water drainage system. At Paykand, the inhabitants simply took advantage of the gentle slopes for gravity-operated drainage, constructing streets with a central ditch that was then maintained through the centuries (Fig. 10). Each time that the occupation level was raised, or the water system required renovation, the streets were arranged with a new accompanying ditch. The city's underground canalisations were also oriented along the natural slopes, making elimination of water more rapid.

5 The Roles of Water in Topography and Organisation of Cities: The Iranian Plateau and Central Asia

Water played a major role in the organisation and topography of cities on the Iranian Plateau and in Central Asia. Eminent scholars have dealt in depth with

Studies Presented to B.I. Maršak on the Occasion of his 70th Birthday (Venice: Cafoscarina, 2006): pp. 555-70; Omel'čenko, Andrej B. and Mirzaakhmedov, D.K., *Raskopki v Pajkende v 2011 godu. Materialy bukharskoj ėkspeditsii* (St Petersburg, 2012), Rante, Rocco, "Les fouilles de Paykend: nouveaux éléments," in Bendezu-Sarmiento, J. (ed.), *Archéologie française en Asie centrale. Nouvelles recherches et enjeux socioculturels* (Paris: de Boccard, 2013): pp. 219-40.

the importance of political decisions in developing the geometrical designs and internal plans of these cities.[48] While recognising the importance of these decisions, the current paper explores how the founding of these cities, and above all the evolution of their urban character, depended first and foremost on the available water resources. In this section, we will examine the ways in which the hydrography and topography of a site influenced urban development. These considerations were in addition to other factors, above all the need for protection. We will also examine the corresponding aspect of how urban development changed the landscape.

In the two cases on the Iranian Plateau studied above, we observe that the city was oriented in relation to the inflow of water. The citadel was the first area to be supplied, whether the water arrived from the foothills via *qanāt*s, canals, rivers or streams. After the citadel, and once the inhabited areas (*šahristān*) were supplied, the remaining water flowed to gardens in and around the urban nucleus, through *qanāt*s and/or surface canals. Here, as in Central Asia, the city plan was thus founded on the principle of supplying water according to the respective importance of the various areas. The principle of urban development was that people were supplied with better quality water than fields and gardens. Was this a general principle which was valid also in other cities, on the Iranian plateau as well as in the central Asian steppe region?

Although our presentation of case studies is limited to Rayy and Nishapur, the same phenomena can be observed at Hamadan and Herat, in the western and eastern parts of the Iranian plateau. At Hamadan, recent archaeological investigations suggest that as early as the Parthian period the political centre, previously situated in Hagmatāna Tepe, was relocated to Muṣallā Tepe (see Fig. 11).[49] If this shift is confirmed by further excavation, it would mean that new urban area was located further to the southeast, on the opposite bank of the Alūsjird River, the main water source, and slightly upstream. The location was therefore supplied with purer water, validating the observed principle of

---

48 See among others, Tskitishvili, O., "Two Questions Connected with the Topography of the Oriental City in the Early Middle Ages", *JESHO*, XIV/3 (1971); pp. 311-20; the interesting suggestions proposed by Grenet, Frantz, "Crise et sortie de crise en Bactriane-Sogdiane aux IVe-Ve s. de notre ère: de l'héritage antique à l'adoption de modèles sassanides", in *La Persia e l'Asia Centrale da Alessandro al X secolo, Atti dei Convegni Lincei*, CXXVII (Rome: Accademia dei Lincei, 1996): pp. 367-90 and Gaube, Heinz, "Iranian Cities", in Jayyusi, S.K., Holod, R., Petruccioli, A. and A. Raymond (eds.), *The City in the Islamic World* (Leiden: Brill, 2008): pp. 159-80, but also Semenov, "Sogdijskij gorod", who explain the urban evolution through historical events and political choices.

49 See Brown, Stuart C., s.v. "Ecbatana", in *EIr*, VIII (1997): pp. 80-4, here pp. 80-81; Āḏarnūš, Mas'ūd, *Ġarāriš-i kāvūšhāy-i lāye-šināḫtī tipi Haġmatānah* (Tehran, ICHHtO-ICAR: 2008): p. 37; Ṣarrāf, M., *Hegmatana* (Tehran, 1392š/2013): p. 11.

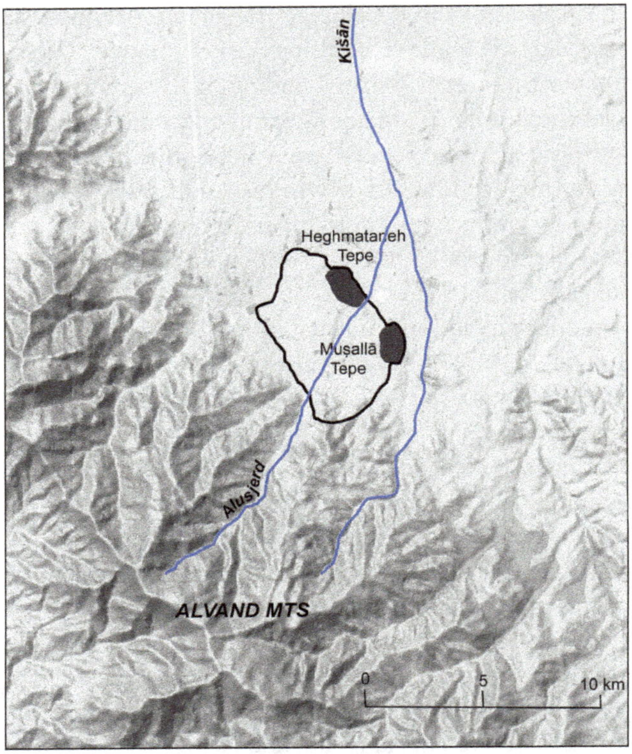

FIGURE 11 Situation map of Hamadan. Drawing D. Durand-Guédy and Rocco Rante (outline of the medieval city according to Sharraf 2013: 144)

urbanisation. Preliminary observations are also that this new urban area received other supplies of water through another branch of the Alūsjird, flowing on the eastern side.

The city of Herat was framed by two important canals, the Jūy-i Injīl to the north and the Karubār to the south, both originating from the Harī Rūd (see Fig. 12), while *qanāt* running from the mountains to the north and the north-east irrigated the vast and fertile lands to the east. Through time the city expanded concentrically, still maintaining the pattern of the medieval eighth/fourteenth century, as mapped by Franke.[50] Since water could be brought from

50 Franke's map, in Franke, Ute, "Ancient Herat Revisited. New Data from Recent Archaeological Fieldwork", in Rante (ed.), *Greater Khorasan*: pp. 63-88, figures pp. 216-41, here p. 234, fig. 27 serves as the basis of Fig. 12, seen here.

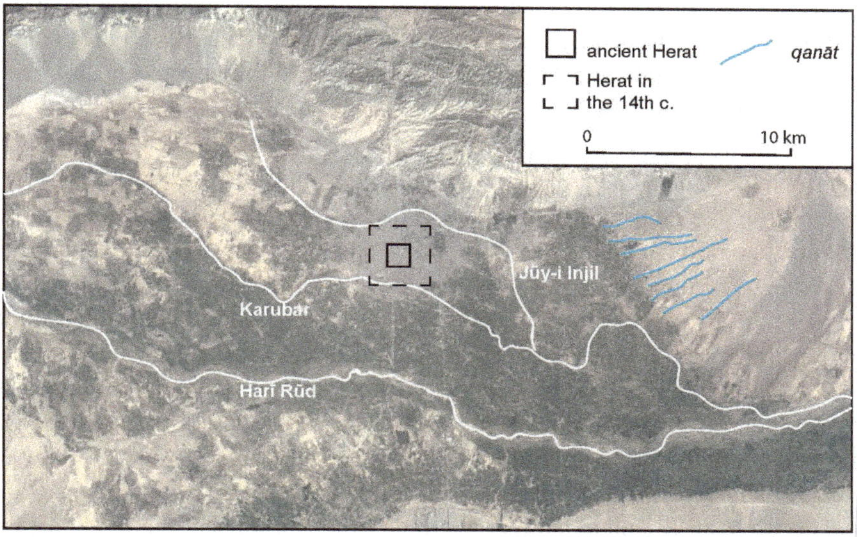

FIGURE 12 Herat in the fourteenth century
SOURCE: IMAGERIE 2015, RECTIFIED BY THE AUTHOR

several directions, the case of Herat validates the principle, because the city's concentric expansion was always toward the inflows of water.

Two factors permitted more varied systems of water supply in the steppe lands of Central Asia. First, it was possible to dig canals fed by a common natural source, shaping the irrigation system in the form of an artificial delta. Second, the water tables were relatively high. Marw al-Šāhijān, like Šādyāḫ on the Iranian Plateau, was built around a main central canal. At Marw, the system of canals was derived from the Murġāb. Walls were not constructed around the city until the fifth/eleventh century, and as shown above, this permitted extensive development along the canal, without precise geometric planning. The south-north axis of the three main water canals, with Mājān in the centre and the Ḫurmūzfarra and Razīq on either side, shaped urban development. The hump-shaped topography, with the central canal on higher ground, represented the best option for supplying the city with water and at the same time draining excess water for preservation of mud-brick structures. However this topography and plan precludes significant accumulation of layers, which would have impeded water supply and drainage.

The more recent site of Šādyāḫ also featured a central water canal. Unlike Marw al-Šāhijān, this city was built in replacement of and at some distance from the old city of Nishapur, and was protected by strong ramparts from the outset. The former location, Old Nishapur, was by that time no longer a city fortress, and continued to decline.[51] Marw al-Šāhijān, on the other hand, still had the ancient city nearby, offering a large strongly fortified space in time of need. The early fortification of Šādyāḫ soon imposed limits, and the city then expanded into other unwalled areas.

Paykand offers a clear example of the citadel being situated close to the main water canal, with the rest of the city (*šahristān*-1 and *šahristān*-2) expanding beyond the citadel over the course of centuries. Archaeological research has evidenced a potter's quarter situated along this main canal. Both this and the *šahristān*-2 locality were abandoned in the seventh/thirteenth century, suggesting that the water supply had progressively decreased, probably beginning in the sixth/twelfth century. This would confirm a contemporary description provided by Qubāwī, the translator of Naršaḫī's *Ta'rīḫ Buḫārā*.[52] The decrease in the water supply evidently forced a reduction in surface size of Paykand and the migration of population, presumably to nearby Bukhara.

Although it is difficult to identify all of Bukhara's early canals, the Šahr-i Rūd must have been the main source of supply from the start.[53] The final part of this main canal is oriented east-west, to the south of the earliest settlement, in the area eventually developed as the city (Fig. 13). Further research is needed to determine exactly when the development of the Bukharan *šahristān* occurred, upstream to the east. In any case, this development ultimately led to the situation that the citadel came to be downstream from the *šahristān*. As we have seen before, this situation is uncommon as usually the citadel got the best water.[54] During the second to fourth centuries CE, Bukhara extended considerably towards the south and west. Since in this dynamic of urban expansion all settlement continued to extend around the ancient nucleus, the city's managers must have supplied water to the newly urbanised areas through several secondary canals, all probably fed from the same main source, the Šahr-i Rūd. By the Medieval period, there had been considerable gains in knowledge in matters of increasing water supply through construction of canals, for purposes of rapid urban development. Semënov and Turebekov surmised a contraction of the city in the fifth century CE.[55] But this contraction did not continue in the early Islamic period.

---

51 Rante and Collinet, *Rayy*: pp. 53-5.

52 Naršaḫī/Qubāwī, *Tārīḫ-i Buḫārā*, ed. Mudarris Rażawī, M.T. (2nd edition, Tehran: Intišārāt-i Ṭūs, 1362/1984); trans. Richard Nelson Frye, *The History of Bukhara* (Princeton: Marcus Wiener, 2007): p. 26, trans.: p. 19.

53 Belenitskij, A.M., Bentovič, I.B., and G.O. Bol'šakov, *Srednevekovyj gorod Srednej Azii* (Leningrad: Nauka, 1973): p. 243, fig. 95.

54 We might add that this city had a (well founded) reputation of having the worst and least healthy water in the entire region.

55 See Grenet, Frantz, "Crise et sortie de crise": p. 374, n. 22.

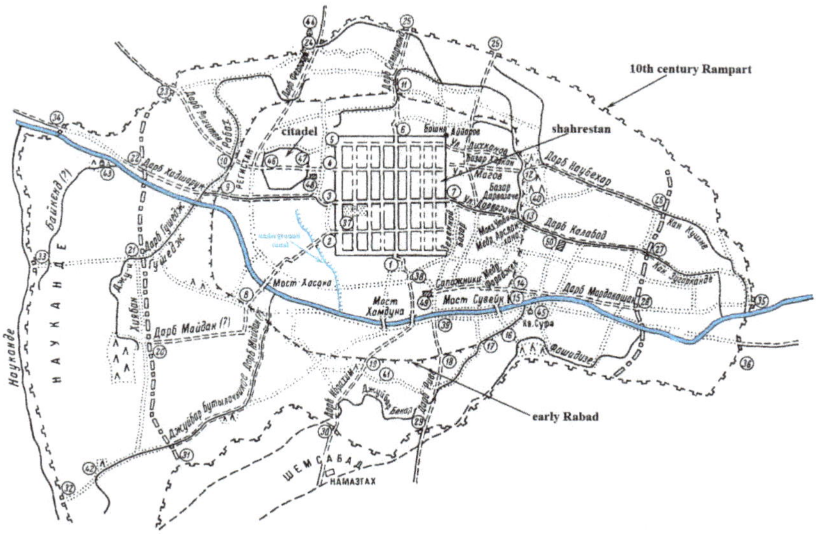

FIGURE 13 Sketch map of medieval Bukhara
AFTER: BELENICKIJ ET AL., SREDNEVEKOVYJ GOROD: P. 95; RECTIFIED BY THE AUTHOR

Although geo-archaeological data are as yet insufficient for firm conclusions, it appears that during the first centuries of Islam the *šahristān* continued to extend southwards, approaching the Šahr-i Rūd until this main canal was ultimately encompassed within an early suburb. Urban expansion at Bukhara continued unabated from the second/eighth to the fourth/tenth century.[56] The southward extension of the secondary canals must have aimed at improving the distribution of water supplies to the city, which had by then doubled in size.

Afrasiyab, or ancient Samarqand, was situated upstream along the Zarafšān and some 250 kilometres to the east of Bukhara. The city was founded during the Achaemenid period, and covered an area of 219 hectares (Fig. 14) on a hill bounded to the north by the Siyāb, the local branch of the Zarafšān. However, it could not be served with water from this branch, because of its higher elevation. Instead, the ancient water system, likely dating to the founding, consisted of basins and canals. The main canal derived from the Zarafšān at a distance, then entered the city from the south, at the Kīsh Gate, and traversed the city

56 Gaube, Heinz, "Bukhara up to the Tenth Century", in Gangler, Anette, Gaube, Heinz, and Attilio Petruccioli (eds.), *Bukhara, the Eastern Dome of Islam: Urban Development, Urban Space, Architecture and Population* (Stuttgart: Axel Menges, 2004): p. 48.

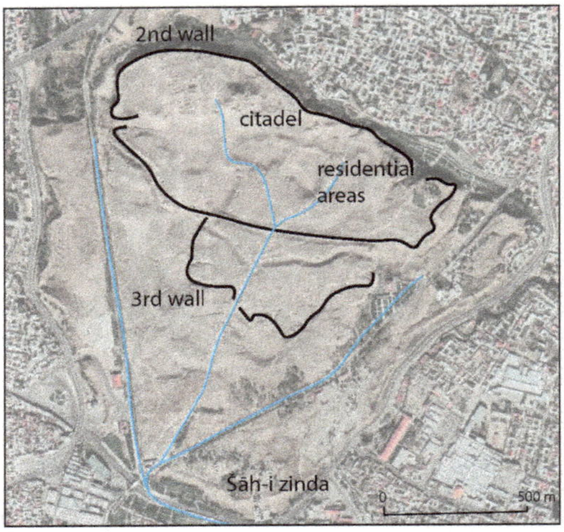

FIGURE 14
Topographic map of Afrasyab
DRAWING D. DURAND-GUÉDY BASED ON HTTP://WWW.ARCHEO.ENS.FR/SPIP.PHP?ARTICLE369; MODIFIED BY RAPIN 2013

centre. Several canals, probably from later periods, supplied water to the different quarters.[57]

In fifth to mid-seventh century CE, Central Asia was subject to a period of sharply decreased precipitation, reaching the level of drought across the north-eastern Tibetan Plateau and as far east as China.[58] Throughout the period under observation, events of population-depopulation-repopulation could have been linked to climatic changes such as this. The varying events of increasing and decreasing population at Samarqand/Afrasiyab clearly had an impact on the water supply system, particularly the event of diminishing population occurring in the fifth century CE.[59] The canals supplying the eastern part of the city centre may have been constructed after events of population decline and curtailment of territorial control,[60] when the political forces of Samarqand, like other cities such as Erkurgan, tried to optimise the military capacity of their sites. Construction of a canal would have been a measure aimed at securing a compact inhabited area, for defensive purposes.[61]

57 Grenet, Frantz and Rapin, Claude, "De la Samarkand antique à la Samarkand médiévale: continuités et ruptures", in Gayraud, R.P. (ed.), *Colloque international d'archéologie islamique* (Cairo: Publications de l'IFAO, 1998): p. 387-402, 387; Rapin, C. and Isamiddinov, M., "Entre sédentaires et nomades: les recherches de la Mission archéologique franco-ouzbèke (MAFOuz) de Sogdiane sur le site de Koktepe", in Bendezu-Sarmiento (ed.), *Archéologie française en Asie centrale*: pp. 113-34, here p. 119.

58 Yang, Bao, Wang, J., Bräuning, A., Dong, Z. and J. Esper, "Late Holocene Climatic and Environmental Changes in Arid Central Asia", *Quaternary International*, CXCIV (2009): pp. 68-78, here p. 76.

59 See Belenickij *et al. Srednevekovyj gorod*: pp. 219-32; Grenet, "Crise et sortie de crise": p. 374.

60 Grenet, "Crise et sortie de crise": p. 374.

61 Semënov, "Sogdijskij gorod".

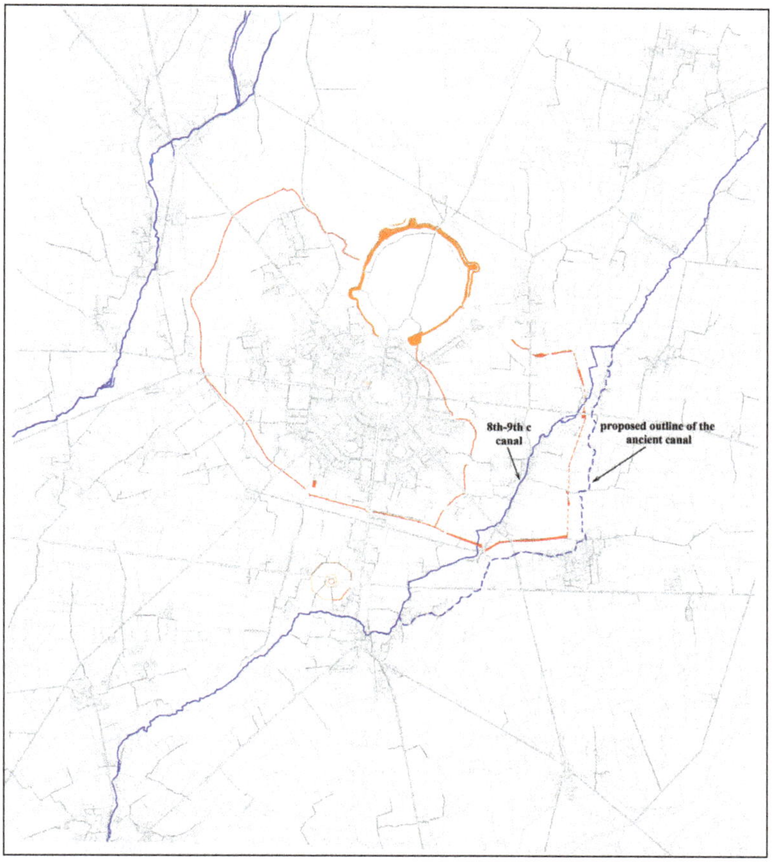

FIGURE 15 Map of Balkh
AFTER MARQUIS 2015, MODIFIED BY THE AUTHOR

The hypothesis would then be that several very large cities, including Bukhara and Samarqand/Afrasiyab shrank because of decreasing water resources.

In Balkh, on the southern side of the Oxus, the first occupation included the entire walled area of Bālā Ḥiṣār, as well Tepe Zargarān through the Bāb Kūh to the south (Fig. 15), essentially corresponding to the Bactres IV level defined by Le Berre and Schlumberger.[62] As at Samarqand, such a large city would need substantial water resources.[63] Supply was achieved through several south/

62 See Le Berre, M. and Schlumberger, D., "Observations sur les remparts de Bactres", *Monuments Préislamiques d'Afghanistan, Mémoires de la Délégation archéologique française en Afghanistan*, XIX (1964): pp. 65-103, fig. 10.

63 I would like to thank Philippe Marquis who provided me with the yet unpublished results of the last researches on Balkh.

north branches of the Balḫāb River.[64] The water table at Balkh is close to the surface, therefore water could also have been obtained by wells. Successive ramparts were built in Balkh, but inside the earlier ones, probably due to the same demographic dynamics characteristic of other large cities in this part of Asia, such as Bukhara, Samarqand and Balkh, between the second century CE and the early Islamic period.

Archaeological investigations conducted by Gardin have shown that following initial occupation, Bālā Ḥiṣār was more or less depopulated, while occupations at Bactres IA and Tepe Zargarān continued to the Timurid period.[65] The Chinese pilgrim Xuangang visited Balkh in 630 CE, and related that although the city was well fortified, it was not heavily populated.[66] These observations indicate that at Balkh, the depopulation of the city occurred prior to the desiccation and drought of the fifth-seventh century CE, however the population may have continued to shrink thereafter.

Historical and archaeological investigations attest that the main water resource for at least the eastern part of the city (Bactres IA-II)[67] was the Nahr-i Muštāq.[68] Recent investigations directed by Philippe Marquis have demonstrated that it was not until the beginning of the Islamic period, probably between the second/eighth and the third/ninth centuries, that the area and its water resource were reorganised. It was then that this main canal crossed Tepe Zargarān (as shown in Gardin's plan).[69] Gardin reports that archaeological tests 1 and 10 in Tepe Zargarān reveal an increment in the ground level, testifying to a reorganisation of the area.[70] The modification of the Nahr-i Muštāq canal, which was reoriented inside the city to cross the southern ramparts and run alongside Tepe Zargarān may have been part of the reorganisation noted

64 For a detailed description of the irrigation system at Balkh in the early Islamic period, see de la Vaissière in this volume.

65 Gardin, Jean-Claude, *Céramiques de Bactres* (Paris: Mémoires de la délégation archéologique française en Afghanistan, 1957): pp. 92-113. Cf. Le Berre and Schlumberger, "Observations": fig. 10.

66 De la Vaissière, Etienne and Marquis, Philippe, "Nouvelles recherches sur le paysage monumental de Bactres", *Comptes-rendus de l'Académie des Inscriptions et Belles-Lettres*, 2013.3: pp. 1155-71, here p. 1157. See also his contribution in this volume. See also Grenet, "Crise et sortie de crise": p. 381, n. 41.

67 It is not impossible that several other more or less important irrigation canals served the city.

68 De la Vaissière, Etienne, "De Bactres à Balkh par le Nowbahār", *JA*, CCXCVIII/2 (2010): pp. 517-33, here p. 518. De la Vaissière and Marquis, "Nouvelles recherches": p. 1163.

69 See Gardin, *Céramiques*: fig. 1; Le Berre and Schlumberger, "Observations": fig. 10; De la Vaissière and Marquis, "Nouvelles recherches": p. 1159, fig. 1.

70 Gardin, *Céramiques*: p. 104.

by Gardin. This reorganisation could be further confirmed by the location of the (now dry) canal running above the site once occupied by the stupa in the Buddhist monastery of Nawbahār.[71] If this was not the result of a full-fledged community reorganisation, then it was at least a socio-religious restructuring of the area. In Tepe Zargarān, a long drainage canal oriented west/east could also be dated to the end of the second/eighth-beginning of the third/ninth century. These elements clarify the important reorganisation identified in the early third/ninth century, when it was probably decided to concentrate the city around its major water source on the eastern side.

6 Conclusion

The cities examined here are situated in regions of different character: the mountainous landscapes of the Iranian plateau, and steppe landscapes of Central Asia. Political, economic and social concerns would be behind the foundation of a settlement at a given place, but in all cases, geomorphological givens also had to be taken into account, and they bore heavily on a city's physical features, in every era. No matter the dynamic of territorial occupation, whether the city was "inherited" (Kennedy's expression)[72] or newly founded, its occupants organised the available urban space in relation to the advantages and restrictions presented by the landscape. The supply of water was always the priority. Thus, the founding locations, and subsequently the form and the urban organisation of these cities, were conditioned by geomorphological considerations, such as the context of mountainous or steppe terrain, water channels, water table, and geological substrate. It is on the basis of substantial geomorphological differences of water and topography that the real distinction between eastern (Central Asian) and western (plateau) Iranian cities should be made. Imperial authorities, in particular the Parthian and Sasanian empires, were surely directly involved in the construction of numerous major irrigation systems. But we need to differentiate among these systems. Recent studies, especially in Mesopotamia concerning the Sasanian empire, have observed the existence of large-scale but short-lived systems alongside long-term core areas of irrigation. This observation could have profound implications in

71 De la Vaissière and Marquis, "Nouvelles recherches": p. 1157. See also De la Vaissière's contribution in this volume.

72 Kennedy, Hugh, "Inherited Cities", in Jayyusi, Holod, Petruccioli, and Raymond (eds.), *The City in the Islamic World*: pp. 93-113, pointed out the different ways in which Muslims inherited cities, offering to the discussion of Iranian cities the useful term of *Decastellamento*, which stresses the dynamic of occupation through socio-economic and political choice.

understanding long-term landscape sustainability: a proposal would be that there was a relationship between the scale and sustainability of the water system and the involvement of imperial rule.[73] A thesis could be that large-scale imperial irrigation schemes were less sustainable than the locally driven systems, which had developed over longer periods of time. Thus, while many hydraulic works in Iranian cities could certainly be attributed to imperial design, local landlords would also have played a fundamental role in providing investment and other forms of capital for construction and operation of public works in the long term.[74]

On the Iranian Plateau, Rayy, and possibly Nishapur as well, would represent examples of the involvement of local actors. Such involvement would be logical, given the attractive opportunities represented by capital-intensive *qanāt* technology, for private investors. The occupation and organisation of areas around Tepe Mīl resulted in large-scale measures for transformation of the landscape, which evolved from a semi-arid, near-desert zone to a well-irrigated area, in which rich residences and a huge fire temple were constructed. Since my research indicates that this area was abandoned for political and social reasons, and not for a decrease of natural resources (i.e., water supply), it could then be repopulated, and in fact is still inhabited today.[75] Thus, the case of Rayy could serve as a model of long-term landscape transformation resulting from a large-scale irrigation system. To confirm this suggestion, deeper geomorphological studies are needed in this zone, where intensive modern urbanisation has unfortunately erased the traces of ancient civilisations.

In Central Asia, Bronze Age settlements occupied areas close to water reservoirs or lakes, while those of the Iron Age and especially of antiquity and late antiquity developed the technology and increased the complexity of water channel networks.[76] In Transoxiana, this dynamic was accompanied by an increase of activities concerning the management of landscapes, often directed by landlords.

In Central Asian cities, water was generally provided by watercourses flowing exterior to the urban nucleus. Topographically, this meant that the inhabited

73 Wilkinson, "From Human Niche Construction": p. 170, including reference to the seminal work of Adams, *Land Behind Bagdad, on* the shaping and decline of the irrigation system built by the Sasanids in Central Mesopotamia.

74 Wilkinson et al., "From Human Niche Construction": pp. 155-76.

75 The buildings around Tepe Mīl were inhabited until the 10th century CE, at least, according to the pottery found during surveys and excavations. See also Rante, *Rayy*: p. 126.

76 Fouache E., Besenval R., Cosandey C., Coussot C., Ghilardi M., Huot S., and M. Lamothe, "Palaeo-Channels of the Balkh River (Northern Afghanistan) and Human Occupation since the Bronze Age Period," *Journal of Archaeological Science*, XXXIX (2012): pp. 3415-27.

areas could grow higher through the accumulation of occupation layers, generating a domed hill (*tepe*). This pattern would be impossible in the case of a surface watercourse flowing through the city centre. Inside the city, the watercourse had to be higher than the dwellings, for distribution of water using gravity. This was unlike regions where *qanāt*s were widely used, and could be tapped like wells. The cities on the Iranian Plateau were supplied with water primarily through *qanāt*s, which crossed the urban space underground or on the surface, facilitating access to the water.

Although the archaeological evidence is limited, Paykand and Balkh appear to present some common features in the matter of elimination of excess water by means of ditches along streets. In Tepe Zargarān there is a drainage canal for elimination of excess water, which according to the stratigraphic sequence would date to the Islamic period. In Paykand, excavation in the *šahristān*-1 area has brought to light street ditches from the beginning of the Islamic occupation. This discovery does not rule out the earlier existence of drainage practices, since urbanisation affected every part of the city during this period, from the large-scale irrigation system to the intricate urban water network.

These different methods of supplying water to the city, and especially of eliminating water, had implications for the urban morphology with respect to water management.

Bibliographical References

Adams, Robert McC., *Land behind Baghdad. A History of Settlement on the Diyala Plains* (Chicago: University of Chicago Press, 1965).

Adams, Robert McC., "Agriculture and Urban Life in Early Southwest Iran", in Caldwell, J.R. (ed.) *New Roads to Yesterday: Essays in Archaeology* (London: Thames and Hudson, 1966): pp. 436-65.

Āḏarnūš, Mas'ūd, *Ġarāriš-i kāvūšhāy-i lāye-šināḫtī tipi Haġmatāna* (Tehran, ICHHtO-ICAR: 2008).

Adle, Chahryar. "Qanāts of Bam: an archaeological perspective", in *Qanāts of Bam* (Tehran: UNESCO, 2006): pp. 33-85.

Allan, Tony, *The Middle East Water Question: Hydropolitics and the Global Economy* (London: Tauris, 2001).

Andrianov, Boris V. et al., "Rol' irrigatsii v sotsial'noj istorii drevnej srednej Azii", in Masson, Vadim (ed.), *Drevnie tsivilizatsii Vostoka* (Taškent: Fan, 1986): pp. 34-44.

Askarov, Akhmed Ali et al., *Rannaja gorodskaja kul'tura ėpokhi bronzy juga Srednej Azii* (Samarqand: Institute of Archaeology of Samarqand, 1993).

Beaumont, Peter, s.v. "Āb iii. The Hydrology and Water Resources of the Iranian Plateau", in *EIr*, I (1982): pp. 28-39.

Belenitskij, A.M., Bentovič, I.B., and G.O. Bol'šakov, *Srednevekovyj gorod Srednej Azii* (Leningrad: Nauka, 1973).

Bosworth, Edmund, s.v. "Nishapur. Historical Geography and History to the Beginning of the 20th century", in *EIr* online, www.iranicaonline.org/articles/Nishapur-i (2010).

Boucharlat, Rémy, "Les galeries de captage dans la péninsule d'Oman au premier millénaire avant J.-C.: questions sur leurs relations avec les galeries du plateau iranien", in Briant, P. (ed.), *Irrigation et drainage dans l'Antiquité, qanāts et canalisations souterraines en Iran, en Egypte et en Grèce* (Paris: Thotm éditions, 2001): pp. 157-83.

Briant, Pierre (ed.), *Irrigation et drainage dans l'antiquité, qanats et canalisations souteraines en Iran, en Egypte et en Grèce*. (Paris: Thotm éditions, 2001).

Brown, Stuart C., s.v. "Ecbatana", in *EIr*, VIII (1997): pp. 80-4.

Bulliet, Richard, *The Patricians of Nīshāpūr: A Study in Medieval Islamic Social History* (Cambridge, Mass.: Harvard University Press, 1972).

Bulliet, Richard, "Medieval Nīshāpūr: A Topographic and Demographic Reconstruction", *StIr*, V (1976): pp. 67-89.

Bulliet, Richard, *Cotton, Climate and Camels in Early Islamic Iran* (New York: Columbia University Press, 2009).

Christensen, Peter, *The decline of Iranshahr. Irrigation and Environments in the History of the Middle East 500 BC to AD 1500* (Copenhagen: Museum Tusculanum Press, 1993).

De la Vaissière, Etienne, "De Bactres à Balkh par le Nowbahār", *JA*, CCXCVIII/2 (2010): pp. 517-33.

De la Vaissière, Etienne and Marquis, Philippe, "Nouvelles recherches sur le paysage monumental de Bactres", *Comptes-rendus de l'Académie des Inscriptions et Belles-Lettres*, 2013.3: pp. 1155-71.

De Planhol, Xavier, s.v. "Cities: Geographical Introduction", in *EIr*, V (1991): pp. 603-7.

De Planhol, Xavier, s.v. "Iran i: Lands of Iran", *EIr*, XIII (2004): pp. 204-12.

Fouache, Eric, Cosandey, Cl., Wormser, P., Kervran, M., and R.A.Kh. Labbaf, "The River of Nishapur", *StIr*, XL/1 (2011): p. 99-119.

Fouache, Eric, Besenval, R., Cosandey, C., Coussot. C., Ghilardi, M., Huot, S., and M. Lamothe, "Palaeo-Channels of the Balkh River (Northern Afghanistan) and Human Occupation since the Bronze Age Period", *Journal of Archaeological Science*, XXXIX (2012): pp. 3415-27.

Fouache, Eric, Francfort, H.-P., Cosandey, C., Adle, Ch., "La crise d'aridité climatique de la fin du 3ème millénaire av. J.-C., à la lumière des contextes géomorphologiques de 3 sites d'Iran Oriental (Bam, Tepe Damghani, Jiroft)", in Rante, R. (ed.), *Greater Khorasan. History, Geography, Archaeology and Material Culture* (Berlin: De Gruyter, 2015): pp. 27-37.

Fouache, Eric et al., "The Role of Catastrophic Floods Generated by Collapse of Natural Dams Since the Neolithic in the Oases of Bukhara and Qaraqöl: Preliminary Results", *International Journal of Geohazards and Environment*, III/2 (2016): pp. 150-65.

Francfort, Henri-Paul and Lecomte, O., "Irrigation et société en Asie Centrale des origines à l'époque achéménide", *Annales. Histoire et Sciences Sociales*, LVII/3 (2002): pp. 625-63.

Franke, Ute, "Ancient Herat Revisited. New Data from Recent Archaeological Fieldwork", in Rante, R. (ed.) *Greater Khorasan. History, Geography, Archaeology and Material Culture* (Berlin: De Gruyter, 2015): pp. 63-88.

Gardin, Jean-Claude, *Céramiques de Bactres* (Paris: Mémoires de la délégation archéologique française en Afghanistan, 1957).

Gaube, Heinz, "Bukhara up to the Tenth Century", in Gangler, Anette, Gaube Heinz, and Attilio Petruccioli (eds.), *Bukhara, the Eastern Dome of Islam: Urban Development, Urban Space, Architecture and Population* (Stuttgart: Axel Menges, 2004): pp. 39-59.

Gaube, Heinz, "Iranian Cities", in Jayyusi, S.K., Holod, R., Petruccioli, A., and A. Raymond (eds.), *The City in the Islamic World* (Leiden: Brill, 2008): pp. 159-80.

Gillmore, Gavin K., Coningham, R.A.E., Fazeli, H., Young, R.L., Magshoudi, M., Batt, C.M., and G. Rushworth, "Irrigation on the Tehran Plain, Iran: Tepe Pardis – the Site of a Possible Neolithic Irrigation Feature?", *Catena*, LXXVIII (2009): pp. 285-300.

Grenet, Frantz and Rapin, Claude, "De la Samarkand antique à la Samarkand médiévale: continuités et ruptures", in Gayraud, R.P. (ed.), *Colloque international d'archéologie islamique* (Cairo: Publications de l'IFAO, 1998): pp. 387-402.

Grenet, Frantz, "Crise et sortie de crise en Bactriane-Sogdiane aux IVe-Ve s. de notre ère: de l'héritage antique à l'adoption de modèles sassanides", in *La Persia e l'Asia Centrale da Alessandro al X secolo, Atti dei Convegni Lincei*, CXXVII (Rome: Accademia dei Lincei, 1996): pp. 367-90.

Herrmann, Georgina, Masson, V.M., Kurbansakhatov, K. et al., "The International Merv Project. Preliminary Report on the First Season (1992)", *Iran*, XXXI (1993): pp. 39-62.

Herrmann, Georgina (ed.), *Monuments of Merv: Traditional Buildings of the Karakum* (London: The Society of Antiquaries of London, 1999).

Honari, Morteza, et al., "Qanāts of Bam. A Multidisciplinary Approach", in *Qanāts of Bam* (Tehran: UNESCO, 2006).

Hourcade, Bernard, s.v. "Jājrud River", in *EIr*, XIV (2008): pp. 399-400.

Hourcade, Bernard, s.v. "Karaj River", in *EIr*, XV (2010): pp. 543-4.

Ibn Ḥawqal, *K. Ṣūrat al-arḍ*, ed. de Goeje (2nd edition revised by J.H. Kramers, Leiden: Brill, 1967). Trans. J.H. Kramers, G. Wiet, *Configuration de la terre*, 2 vols. (Paris and Beirut: Maisonneuve et Larose, 1964).

Ibn Rusta, *K. al-Aʿlāq al-nafīsa*, ed. de Goeje (2nd edition, Leiden: Brill, 1967). Trans. G. Wiet, *Les atours précieux* (Cairo: Société de géographie d'Egypte, 1955).

Karīmān, Ḥusayn, *Rayy-i bāstān*, 2 vols. (Tehran, Anjuman-i Āṯār-i millī, 1345-1349/ 1966-70).

Kennedy, Hugh, "Islamic Geographies", in Herrmann, Georgina (ed.), *Monuments of Merv: Traditional Buildings of the Karakum* (London: The Society of Antiquaries of London, 1999): pp. 121-6.

Kennedy, Hugh, "Inherited Cities", in Jayyusi, S.K., Holod, R., Petruccioli, A. and A. Raymond (eds.), *The City in the Islamic World* (Leiden: Brill, 2008): pp. 93-113.

Labbāf, Khānikī Rajab'alī, *Reports of excavation* (Tehran: ICHHTO-ICAR, 2004).

Le Berre, M. and Schlumberger, D., "Observations sur les remparts de Bactres", *Monuments Préislamiques d'Afghanistan, Mémoires de la Délégation archéologique française en Afghanistan*, XIX (1964): pp. 65-103.

Le Strange, Guy, *The Lands of the Eastern Caliphate* (3rd edition, London: Frank Cass, 1966).

Lioubimtseva, Elena, "Climate and Environmental Change in Arid Central Asia: Impacts, Vulnerability, and Adaptations", *Journal of Arid Environments*, LXXIII (2009): pp. 963-77.

Lisitsyna, Gorislava N., *Orošaemoe zemledelie ėpokhi ėneolita na juge Turkmenii* (Moscow: Nauka, 1965).

Lisitsyna, Gorislava N., "The Earliest Irrigation in Turkmenia", *Antiquity*, XLVII (1969): pp. 279-88.

Magee, Peter, "The Chronology and Environmental Background of Iron Age Settlement in Southeastern Iran and the Question of the Origin of the qanat Irrigation System", *Iranica Antiqua*, XL (2005): pp. 217-31.

Margueron, Jean-Claude, *Cités invisibles* (Paris: Geuthner, 2013).

Melville, Charles, "Earthquakes in the History of Nishapur", *Iran*, XVIII (1980): pp. 103-20.

Mukhamedzhanov, A.R., Semenov, G.L., Mirzaakhmedov, D.K., and Š.T. Adilov, *Gorodišče Pajkend* (Taškent: Fan, 1988).

Naršaḫī/Qubāwī, *Tārīḫ-i Buḫārā*, ed. M.T. Mudarris Rażawī (2nd edition, Tehran: Intišārāt-i Ṭūs, 1362/1984); trans. Richard Nelson Frye, *The History of Bukhara* (Princeton: Marcus Wiener, 2007).

Omel'čenko, Andrej B. and Mirzaakhmedov, D.K., *Raskopki v Pajkende v 2011 godu. Materialy bukharskoj ėkspeditsii* (St Petersburg, 2012).

Prickett, Martha, "Settlement during the Early Periods", in Beale, T.W. (ed.), *Excavations at Tepe Yahya, Iran 1967-1975. The early periods* (Cambridge, Mass.: Harvard University Press, 1986): pp. 215-46.

Rante, Rocco, "Les fouilles de Paykend: nouveaux éléments," in Bendezu-Sarmiento, J. (ed.), *Archéologie française en Asie centrale. Nouvelles recherches et enjeux socioculturels* (Paris: de Boccard, 2013): pp. 219-40.

Rante, Rocco, *Rayy: from its Origins to the Mongol Invasion. An Archaeological and Historiographical Study* (Leiden: Brill, 2015).

Rante, Rocco and Di Pasquale, S.C., "The Urbanisation of Rayy at the Seljuq epoch", in Tor, D., Schmidtke, S., and H. Ansari (eds.), *The Religious and Intellectual History of Rayy from 900 Through the Seljūq Period* [*Der Islam*, XCIII/2 (2016)]: pp. 413-32.

Rante, Rocco and Collinet, A., *Nīshāpūr Revisited. Stratigraphy and Ceramics of the Qohandez*, (Oxford: Oxbow Books, 2013).

Rapin, C. and Isamiddinov, M., "Entre sédentaires et nomades: les recherches de la Mission archéologique franco-ouzbèke (MAFOuz) de Sogdiane sur le site de Koktepe", in Bendezu-Sarmiento, J., (ed.), *Archéologie française en Asie centrale. Nouvelles recherches et enjeux socioculturels* (Paris: de Boccard, 2013): pp. 113-34.

Sarrāf, Muḥammad-Raḥīm, "Hagmatāna", in Kiāni, M.Y. (ed.), *Šahrhā-yi Īrān* (Tehran: ICHHTO-ICAR, 1989): pp. 290-311.

Ṣarrāf, Muḥammad-Raḥīm, *Hagmatāna* (Tehran, 1392š/2013).

Semenov, Grigorij L., "Sogdijskij gorod v rannem srednevekov'e: formirovanie plana", in Smirnova, G.I. (ed.), *Itogi rabot arkheologičeskikh ėkspeditsij Gosudarstvennogo Ėrmitazha* (Leningrad: Gosudarstvennyj Ėrmitazh, 1989): pp. 128-140.

Semenov, G.L., "Dwelling Houses of Bukhara in the Early Middle Ages", in Compareti, M., Raffetta, P., and G. Scarcia (eds.), *Ērān ud Anērān. Studies Presented to B.I. Maršak on the Occasion of his 70th Birthday* (Venice: Cafoscarina, 2006): pp. 555-70.

Semenov, G.L. and Mirzaakhmedov, D.K., *Raskopki v Pajkende v 2000 godu. Materialy bukharskoj ėkspeditsii* (St Petersburg, 2004).

Spooner, B., "City and River in Iran: Urbanization and Irrigation of the Iranian Plateau", *IrSt*, VII (1974): pp. 681-713.

Torgoev, A.I. and Mirzaakhmedov, D.K., *Raskopki v Pajkende v 2006 godu. Materialy bukharskoj ėkspeditsii* (St Petersburg, 2009).

Tskitishvili, O, "Two Questions Connected with the Topography of the Oriental City in the Early Middle Ages", *JESHO*, XIV/3 (1971): pp. 311-20.

Valles, Vincent, Gholami, M., and R. Lambert, "Chimie des eaux et alimentation du bassin versant du Djajerud (Iran)", *Hydrologie Continentale*, V/1 (1990): pp. 61-9.

Wilkinson, Charles K., *Nīshāpūr, Some Early Islamic Buildings and Their Decoration* (New York: Metropolitan Museum of Arts, 1986).

Wilkinson, Tony J., *Archaeological Landscapes of the Near East* (Tucson, Arizona: The University of Arizona Press, 2003).

Wilkinson, Tony J. et al., "From Human Niche Construction to Imperial Power: Long-Term Trends in Ancient Iranian Water Systems", *Water History*, IV/2 (2012): pp. 155-76.

Williams, Tim, "The City of Sultan Kala, Merv, Turkmenistan. Communities, Neighbourhoods and Urban Planning from the Eight to the Thirteenth Century", in Bennison, A.K. and Gascoigne, A.L. (eds.), *Cities in the Pre-Modern Islamic World. The Urban Impact of Religion, State and Society* (London and New York: Routledge, 2007): pp. 42-62.

Wittfogel, Karl August, *Oriental Despotism. A Comparative Study of Total Power* (New Haven: Yale University Press, 1957).

Wordsworth, Paul, "Merv on Khorasanian trade routes from the tenth-thirteenth centuries", in Rante, R., (ed.), *Greater Khorasan. History, Geography, Archaeology and Material Culture* (Berlin: De Gruyter, 2015): pp. 51-62.

Yang, Bao, Wang, J., Bräuning, A., Dong, Z. and J. Esper, "Late Holocene Climatic and Environmental Changes in Arid Central Asia", *Quaternary International*, CXCIV (2009): pp. 68-78.

Yaʿqūbī, *Kitāb al-buldān*, ed. de Goeje (Leiden: Brill, 1892). Trans. G. Wiet, *Les pays* (Cairo: Institut français d'archéologie orientale, 1937).

Yāqūt, *Muʿjam al-buldān*, ed. F. Wüstenfeld, *Jaqut's Geographisches Wörterbuch*, 6 vols. (Leipzig: Brockhaus, 1866-73). Partial trans. C. Barbier de Meynard, *Dictionnaire géographique, historique et littéraire de la Perse et des contrées adjacentes* (Paris: Imprimerie Impériale, 1861).

Biographical Note

Rocco Rante is Researcher-Archaeologist at the Musée du Louvre, Paris. He specialises in civilisation and culture of the Middle East between antiquity and the medieval Islamic period. He is director of the French components of the Archaeological Missions in the Bukhara Oasis (Uzbekistan) and Khorasan (Iran). His publications include *The Bukhara Oasis, vol. 1*, Brill 2018; *Rayy: From its origins to the Mongol invasion*, Brill 2015; *Greater Khorasan*, De Gruyter 2015; *Nishapur Revisited* (coauthored with Annabelle Collinet), 2013.

“From Shahristān to Medina” Revisited

Donald Whitcomb
The Oriental Institute, The University of Chicago
d-whitcomb@uchicago.edu

Abstract

The title is taken from an article published in 2006 by Hugh Kennedy that is an examination of the Sasanian city. This subject is re-examined with a presentation of archaeological information on Sasanian cities in Fars province and the city of Jundi Shapur in Khuzistan. The urban form and key institutions are presented with consideration of changes introduced with the Islamic period. In contrast to the Classical *polis*, the problem is limited information on urban element in Sasanian cities and some of these elements are described.

Keywords

Sasanian – urban plan – Fīrūzābād (Jūr) – Bīšāpūr – Istakhr – Shiraz – Jundi Šāpūr

1 Introduction

In 1985, Hugh Kennedy wrote the article, “From Polis to Medina”, which set a new paradigm for understanding the urban transition from Late Antiquity to Early Islam in Syria. This was the same year of publication of my monograph on the excavations at Qaṣr-i Abū Naṣr, giving evidence of a Sasanian town and fortress and its transformation in the early Islamic period. Somehow Qaṣr-i Abū Naṣr does not seem paradigmatic, at best offering tangential evidence for the development of Shiraz. Some twenty years later, Kennedy has returned to the study of the urban transition of Iranian cities, with far less impact. One reason could be the state of understanding the Sasanian city.[1] His research reflects the common assumption that the urban form of both late antique and Sasanian

1 Kennedy, Hugh, “From Polis to Madina: Urban Change in Late Antique and Early Islamic Syria”, *Past and Present* 106 (1985): p. 3-27; Whitcomb, Donald, *Before the Roses and Nightingales: Excavations at Qasr-i Abu Nasr, Old Shiraz*, (New York: The Metropolitan Museum of Art, 1985); Kennedy, Hugh, “From Shahristān to Medina”, *StIs*, CII-CIII (2006): pp. 5-34.

 | DOI:10.1163/9789004434332_006

cities are based on the classical city. In other words, the structures of classical cities in the Levant may have been imported into Iran. The descriptions of Bīšāpūr in Ghirshman's reports would be good examples of this process.[2] The perhaps quixotic intent of this current paper is to take a first step toward definition of an Iranian urban tradition and its successor in the Islamic city.

Patterns of Sasanian cities can be readily found in aerial photographs, particularly the spectacular results of Erich Schmidt in the 1930s.[3] Schmidt recorded views of famous cities and flew over many other large settlements that remain nameless and uninvestigated to this day. He often landed his plane and collected some sherds for dating, in what today might be called 'ground-truthing'. This process continues today with ready access to rather more remote satellite images, such as the Corona series, which offer fine details.[4] Nevertheless, understanding cities requires archaeological field research, always on a large scale. The long tradition of research on classical cities places knowledge of the Iranian phenomenon at a great comparative disadvantage. Kennedy could observe change from Byzantine to early Islamic cities and discuss details in patterns of development. The Sasanian city poses a completely different situation, where broad patterns remain undefined and changes may in fact often be continuities. The lack of classic urban attributes has too often led to the suggestion that the Sasanians had no urban tradition. Some could even ask: "Did they have cities at all?"

The vast Sasanian empire was in fact a region of well-established and often extremely ancient cities (Fig. 1). Sasanian occupation may well have involved changes, or indeed development of new cities, and one must always consider regional influences. The cities of Mesopotamia offer good examples of specific Sasanian projects on a new scale. A Sasanian proclivity toward 'hyper-development' produced Ctesiphon, an urban phenomenon so vast that it received the moniker *al-Madā'in* (the cities). However, analysis of the site is complicated because of interactions with earlier (and later) urban structures.[5]

---

2 Ghirshman, Roman, *Bichapour*, II: *Les mosaïques sassanides*, (Paris: Geuthner, 1956) and Ghirshman, Roman, *Bichapour*, I, (Paris: Geuthner, 1971). It should be noted that Ghirshman always recognized the possibility of continuation of ancient urban forms; *Bichapour*: I, p. 24.

3 Schmidt, Erich, *Flights over Ancient Cities of Iran*, (Chicago: Chicago University Press, 1940).

4 Corona images are satellite photographs produced in the 1960s, declassified in 1995. They often reveal landscapes before extensive modernization. See http://scholar.harvard.edu/jasonur/pages/corona-photography-1.

5 Invernizzi, Antonio, "Ten Years' Research in the al-Mada'in Area: Seleucia and Ctesiphon", *Sumer*, XXXII (1976): pp. 167-76 provides a general overview of Italian research. On the early Sasanian contribution, see Morony, Michael G., "Beh-Ardašīr", *EIr*: IV, pp. 93-94. Maps still

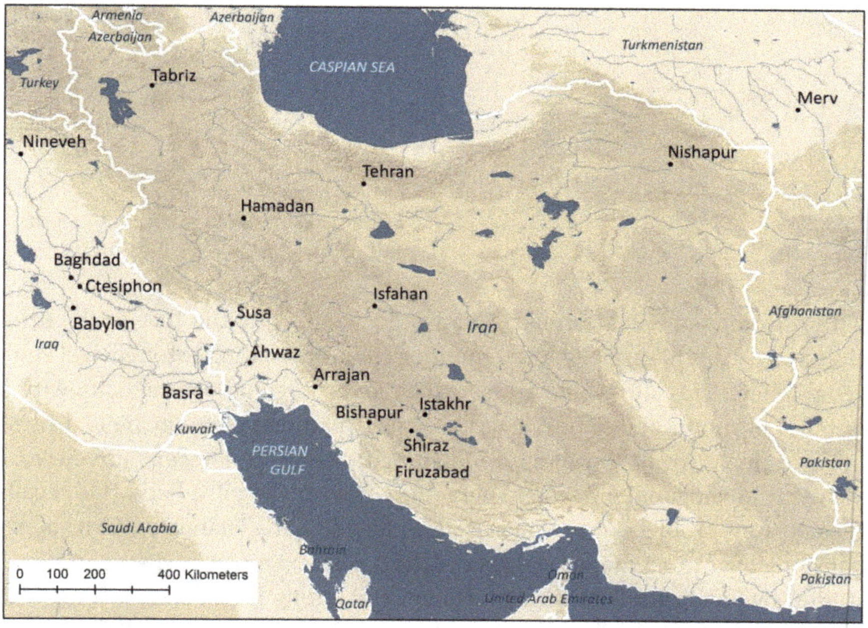

FIGURE 1 Situation map with Sasanian (and modern) cities
DRAWING K.J. FRANKLIN, CENTER FOR ANCIENT MIDDLE EASTERN LANDSCAPES, THE ORIENTAL INSTITUTE

The nature of the Sasanian contribution will remain clouded, though new breakthroughs are always possible. One factor noted in Kennedy is the extensive use of mud-brick and plaster for construction, a source of frustration for one more familiar with the fine stones used in the Levant and Mediterranean in general; this is less problematic for one familiar with the challenges of Mesopotamian and more eastern cities (e.g. Simpson, see below).

This paper will focus on the towns and cities of Fars province (Fig. 2), a district and subject explored by the author in his dissertation in 1979. The philological and archaeological characteristics of this core province have more recently been compiled in a detailed dissertation and monograph by Negin Miri. As an example of her contributions, she discusses the town of Abruwān in the district of Dašt-i Barīn. Her geographical study has an excellent complement in the archaeological investigations by Parsa Ghasemi. He has discovered and described a vast field of ruins of an urban settlement, in association

refer back to that of Reuther, Oscar, "German Excavations at Ctesiphon", *Antiquity* 3 (1929): pp. 434-52, 435.

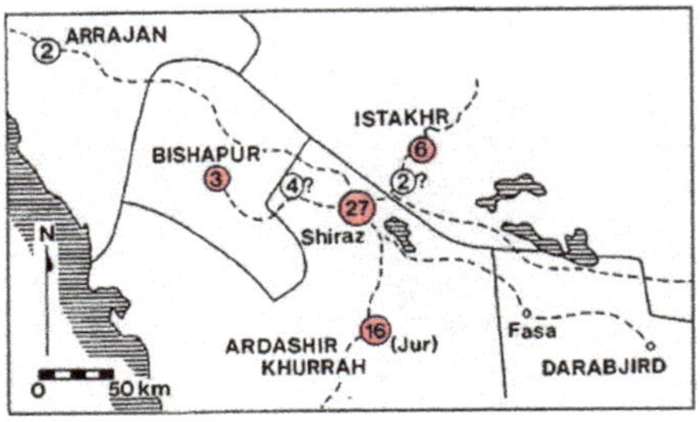

FIGURE 2
Central cities of Fars province
SOURCE: WHITCOMB *QASR-I ABU NASR*, FIG. 1

with the ruins at Sar Mashhad. This would seem to be the first evidence of Ġondekān/Ġundijān, a city often mentioned in the writings of medieval Arab geographers. This site would appear to be an orthogonal planned city with a number of monumental buildings.[6] A similar planned city may be identified in the ruins of Arrajān (Veh-az-Amīd-Qubād; see below). Recent survey work shows an orthogonal city with a large mosque on the periphery; this recalls the constant problem of separating Sasanian and early Islamic patterns of expansion concerning the same urban entity. These examples of discoveries and studies by young Iranian scholars convey the strong levels of activity in this field of investigation. Though it may be premature to summarize this active research, we can illustrate the recent advances by describing the new understandings of a few of the more prominent cities. The descriptions that follow attempt to provide a broad outline of urban development during Sasanian rule, in each case beginning with general descriptions and where appropriate, adding results from new research.

1.1 *Jūr (Fīrūzābād)*

The city of Jūr, later known as Fīrūzābād, is taken as the earliest in the typology of Sasanian foundations. It was approached from the north through a narrow valley with a fortress (Qalʿa-yi Duḫtar), passing magnificent reliefs in the gorge, and a well-preserved palace (Fig. 3). Beyond these features was a circular city, perhaps the most famous of the cities constructed by Ardašīr I (227-241).[7] The surrounding landscape is divided by radial roads, some extending to more than

---

6 Miri, Negin, "Historical geography of Fars during the Sasanian period", *e-Sasanika* 10, www.sasanika.org/esasanika/historical-geography-of-fars-during-the-sasanian-period, posted 03/25/2013: p. 5-6, 44-45. Ghasemi, Parsa, "Preliminary Report on the Identification of Ghandejan 'Dasht-e Barin City Center in Sar Mashhad Plain, north of the Persian Gulf", *Iranian Archaeology* 3 (2012): p. 65-71; his work and others on Sar Mashhad is soon to be published.

7 As many as 14 cities have been named in association with Ardašīr I, however the locations of many of these have not been identified. One set of lists appears in al-Ṭabarī, though the

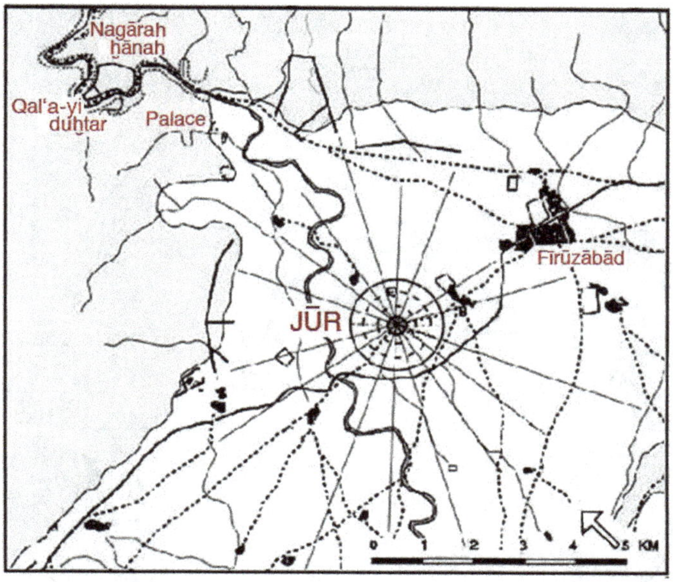

FIGURE 3 Region of Jūr (Fīrūzābād)
SOURCE: HUFF, "IRAN", FIG. 2

5 km beyond the double walls of the city. Within the city walls (ca. 2000 m in diameter), the roads divide the town into twenty sectors, and there are three concentric ring-roads.

The central circle abandons the pattern of radiating streets and contains buildings in a rectilinear grid, oriented in relation to four axial streets and gates (Fig. 4). There have been no excavations within this urban core. Attention has instead focused on the central tower called the *tirbāl* or *manār*, and a large platform called the Taḫt-i Nišīn, assumed to contain important structures. Air photographs and satellite images suggest that most occupation mounding seems to be around the central orthogonal arrangement of buildings, with some deviation toward the southwest. This may reflect an Islamic reoccupation more aligned with *qibla*. This central district has a diameter of 400 m, large enough to include administrative, religious and core residential functions.[8]

urban form is not specified; see Karimian, H., "Cities and Social Order in Sasanian Iran – the Archaeological Potential", *Antiquity* 84 (2010): pp. 453-66, 456, for a discussion.

8 Huff has conducted architectural studies: Huff, Dietrich, "Der Takht-i Nishin in Firuzabad", *ArchäologischerAnzeiger*, 1972: pp. 517-40; Huff, Dietrich, "Archaeology: IV. Sasanian", *EIr*, II (1986): pp. 302-8. Whitcomb, Donald, *Trade and Tradition in Medieval Southern Iran*. PhD dissertation, University of Chicago, 1979: pp. 64-6, proposes a hypothesis on the Islamic reoccupation (though the plan there contains an erroneous scale), specifically positing that

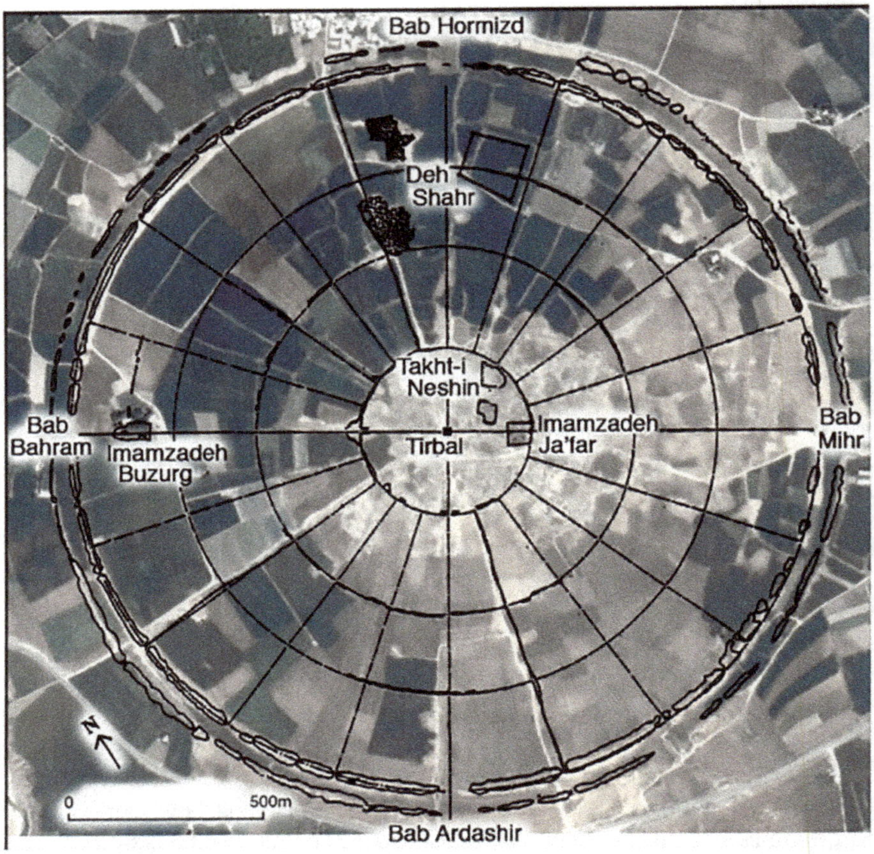

FIGURE 4 Circular city of Jūr
SOURCE: HUFF, "ARCHAEOLOGY", P. 192, WITH GOOGLE EARTH IMAGE

1.2 *Bīšāpūr*

The second city in our typological discussion is Bīšāpūr, founded by Šāpūr I (241-272). Here there is a famous series of reliefs carved in the Tang-i Čugān, and a high fortress guards the opening, where the Šāpūr river leads onto the plain. The eastern sector of the city was excavated by Roman Ghirshman, who discovered a temple, probably to Anahita, and a palace (now often considered

Islamic occupation often involved some reorientation toward *qibla*, and a plan of 400 m in diameter or a square of 400 m per side. Huff and Niakan have conducted new excavations in 2005 that may lead to better ideas of urban development.

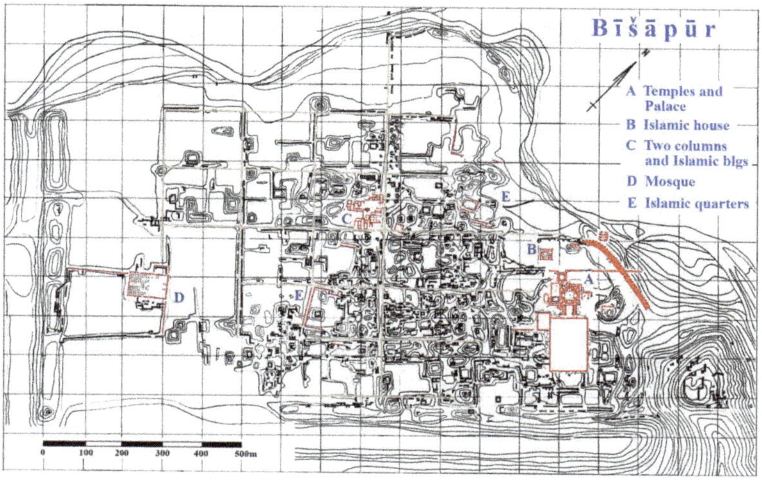

FIGURE 5 City of Bīšāpūr
SOURCE: GHIRSHMAN, *BICHAPOUR*, PLAN I, WITH ADDITIONS BY DONALD WHITCOMB

a fire temple).[9] More recently Ali Sarfaraz discovered an early Islamic residence and a multi-towered wall. This wall may partially enclose this 'royal quarter,' which once held administrative buildings.[10]

The city is laid out in an orthogonal grid of blocks, each 200 × 160 m, most clearly defined in the center square of four blocks per side, with an apron on the periphery of the wall and ditch to the south, and the river on the north. A pair of columns dedicated to the memory of Šāpūr marks the crossing near the center. This monument was thoroughly engulfed by buildings of the Islamic town, where a bath and reception hall have been found.[11] These latter structures are dated to the Saljuq period (fifth/eleventh-sixth/twelfth centuries) but could represent an occupation continuing from the earlier Islamic period. Al-Muqaddasī describes the city, in the tenth century, as having four

9 Huff, Dietrich, "Iran in Sasanian and Medieval-Islamic Periods", in T. Stöllner, et al. (eds), *Persiens antike Pracht*, (Bochum: Deutsches Bergbau-Museum, 2004): p. 416-39, 424; fig. 5A.

10 Ghirshman, *Bichapour* I and II; Sarfaraz, Ali A. "Bichapour grande ville Sassanide", *Bastan Chenassi va Honar-e Iran*, 2 (1969): p. 26. The standard plan of figure 5 is an adaptation of Ghirshman, *Bichapour*, II: plan I. An improved plan in figure 7, utilizing new surveys and satellite images, is the result of research by Shahmohammadpour, "*Ulgū-yi ṭarāḥī-yi šahrī-yi Bīšāpūr* [Patterns of Urban Design of Bishapur]", *Muṭāliʿāt-i miʿmāri-yi Īrān*, 6 (2014): pp. 107-25: fig. 9 (with additions by the current author).

11 Mihryār, Mihdī, "Sīmā-yi šahr-i Bīšāpūr dar dawrān-i islāmī", *Second congress on the history of architecture and urbanization of Iran*, III (2000), Tehran: 11-138.

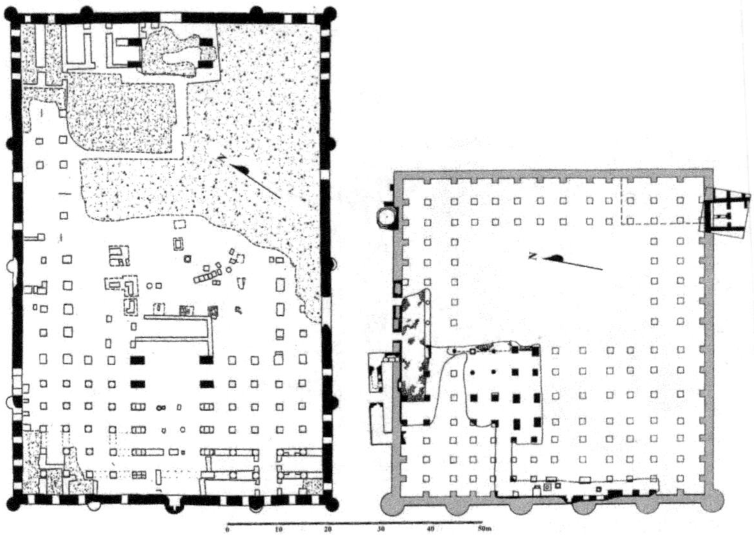

FIGURE 6 Comparison of mosques at Bīšāpūr (right) and Iṣṭaḫr (left)
SOURCES: NAWRŪZĪ, "BĪŠĀPŪR", P. 420 AND WHITCOMB, *ISTAKHR*, P. 365

gates, a trench, two fire temples, and at least two mosques.[12] He states that "the mosque is outside the town amidst the gardens, handsome and delightful." This mosque has been excavated and would seem early Islamic in date (cf. building D and in fig. 6); its characteristics and proportions are very similar to the mosque at Istakhr, though the *ṣaḥn* or court is much longer.[13]

Perhaps more puzzling is al-Muqaddasī's statement that "the town ... has two sections, both having gardens, trees, villages." He describes the "town as fallen into disrepair, near a major bridge, and having an old market (*al-sūq al-ʿatīq*)." One could wonder whether the oldest part of the site was a circular city,

12 He names the four gates as Bāb Hurmizd, Bāb Mihr, Bāb Bahrām, and Bāb Šahr; see Muqaddasī, translated by B. Collins, *The Best Divisions for Knowledge of the Regions: Aḥsan al-Taqāsīm fī Maʿrifat al-Aqālīm* (Reading: Garnet, 2001): p. 433. Since the habit of this author was to name gates beginning from the southwest and proceeding counterclockwise around the city, we insert these names in similar fashion in the plan offered here. Note that the Bāb Šahr opens onto the circular area (on Fig. 7), which could possibly be recognized as the older part of the town.

13 Mihryār, "Sīmā-yi šahr-i Bīšāpūr", and Whitcomb, Donald, "The City of Istakhr and the Marvdasht Plain", *Akten des VII. Internationalen Kongresses für Iranische Kunst und Archäologie*, (Berlin: D. Daimer, 1979): pp. 363-70, 365. For the Bīšāpūr mosque, see Nawrūzī, Riżā, "Masjid-i Jumʿa-yi Bīšāpūr: Panjumīn faṣl az dawra-yi sivum muṭālaʿāt-i bāstānšināsī-yi Bīšāpūr," *The Second symposium of Iranian Young Archaeologists, Iran*, Teheran 2003: pp. 411-30, 420.

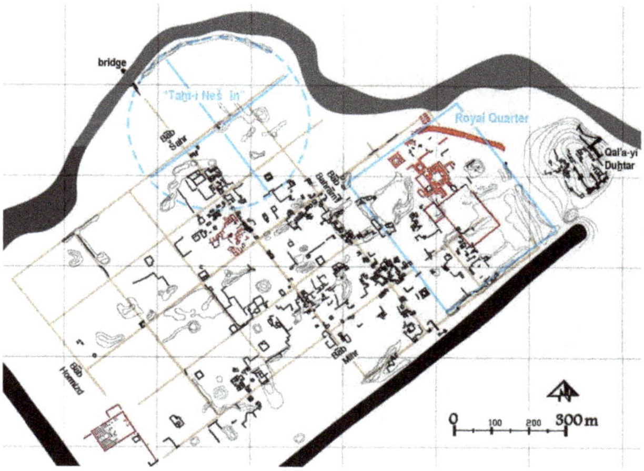

FIGURE 7 Bīšāpūr
SOURCE: SHAHMOHAMMADPOUR, BISHAPUR, FIG. 9, WITH ADDITIONS BY DONALD WHITCOMB

with its center on the northwest edge of the later city grid (Fig. 7). With a diameter of 400 m, this circle is the same size at the 'Taḫt-i Nišīn,' or administrative center of Jūr. This relationship of urban phases is remarkably similar to Jundi Šāpūr (see below).[14] This sector would have been separate from Šāpūr's 'royal district,' where Muqaddasī found a "prison with marble walls".

2 Istakhr

The third city considered here is perhaps the most interesting for early Islamic city planning. Istakhr was founded near the ancient monuments of Persepolis, and was reputed in the sources as the urban center for the Sasanian dynasty, continuing through the early Islamic period.[15] As al-Muqaddasī notes "I found it similar to Mecca, since it also lies in two valleys and two mountains immediately close around it."[16] Once again, access to the city leads past Sasanian stone

14 This hypothesis is added to a base map by Šāh-Muḥammad-Pūr Salmānī, "Patterns": fig. 9.

15 The role of Persepolis in later Islamic thought is explored in Whitcomb, Donald, "Istakhr in the Eyes of Muqaddasi and Modern Excavations", *Bāstānpažūhī*, III/ 5 (2008): p. 10-11, and more extensively in Mottahedeh, R. P., "The Eastern travels of Solomon: Reimagining Persepolis and the Persian Past", in M. Cook et al. (eds.), *Law and Tradition in Classical Thought: Studies in honor of Professor Hossein Modarressi*, (New York: Palgrave Macmillan, 2013): pp. 247-67.

16 Muqaddasī, *Aḥsan al-taqāsīm*: p. 436.

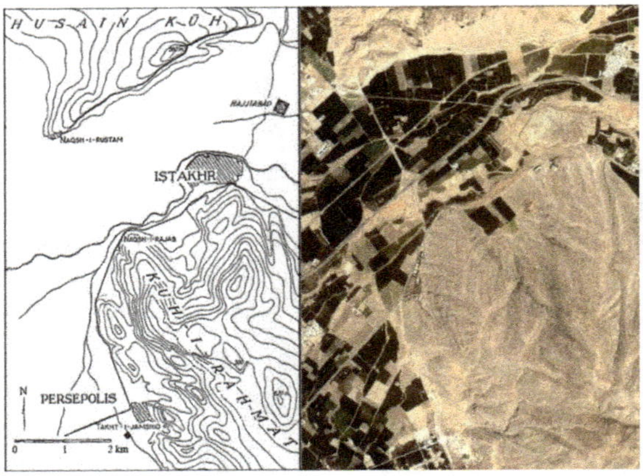

FIGURE 8 Region of Iṣṭaḫr
SOURCES: SCHMIDT, *FLIGHTS*, MAP 1, AND GOOGLE EARTH IMAGE

reliefs and a fortress, both at Naqš-i Rustam. The river bends around the city before entering the Marvdašt plain, and it is itself enclosed with an apparently irregular brick wall (Fig. 8).

The ruins of Istakhr were investigated by Herzfeld and then Schmidt, during the Persepolis excavations of 1930 to 1938.[17] Analysis of the early Islamic city is possible using the air photographs by Schmidt, which he supported with trench testing.[18] The pattern of streets and buildings indicate the Islamic city formed a square, 400 m per side, with external streets radiating out of gate areas (Fig. 9). Suddenly, the details of description in Muqaddasī fit the plan:

> The Friday mosque stands in the bazaar. It is built according to the style of Syrian mosques with round columns, each of which carries a cow [a Persepolitan bull capital]. They say that it was in older times a fire temple.[19] The bazaar surrounds the mosque on three sides. In the middle of the city is a valley-like depression.

17 These excavations have not been published, though information has been used in Whitcomb, *Trade and Tradition*: p. 97-101, and Whitcomb, "The City of Istakhr": pp. 363-68.

18 Schmidt, Erich, *The Treasury of Persepolis and Other Discoveries in the Homeland of the Achaemenians*, (Chicago: Chicago University Press, 1939): pp. 105-121, and Schmidt, *Flights*: pp. 12-16, pl. 8-12.

19 Muqaddasī, *Aḥsan al-taqāsīm*: p. 436. Kennedy states that "Muqaddasī ... never mentions a fire temple being converted", "From Shahrestān": p. 26. This shows that Muqaddasī is always full of surprises and was not always correct.

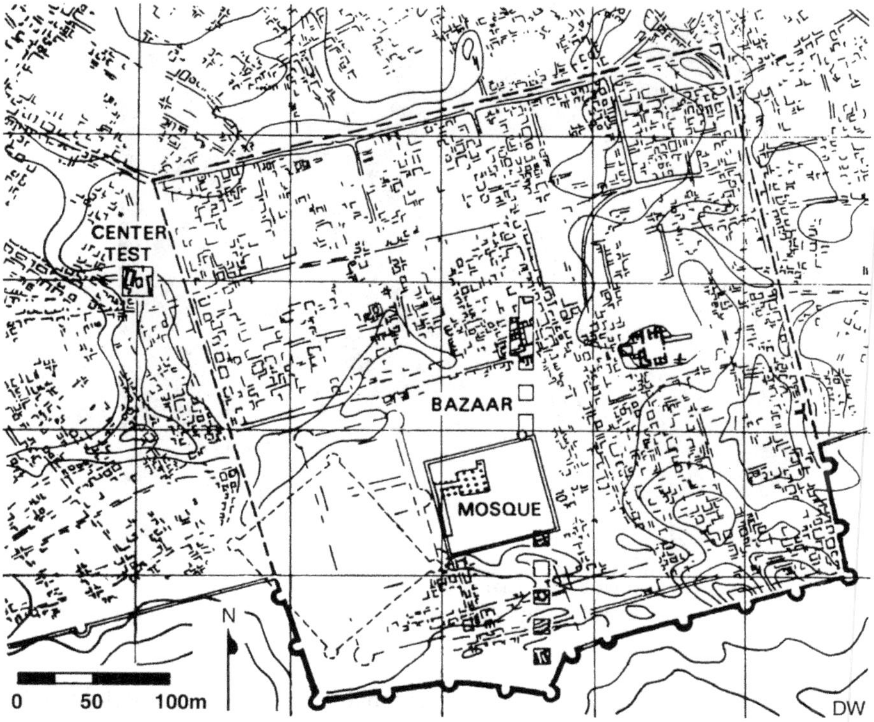

FIGURE 9 Plan of the early Islamic city of Iṣṭaḫr
SOURCE: WHITCOMB, *ISTAKHR*, P. 367

Further trenches to the north and west show a continual growth of urban accumulation in the tenth century and into the eleventh.

However, the so-called "west test", a 30 × 30 m trench, revealed architecture of a different style: heavy mud-brick walls delineate rooms with corridors, and rooms of a formal (perhaps ceremonial) nature. The ceramics seem to be more clearly Sasanian than early Islamic.[20] This is admittedly minimal evidence to suggest that a Sasanian city existed next to the later early Islamic foundation (Fig. 10). This Sasanian city would seem to have formed an orthogonal plan about 550 × 350 m with a distinctive orientation. The area of the city would at maximum have been half that of Bīšāpūr. Given the importance of Istakhr as the original dynastic capital, one must expect the true plan to be somewhat different, or at least more complex. Based on the description of this Sasanian

20 Analysis of this architecture and the artifacts is incomplete. It is interesting that a kiln for ceramics was inserted at a later date, indicating that the western periphery was outside the later occupation area.

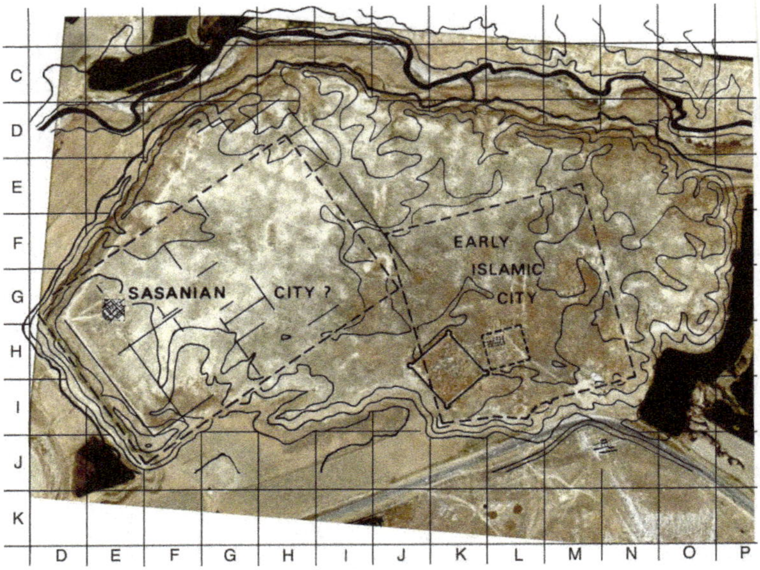

FIGURE 10 Plan of Iṣṭaḫr with early Islamic and hypothesized Sasanian city
SOURCE: WHITCOMB, *ISTAKHR*, P. 364

city, James Breasted[21] understood it as evidence of the last great 'ancient civilization,' which would have also provided a transition to the Islamic city.

2.1 *… and Shiraz in the Center*

Fars province is the ancient home of Iranian culture, but Shiraz was an urban foundation of the early Islamic period. The city was located at the center of the triangle of the three royal Sasanian cities of the province, Bīšāpūr, Jūr, and Istakhr, each of which was prone to revolt during the first decades of Islamic rule (Fig. 2).[22] This central location was noted by medieval geographers and leads to the claim of Shiraz as one of the earliest Islamic cities.[23]

There are various accounts for the foundation of Shiraz as a Muslim control point, as a *miṣr* or garrison town. According to al-Balāḏurī (end of third/ninth century): "The city of Shiraz is Islamic and not old; it was the first city to be built in Islam.… It was the camp of the Muslims when they attempted to conquer Istakhr." This action was under Muḥammad ibn al-Qāsim, cousin of the powerful governor of Iraq al-Ḥajjāj ibn Yūsuf in 693/4. The influence of this figure in founding the city is always noted, although there are also other

21 Director of the University of Chicago Oriental Institute, 1919-1935.

22 See Hinds, Michael, "The First Arab Conquests in Fars", *Iran* 22 (1984): pp. 39-53.

23 See Karimian, "Cities".

traditions. However Ḥamd-Allāh Mustawfī states there was an older foundation called "Fars".[24] For al-Muqaddasī, Shiraz was a *miṣr* in another sense, in that it was a metropolis, an administrative center and the principal seat of the ruler. In the late fourth/tenth century, this ruler was ʿAḍud al-Dawla (d. 983), who created another city called Kard Fanā Ḫusraw, within which were gardens, fine houses, and textile manufactures.[25]

While the archaeology of Shiraz has not been attempted in this thriving city of today, the site of Qaṣr-i Abū Naṣr, often known as Old Shiraz, lies about five kilometers to the southeast. The site was excavated by Upton and Hauser in 1932 to 1935, the same years that Herzfeld was at Persepolis and Istakhr.[26] Their attention focused first on the so-called "Western Area", the location of a church of the Sasanian period where the baptistery was later reworked in the form of an Achaemenian pavilion.[27] Next, they completely excavated the fortress, a high, natural spur of the mountain. Though the stratigraphy is difficult to discern, there are early Sasanian (and perhaps Parthian), later Sasanian, and early Islamic occupations. There was a main street with branching lanes, covering an area of 120 × 50 m.[28]

About five kilometers south of this fortress is a spring with Sasanian reliefs, called the Barm-i Dilak. Below the fortress, a Sasanian town spreads out to the southwest. The few trenches excavated within the town have an orientation that aligns closely with the Western Area, and a projection from towers on the northern wall gives a regular interval.[29] An urban space of about 1000 × 700 m may have enclosed blocks of perhaps 80 × 60 m. This would suggest a city of approximately the same size as Bīšāpūr, but with a smaller grid (Fig. 11). The

24 Le Strange, Guy, *The Geographical Part of the Nuzhat-al-Qulūb Composed by Ḥamdallāh Mustawfī of Qazwīn in 740 (1340)*, (Leiden: Brill, 1919): p. 113.

25 Muqaddasī, *Aḥsan al-taqāsīm*: p. 431; he also seems to mention production of *ṭirāz* cloth there. This town also held a library, probably where Muqaddasī wrote his geography. This suburb was probably located between the old Shiraz airport and Qaṣr-i Abū Naṣr.

26 Whitcomb, *Before the Roses and Nightingales*. The fortunes of these expeditions were ironically reversed: whereas Herzfeld sought the "last Oriental empire" of Sasanian, in vain (noted above), Upton was charged with finding an Achaemenian site for the Metropolitan Museum of Art, in competition with Persepolis, and in fact reluctantly settled for major excavations on a Sasanian site.

27 Whitcomb, *Before the Roses and Nightingales*: pp. 32-86. This Achaemenid structure was no doubt part of Kard Fanā Ḫusraw, documenting ʿAḍud al-Dawla's well-attested interest in this ancient kingship. The Achaemenid stones have been returned to their original locations at Persepolis.

28 Whitcomb, *Before the Roses and Nightingales*: pp. 87-209, fig. 30. In spite of a lack of exact right angles, the fortress seems carefully planned, though likely not large enough to accept wheeled vehicles, Kennedy, "From Shahrestān": p. 21.

29 Whitcomb, *Before the Roses and Nightingales*: pp. 220-21, fig. 80.

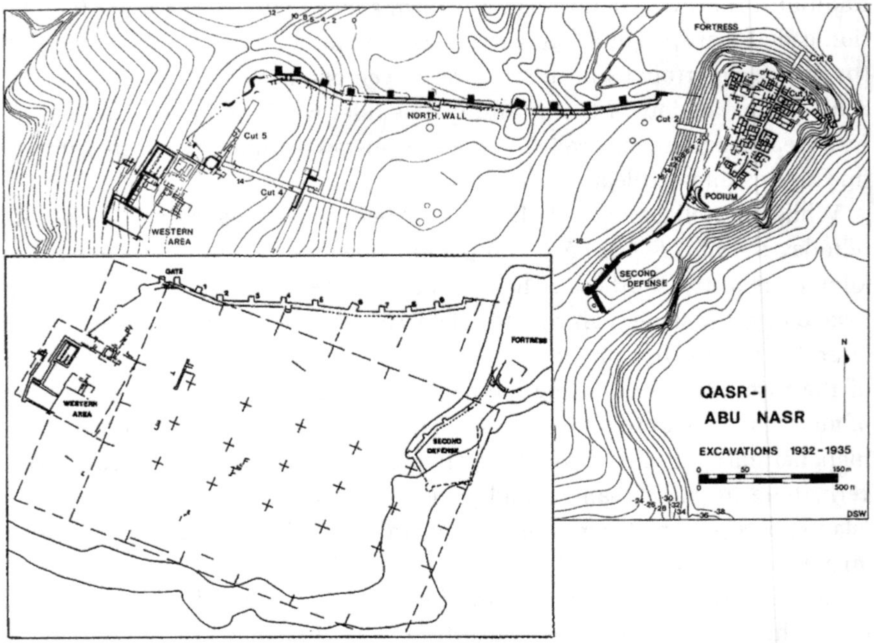

FIGURE 11 Plan of Qaṣr-iAbū Naṣr
SOURCE: WHITCOMB, *QASR-I ABU NASR*, FIGS. 2 AND 80

city must have had some importance, but scholarship has yet to advance a name for it, apart from a possible antecedent of 'Shiraz'.[30]

2.2 *Two Cities of Jundi Šāpūr*

As a last example of Sasanian cities we consider a location beyond Fars province, in Khuzistan. This great alluvial plain, in some ways an extension of Mesopotamia, was the most fertile part of Sasanian Iran and thus of economic and political importance. During the Sasanian period four cities dominate this province: Susa, Šuštar, Īwān-i Karḫah, and Jundi Šāpūr. Each of these cities is an archaeological site with respectable antiquity and investigations, but perhaps the most comprehensive plan is found for Jundi Šāpūr.[31] In 1963, Robert Adams

30 The name of Shiraz appears frequently on sealings; Frye, Richard N., (ed.) *Sasanian Remains from Qasr-i Abu Nasr: Seals, Sealings, and Coins*, (Cambridge, MA.: Harvard Univ. Press, 1973): pp. 1-2, 52; discussion in Whitcomb, *Before the Roses and Nightingales*: pp. 229-33, and Miri, "Historical Geography": pp. 34-39.

31 There is a frustrating lack of consistency in rendering this name. As described below, Gondēšāpūr may be the most accurate reflection of the Sasanian name while Jundaysābūr reflects the later, Islamic (Arabic) modification. The present article uses Jundi Šāpūr for

surveyed this site and carried out limited test excavations. Though he decided not to continue on this site, his report is a valued testimony, since modern farming has now obscured the visible traces.

Surveys by Adams and his team revealed an orthogonally planned city, a walled area of 3 × 1.5 km, a rectilinear system of city walls, axial streets, and a cityscape divided into blocks of buildings, residential areas and park lands. Their limited excavations were less successful, without exploration of monuments known from historical sources – the medical academy, astronomical observatory, Nestorian cathedral and monastery, congregational mosque and madrasas. The vast grid imposed on the landscape was still visible in the eleventh century, when it was described as a "chess board", and it was later noted by Yāqūt and Mustawfī.[32]

This layout of the city would seem to confirm the interpretation of the name as '*bih az Andaw-i Šāpūr*' (the better than Antioch of Šāpūr), and the thesis of construction by Roman captives, brought from Antioch by Šāpūr to build his new city in the mid-third century.[33] The hypothesis can be tested by comparing the plan of Jundi Šāpūr with that of Antioch. Setting aside consideration of the Palatial Island, the urban layout of Antioch has a broad main street with a slight bend in its axis at the crossing of a major street in the middle. The overall size and configuration, including the concentration of mounding analogous to the citadel of Antioch, appears to confirm an influence on the Sasanian city (Fig. 12). Ghirshman would use this planning for confirming the influence of Roman city planning at Bīšāpūr, as well as Jundi Šāpūr.[34]

The study of Jundi Šāpūr has revealed the presence of a second city. Adams used new air photographs and the 1960s Corona images to prepare a detailed plan. More recent analysis of these images reveals that this plan is a dramatic palimpsest covering an early city plan (Fig. 14). The central point along the north wall, where it intersects with an axial north-south street, now appears to be the central point of a great circle. The arc of this circle curves north of the city wall, clearly visible in the countryside. The radius measures about a

---

the archaeological site, respecting usage in the major papers by Adams, Robert McC., and Donald P. Hansen, "Archaeological Reconnaissance and Soundings in Jundi Shahpur", *Ars Orientalis*, VII (1968): pp. 53-70, and Abbott, Nabia, "Appendix: Jundi Shahpur: A Preliminary Historical Sketch", *Ars Orientalis*, VII (1968): pp. 71-73. Abbott seems to follow Adams in adding an unjustifiable second 'h'.

32 Wheatley, Paul, *The Places where Men Pray Together: Cities in Islamic Lands Seventh through the Tenth Centuries*, (Chicago: University of *Chicago* Press, 2001): p. 501, n. 593.

33 See Frye in Abbott, "Appendix": p. 71 n. 1.

34 Ghirshman, *Bichapour* I, p. 28. It would seem that, even after so many years of interest and excavations, further study of Jundi Šāpūr would assist in understanding details of the city of Antioch.

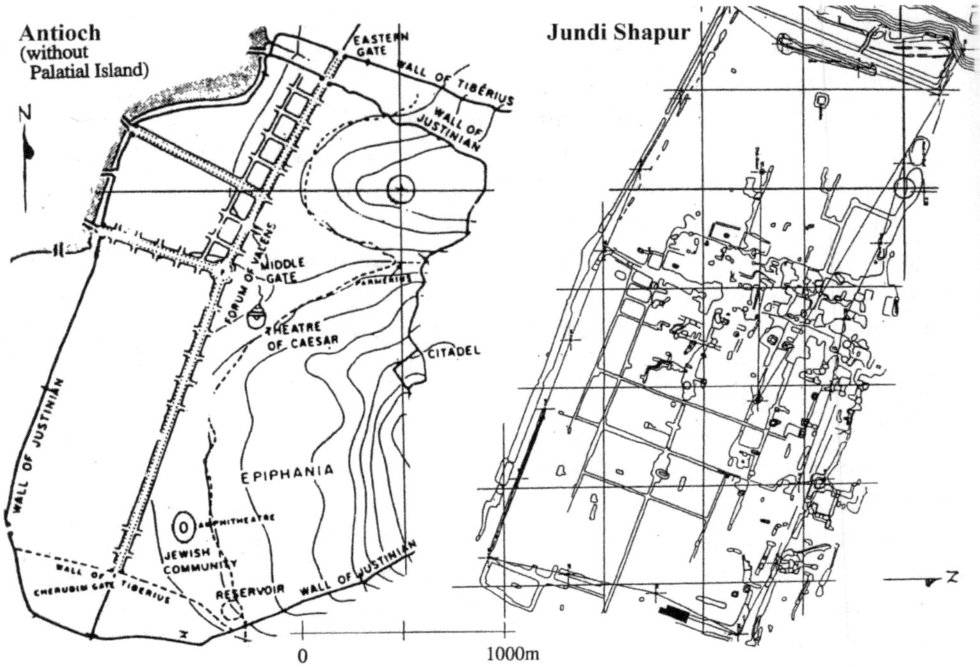

FIGURE 12–13 Plans of Antioch (left) and Jundi Šāpūr (right)
SOURCES: ADAMS AND HANSEN, JUNDI SHAHPUR, FIG. 1, AND MAPPERY.COM

kilometer in length, describing a city of about two kilometers in diameter, within which we find virtually all the mounding of buildings (Fig. 14). This circular occupation is the same size as Jūr (Fīrūzābād), and that city could provide a model for the features to be expected at Jundi Šāpūr: the axial gates and streets, the built-up urban core, the Taḫt-i Nišīn platform, and especially the great pivotal tower (the *manār* or *tirbāl*) in the center.

The circular city at Jundi Šāpūr takes its place with other circular cities, all traditionally associated with Ardašīr, the father of Šāpūr and founder of the Sasanian empire. There is a report that Šāpūr found a number of his father's foundations in decline and renewed them.[35] Jundi Šāpūr may offer a special set of archaeological circumstances: the first was an imperial foundation of

35 The Chronicle of Seert composed during the Abbasid era records that Šāpūr "rebuilt Gondēsābūr which had fallen into ruins". See Scher, Addai, *Histoire nestorienne inédite (Chronique de Séert), première partie (I)*, (Paris: Firmin-Didot, 1907; reprint Turnhout: Brepols, 1993): p. 221.

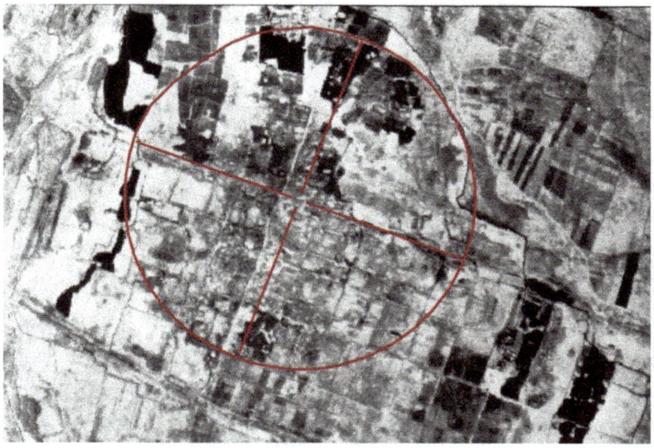

FIGURE 14 Map of Jundi Šāpūr
SOURCE: CORONA SATELLITE IMAGE (COURTESY OF ABBAS ALIZADEH) WITH ADDITION OF CIRCULAR FEATURE

Ardašīr reflecting Sasanian (and older) urban concepts; and second that Šāpūr replaced this city with another urban form, meaning a Western rectilinear city and its institutions (Fig. 15). This is not the close of urban history at the site, since an important Middle Islamic city developed on these foundations.[36]

3 Conclusions

This paper began with the rhetorical question of "whether there was ever a Sasanian city". This is an obvious matter for archaeological research, and only a small fraction of recent Iranian research has been cited here. Perhaps more important than fieldwork are the background conceptions and assumptions that scholars bring to the subject.

We could begin with consideration of long-term excavations by the British Museum, as part of the International Merv Project. St. John Simpson reports on a mid-Sasanian (fourth-fifth century CE) residential quarter in Gyaur-kala, the lower city (MGK5). He systematically discusses the domestic architecture in terms of building materials, house patterns, surrounding streets and lanes, and

36 One must also note that Jundaysābūr once again became a major city under the Saffarid Ya'qūb ibn Layṯ (840-879), whose tomb lies at the city's edge. Muqaddasī states the city produced "most of the sugar of Khurasan and al-Jibāl", (Muqaddasī, *Aḥsan al-taqāsīm*: p. 408). The Oriental Institute has sherds from this site that seem to come from Nishapur, which may bear out this economic interaction; Whitcomb, Donald, "Bull Spout from Jundi Shapur", *Oriental Institute News & Notes*, CLXXIX (2003): pp. 6-7.

FIGURE 15 Plan of Jundi Šāpūr, with two suggested cities
SOURCE: WHITCOMB, "JUNDISHAPUR" AND "IRANIAN CITIES", P. 93

patterns of refuse deposition. He compares these housing patterns to those at Ctesiphon (Veh Ardašīr) and then numerous other excavations, resulting in "a vivid illustration of the longevity of urban tradition". He finds variations in vernacular architecture, common occurrence of "major religious buildings", descriptions of bazaars and market fairs, and regular main streets and alleys.[37]

This is a very different picture from that painted by Kennedy, who states that "most Sasanian towns were no doubt busy little huddles of mud-brick houses, looking onto streets that were little more than wide footpaths." His next sentence reveals the core problem, that "there was probably no monumental planning and certainly no great baths or theatres."[38] The Sasanian city presents problems of interpretation because it was "oriental", whatever that

37 Simpson, St John, "Suburb or Slum? Excavations at Merv (Turkmenistan) and Observations on Stratigraphy, Refuse and Material Culture in a Sasanian City", in Derek Kennet and Paul Luft, (eds.), *Current Research in Sasanian Archaeology, Art and History* (Oxford: Archaeopress, 2008): pp. 65-78, 70, 72-73, and Simpson, St John, "Merv, an Archaeological Case-Study from the Northeastern Frontier of the Sasanian Empire", *Journal of Ancient History*, 11/2 (2014): pp. 116-43. Given his experience in sites of Iraq, he is perhaps sensitive to the interpretation of urban architecture on the basis of brick remains.

38 Kennedy, "From Shahrestān": p. 22.

might connote. It is curious that many features, such as narrow, twisting roads, unplanned neighborhoods, etc. are aspects that older scholars used to characterize the 'Islamic city'. Clearly there is a need for more fieldwork and archaeological models for understanding this apparently alien urban phenomenon.

One complication stems from the problem that drew Kennedy to the transition of the Late Antique (Byzantine) city into the Islamic. As he was able to deduce, the sixth-century cities of Syria-Palestine were no longer 'classical' cities: they had begun to change into a new urban form.[39] The evidence for Sasanian cities produced in this paper uses the foundations of Ardašīr and Šāpūr, that is, the *early* Sasanian city of the third century. Rezakhani raises the important point that reforms and urban foundations of Kavād I and Ḫusraw I (late fifth/sixth centuries) need to be studied as the antecedent for continuities with early Islamic patterns.[40] Ḫusraw I is associated with foundation of numerous cities and even Veh Antiok Ḫusraw at Ctesiphon.[41] His father Kavād I (Qubād I) founded at least four cities in Fars province, including Veh az-Amīd Kavād, better known as Arrajān.[42]

The implication of these later urban foundations is that Sasanian cities may have had more than one model of form and institutions, a complexity that requires consideration before addressing changes and continuities with the early Islamic occupations. The situation is analogous to the state of research for Syro-Palestinian cities, both those used by Kennedy in 1985, and the extensive studies made available by Gideon Avni in 2014. Indeed, Avni states that archaeological research during the past twenty years enabled him to take new

39 Kennedy, "From Polis"; this study has the important addendum that this change was not a matter of 'decline'. The archaeological evidence for this problem in the Palestine region is carefully addressed by Gideon Avni, first in articles and recently in an excellent monograph: Avni, Gideon, "'From Polis to Madina' revisited – Urban change in Byzantine and Early Islamic Palestine", *JRAS*, XXI (2011): p. 301-329; Avni, Gideon, *The Byzantine-Islamic Transition in Palestine: An Archaeological Approach*, (Oxford: Oxford University Press, 2014).

40 Rezakhani, Khodadad, "Continuity and change in Late Antique Iran: An Economic view of the Sasanians", *International Journal of the Society of Iranian Archaeologists*, I (2015): pp. 95-108, 97. This point is also made in Whitcomb, Donald, "Landscape Signatures in Sasanian Archaeology", *Journal of Ancient History*, II/2 (2014): p. 209-15, 213.

41 Rather like his predecessor, Šāpūr I, founding Jundi Šāpūr some 300 years earlier, Ḫusraw is also associated with the foundation of the Academy at Gondēšāpūr, yet another institution that may be sought at that site (Abbott, "Appendix": p. 72).

42 This foundation was presumably after his capture of Amida in 502 CE. For identifications of Sasanian place names, see Gyselen, Rika, *La géographie administrative de l'empire sassanide. Les témoignages sigillographiques* (Paris and Leuven: Groupe pour l'étude de la civilisation du Moyen-Orient, 1989).

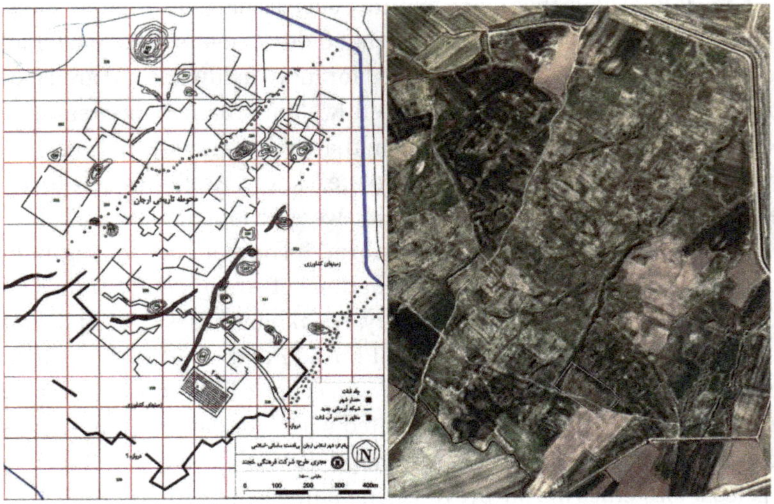

FIGURE 16 Ruins of Arrajān: map (drawing: Širkat-i farhangī-yi Ḫujand) and photo (Google earth image)

approaches in considering this urban development.[43] The study of the city of Arrajān could offer a useful example (Fig. 16). Recent Iranian planning of the ruins, combined with satellite images, serves in outlining a sixth century, late Sasanian foundation.[44] The new satellite images now capture unknown sites with astonishing detail, just as we previously benefitted from the aerial photography by Schmidt. An example would be his record of Dar-i Šahr, a Sasanian and early Islamic site in western Iran (Fig. 17).[45] "The cities in the land of Ērānšahr which have been built in different days"[46] remain to be studied, and not just in Fars province. New archaeological studies on Sasanian and early Islamic urban sites will bring the revisitation that the *šahristān* deserves.

43 Avni, *The Byzantine-Islamic Transition*: p. 299.

44 The author wishes to thank Dr Abbas Moghaddam for images and discussion on the apparent mosque at Arrajan.

45 Schmidt, *Flights*: p. 43, pl. 53.

46 See Daryaee, Touraj, *Šarestānīha ī Ērānšahr: A Middle Persian Text on Late Antique Geography, Epic, and History*, (Costa Mesa: Mazda, 2002): p. 17, for a reinterpretation of a classic text on Iranian cities.

FIGURE 17 Air photo of Dar-i Šahr, Saymarra
SOURCE: SCHMIDT, FLIGHTS, P. 43, PL. 53

Bibliographical References

Abbott, Nabia, "Appendix: Jundi Shahpur: A Preliminary Historical Sketch", *Ars Orientalis*, VII (1968): pp. 71-73.

Adams, Robert McC., and Donald P. Hansen, "Archaeological Reconnaissance and Soundings in JundiShahpur", *Ars Orientalis*, VII (1968): pp. 53-70.

Avni, Gideon, "'From Polis to Madina' revisited – Urban change in Byzantine and Early Islamic Palestine", *JRAS*, XXI (2011): pp. 301-329.

Avni, Gideon, *The Byzantine-Islamic Transition in Palestine: An Archaeological Approach* (Oxford: Oxford University Press, 2014).

Daryaee, Touraj, *Šarestānīha ī Ērānšahr: A Middle Persian Text on Late Antique Geography, Epic, and History* (Costa Mesa: Mazda, 2002).

Frye, Richard N., (ed.) *Sasanian Remains from Qasr-i Abu Nasr: Seals, Sealings, and Coins* (Cambridge, MA.: Harvard Univ. Press, 1973).

Ghasemi, Parsa, "Preliminary Report on the Identification of Ghandejan Dasht-e Barin City Center in Sar Mashhad Plain, north of the Persian Gulf", *Iranian Archaeology*, III (2012): pp. 65-71.

Ghirshman, Roman, *Bichapour*, II: *Les Mosaïques sassanides* (Paris: P. Geuthner, 1956).

Ghirshman, Roman, *Bichapour*, I (Paris: Paul Geuthner, 1971).

Gyselen, Rika, *La géographie administrative de l'empire sassanide. Les témoignages sigillographiques* (Paris and Leuven: Groupe pour l'étude de la civilisation du Moyen-Orient, 1989).

Hinds, Michael, "The First Arab Conquests in Fars", *Iran*, XXII (1984): pp. 39-53.

Huff, Dietrich, "Der Takht-i Nishin in Firuzabad", *Archäologischer Anzeiger* (1972): pp. 517-40.

Huff, Dietrich, "Archaeology: IV. Sasanian", *EIr*, II (1986): pp. 302-308.

Huff, Dietrich, "Iran in Sasanian and Medieval-Islamic Periods", in T. Stöllner, et al. (eds), *Persiens antike Pracht* (Bochum: Deutsches Bergbau-Museum, 2004): pp. 416-39.

Invernizzi, Antonio, "Ten Years' Research in the al-Mada'in Area: Seleucia and Ctesiphon", *Sumer*, XXXII (1976): pp. 167-76.

Karimian, H., "Cities and Social Order in Sasanian Iran – the Archaeological Potential", *Antiquity*, LXXXIV (2010): pp. 453-66.

Kennedy, Hugh, "From Polis to Madina: Urban Change in Late Antique and Early Islamic Syria", *Past and Present*, CVI (1985): pp. 3-27.

Kennedy, Hugh, "From Shahristān to Medina", *StIs*, CII-CIII (2006): pp. 5-34.

Le Strange, Guy, *The Geographical Part of the Nuzhat-al-Qulūb Composed by Ḥamd-allāh Mustawfī of Qazwīn in 740 (1340)* (Leiden: Brill, 1919).

Mihryār, Mihdī, "Sīmā-yi šahr-i Bīšāpūr dar dawrān-i islāmī", *Second congress on the history of architecture and urbanization of Iran*, III (2000), Tehran: pp. 11-138.

Miri, Negin, "Historical geography of Fars during the Sasanian period", *e-Sasanika* 10, www.sasanika.org/esasanika/historical-geography-of-fars-during-the-sasanian-period, posted 03/25/2013.

Morony, Michael G., "Beh-Ardašīr", *EIr* IV: pp. 93-94.

Mottahedeh, R. P., "The Eastern travels of Solomon: Reimagining Persepolis and the Persian Past", in M. Cook et al. (eds.), *Law and Tradition in Classical Thought: Studies in honor of Professor Hossein Modarressi* (New York: Palgrave Macmillan, 2013): pp. 247-67.

Muqaddasī, translated by B. Collins, *The Best Divisions for Knowledge of the Regions: Aḥsan al-Taqāsīm fī Maʿrifat al-Aqālīm* (Reading: Garnet, 2001).

Nawrūzī, Rižā, "Masjid-i Jumʿa-yi Bīšāpūr: Panjumīn faṣl az dawra-yi sivum muṭālaʿāt-i bāstānšināsī-yi Bīšāpūr", *The Second symposium of Iranian Young Archaeologists, Iran*, (Teheran: 2003): pp. 411-30.

Reuther, Oscar, "German Excavations at Ctesiphon", *Antiquity*, III (1929): pp. 434-52.

Rezakhani, Khodadad, "Continuity and change in Late Antique Iran: An Economic view of the Sasanians", *International Journal of the Society of Iranian Archaeologists*, I (2015): pp. 95-108.

Sarfaraz, Ali A., "Bichapour grande ville Sassanide", *Bastan Chenassi va Honar-e Iran*, II (1969): p. 26.

Scher, Addai, *Histoire nestorienne inédite (Chronique de Séert), première partie (I)* (Paris: Firmin-Didot, 1907; reprint Turnhout: Brepols, 1993).

Schmidt, Erich, *The Treasury of Persepolis and Other Discoveries in the Homeland of the Achaemenians* (Chicago: Chicago University Press, 1939): pp. 105-21.

Schmidt, Erich, *Flights over Ancient Cities of Iran* (Chicago: Chicago University Press, 1940).

Shahmohammadpour, A. [Šāh-Muḥammad-Pūr Salmānī, 'Alī-Riḍā], "*Ulgū-yi ṭarāḥī-yi šahrī-yi Bīšāpūr* [Patterns of Urban Design of Bishapur]", *Muṭāli'āt-i mi'māri-yi Īrān*, VI (2014): pp. 107-25.

Simpson, St John, "Suburb or Slum? Excavations at Merv (Turkmenistan) and Observations on Stratigraphy, Refuse and Material Culture in a Sasanian City", in Derek Kennet and Paul Luft, (eds.), *Current Research in Sasanian Archaeology, Art and History*, (Oxford: Archaeopress, 2008): pp. 65-78.

Simpson, St John, "Merv, an Archaeological Case-Study from the Northeastern Frontier of the Sasanian Empire", *Journal of Ancient History*, II/2 (2014): pp. 116-43.

Wheatley, Paul, *The Places where Men Pray Together: Cities in Islamic Lands Seventh through the Tenth Centuries* (Chicago: University of Chicago Press, 2001).

Whitcomb, Donald, *Trade and Tradition in Medieval Southern Iran*. PhD dissertation, University of Chicago, 1979.

Whitcomb, Donald, "The City of Istakhr and the Marvdasht Plain", *Akten des VII. Internationalen Kongresses für Iranische Kunst und Archäologie* (Berlin: D. Daimer, 1979): pp. 363-70.

Whitcomb, Donald, *Before the Roses and Nightingales: Excavations at Qasr-i Abu Nasr, Old Shiraz*, (New York: The Metropolitan Museum of Art, 1985).

Whitcomb, Donald, "Bull Spout from Jundi Shapur", *Oriental Institute News & Notes*, CLXXIX (2003): pp. 6-7.

Whitcomb, Donald, "Istakhr in the Eyes of Muqaddasi and Modern Excavations", *Bāstānpažūhī*, III/5 (2008): pp. 10-1.

Whitcomb, Donald, "Landscape Signatures in Sasanian Archaeology", *Journal of Ancient History*, II/2 (2014): pp. 209-15.

Biographical Note

Dr. Donald Whitcomb teaches Islamic Archaeology at the University of Chicago, where he took his doctorate in the Islamic archaeology of Iran. He has conducted excavations at Quseir al-Qadim (a port on the Egyptian coast of the Red Sea), at the city of Luxor, at Aqaba in Jordan, at Hadir Qinnasrin (the early Islamic capital of north Syria near Aleppo) and most recently at Khirbet al-Mafjar (near Jericho). He has conducted surveys in Iran, Oman, Syria, Jordan and Egypt. He is the author of *Before the Roses and Nighingales. Excavations at Qasr-i Abu Nasr, Old Shiraz* (1985), and monographs on his work on the Red Sea. He has edited a seminar in Chicago on *Changing Social Identity with the Coming of Islam: Archaeological Perspectives*.

Why Nishapur?

Richard W. Bulliet
Columbia University
rwb3@columbia.edu

Abstract

Given the high cost of premodern land transport, inland cities not serviced by any sort of water transport were quite rare before the appearance of motorized vehicles. The emergence of Nishapur in northeastern Iran as the Abbasid caliphate's second largest metropolis, with a population over 150,000 in the year 1000, thus presents a problem. That it was neither a major governing center during most of its growth period, nor a transshipment node or pilgrimage site, compounds the problem. This article proposes a set of economic, geographic, and social circumstances that taken together may account for its flourishing. It further suggests that these same factors contributed to the city's emerging distinction as a Muslim religious and cultural center.

Keywords

Nishapur – urbanization – camels – Islamic society – Khurasan

In 1965 I decided to devote my doctoral dissertation to the history of Nishapur, a once flourishing city in northeastern Iran. Through the good offices of Professor Richard N. Frye I acquired copies of three manuscripts specific to the city, one being little more than a listing of scholars once accorded biographical entries in al-Ḥākim al-Bayyiʿ al-Naysābūrī's biographical dictionary (*ṭabaqāt*) of the city, and the other two being separate abridgements of a work by ʿAbd al-Ġāfir al-Fārisī, which extended al-Ḥākim's biographical coverage through the eleventh century.

Personal computers were not yet available, but I had learned from my father about the Royal-McBee Keysort system, which featured a type of card with holes around the edges that could be converted into notches by a special punch. Push a stylus or knitting needle through a hole in a stack of cards, and any card on which that hole had been converted into a notch dropped out. Four-hole groupings allowed for sixteen permutations of holes and notches,

 | DOI:10.1163/9789004434332_007

which meant that I could divide the Arabic alphabet into sixteen segments for coding purposes. By this means, I was able to transcribe the data from each biography onto a separate card, and then code that data with my notch punch, thus making it randomly retrievable.

Having devised what was at that time a unique means of quantifying the data in Arabic biographical dictionaries, a world of quantitative history opened up to me, along with another world of multigenerational family history. Half a century has passed since I devised my method for dealing with quantifiable *ṭabaqāt* data, and I still go back to my original cards when questions occur that might be amenable to quantitative treatment.

My first study based on the Nishapur data was published in 1972, as *The Patricians of Nishapur: An Essay in Medieval Islamic Social History.*[1] Three subsequent books expanded my method and addressed a broader array of historical questions. In *Conversion to Islam in the Medieval Period: An Essay in Quantitative History*, using additional sets of cards derived from the biographical dictionaries of Jurjān by al-Sahmī, and of Isfahan by Abū Nuʿaym al-Iṣfahānī, I addressed the pace and pattern of conversion to Islam.[2] *Islam: The View from the Edge* advanced a fresh interpretation of medieval Islamic social history based on the chronology of conversion.[3] More recently, *Cotton, Climate, and Camels in Early Islamic Iran: A Moment in World History* focused on the economic developments triggered by the Arab conquest of Iran in the seventh century.[4]

In reflecting on the economic matters addressed in the last book, I realized that certain important aspects of the city's history had hitherto escaped my grasp. The purpose of this paper is to properly consider these aspects.

In the year 1000 CE, the northeastern Iranian metropolis of Nishapur was the second largest city in the Muslim world, surpassed only by Baghdad. In that same year, Nishapur may also have been the largest city in the world lacking any access by water transport. These two distinctions present different historical challenges. Why was the city so big? And how did it sustain its population on a great arid plateau?

---

1 *The Patricians of Nishapur: An Essay in Medieval Islamic Social History* (Cambridge, MA: Harvard University Press, 1972).

2 *Conversion to Islam in the Medieval Period: An Essay in Quantitative* History (Cambridge, MA: Harvard University Press, 1979).

3 *Islam: The View from the Edge* (New York: Columbia University Press, 1994).

4 *Cotton, Climate, and Camels in Early Islamic Iran: A Moment in World History* (New York: Columbia University Press, 2009).

The city of Ṭūs, 75 miles east of Nishapur, would seem to be a more logical location for a regional centre representing a caliphate centered on Baghdad. Strategically, Ṭūs commanded the end of a mountain range that potential invaders from the northeast would have to round to access the central Iranian plateau. Ṭūs had also been a major governing center under the Sasanid empire, prior to Arab conquest, and in its modern incarnation as the city of Mashhad it is now the regional metropolis of northeastern Iran. Perhaps it was the close association between Ṭūs and the Persian empire that slowed its absorption into the Muslim world. In fact recollections of that past still remained so important in the late tenth century that two city residents compiled collections of pre-Islamic royal tales: an unfinished *Šāh-nāma* (Book of Kings) by Abū Manṣūr Daqīqī and a much more famous *Šāh-nāma* by Firdawsī.

If an Arab military presence, as opposed to Sasanid imperial continuity, had been key to caliphal rule in northeast Iran, then the logical site for a metropolis might have been neither Nishapur nor Ṭūs, but the city of Jurjān, modern Gunbad-i Qābūs, 300 miles to the west of Nishapur.[5] Jurjān was situated at the southeastern corner of the Caspian Sea, a spot so vulnerable to invasion that earlier rulers had constructed an immense protective wall to fend off raiders arriving via the deserts to the north. Jurjān then became a major Arab military encampment in the course of the conquest, and over subsequent centuries Muslim warriors would continue to dedicate themselves to guarding the Caspian frontier. By contrast, the Arabs did not install a major garrison in Nishapur. Sheltered to the north by the rugged Bīnālūd mountain range, Nishapur seldom experienced direct invasion, and the surrounding terrain did not afford any tactical strongpoints.

From a dialectical point of view, one might opine that Ṭūs represented the thesis of Persian royal and cultural tradition and Jurjān the antithesis of Arab Muslim dominion. This would make Nishapur the synthesis, a Muslim city that was in thrall to neither the Persian nor the Arab traditions. But a dialectical approach has little to recommend it. To be sure, medieval geographers credit Nishapur with a large and well-constructed congregational mosque, supplemented by a second such mosque in the eleventh century; and a number of eminent Muslim scholars came to be buried there. But it was not a pilgrimage site or a shrine center. Nishapur's growth, in other words, did not stem from some military role or popular religious ritual particular to Islam.

5 Though the modern city of Gurgān takes its name from the medieval city, it is instead located on the site of medieval Astarābād.

Yet it did grow prodigiously.[6] The citadel and the walled inner city date from the pre-Islamic period, and together cover an area of 17.6 hectares. Calculating from modern estimates that medieval Muslim population densities ranged from 100 to 200 people per hectare, this area might have housed 1760 to 3520 inhabitants. A list of 65 villages that were incorporated into the city as it grew indicates a fairly dense cluster of settlements outside the walls, presumably taking advantage of the security provided by the citadel's garrison and the refuge afforded by the city walls. Judging from tabulations of Iranian village life made in the 1950s, when most villages still drew their water from a single underground channel, most of those villages would have numbered around 160 inhabitants, and some would have had as many as 350.[7] These sizes reflect the food-growing potential of the land irrigated by the average underground channel, called a *qanāt* or *kārīz*. Virtually all of the Nishapur area villages relied on water from a *qanāt*. Yet even if all of the villages fell in the range of 350 inhabitants, their collective population would have amounted to only 22,750 souls. Added to those who lived within the walled precincts, this suggests a maximum urban and semi-urban population of 25,000 at the time of the Arab conquest in the seventh century.

By the year 1000, the built-up area of the city, as estimated from aerial photographs and on-site observations, had grown to approximately 1680 hectares.[8] At 100 residents per hectare, this translates into a population of 168,000; at 200 per hectare it comes to 336,000. If one reduces these estimates by one-third on the assumption that the neighborhoods farthest from the city center had a good deal of space devoted to growing fruits and vegetables, it seems reasonable to suppose that the population of Nishapur at its peak numbered between 110,000 and 220,000. In other words, during the three centuries following the definitive Arab conquest of northeast Iran, Nishapur saw a roughly ten-fold increase in its population. To be sure, other Iranian cities, such as Shiraz, Isfahan,

6 For an extensive and detailed discussion of the topography and demography of Nishapur see my article "Medieval Nishapur: A Topographic and Demographic Reconstruction," *StIr*, V/1 (1976): pp. 67-89. For maximum population estimates see pp. 87-9.

7 These are two size ranges, largely determined by water supply, suggested for the central Iranian plateau by J. Behnam in "Population", Fisher, W. B. ed., *The Cambridge History of Iran*, 1st ed., I (Cambridge: Cambridge University Press, 1968): p. 449.

8 In walking outward in different directions from the walls of the citadel and inner city, the boundary between a built zone (strewn with mounds, potsherds, bricks) and the surrounding farmland and desert becomes strikingly apparent.

Rayy, and Hamadan, also experienced extraordinary growth during those centuries, but Nishapur surpassed them all.[9]

The city's enormous population was sustained exclusively by land transport, and this in itself poses questions. Worldwide, cities without access to sea-, lake-, river-, or canal-borne transport seldom grew to great size prior to the advent of motor transport in the nineteenth century. Without bridges or port facilities to create traffic and transshipment bottlenecks, inland cities without access to water transport did not normally have a logistical or economic reason to expand. A major pilgrimage or shrine center, a border crossing, or a road junction might also provide such a reason. But Nishapur had none of these.

As for provisioning, the populous medieval city of Florence in the fourteenth century imported sixty percent of its annual grain requirements by sea.[10] The immediate hinterland was not productive enough to support the city's population, and the cost of importing food overland from any great distance was simply too expensive to compete with water transport. For Nishapur, however, all transport was by pack animal.[11] Therefore, if the city had 150,000 non-farming inhabitants and consumed food at the same rate as has been estimated for medieval European working people, namely, 2.5 lbs. of bread or its equivalent per person per day, supplying the city with the staff of life would have required more than a quarter of a million camel-loads of grain per year, or several thousand camel-loads per day during the harvest season. Additional thousands of loads would have been required for other foodstuffs, charcoal, and building supplies, not to mention the many animals needed for the long-distance caravan trade carrying higher value goods along the Silk Road, and for the local requirements of Nishapur's main industry, the production of cotton cloth. Raw cotton had to be transported to ginning, dyeing, and weaving facilities, and finished cloth had to be shipped to markets.

This raises a question: Where did all those camels come from?

Iran being part of the Middle East, and the camel being a time-honored stereotype of Middle Eastern life, this may seem like a trivial question. But it is not. Great strength, endurance, and the ability to continue working with minimal rations of food and water make camels ideal laboring animals. In Roman

---

9 Demographic study has been more thorough for Isfahan than for the other cities listed. See David Durand-Guédy's *Iranian Elites and Turkish Rulers: A History of Iṣfahān in the Saljūq Period* (Abingdon: Routledge, 2010): pp. 24-26.

10 Bulliet, Richard W., *Islam: The View from the Edge* (New York: Columbia University, Press, 1994): pp. 136-8.

11 For an exploration of the thesis that wheeled transport vanished from the Middle East before the rise of Islam, see my book *The Camel and Wheel* (Cambridge, MA: Harvard University Press, 1975).

times, a pack camel carrying half the load pulled by two oxen in a cart could travel as many as twenty miles per day, as opposed to the oxcart's ten.[12] Yet despite many efforts over the centuries to acclimate them to different lands, camels have never flourished as beasts of burden in the absence of wastelands in which they can graze, and of the pastoralists who are willing to live in those wastelands and keep track of their far-ranging livestock.

The reason for this relates to the camel's age of maturation. Unlike horses, oxen, or donkeys, which mature and can be put to work in their second year, camels reach sexual maturity at the age of six. Camel-breeding groups in Iran, which do not normally geld their male animals, begin to work their camels with a light load at age three, and load them fully only when they reach four.[13] Given at least three juvenile years during which the growing camels are not worked at all, it therefore becomes essential that the cost of feeding be kept to a minimum. (So-called "domestic" elephants in South Asia are born in the wild and grow to maturity at age 14. This saves the expense of feeding juveniles that have little economic use in captivity.) If, like horses and oxen, working camels were commonly fed grain to supplement their grazing or were set to graze on land that could otherwise be used to grow crops, particularly during their juvenile non-productive years, ground transport would have been much more expensive. Here is one illustration of this fact developed in a study of the economics of camel use in contemporary Pakistan (see Table 1).

If this illustration is representative, as the study asserts, the person putting a camel to work at a task like crushing sugar cane, operating an irrigation wheel, pulling a cart, or carrying a load, does not have to budget for any food-related costs. Consequently, even today the camel can compete economically with an electric motor or internal combustion engine. Thus it is not surprising that despite the modernization of Pakistan's transport system, the country's camel population rose from 719,000 in 1955 to a million by 2006, and quintupled in the desert province of Baluchistan. The authors of the 1985 study concluded that:

> … the camel can compete successfully against other work animals and/or machinery in several ecological situations. Wherever there are no direct feeding costs for camels, their number is likely to dominate that of all other work animals, except perhaps that of donkeys … It is our conjecture

12 Ibid.: p. 20.

13 Tapper, Richard, "One hump or two? Hybrid camels and pastoral cultures: an update," www.soas.ac.uk/camelconference2011/file74604.pdf (recovered October, 2015) This paper is a slight revision of one that Tapper originally presented in Paris in 1975 on "L'acculturation turque dans l'Orient et la Méditerranée".

TABLE 1 The economics of camel use in contemporary Pakistan

Feeding costs	
200 working days	0
165 nonworking days	0
Medicines	200
Cost of harness	50
Amortization of camel cost	1350
Revenues	12,000
Return to labor/owner	10,400

SOURCE: ALAN HESTON, H. HASNAIN, S. Z. HUSSAIN, AND R. N. KHAN, "THE ECONOMICS OF CAMEL TRANSPORTATION IN PAKISTAN," *ECONOMIC DEVELOPMENT AND CULTURAL CHANGE*, 34/1 (OCT. 1985): PP. 121-141.

> that if it is necessary to incur feed costs for camels, they cannot compete in traditional tasks like sugarcane crushing and drawing a Persian wheel for irrigation. In these areas the machine is likely to be more cost effective in view of the pricing of machinery and fuels.[14]

Extrapolating from a single study is hazardous but in this case the economic logic seems clear, whether referring to the twentieth century or to the tenth. In non-desert habitats, the costs of using rapidly maturing work animals like horses, mules, oxen, and donkeys are going to be less than those associated with late maturing camels, if the juveniles require fodder grown specifically for them for three or four years before they can be put to work. This, along with a lack of pastoral nomads dedicated to camel breeding, is the principal reason for the failure over the centuries of dozens of efforts to export camels as work animals to non-desert environments around the world. As for desert environments, such as the interior of Australia, the existence today of tens of thousands of feral animals descended from turn of the twentieth century working camels conceals the fact that the livestock censuses of that time rarely

14 Heston, Alan, Hasnain, H., Hussain, S. Z. and R. N. Khan, "The Economics of Camel Transportation in Pakistan," *Economic Development and Cultural Change*, XXXIV/1 (Oct. 1985): pp. 130-1.

list more than 10,000 animals. The Australian working camels were mostly imported from India rather than bred locally.[15]

To return to Nishapur, availability of arid grazing was never a problem since, like other cities ringing the interior plateau of Iran, it was situated in a desert environment where irrigation was needed to grow crops. Yet the same cannot be said for the availability of pastoralist camel-breeders. Over the centuries, camel-herding societies have indeed grown up in Iran, particularly the Baluch, who live in the arid southeastern part of the country. But pastoral groups receive almost no mention in the chronicles that relate to Nishapur as it was before the year 1000. Moreover, though nothing like a medieval livestock census exists, one-humped camels do not seem to have been common in Iran before the Arab conquest. The bas-reliefs of tribute delegations at Persepolis dating to the fifth century BCE depict several Iranian peoples accompanied by two-humped camels, but only the Arabs, a non-Iranian subject population then living in Iraq, are shown with one-humped animals. This situation seems not to have changed over the next 700 years, but under the Sasanid dynasty that immediately preceded the Arab conquest it is apparent that one-humped camels did become known to some degree in Iran.[16] To be sure, the fifth-century CE emperor Bahrām Gūr, who is frequently portrayed hunting from the back of a one-humped animal, was raised among Arabs in Iraq. But numerous sealings survive that also bear the image of one-humped beasts. Even as far from Arabia as Central Asia, a pre-conquest seventh-century wall painting from Samarqand depicts two men riding one-humped camels.

Nevertheless, the history of various camel acclimatization projects in such places as the American southwest shows that importing a few animals does not lead to the formation of large breeding herds. Considering the transport needs of a city the size of Nishapur, it is hard to imagine a regional camel population of fewer than 10,000 animals. This figure assumes that 2000 female camels give birth to one foal every two years, or 1000 per year. Four years later, the males and some of the females – maybe 600 of the birth cohort – enter the labor market, where they remain of service, on average, for the next dozen years. If 2000 females, therefore, could sustain a workforce of 7200 animals (600 per year for 12 years), which might have been enough for the city's needs, then it is not unreasonable to assume an overall camel population of 7200 working animals,

---

15 Richard W. Bulliet, "Afterword: Camels and Deserts," in Knoll, Eva-Maria and Pamela Burger, eds., *Camels in Asia and North Africa. Interdisciplinary Perspectives on their Significance in Past and Present* (Vienna: Austrian Academy of Sciences Press, 2012): pp. 231-36.

16 *The Camel and the Wheel*: pp. 141-75.

along with 3000 non-working juveniles, for a rough total of about 10,200 animals. These estimates, of course, are highly conjectural, but they clearly suggest a very large number of camels roaming the deserts around Nishapur, and thus a substantial pastoral, or at least camel-savvy, population handling breeding, keeping track of juveniles, servicing the needs of working animals, and trekking on caravans, both local and trans-regional.

Who were these pastoralists? Medieval sources identify two districts in the vicinity of Nishapur that produced camels, specifically Ustuvā to the northwest of the city beyond a range of hills, and implicitly Bušt some 50 miles to the southeast.[17] The latter district, which was sometimes referred to as the Arabistan of Khurasan, would also seem to have had a substantial Arab presence. The epithet Arabistan, of course, could simply refer to the numerous Arab litterateurs who came from Bušt; but it more likely reflects the presence of Arabs who relocated there in the conquest era and maintained their pastoral lifestyle, including the herding of their traditional livestock and the cultivation of poetry, a classic Bedouin art form.[18] If this is the case, then Bušt might have been to Nishapur what the deserts near the huge Arab military encampments in Balkh in northern Afghanistan and Marw in Turkmenistan were to those burgeoning cities.[19] In the case of Nishapur, the camel industry (whether in Ustuvā and Bušt, or only in Ustuvā) would have originated in the relocation of Arab tribes to provide logistical support for the Arab conquerors. Around Balkh and Marw, where two-humped camels were bred in the pre-conquest era, a thriving business arose to supply strong hybrid animals – two-humped sires and one-humped dams – for the caravan trade and military mounts, but this seems not to have occurred around Nishapur. Presumably, goods traveling from the east in winter caravans would have already been shifted from the two-humped camels of Central Asia onto hybrids or one-humped camels in Transoxania before reaching Nishapur. If these conjectures are more or less sound, then the post-conquest arrival of Arabs with their camels would have been a *sine qua non* for the phenomenal growth of Nishapur between 700 and 1000.

17 Bulliet, Richard W., *The Patricians of Nishapur* (Cambridge, MA: Harvard University Press, 1972): p. 16.

18 The association of Arab tribes with Bušt (later "Turaiṯīṯ" or "Turšīz") and other central Iranian desert locales survived into the twentieth century. Anthropologist Henry Field, writing in 1939, remarks that "tribes of Arab origin were formerly found in Turshiz, Qain, Firdaus, and Tabas". *Publications of the Field Museum of Natural History*, Anthropological Series, XXIX (Chicago: Field Museum of Natural History, 1939): p. 249.

19 Richard W. Bulliet, *Cotton, Climate, and Camels in Early Islamic Iran: A Moment in World History* (New York: Columbia University Press, 2009): pp. 96-117.

The two stories we began with now converge. Nishapur grew into a great metropolis that produced religious scholars who were known throughout the Muslim world, but this was largely a characteristic of the tenth century after most of the growth had already taken place. The basis of the city's growth came not from religion per se but because of its being situated on a fertile, irrigable plain adjoining a vast arid wasteland that could sustain the thousands of camels brought by Arab migrants, some of whom continued to pursue their pastoral lifestyles. In transforming Iran's transport economy, the Arabs' herds of inexpensive pack animals made it possible for more people to live in urban agglomerations on the desert fringe than would otherwise have been feasible without water transport.

All that is missing is a linkage between camel use and religious scholarship. That linkage is supplied by the rise of an export-oriented cotton industry, another consequence of Arab immigration. Arabs in Yemen grew cotton and exported it in the pre-Islamic period, but Iranians did not. Though cotton *was* grown in river valleys and deltas in Central Asia, and the long, hot summers of the Iranian plateau were ideal for cotton production, the water required for the thirsty crop was severely lacking in the latter region. This changed when Arab settlers and Iranian converts to Islam, using indigenous Iranian hydraulic technology, began to finance underground canals that channeled ground water to newly created villages in the desert.[20] This produced a double benefit for the entrepreneurs. By bringing desert land under production, they obtained, under Islamic law, freehold ownership of the canal, the village, and the irrigable croplands. And by producing a crop that supplied a previously unknown cotton textile industry, they tapped into a vast market for lightweight fabrics.

Compounding these benefits was a religious doctrine contained in the sayings of Muḥammad, which spread throughout Iran: A Muslim man should not wear silk. Silk is a reward for Muslims in the afterlife, but in this world it is only for unbelievers, meaning for the non-Muslim Iranian elite who continued to favor the clothing preferences of Sasanid times. The Muslim religious scholars who repeated this anti-silk doctrine came disproportionately from families that either grew cotton or fabricated and sold cotton cloth. Of the Nishapuri religious scholars whose occupational backgrounds can be inferred from their surnames, some forty percent bore the names Qaṭṭān, Bazzāz, or Karābīsī, which were associated exclusively with the cotton industry.

Being situated on a perfectly flat arable plain running along the southern foothills of the Bīnālūd mountain range, whose water table was refreshed every year by winter rain and snow, Nishapur was ideal cotton country. By

20 Ibid., pp. 1-68.

comparison, the Ṭūs farming area was physically constrained between mountain ranges, and the land north of the Bīnālūd mountains was chopped up by deep streambeds making it unsuited for extensive agriculture. As for the Arab settlement in Jurjān, a well-established silk industry there could have retarded the spread of cotton.

A Muslim doctrinal preference for plain white, black, or striped cotton garments contrasted with the brightly colored silk brocades with figurative designs worn by the Iranian elite in Sasanid times and still preferred by non-Muslim nobles after the Arab conquest.[21] These silks were so attractive that

21 A reviewer says:

You write that before the Samanids plain white cotton signified Islam and silk brocade marked one as a non-Muslim. This is a false antithesis since there is no evidence about non-silk brocade and no clear evidence that plain white cotton signified anything except the official dress of the Umayyads.

I respectfully disagree with this argument. The same writes: "Despite a few canonical *ḥadīṯs* which attribute to Muḥammad the statement that 'the best of your clothes are the white' and non-canonical ones such as 'God loves white clothing and he created Paradise white,' *which reflect a later pious preference*, it is clear that the Prophet wore many different colored garments, including red, green, and black, in addition to white". [Yedida Kalfon Stillman, *Arab Dress* (Leiden: Brill, 2003): pp. 23-24 (italics added)] It is precisely this later pious preference that interests me. For example, 'Abd al-Karīm al-Sam'ānī in his *Kitab adab al-imlā' wa'l-istimlā'* [*Die Methodik des Diktatkollegs*, ed. Max Weisweiler (Leiden: Brill, 1952): p. 29] stipulates that a scholar preparing to hold a class to transmit *ḥadīṯ* should put on white clothes. Adding to the piety of white is the specification of white Yemeni cotton for Muḥammad's winding sheet and the undyed nature of *iḥrām* garments. For women, "white garments of any kind – except for the *šīrwāl* – were considered masculine and were to be avoided". [Ibid.: 45, quoting al-Waššā'].

With regard to brocade, "the iconoclasm of early Islam extended to garments with images embroidered upon them. Coptic, Byzantine, and Persian garments of the period frequently had human, animal, and vegetal figures on them on the decorative bands and patches ... Muḥammad found such clothing objectionable, removed all such items from his wives' living quarters and refused to wear garments on which images were woven". [Stillman, *Arab Dress*: pp. 25-26].

Concerning silk, most surviving specimens of early *ṭirāz* consist of plain or striped linen or cotton – not necessarily white – with 2-3 inches of silk embroidery. As late as the twelfth century, Abū al-Faḍl al-Dimašqī remarked that "people's tastes vary in regard to the *ṭirāz* borders and the ornamented embroideries, but they are agreed in the preference of that which is of the finest thread, and closest of weave, of the purest white". [Quoted in Stillman, *Arab Dress*: p. 133].

My assertion of a doctrinal preference, what Stillman calls a "pious preference," does not imply that every Muslim wore clothing of uniform color and ornament. But plain cloth seems to have been preferred to figured brocades. These were not necessarily woven of silk, but surviving specimens are normally of silk, and the word for "brocade," *dībāj*, is twinned with "silk" in some condemnations of silk wearing. It is hard

they resumed their place in the elite Iranian wardrobe after the year 1000, by which time the conversion of the country to Islam was so complete that no one could mistake a brocade-wearing gentleman for a non-Muslim reactionary.[22] By that time, however, northern Iran was on the cusp of a climatic chilling that reduced agricultural production and thus stemmed and eventually reversed the urban growth that had marked the ninth and tenth centuries.[23]

Thus Nishapur's development into a great metropolis can be seen as coinciding with a consumer rivalry between plain Muslim cotton garments and elegant non-Muslim figured silk. The fact that pre-Islamic Nishapur had been a small city which could not have been home to more than a handful of elite families probably made Muslims more visible on the street and in the markets of the growing metropolis than they would have been in a place like Ṭūs, which had been a bastion of Sasanid aristocracy and did not experience the same burgeoning of its population.[24] Conceivably, if the costumes preferred by Muslims became visible more rapidly in Nishapur than elsewhere, converts to Islam migrating from smaller towns and villages may have been disproportionately drawn to what they perceived as a growing Muslim city.

Be that as it may, Nishapur certainly became a major exporting center for cotton cloth, and just as certainly it became, along with Samarqand, which is situated in a cotton-producing river valley in Central Asia, a production center

---

to imagine prohibiting silk and *dībāj* together if the latter word was not understood to denote a silken fabric.

22 The return of figured brocade as an elite class preference after 1000 is best indicated by Saljuq and post-Saljuq depictions of rulers wearing brocade tunics with *ṭirāz* embroidery limited to plain cloth bands around the upper arms.

23 Bulliet, *Cotton*: pp. 69-95.

24 The possibly questionable assertion of a limited elite population for pre-conquest Nishapur arises primarily from the small size of the city compared with what it would become two centuries later. To be sure, under the Sasanids, Nishapur was an administrative center that had a mint and a bishopric. Moreover, a son of Yazdegerd, the last Sasanid emperor, took refuge in Nishapur. Yet the only eminent Muslim families in later times whose distinction was clearly based on a high level of pre-Islamic eminence were the Māsarjisīs [see my *Patricians of Nishapur*: 106-114] and the Mīkālīs [Bayhaqī, *Tārīḫ-i Bayhaqī*, ed. Saʿīd Nafīsī, III: pp. 969-1008; and Bulliet, *Patricians*: p. 67, n. 23].If there had been a large number of families boasting similarly august Sasanid – era lineages, I think they would have left a trace. Having a mint and collecting taxes from surrounding lands does not necessarily imply a sizable elite. Nor does the use of the citadel and its garrison as a temporary protection for fleeing royalty. True, the important fire temple of Ādur Burzēn-Mihr was situated somewhere in the vicinity of Nishapur and would have had a significant number of Zoroastrian staff. But its locale has not been identified archaeologically, and the fire temple in the inner city seems to have been quite small since the congregational mosque that was built above it, after the conquest, was quickly outgrown by the increasing Muslim community.

for enormous quantities of plain white (and occasionally black) pottery. Some of this pottery was decorated with elaborate, and often unreadable, Arabic calligraphy, usually of a black or aubergine tint, though sometimes including red or yellow pigment. The significance of this writing is debatable. On the one hand, the stark contrast between plain white background and ornamental Arabic writing is suggestive of the cotton (or linen) cloth called *ṭirāz* that was woven in government factories, including one or more in Nishapur, to be fabricated into robes of honor. *Ṭirāz* garments featured a single two-to-three-inch band of ornate silk embroidery near the hem of a large piece of plain fabric. The inscriptions usually included the name of the caliph and the date of manufacture. This distinguishes *ṭirāz* inscriptions greatly from the bland adages characteristic of Nishapur's pottery.

The date of production of the pottery is in dispute. Art historians favor the tenth century, not for archaeological reasons but because they associate such beautiful items with the flourishing of the regional Samanid dynasty in Bukhara. However, since the Samanid court was famed for sponsoring the writing of Persian in Arabic script, the absence of any Persian adages on the pottery points to an earlier date, namely the ninth century, a time when Arabic was the exclusive written language of Muslims, and plain (white) cotton was competing in the aesthetic marketplace with silk brocade. Moreover, as I have argued elsewhere on the basis of reports from the Metropolitan Museum of Art's pre-World War II excavations of the Nishapur's ruins, the white Nishapur/Samarqand pottery type with dark calligraphy, the so-called 'slip-painted ware (see Fig. 1 and 2)', appears somewhat more frequently in parts of the ruins that are pre-Samanid than does an alternative type featuring motifs of brocade-clad Iranian noblemen feasting and hunting. The latter style imitates Sasanid silver plates, which continued to be produced after the conquest.[25] Evidently

25 Bulliet, Richard W., "Pottery Styles and Social Status in Medieval Khurasan", in A. Bernard Knapp, ed., *Archaeology*, Annales, *and Ethnohistory*, (Cambridge: Cambridge University Press, 1992): pp. 131-4. My dating proposal stems from archeological findings. Charles K. Wilkinson's study of the pottery retrieved by the excavations of the Metropolitan Museum of Art provides the basis for my dating. Most of the pottery with inscriptions falls into two categories that Wilkinson labels "Black on White Ware" and "Polychrome on White Ware". He illustrates, describes, and gives the find spot for 90 items in the first category and 59 in the second. Seventy-two percent of these pots come from three excavated mounds. Of the 312 Islamic coins from the mound called Sabz Pūšān, 154, or 49 percent, date to before the tenth century. For Tepe Madrase, 80 out of 123 coins, or 65 percent, are pre-tenth century; and for Qanāt Tepe the numbers are 121 pre-tenth century out of 145, or 83 percent. I do not know how it can be claimed that the pottery in question dates from the Samanid era in the tenth century when an aggregate 61 percent of the concretely datable material from the find spots dates to the eighth and ninth centuries, and only 24 percent

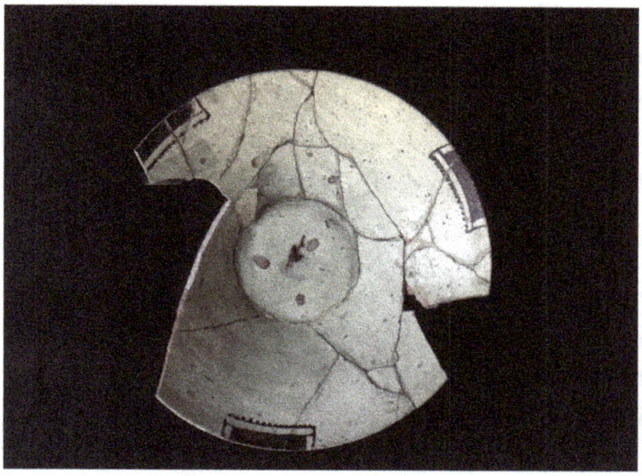

FIGURE 1
Nishapur slip-painted ware without inscription (author's personal collection)

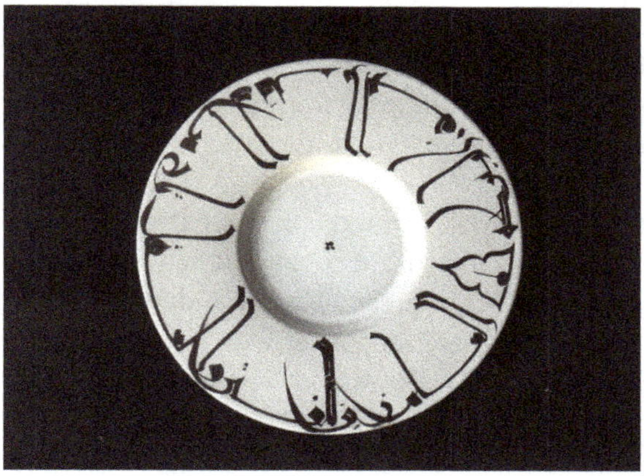

FIGURE 2
Nishapur slip-painted ware with inscription (modern commercial reproduction, Metropolitan Museum of Art, New York)

both wares were produced simultaneously for a lengthy period, as one would expect if the market consisted of some consumers who wanted to be visibly oriented toward the Muslim Arab rulers and other consumers who preferred the pre-Islamic culture.

Robert Hillenbrand, who has published a lengthy analysis of the inscriptions on Nishapur/Samarqand ware, acknowledges the possible cogency of this preference for the ninth century over the customary art-historical dating to a century later. Nevertheless, he remains committed to the Samanid identification.[26]

of the coins belong to the tenth century. For the archaeological details see Charles K. Wilkinson, *Nishapur: Pottery of the Early Islamic Period* (New York: Metropolitan Museum of Art, 1973): pp. xxx-xxxiii, 90-157.

26 Hillenbrand, Robert, "Content Versus Context in Samanid Epigraphic Pottery," in A. C. S. Peacock and D. G. Tor, eds., *Medieval Central Asia and the Persianate World* (London: I. B. Tauris, 2015): pp. 56-106.

As for the texts of the inscriptions, he dismisses them as trivialities. "The specious wisdom and empty rhetoric of the fortune cookie, or the trite message of the Christmas cracker, quickly come to mind".[27] This dismissal is a bit off-base since the buyer of a fortune cookie or holiday cracker does not know what text will be revealed while the buyer of a plate ornamented with calligraphy sees the text before buying it.

Given that customers actively chose these calligraphic wares, dismissing them as trivial seems unwise. Yet what is one to make of the blandness of their inscriptions? Basing his discussion on his own investigations and on a corpus of 140 texts deciphered by the Iranian epigrapher ʿAbdullāh Qūčānī,[28] Hillenbrand confirms that all of the inscriptions are in Arabic, that none of them come from the Qur'ān, and that many are so ornate as to be scarcely legible. One characteristic of the pottery escapes detailed comment, however, namely, the fact that most items do not have any calligraphy at all. They are plain white plates and dishes with spare, dark, non-figurative, markings. This means that many buyers, perhaps those with less money to spend, opted for plain white with dark ornament, a visual style that appears most notably in Nishapur/Samarqand ware.

To my mind, this amounts to the pottery being the ceramic counterpart of the cotton fabrics being produced in the eighth through and tenth centuries for Muslim consumption in places like Nishapur and Samarqand. Wealthier, or more pretentious, customers may have preferred pieces with fancy Arabic inscriptions because they called to mind the *ṭirāz* robes of honor, without in any way touching on the wording and religious overtone of actual *ṭirāz*. But common buyers, of which there must have been a great many given the ubiquity of diagnostic sherds in the rubble of Nishapur, simply liked the look of plain white with bits of dark ornament. Inasmuch as the Samanid period ushered in a period of nostalgia for the motifs of pre-Islamic Iran, including the wearing of silk brocade, the taste for plain white seems particularly appropriate to an earlier, pre-Samanid period when plain white cotton suggested Islam, while silk brocade might mark one as a non-Muslim.[29]

---

27 Ibid.: p. 57.

28 ʿAbdullāh Qūčānī, *Kātibahā-yi sufāl-i Nīšāpūr* (Tehran: Reza Abbasi Museum, 1986).

29 This is not to say that everyone in Nishapur necessarily wore one or the other of these styles, but rather that the choice between them derived from early Arab efforts to keep themselves visually distinct from the populations they conquered. The Covenant of ʿUmar, a late document of purportedly pre-Umayyad origin, expressly prohibits non-Muslims from wearing Arab headgear and military garments. As for Muslims dressing like non-Muslims, soldiers are reported to have been punished if they donned an Iranian cuirass or leggings. [Stillman, *Arab Dress*: p. 39] Costume specialist Yedida Stillman observes

In other words, the transition by way of conversion of most Iranians from non-Muslim communities into the burgeoning Muslim society, a three-century wave of conversion that was largely accomplished by the time the Samanids gained prominence at the end of the ninth century, created a visual contrast between plain cotton Muslim gowns and turbans and the figured silk tunics and trousers of non-Muslims. But as conversion progressed, that contrast gradually became religiously irrelevant. This paved the way for an intensifying elite preference for silk brocade after the year 1000, as well as for growing popularity of the pottery designs featuring feasting and hunting that reflected a return to a Sasanid aesthetic.

Revising the date of Nishapur/Samarqand ware to the ninth century makes the inscriptions on the pottery germane to our inquiry since they would have been simultaneous with, rather than subsequent to, the period of Nishapur's greatest growth. Therefore, trivial though they may seem, a reconsideration of these texts may yet yield some insights into the period. Hillenbrand's affirmation that the texts were all in Arabic but not specifically Muslim in content provides a starting point. However, unlike buyers of fortune cookies or holiday crackers, the buyers of Nishapur pots knew what they were getting and did not demand religious or Persian-language alternatives to the Arabic motifs available in the market. And unlike the purchasers of souvenir T-shirts, who typically browse a range of items catering to different tastes – humorous, respectful, familiar, punning, outrageous, what have you – people in the market for white pottery had but two choices, plain plates or plates with ornate Arabic calligraphy. Thus even though the adages are bland, they define a particular taste. In negative terms, Qūčānī's corpus of translations indicates that buyers in Nishapur did not want either Muslim religious sentiments or adages couched in the Persian language.

But what do they convey in a positive sense? Hillenbrand observes that the texts are not only barely legible, but sometimes stop short when an adage fills the space available before it is finished. One can surmise from this that some (perhaps many) buyers interested in fancy Arabic writing could not read the

---

that attempts to enforce such clothing differentiation "were meant to still the rising tide of assimilation in the eastern provinces. Such attempts proved futile, however. By the end of the Umayyad period, the Arabs who had settled in Khurasan were becoming increasingly assimilated into the local culture, and, *it may be assumed*, had adopted many elements of native attire which was part and parcel of the continuing fusion of Persian modes into the evolving Islamic vestimentary system". [Ibid.: 40 (italics added)] Despite Stillman's vague assumption about dating, the ninth century fits better than the Samanid tenth as a period of transition from costume elements that were associated with pre-Islam to ones that signaled Muslim identity.

inscriptions they purchased, but wanted them anyway. The absence of texts using contemporary non-Arabic alphabets – Hebrew, Pahlavi, Soghdian – combined with the difficulty of deciphering the texts indicates that a desire to be identified with the Arabic alphabet, as opposed to an actual ability to read the alphabet, was a common consumer characteristic. That the Arabic alphabet always conveys a text in the Arabic language further suggests, though it does not prove, that the pottery was produced before the Nishapuri population had gained familiarity with that alphabet being used for their own Persian language.

Moving on to the 121 texts translated by Qūčānī,[30] 58 are unique, and 33 are repeated at least twice.

TABLE 2 Texts on Nishapur ceramics (based on the translation of 121 translations by Qūčānī)

Number of occurrences	Text	%
12	Blessing; – to the owner; – and prosperity to the owner; – and good fortune and joy; – and good fortune and delight and happiness; – and exultation and prosperity and good health and happiness	10
10	May everything be wholesome; – it is permissible and dainty to eat from a bowl that is always full	8
8	The free man is free even if touched by harm	7
8	Generosity is a quality of the people of paradise	7
6	Knowledge is the noblest of values and manliness is the more intricate heritage	5
5	Deliberation before action protects you from regret	4
5	The noblest of riches is the abandonment of desire	4
3	Patience before opinions	2
64	[other sentences]	52

As table 2 shows, the most common texts, recurring twelve times, begin with "Blessing". Following this cluster comes the phrase "May everything be wholesome," with some occurrences adding the words "it is permissible and dainty

30 Qūčānī's specimens number 140, but I have subtracted 19 in this analysis because they have a different fabric and contain only the potter's name.

to eat from a bowl that is always full". The third and fourth most common texts read "The free man is free even if touched by harm," and "generosity is a quality of the people of paradise". Then come adages that extol virtues:

> "Knowledge is the noblest of values and manliness is the more intricate heritage".
> "Deliberation before action protects you from regret".
> "The noblest of riches is the abandonment of desire".

Looked at in terms of the virtues being extolled, the most common sentiments praise generosity (24 times), knowledge (12 times), free status (11 times), patience (8 times), and loyalty (4 times). But silence and reticence are also praiseworthy:

> "Patience before opinions" (3 occurrences).
> "A person who talks too much has many faults" (2 occurrences).
> "Safety lies in silence, only speech will reveal the inner side of the man with faults" (2 occurrences).
> "Yea! A person who jokes too much, his worth is contemptuous".
> "He who is covered with shyness, his faults are hidden from the eye".
> "He who speaks, his speech is silver, but silence is a ruby".
> "Your modesty keep to yourself, because it proves the action of a generous man".

Speculation about the semantic overtones of the pottery inscriptions can proceed endlessly without firm conclusions. Yet combining their litany of commendable virtues with the other considerations raised in this essay, a plausible picture of Nishapur's rise as a Muslim commercial and religious center begins to emerge:

In the course of the late eighth and ninth centuries, descendants of Arab migrants and an increasing number of Iranian converts to Islam paid for the digging of new *qanāt*s and the creation of villages in the broad, arid plain around the compact urban nucleus of Nishapur. The lands brought under production were devoted to growing cotton as a summer crop. The entrepreneurs thus became prosperous landowners, and some of them branched out into the various industrial processes involved in the production of cotton cloth: ginning, spinning, weaving, and dyeing.

As Muslims, they may have required a profession of allegiance to Islam from the farmers they recruited as a work force for the new cotton villages, whether or not these workers knew Arabic or anything substantial about the religion.

This sort of linkage between the development of new farming villages and religious conversion has been studied by Richard Eaton concerning the spread of Islam in eastern Bengal. It seems unlikely that the Muslim owners of new cotton lands would have followed equal employment opportunity guidelines in their labor recruitment, more so that they would favor fellow Muslims.[31]

It is also likely that the cotton entrepreneurs, as well as other Muslims, preferred to transact business with fellow Muslims, just as Jews and Armenians tended to form business associations with their respective coreligionists. Distinguishing Muslims from non-Muslims thus became socially and commercially important as the city attracted migrants from the countryside and other towns. This distinction was facilitated by male Muslims wearing distinctive garb. Those aspiring to a pious reputation are likely to have preferred plain cotton garments and to have eschewed the figural silk brocades of the Sasanid period. There is no contemporary evidence of fixed or severely limited costume rules for Muslims, but the wearing of plain white cloth and avoidance of fabrics with human, animal, or vegetal designs were both supported by traditions attributed to the Prophet Muḥammad and taught to people seeking knowledge about the faith.[32]

Assuming that the buyers of inscribed plates were either literate or wanted to be thought of as literate, it is reasonable to surmise from the content of the adages that they also wanted to be thought of as generous, knowledgeable, patient, loyal, reserved in their opinions, and opposed to idle talk – all of this while endorsing a style of black calligraphic ornament on plain white pottery. Is it too much to suggest that Nishapur ware (and presumably Samarqand ware in its own economic context) appealed especially to Muslim merchants who were involved in the religiously praiseworthy cotton trade? Such a clientele would both explain the popularity of the ware in the ninth century and its disappearance after the year 1000 when, on the one hand, worsening climatic conditions seem to have wiped out much of the cotton industry, and, on the other, the return of Sasanid design motifs made the traditional *ṭīrāz* fabric obsolete. Judging from Saljuq era imagery, later robes of honor were often made of silk brocade with the caliphal *ṭirāz* inscription confined to strips of cloth around the upper arms.

In conclusion: Why Nishapur?

31 Eaton, Richard M., *The Rise of Islam and the Bengal Frontier, 1204-1760* (Los Angeles: University of California Press, 1993).

32 Still today, religious scholars in Iran customarily wear a plain white gown with a plain black or brown cloak and avoid fabrics with figural motifs. Turbans are also of plain rather than figured fabric.

The improbable growth of a great metropolis where there had been none before, and where there has been no successor since the twelfth century, depended upon a concatenation of factors.

First, Nishapur was situated on a broad, flat piedmont plain that was naturally arid but readily irrigable by means of underground channels (*qanāts*) bringing ground water from the foothills to individual villages.

Second, the opportunity for Arabs to herd camels in deserts free of previous heavy pastoral exploitation opened the way to the inexpensive transport necessary for the commerce of a large city, remote from any navigable water. Third, speaking of Iran's central plateau in general, the need of Arab immigrants to find a way to become part of Iran's farming economy without seizing the lands of non-Muslims sparked the building of new *qanāts* and villages and the growing of cotton, a crop known to the Arabs from Yemen that uniquely benefited from summer irrigation in a land where the agricultural sector was traditionally based on winter wheat and barley. A relatively small number of entrepreneurs, whether Arabs or Iranian converts to Islam, would have been sufficient to redirect the agricultural economy.

Fourth, the development of a cotton industry required the ginning, spinning, dyeing, weaving, and transport facilities of the sort normally found only in cities. So Nishapur became a manufacturing and exporting center.

Fifth, to build a market for their new crop, Muslim cotton growers and cloth exporters – first probably Arabs, but increasingly Iranian Muslims as conversion accelerated in the ninth century – used their position as expositors of the new faith to disparage the Sasanid taste for silk brocade and proclaim plain cotton as the fabric favored by Muḥammad. Arabic was the language of this population, and writing, or exhibiting writing, in Arabic script both gave evidence of piety and enabled merchants to communicate in a medium that non-Muslim competitors could not understand.

Sixth, the rapidly growing population of converts to Islam, both in Iran and Iraq, provided a growing market for cotton; and the taxation rate for cotton, which was consistently higher than that for grain, suggests that great profits were earned from manufacturing and selling cloth.[33] Judging from the high frequency of names for 'cotton grower' and 'cotton cloth dealer' among the religious scholarly elite, the new Muslim economy evidently fostered a leisure class that had a vested interest in the spread of Islam, not as a religion of conquest, as might be expected of a garrison community, but as a religion of peaceful commerce. Populating cotton villages with Muslim workers could

33 Bulliet, *Cotton, Climate, and Camels*: pp. 34-39.

have served this interest, though dispositive evidence of conversion being required is lacking.

Seventh, the street-level contrast between a Muslim aesthetic based on a plain or striped fabric and a non-Muslim aesthetic associated with colorful silk brocades, enhanced Nishapur's attraction as a religious center to which converts to Islam might migrate. Since the pre-Islamic city was quite small, the non-Muslim population continuing in residence from before the Arab conquest could not have been very great. "Two villages of Jews" are mentioned as being absorbed in the city's expansion, but aside from occasional mention of interfaith prayers for rain during droughts, non-Muslims show up only rarely in texts devoted to Nishapur. A pilgrimage site called Qadamgāh, "the place of the footprint," located some distance away on the road to Ṭūs, may have originated as a shrine sanctified by the Buddha's footprint; but it was converted to Muslim use. Otherwise, the Nishapur area had no significant shrines, either Muslim or non-Muslim, in the early Muslim centuries.

Eighth, the tenor of life was dominated by a commercial and religious elite whose way of life was consonant with the values embodied in the Arabic inscriptions on ninth-century Nishapur pottery crafted to appeal to literate, or would-be-thought literate, Muslim buyers. While a sentiment like "Generosity is a quality of the people of paradise" is indeed bland, its tone resembles such homely American sayings as "Home sweet home" and "Home is where the heart is". Today these sayings are cloyingly trite, but past generations of sober and genteel Americans felt warmed by hanging samplers embroidered with those words on the walls of their homes. Nishapur's patrician class, in other words, had would have been called "bourgeois tastes".

A skeptic might object that this eight-part formula does not involve, much less depend on, political rule or the attraction of Nishapur being a capital city. Thus it does not include a discussion of the Tahirids, a non-Nishapuri Iranian family that played an important military role in the history of the Abbasid caliphate and supplied northeastern Iran with governors between 821 and 873. Clifford Edmund Bosworth declared: "It is with the advent of the Ṭāherid governors in Khorasan in the early 9th century that Nishapur rises to prominence as the political and cultural center of the East ... Under the patronage there of a governor like ʿAbd-Allāh b. Ṭāher (828-45), the city became a lively intellectual and literary center for Arabic scholarship, with *adab* and poetry especially prominent".[34]

34 Cited *in EIr* article "Nishapur i: Historical Geography and History to the Beginning of the 20th Century" http://www.iranicaonline.org/articles/nishapur-i (retrieved January 2016).

The Tahirid period does indeed coincide with Nishapur's flowering, but aside from some pious building projects, evidence of significant Tahirid impact on the city is lacking. One wealthy patrician family is said to have loaned (been forced to loan?) a Tahirid governor one million dirhams,[35] but otherwise the biographical dictionaries rarely mention them. They established their governing center at Šādyāḫ, a site some two to three kilometers from the city center, with a stream nearby or running through. The city proceeded to expand in the direction of Šādyāḫ, but it did not reorient toward the ruling compound.

In 1037, the Saljuq army camped in a portion of Šādyāḫ when Ṭoġrïl Beg was negotiating for the city's surrender, indicating that the district was still not fully urbanized. As with the Tahirids, the Saljuqs were probably happy to find an area where their horses and men could be watered from a surface stream and not be pestered by the hustle and bustle of city life.

In the twelfth century, following a horrendous series of disasters – an earthquake, nomadic depredations, and intra-urban factional warfare – which destroyed most of the pre-Saljuq city, Šādyāḫ became Nishapur's new center and acquired a city wall that can still be easily traced. However, the relative paucity of mounds and sherd scatters within this wall speaks to the new Nishapur being less intensely urbanized than its predecessor.

Prior to the twelfth century, the leading clerical-cum-commercial clans, what I call the patriciate, dominated Nishapur in the areas of religion, education, commerce, and land-holding. Their collective reputation and the institutional changes they fostered in education (the *madrasa*), theology (the Aš'arī school of thought), and Sufism vaulted the city to its preeminent place in medieval Islam. On the political side, though Nishapur was an occasional object of contention, its fame did not depend on a single dynasty or potentate. Indeed, one can quite readily imagine Nishapur as a great city, without frequent political role; more difficult is the proposal of its growth as a giant metropolis, without the support of pastoralists in Ustuvā and Bušt, raising the camels needed to keep the population supplied with food.

My surmise is that the Tahirids were drawn to place their capital in Nishapur precisely because it was becoming such a dynamic economic and cultural center, rather than that the Tahirid court became a magnet attracting wealth and religious distinction to the city. Though this conclusion runs counter to conventional approaches to Iranian history that focus almost exclusively on who is ruling at any given point of time, from the point of view of economic and social history, it is apparent that the putative charisma of the Tahirids, to whatever degree it affected the city's growth, would have gone for naught if the

35 Bulliet, *Patricians*: 107.

conditions described in my eight-fold non-political assessment of Nishapur's growth had not been fulfilled by processes that the Tahirids had little to do with.

Bibliographical References

Behnam, J., "Population," in Fisher, W. B. ed., *The Cambridge History of Iran*. 1st ed., I (Cambridge: Cambridge University Press, 1968).

Bulliet, Richard, *The Patricians of Nishapur: An Essay in Medieval Islamic Social History* (Cambridge, MA: Harvard University Press, 1972).

Bulliet, Richard, *The Camel and Wheel* (Cambridge, MA: Harvard University Press, 1975).

Bulliet, Richard, "Medieval Nishapur: A Topographic and Demographic Reconstruction," *StIr*, V/1 (1976): pp. 67-89.

Bulliet, Richard, *Conversion to Islam in the Medieval Period: An Essay in Quantitative History* (Cambridge, MA: Harvard University Press, 1979).

Bulliet, Richard W., "Pottery Styles and Social Status in Medieval Khurasan", in Knapp, A. Bernard ed., *Archaeology*, Annales, *and Ethnohistory* (Cambridge: Cambridge University Press, 1992): pp. 131-4.

Bulliet, Richard, *Islam: The View from the Edge* (New York: Columbia University Press, 1994).

Bulliet, Richard, *Cotton, Climate, and Camels in Early Islamic Iran: A Moment in World History* (New York: Columbia University Press, 2009).

Bulliet, Richard W., "Afterword: Camels and Deserts," in Eva-Maria Knoll and Pamela Burger, eds., *Camels in Asia and North Africa. Interdisciplinary Perspectives on their Significance in Past and Present* (Vienna: Austrian Academy of Sciences Press, 2012): pp. 231-36.

Durand-Guédy, David, *Iranian Elites and Turkish Rulers: A History of Iṣfahān in the Saljūq Period* (Abingdon: Routledge, 2010).

Eaton, Richard M., *The Rise of Islam and the Bengal Frontier, 1204-1760* (Los Angeles: University of California Press, 1993).

Field, Henry, *Publications of the Field Museum of Natural History*, Anthropological Series, XXIX (Chicago: Field Museum of Natural History, 1939).

Heston, Alan; H. Hasnain, S. Z. Hussain, and R. N. Khan, "The Economics of Camel Transportation in Pakistan," *Economic Development and Cultural Change*, XXXIV/1 (Oct. 1985).

Hillenbrand, Robert, "Content Versus Context in Samanid Epigraphic Pottery," in A. C. S. Peacock and D. G. Tor, eds., *Medieval Central Asia and the Persianate World* (London: I. B. Tauris, 2015): pp. 56-106.

Qūčānī, ʿAbdullāh, *Kātibahā-yi sufāl-i Nīšāpūr* (Tehran: Reza Abbasi Museum, 1986)

al-Samʿānī, ʿAbd al-Karīm, *Kitāb adab al-imlāʾ wa'l-istimlāʾ* [*Die Methodik des Diktatkollegs*], ed. Max Weisweiler (Leiden: Brill, 1952).

Stillman, Yedida Kalfon, *Arab Dress* (Leiden: Brill, 2003).

Tapper, Richard, "One hump or two? Hybrid camels and pastoral cultures: an update," www.soas.ac.uk/camelconference2011/file74604.pdf.

Wilkinson, Charles K., *Nishapur: Pottery of the Early Islamic Period* (New York: Metropolitan Museum of Art, 1973): pp. xxx-xxxiii, 90-157.

Biographical Note

Richard W. Bulliet retired in 2015 after 39 years as Professor of Middle Eastern History at Columbia University, where he also directed the Middle East Institute of the School of International and Public Affairs for twelve years. He is a specialist on Iran, the social history of the Islamic Middle East, the history of human-animal relations, and the history of technology.

His most recent book is *The Wheel: Inventions and Reinventions* (2016). His earlier works include *Cotton, Climate, and Camels in Early Islamic Iran: A Moment in World History* (2009), *Hunters, Herders, and Hamburgers: The Past and Future of Human-Animal Relationships* (2005), *The Case for Islamo-Christian Civilization* (2004), *Islam: The View from the Edge* (1994), *Conversion to Islam in the Medieval Period: As Essay in Quantitative History* (1979), *The Camel and the Wheel* (1975), and *The Patricians of Nishapur: A Study in Medieval Islamic Social History* (1972).

Inherited Landscapes in Muslim Bactra

Etienne de la Vaissière
EHESS (Paris)
Etienne.De-La-Vaissiere@ehess.fr

Abstract

This article presents recent archaeological discoveries on the Buddhist monasteries, the irrigation system and defence systems of the oasis of Balkh, intending to explore the depth of the gap between these central features of the oasis landscape and the transmitted early Islamic texts describing the site. The *Fażā'il-i Balḫ* and the *Ḥudūd al-ʿĀlam* do provide us with some data but the archaeology allows a detailed demonstration of how many parts of the pre-Islamic past were quickly forgotten or reinterpreted in the following centuries.

Keywords

Balḫ – Central Asia – Landscape Archaeology – Buddhism – Samanids – Irrigation

1 Introduction

Bukhara, Samarqand and Marw have been excavated by various teams since the Second World War, whereas Bactra remains the least known of all the major cities of western Central Asia. After a first attempt to excavate part of the city by Alfred Foucher in 1923, his description of the remains as a vast ocean of mud might have deterred further attempts, while the newly created Délégation archéologique française en Afghanistan (DAFA) focused on Antiquity.[1] Some surveys were undertaken in the oasis, as in 1948 by Marc Le Berre, an architect working with the DAFA, and there were some attempts to describe the numerous walls of the city.[2] The Soviets, operating in the 1970s, surveyed mainly

1 Foucher, Alfred, and Mme Eugénie Bazin-Foucher, *La Vieille Route de l'Inde de Bactres à Taxila* (Paris: Éditions d'Art et d'Histoire, 1982).

2 The 1948 survey was not published but its results are incorporated in Ball, Warwick, and Jean-Claude Gardin, *Archaeological Gazetteer of Afghanistan: Catalogue des sites archéologiques*

 | DOI:10.1163/9789004434332_008

the deserted north-western part of the alluvial fan beyond the Islamic oasis.[3] War then prevented any excavations other than pillaging for twenty-five years. However, since 2005, the DAFA has again worked on Balkh, with important results for both Antiquity[4] and the early Islamic period. As of 2015 the members of the DAFA and myself are still conducting surveys in the oasis, although the worsening of the security situation has put a halt to the actual excavations.[5]

This article aims to present some of the main hypotheses and results of the last 10 years, which significantly modify the accepted narrative of the transition from pre-Islamic Bactra to Islamic Balkh.[6] A convenient starting point is the depiction of the town and the oasis provided by the local Balkhi authors. A local history of the town from the thirteenth century, the *Fażā'il-i Balḫ*, reinterprets this landscape in religious terms, as demonstrated by Arezou Azad.[7] Unfortunately, this text also demonstrates the extremely heavy toll paid with the passing of time. Apart from a few exceptions, it contains extremely little data on the transformation of the town from the pre-Islamic to the Islamic period. We are therefore forced to rely on texts written by earlier non-Balkhi authors, geographers, historians and encyclopaedists of the tenth century,

---

d'Afghanistan (Paris: Éditions Recherches sur les Civilisations, 1982); for the walls, see Le Berre, Marc, Daniel Schlumberger, "Observations sur les remparts de Bactres", in Bruno Dagens, Marc Le Berre, and Daniel Schlumberger, *Monuments Préislamiques D'Afghanistan*, (Paris: Klincksieck, 1964).

3 Pugačenkova, Galina, "K poznaniju antičnoj i rannesrednevekovoj arkhitektury severnogo Afganistana", in U. T. Kruglikova, V. I. Sarianidi (eds.), *Drevnjaja Baktrija*, (Moscow: Nauka, 1976): pp. 125-62.

4 Besenval, Roland, and Philippe Marquis, "Les travaux de la délégation archéologique française en Afghanistan (Dafa): Résultats des campagnes de l'automne 2007-printemps 2008 en Bactriane et à Kaboul", *Comptes rendus des séances de l'Académie des Inscriptions et Belles-Lettres*, 152/3 (2008): pp. 973-95.

5 The archaeological part of these results was published in La Vaissière, Etienne de, Philippe Marquis, "Nouvelles recherches sur le paysage archéologique de Bactres," *Comptes-Rendus de l'Académie des Inscriptions et Belles-Lettres*, 2013 (sic)-3 (juillet-octobre): pp. 1155-71 (published 2015).

6 This article does not deal with the early Abbassid mosque of Hāji Piyāda. See most recently Adle, Chahryar, "La Mosquée Hâji-Piyâdah / Noh Gonbadân À Balkh (Afghanistan). Un Chef D'Œuvre De Fazl Le Barmacide Construit en 178-179/794-795 ?", *Comptes-rendus des séances de l'Académie des Inscriptions et Belles-Lettres*, 1 (2011): pp. 565-625. My article is dedicated to the memory of Chahryar, a lively, joyful and learned friend and colleague, who devoted the last years of his life to this excavation.

7 Azad, Arezou, *Sacred Landscape in Medieval Afghanistan: Revisiting the Faḍā'il-i Balkh*, (Oxford: Oxford University Press, 2013).

writing in Arabic and Persian. These texts dwell on the main landmarks, composing an inherited landscape of walls and monuments, which we are now able to date. The relevant paragraph of the *Ḥudūd al-ʿĀlam*, written in nearby Gūzgānān in 982 constitutes a good introduction:

> Balkh, a large and flourishing town which was formerly the residence of the Sasanian kings. In it are found buildings of the Sasanian kings with paintings and wonderful works, [which] have fallen into ruins. [That place] is called Nawbahār. (Balkh) is the resort of merchants and is very pleasant and prosperous. It is the emporium (*bār-kaḏa*) of Hindūstān. There is a large river in Balkh that comes from Bāmiyān and in the neighbourhood of Balkh is divided into twelves branches; it traverses the town and is altogether used for agriculture of its districts. Balkh produces citrons and sour oranges, sugarcane, and waterlilies. Balkh has a *šahristān* surrounded by a mighty wall. In its suburb there are numerous marshes.[8]

Like many of his contemporaries, the anonymous author focuses on the Nawbahār just south of the town, which was the principal Buddhist monastery in ancient Bactria. This site seems to have greatly impressed all these authors due to the splendour of its remains. But the text is also focused on the river and its irrigation channels as well as on the complex system of walls surrounding the citadel, the town, and indeed the whole oasis. On these three topics – the Nawbahār, the river and the walls – recent excavations have provided new data.

2 The Monastery

In the second edition of the *Encyclopaedia of Islam*, R. Frye could still write that because of the Arabic conquest, Balkh "seems to have suffered considerably from warfare, and there are indications in Tabari that the city was in ruins about 705 AD."[9] However, it is now clear that ancient Bactra was subject to destruction not only during the Islamic conquest, but also well before.[10] The city had actually been in ruins since the beginning of the seventh century, when

8 *Ḥudūd al-ʿĀlam*, ed. V. V Bartol'd, (Leningrad, 1930); trans. VladimirMinorsky, Clifford Edmund Bosworth, *Ḥudūd al-ʿĀlam; "the Regions of the World"; a Persian Geography, 372 AH-982 AD*, (London: Gibb Memorial Trust, 1970): § 67 trans. 108. Transliteration has been edited for the sake of consistency.

9 Frye, Richard, "Balkh", *EI*^{2}, I (1960): pp. 1000-2, 1001.

10 La Vaissière, Etienne de, "De Bactres à Balkh par le Nowbahār", *Journal Asiatique*, 298.2 (2010): pp. 517-33.

its position made it a buffer zone between Sasanian and subsequently Arab power, centred on Marw, and the Nomadic power of a sub-group of Western Turks centred on Qunduz. As early as 630, the Chinese pilgrim Xuanzang describes Balkh as an empty shell: "although strongly fortified the city has few inhabitants (其城雖固居人甚少 qi cheng sui gu ju ren shenshao.)"[11] During the seventh century the effective economic and political centre of the oasis was no longer the town itself but the gigantic monastery of the Nawbahār, immediately to the south. Balkh was a monastic estate, without king or aristocratic power, by the time of arrival of the Arab armies. A key source for this period is Ibn al-Faqīh's *K. al-Buldān*. He drew on the lost *Aḫbār al-Barāmika wa-faḍā'iluhum* of al-Kirmānī, written during the heyday of the Barmakids in Baghdad, in the second/eighth century. According to the relevant passage, the area of the oasis owned by the monks was seven by seven *farsaḫs* square, meaning about 1600 square kilometres, comprising most of the useful central part of the oasis. If the monks owned most of the agricultural land, this means that they had either extinguished the local royal power in Balkh, as they had done in some parts of India and would later do in Tibet, or that they had simply been awarded local power by the Turks, to the detriment of the resident aristocracy.

Recent surveys have produced results confirming such an assumption of power, but before turning to these we should first consider to two depictions of the Nawbahār surviving from the seventh and eighth century sources.

The first is the previously mentioned book of al-Kirmānī, as quoted in Ibn al-Faqīh:

> The Iranians used to glorify it [the Nawbahār] and make pilgrimage to it, send gifts to it, drape it in different kinds of cloths, and fix flags on top of its dome, the name of which among them was al-Ustup. The dome was one hundred cubits in diameter, its height over one hundred cubits, with a circular arcade around it. Around the house (*ḥawla al-bayt*) there were 360 compartments (*maqṣūra*) in which dwelt servants, attendants and custodians.[12]

---

11 Xuanzang, *Da Tang Xiyu Ji*, ed. *Taishô*, LI, n.°2087; trans. Li Rongxi, *The Great Tang dynasty record of the Western regions*, (Berkeley: BDK America, 1996): 872c.

12 van Bladel, Kevin, "The Bactrian Background of the Barmakids", in Charles Burnett Anna Akasoy, Ronit Yoeli-Tlalim (eds) *Islam and Tibet: Interactions Along the Musk Routes*, (Farnham: Routledge, 2010): pp. 45-87, 64, translating folio 321 of the Mashhad manuscript (reproduced in Ibn al-Faqīh, *Collections of Geographical Works by Ibn al-Faqīh, Ibn Faḍlān, Abū Dulaf al-Khazrajī*, ed. Fuat Sezgin, (Frankfurt/M: Johann Wolfgang Goethe Universität Frankfurt, 1987).

The second is by Xuanzang:

> To the north of the monastery there is a stupa over two hundred feet in height (伽藍北有窣堵波。高二百餘尺 qielan bei you sudubo. Gao erbai yu chi).[13]

These texts are very important first-hand accounts. Xuanzang is known to have stopped in the Nawbahār in 630, for several months, while al-Kirmānī, writing at the end of the eighth century, was clearly transcribing accounts from eyewitnesses bearing various Bactrian *nisbas*, given the detailed and accurate descriptions of the ceremonies at the Nawbahār in the *Aḫbār al-Barāmika*.[14]

However, the two accounts differ on a crucial architectural point: in Xuanzang the monastery is south of the stupa, while in al-Kirmānī the monastery surrounds it. These discrepancies between two reliable accounts should be interpreted chronologically, which means that between 630 and the end of the eighth century, the monastery must have been reconstructed around the stupa.

The Nawbahār stupa, which is the only well-identified monument for Bactra, corresponds to the Top-e Rustam excavated by Alfred Foucher in the 1920s. More recently, working in collaboration with Philippe Marquis, then head of the DAFA, and using aerial photographs supported by on-site surveys, we were able to identify a huge octagon with sides of 140 metres and a diameter of around 400 metres around Top-e Rustam. This octagon was centered on the stupa, thus confirming al-Kirmānī's description of the cells "around" (*ḥawla*) the central structure (Fig. 1). The octagon is no longer complete but was still fully extant in the 1950s, as evidenced by aerial photographs held in the DAFA archives (Fig. 2).[15] Two of the octagon's eight sides are now invisible, however six sides remain in micro-topographic relief, marked by earthen paths and limits of fields. Excavation would be the only means of confirming or excluding the existence of the 360 cells presumed to lie along the octagonal walls. However we should note that monasteries with 180 cells around a central stupa (possibly double that number with a second floor) have been found in the Indian world, for instance at Paharpur.

---

13 Xuanzang, *Da Tang*: p. 34.

14 Melikian-Chirvani, Assadullah Souren, "L'évocation littéraire du bouddhisme dans l'Iran musulman", in Jean Aubin (ed) *Le Monde Iranien et l'Islam. 2: Sociétés et Cultures*, (Paris, Genève: EPHE-Droz, 1974): pp. 1-72; Bosworth, C. E., "Abū Ḥafṣ 'Umar Al-Kirmānī and the Rise of the Barmakids", *BSOAS*, 57.2 (1994): pp. 268-82, 269.

15 Reproduced in Dagens, Bruno, Marc Le Berre, and Daniel Schlumberger, *Monuments Préislamiques d'Afghanistan*, (Paris: Klincksieck, 1964): pl. XXXII.

FIGURE 1 Aerial photography of the octagon in Balkh
PHOTO DAFA

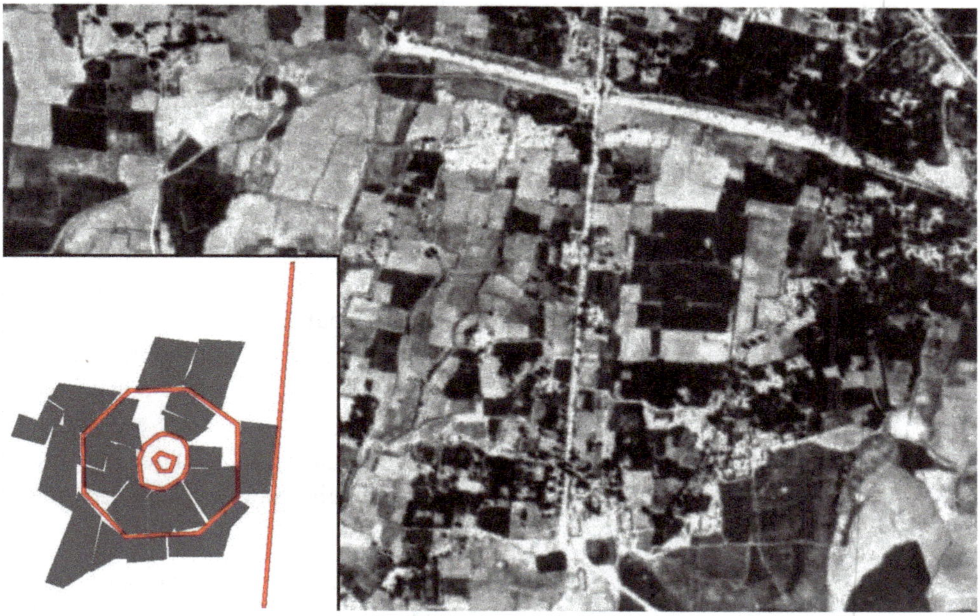

FIGURE 2 Aerial photography of the octagon in Balkh, in the 1950's.
PHOTO DAFA

Octagonal shapes are well-known in Buddhist architecture and could reflect a devotion to the eight bodhisattvas. There are numerous octagonal stupas throughout Eastern Asia, as well as octagonal pillars engraved with prayers and texts. Monasteries built on an octagonal plan are unknown, although some octagonal rooms surrounding small stupas have been discovered, such as in excavations of the first century CE Kālawān monastery in Taxila.[16] The design of an octagonal monastery was a striking innovation by the monks of Bactra, but one leaving no trace in architectural history due to the demise of Iranian Buddhism under Muslim rule. Such capacities for grand innovation resonate soundly with what is known of the wealth and influence of the Nawbahār monks, then controlling an entire oasis. The octagon would represent the final phase of architecture based on the monastery's immense wealth in the seventh century.

The date of construction for the octagon can be placed fairly precisely as occurring after 630 (the moment of Xuanzang's sojourn at the site) but before the Arab seizure of the oasis in 704. The structure should also predate the rule of Nīzak Tarḫan over north-western Afghanistan, in the last two decades of the seventh century, which involved the expulsion of the Barmak from the monastery.[17] Al-Balāḏurī also mentions the pillage of the monastery by the Arabs in 663 as occurring during this period (630 to the end of the century).[18] The construction of the octagonal monastery, extant and functioning in the descriptions of al-Kirmānī, and still visible in the landscape today, was therefore probably accomplished between the pillage of 663 and the 680s.

This discovery confirms the hypothesis derived from the texts. The ability of the Barmak of the Nawbahār to construct their monastery in an entirely new, gigantic and heretofore unseen design demonstrated extreme power and wealth. We should not be surprised that the Nawbahār, along with its design and decoration, left such lasting traces in Persian poetry,[19] nor by the exceptional status acquired by Barmakids at the Abbasid court at the end of the eighth century.[20]

Having dealt with this first octagon we can now turn to a second one known at the oasis. Near Siyāhgird, at the site of Dih-i Naw, in a deserted area

16 Behrendt, Kurt Allen, *The Buddhist Architecture of Gandhara*, (Leiden: Brill, 2004): fig. 20.

17 Esin, Emel, "Tarkhan Nīzak or Tarkhan Tirek? An Enquiry Concerning the Prince of Badghīs who in a.h. 91/a.d. 709-710 opposed the 'Omayyad Conquest of Central Asia", *JAOS*, 97.3 (1977): pp. 323-32.

18 Balāḏurī, *Liber expugnationis*: p. 409.

19 Melikian-Chirvani, "L'évocation": pp. 11-23.

20 See van Bladel, "Bactrian Background", on the Barmakid family. Van Bladel's article predates the scholarly discovery that most of the oasis was owned by the monastery during the seventh century.

FIGURE 3 Aerial photography of the octagon (in right down quarter) and the settlement in Siyāhgird
PHOTO DAFA

northeast of the oasis, another octagon was discovered. This second form is complete, with inner and outer limits, and may include subdivisions and an octagonal central part (Fig. 3). Some lateral walls remain, however these could be later elements based on the final phase of the first octagonal structure. The 'Siyāhgird octagon' is very similar in general dimensions to that of the Nawbahār, but whereas at Balkh the sides are oriented towards the cardinal points, at Siyāhgird, it is the angles which point south-west-north-east. A small test trench executed at the centre of the overall feature has revealed a very hard octagonal base or terrace, which could have served as base for further construction, although no stupa remains.

This second octagon is associated with a complex of regularly patterned streets and field borders composing a very substantial settlement. Based on the examination of images by Prof R. Hillenbrand, in which he identifies the architectural style of one of the main buildings with typically Saljuq forms, we can conclude that some habitation survived at the settlement up to at least the twelfth century (Fig. 4).[21]

Several hypotheses can be proposed. Among these is the proposal that the Siyāhgird octagon would be a second monastery modelled on the first, the construction of which would then have been interrupted by the troubles at the

21 Email of June 13, 2013.

FIGURE 4 Seljuk building in Siyāhgird
PHOTO E. DE LA VAISSIÈRE

end of the seventh century. The settlement plan calls to mind another passage by al-Kirmānī, in which he explains that the peasants at the oasis were the slaves of the monastery, a situation not uncommon in Buddhist lands of the time. If this were the case then Dih-i Naw might represent the monastery village, with dwellings for its enslaved peasants, planned but never finished.

Such a tempting hypothesis must be balanced by another. It is well known that the Buddhist origin of the Nawbahār was soon forgotten in the Arabic and Persian literature. As an example, the tenth century polygraph al-Mas'ūdī describes the Nawbahār as a fire temple founded by the mythical Iranian king Manūčihr.[22] In the *Fażā'il-i Balḫ*, the site is introduced as a "temple of the Magians":

> In 'Ajam [i.e. Iran] during the Time of Ignorance, [the Nawbahār] was the sanctuary (*ka'ba*) and temple (*ma'bad*) of the Magians (*muġān*). It is attributed to Ibn Šawḏab, that the Devil (*Iblīs*) has his home in Ḫurāsān, which they call 'the Naw Bahār of Balkh'. Every year, he dons the garb of a pilgrim (*iḥrām gīrad*) and visits that house. And the sanctuary of the Magians and the temple of the *'ajamī*-s was of this ilk. They also transmit from Ibn Šawḏab that at the beginning of each solar year the great and noble [men] of Ṭuḫāristān, Hindūstān and Turkistān and the lands

22 Mas'ūdī, *Murūj al-Ḏahab* trans. Charles Pellat *Les prairies d'or*, II, (Paris: Société asiatique, 1965): II; p. 525 § 1374.

> of ʿIrāq, Syria and Greater Syria (Šām wa Šāmāt) would come to and celebrate for seven days at Nawbahār.[23]

However, it seems that the story went one step further. The *Ḥudūd al-ʿĀlam* states that the Nawbahār was constructed by the ancient Sasanian kings, and praises the wonderful work. Should a Samanid or Saljuq prince have wanted to build himself a whole new domain on the watered periphery of the oasis, with a part for the peasants and a palatial part for himself, he might well have imitated the building that for him would have represented the quintessential palace of Iranian kings, namely the octagonal Nawbahār of the Sasanians. In that case, this second octagon at Siyāhgird would have nothing to do with Buddhism itself but would represent a false assimilation to the kingly Iranian past, based on what was really a Buddhist model. In this instance, Siyāhgird would be an extremely interesting example of an architectural reappropriation of the past.

Still other hypotheses could certainly be advanced, however a useful beginning would be to undertake excavations serving towards confirmation or denial of those advanced here.

Balkh is clearly unique for its Buddhist past and survivals, compared to the other main towns of Greater Khurāsān. In the cases of Samarqand and Bukhara, the only memory of the Buddhist past was toponymic: both had a gate of the Nawbahār. For Marw, once graced with two monasteries and stupas, discovered by Soviet excavations, there is no mention of this past in the surviving Islamic texts.[24] The only testimony is Chinese, from 751.[25]

3 The Irrigation Network

Dih-i Naw provides other important clues to the history of the oasis, in particular concerning the irrigation network. Given the convex surface of an alluvial fan, watercourses will continually change course. The challenge of irrigation becomes to restrict the water to established channels, rather than diverging and spreading. A significant flood or geological disaster can change the

23 Balḫī, *Fażāʾil*. Trans. in Azad, *Sacred Landscape*: p. 79.

24 The single exception of Gurgānī adduced by Melikian-Chirvani, "L'évocation": p. 7 is not fully convincing: the poem describes a cupola (*gumbad*) with idols (*but*) painted in gold. Its model might be Buddhism, but the paintings and idols of the Central Asian iconodule variant of Zoroastrianism cannot be excluded.

25 Akin, Alexander, "The *Jing Xing Ji* of Du Huan: Notes on the West by a Chinese Prisoner of War", *Harvard Middle Eastern and Islamic Review* 5 (1999): pp. 77-102, 87.

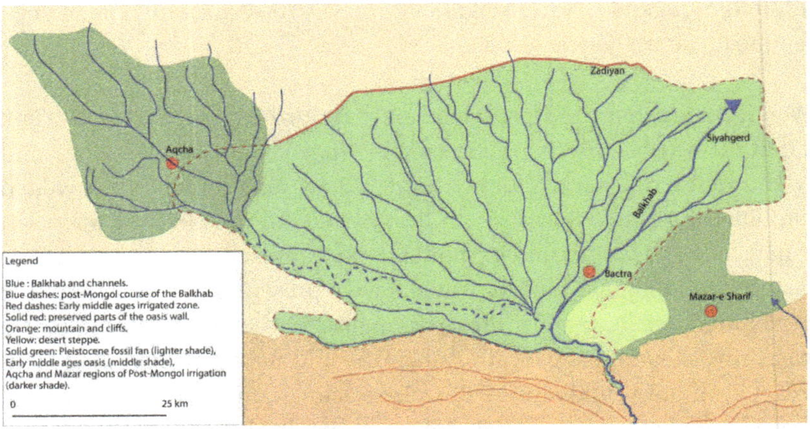

FIGURE 5 Schematic map of the oasis of Balkh
DRAWING E. DE LA VAISSIÈRE

entire morphology of an alluvial fan, as well as the economy. Such an event definitely occurred at Balkh sometime in the Islamic period (Fig. 5), when the Balḫāb River shifted to the west, creating the alluvial fan of Aqča. The pillars of an early Muslim bridge at Tepe-yi Zargarān, the eastern limit of pre-Islamic Bactra, represent archaeological proof that the course of the Balḫāb River passed further to the northeast in pre-Islamic and early Islamic times.[26] This is confirmed by texts on Balkh: the tenth century geographer Ibn Ḥawqal wrote that the River of Balkh flowed outside the gate of the Nawbahār and as far as Siyāhgird.[27] Whoever controlled the Nawbahār also clearly controlled the irrigation network, as frequently occurred for the oases of Central Asia. The Khotan and Kucha oases represent further cases, where the monasteries are situated at the apex of the alluvial fans in places devoid of agriculture, and the only potential strategy for the location would be to control the head of the irrigation system.[28]

The secondary Aqča fan was later formed by a radical shift of the river to the west. Indeed, the toponymy of the Aqča region is purely New Persian and Muslim, reflecting the timing of this shift. In a very interesting article on the long-term evolution of the river course and its channels, our colleagues Roland

26 Besenval, Marquis, "Les travaux": p. 975.

27 Ibn Ḥawqal, *Kitāb Ṣūrat al-Arḍ*, ed. M. Goeje, 2nd edn revised by Kramers, J.H., (Leiden: Brill, 1967): p. 448.

28 La Vaissière, Etienne de, "The Monks and the Control of the Irrigation Networks", in Frantz Grenet, Étienne de la Vaissière, Simone Mantellini (eds.) *L'irrigation en Asie centrale*, (Paris: Collège de France, forthcoming 2019).

Besenval and Eric Fouache have recently suggested an early Islamic date for this shift, deriving from a major earthquake in 818.[29] However, the Saljuq dating of some buildings at Siyāhgird, as well as the text of Ibn Ḥawqal, prove that this cannot be true.[30] Instead, it is most likely the river changed its course in the period of Mongol control, leading to desiccation of the immense zone northeast of Balkh. The geophysical data provided by our colleagues confirms this hypothesis, as the medieval sedimentary beds sampled for luminescence dating in their Aqča test-well are dated between around 1100 and 1600 CE.[31]

4 The Wall

A last landmark described by the Muslim authors, which the DAFA surveys have helped clarify, is the oasis wall itself. The most detailed description is that of al-Yaʿqūbī:

> a large wall encloses the villages, farms and cultivated lands of Balkh, with 12 *farsaḫ*s between one gate to this wall and the opposite one. Outside of the wall there are no buildings or villages or farms, only sands. This biggest wall surrounding the land of Balkh has 12 gates while the wall surrounding the suburbs has 4 gates. There are 5 *farsaḫ*s separating the biggest wall from the second one.[32]

A Soviet team led by Galina Pugačenkova surveyed the oasis wall in the Northwestern part of the oasis in the early 1970s.[33] Subsequently, in 2013, we surveyed along the sector of the oasis wall close to the desert edge, at Zādiyān, a village

29 Fouache, Eric, Roland Besenval, Claude Cosandey, Céline Coussot, Matthieu Ghilardi, Sébastien Huot, and Michel Lamothe, "Palaeochannels of the Balkh River (Northern Afghanistan) and Human Occupation Since the Bronze Age Period", *Journal of Archaeological Science*, 39.11 (2012): pp. 3415-27, 3424 and especially 3426 fig. 9.

30 Wordsworth, Paul, "The hydrological networks of the Balkh Oasis after the arrival of Islam: a landscape archaeological perspective", *Afghanistan*, 1.1 (2018): pp. 182-208, 186-188 followed Fouache *et al.*, "Paleochannels", and did not take into account the text of Ibn Ḥawqal. His conclusions should be modified accordingly.

31 Fouache *et al.*, "Paleochannels": pp. 3420, 3422. There have been 2 m. of deposits in the Aqča delta between the 11th c. and the 16th c. CE while only 1 m. of sediments were deposited there in the five millennia before the 11th c.

32 Yaʿqūbī, *Kitāb al-Buldān*, ed M.J. de Goeje, (Leiden: Brill, 1892): p. 288.

33 Pugačenkova, Galina, "K poznaniju antičnoj i rannesrednevekovoj arkhitektury severnogo Afganistana", in U. T. Kruglikova, V. I. Sarianidi (eds.), *Drevnjaja Baktrija*, (Moscow: Nauka, 1976): pp. 125-62, 137-40.

FIGURE 6 The oasis wall at Zādiyān seen from the desert
PHOTO PH. MARQUIS

directly north of Balkh (Fig. 6). Here we made an important discovery: a gigantic military camp dating to the earliest years of the first millennium CE, created by the Kushans (first to third century). The camp is four by four kilometres in size, with an entirely regular grid plan, surrounded by a wall of sixteen kilometres total length.[34] At the centre of the walled compound and camp there stood a huge Kushan citadel, covering four hectares. The Kushan military camp was evidently never reused, however the citadel was: carbon-14 testing has shown that the upper construction of the citadel was executed using large bricks of the Samanid or early Ghaznavid period. The straw of the bricks has been dated at 892 to 1031 CE with 95.4% probability.[35] Thus for a span on the order of 1000 years, the Kushan citadel continued to protect Balkh from the north, an orientation that was probably more strategic during the wars between the Ghaznavids and Qarakhanids than during the Samanid period. The repairs to the citadel would then likely be linked to the disintegration of the Samanid empire into the rivalry between Qarakhanid and Ghaznavid empires, separated by the Amu Darya, rather than to the Samanid heyday, when the southern and northern banks of the Amu Darya were unified. It is possible that the Kushan citadel remained in long use after construction, given that some of its architectural features seem to be Sasanian, in particular several vaults. To my

34 La Vaissière, Etienne de, Philippe Marquis, Julio Bendezu Sarmiento, "A Kushan military camp near Bactra", in Harry Falk (ed.), *Kushan Histories. Literary Sources and Selected Papers from a Symposium at Berlin, December 5 to 7, 2013*, Bremen: Hempen, 2016): pp. 241-54.

35 BP 1055 ±40, calibrated. Laboratory Code: Lyon-11319(GrA). The samples were prepared at the Centre National de la Recherche Scientifique C-14 laboratory and Université Claude Bernard Lyon 1, Centre de Datation par le RadioCarbone, and dated using an accelerator at Heidelberg, applying the IntCal13 atmospheric curve.

knowledge this structure is not mentioned in the sources, although it remains an extremely impressive presence (Fig. 7).

The northern wall of the gigantic Kushan camp was incorporated into the oasis walls, when the latter were constructed (Fig. 8). One of the twelve gates of the oasis walls, as mentioned in al-Yaʿqūbī, still exists at Zādiyān. A major result of our survey has been to prove that this feature dates to the early Sasanian period, based on two carbon-14 samples, rather than to the Greek period, as previously supposed by the Soviet team.

The text by al-Yaʿqūbī provides no indication of either the builder or dates for the wall construction. They were certainly in ruins by the end of the ninth century, which explains why later geographers such as al-Iṣṭaḫrī, Ibn Ḥawqal and al-Masʿūdī offer no descriptions. However, the *Fażāʾil-i Balkh* suggests a faint allusion to the Sasanian past of the oasis wall, in indicating that a Qubād son of Ardašīr is supposed to have rebuilt the city and its walls.[36] In fact, the son of Ardašīr was not Qubād, but Šāpūr. The latter could indeed have been the builder of the wall, considering the carbon-14 analyses as correct. However, it is also possible that the *Fażāʾil* simply preferred to ascribe anything ancient to the Sasanians.

Step by step, our knowledge of early Islamic Balkh is becoming more precise. It is now clear that the frequently quoted text of al-Ṭabarī describing how the Arab governor, Asad b. ʿAbd-Allāh, decided to *build* the city of Balkh in 725 should be taken at face value. Contrary to the situation at Marw, Samarqand or Bukhara, Balkh and its oasis had to be totally reorganised due to the highly uncommon monastic system prevailing just before the conquest. Balkh was an Islamic city built almost from scratch, except for a few landmarks: the old Achaemenid citadel, the associated city walls and the Buddhist monasteries. Concerning the oasis system itself, the Muslim period inherited a complete irrigation network, which does not seem to have been modified during the first centuries of rule. The wall surrounding the oasis, also inherited, soon fell into decay. The various discoveries demonstrate the course of chronological development, involving a very rapid pace of assimilation during the ninth century. Even though al-Kirmānī could still very accurately describe the Nawbahār Buddhist monastery at around 800, when the octagon structure was still of quite recent construction, this knowledge would soon be entirely lost. Just one century later, al-Yaʿqūbī wrongly describes the monastery as the residence (*manāzil*) of the Barmakids.[37] As I have previously demonstrated

36 Schefer, Charles Henri Auguste, *Chrestomathie Persane à l'usage des élèves de l'école spéciale des langues orientales vivantes*, (Paris: E. Leroux, 1883): p. 73, Persian text: p. 70.

37 Yaʿqūbī, *Buldān*: p. 288.

FIGURE 7 The citadel at Zādiyān
PHOTO R. BESENVAL

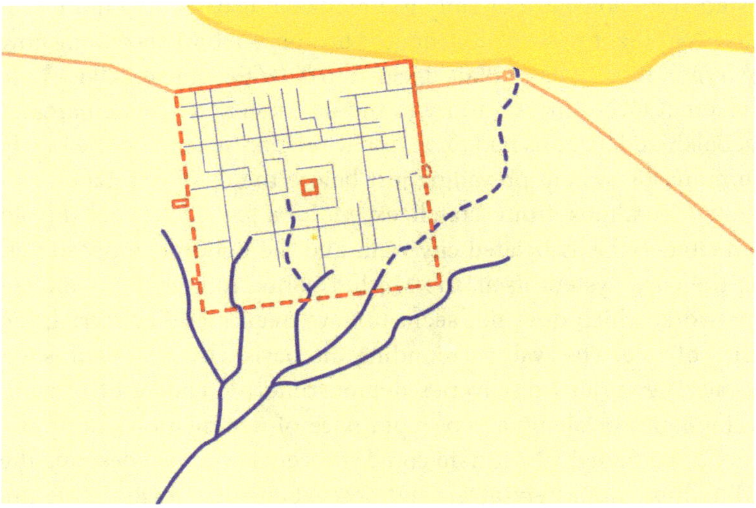

FIGURE 8 Schematic plan of the Kushan military camp at Zādiyān (preserved parts of the camp in solid red; oasis walls in orange)
DRAWING E. DE LA VAISSIÈRE

for Sogdiana,[38] the middle of the ninth century, meaning the Tahirid period, seems to have been key to the acculturation of non-Sasanian Central Asia into an Iranian and Muslim culture, and the transformation of the non-Sasanian past as a 'Sasanian' one.

Bibliographical References

Adle, Chahryar, "La mosquée Hâji-Piyâdah / Noh Gonbadân à Balkh (Afghanistan). Un chef-d'oeuvre de Fazl le Barmacide construit en 178-179/794-795?", *Comptes-rendus des séances de l'Académie des Inscriptions et Belles-Lettres*, I (2011): pp. 565-625.

Akin, Alexander, "The *Jing Xing Ji* of Du Huan: Notes on the West by a Chinese Prisoner of War", *Harvard Middle Eastern and Islamic Review* 5 (1999): pp. 77-102.

Azad, Arezou, *Sacred Landscape in Medieval Afghanistan: Revisiting the Faḍā'il-i Balkh*, (Oxford: Oxford University Press, 2013).

Balāḏurī, *Liber expugnationis regionum / auctore Imámo Ahmed ibn Jahja ibn Djábir al-Beládsorí*, ed M.J. de Goeje, (Leiden: Brill, 1866).

Balḫī, 'Abd Allāh Muḥammad Ḥusaynī, *Fażā'il-i Balḫ*, ed. 'Abd al-Ḥayy Ḥabībī, (Tehran: Intišārāt-i bunyād-i farhang-i Īrān, 1350sh./1971.)

Ball, Warwick, and Jean-Claude Gardin, *Archaeological Gazetteer of Afghanistan: Catalogue des sites archéologiques d'Afghanistan*, (Paris: Éditions Recherches sur les Civilisations, 1982).

Behrendt, Kurt Allen, *The Buddhist Architecture of Gandhara*, (Leiden: Brill, 2004).

Besenval, Roland, and Philippe Marquis, "Les travaux de la délégation archéologique française en Afghanistan (Dafa): Résultats des campagnes de l'automne 2007-printemps 2008 en Bactriane et à Kaboul", *Comptes rendus des séances de l'Académie des Inscriptions et Belles-Lettres*, 152/3 (2008): pp. 973-95.

Bosworth, C. E., "Abū Ḥafṣ 'Umar Al-Kirmānī and the Rise of the Barmakids", *BSOAS*, 57.2 (1994): pp. 268-82.

Dagens, Bruno, Marc Le Berre, and Daniel Schlumberger, *Monuments Préislamiques d'Afghanistan*, (Paris: Klincksieck, 1964).

Esin, Emel, "Tarkhan Nīzak or Tarkhan Tirek? An Enquiry Concerning the Prince of Badghīs who in AH 91/AD 709-710 opposed the 'Omayyad Conquest of Central Asia", *JAOS*, 97.3 (1977): pp. 323-32.

Fouache, Eric, Roland Besenval, Claude Cosandey, Céline Coussot, Matthieu Ghilardi, Sébastien Huot, and Michel Lamothe, "Palaeochannels of the Balkh River (Northern

38 La Vaissière, Etienne de, *Samarcande et Samarra. Élites d'Asie centrale dans l'empire abbasside*, (Paris, Louvain: Association pour l'avancement des Études iraniennes, 2007).

Afghanistan) and Human Occupation Since the Bronze Age Period", *Journal of Archaeological Science*, 39.11 (2012): pp. 3415-27.

Foucher, Alfred, and Mme Eugénie Bazin-Foucher, *La Vieille Route de l'Inde de Bactres à Taxila*, (Paris: Éditions d'Art et d'Histoire, 1982).

Frye, Richard, "Balkh", *EI*², I (1960): pp. 1000-2.

Ḥudūd al-ʿĀlam, ed. V. V Bartol'd, (Leningrad, 1930); trans. Vladimir Minorsky, Clifford Edmund Bosworth, *Ḥudūd al-ʿĀlam; "the Regions of the World"; a Persian Geography, 372 AH-982 AD*, (London: Gibb Memorial Trust, 1970).

Ibn al-Faqīh, *Collections of Geographical Works by Ibn al-Faqīh, Ibn Faḍlān, Abū Dulaf al-Khazrajī*, ed. Fuat Sezgin, (Frankfurt/M: Johann Wolfgang Goethe Universität Frankfurt, 1987).

Ibn Ḥawqal, *Kitāb Ṣūrat al-Arḍ*, ed. M. Goeje, 2nd edn revised by Kramers, J.H., (Leiden: Brill, 1967).

La Vaissière, Etienne de, *Samarcande et Samarra. Élites d'Asie centrale dans l'empire abbasside*, (Paris, Louvain: Association pour l'avancement des Études iraniennes, 2007).

La Vaissière, Etienne de, "De Bactres à Balkh par le Nowbahār", *JA*, 298.2 (2010): pp. 517-33.

La Vaissière, Etienne de, Philippe Marquis, "Nouvelles recherches sur le paysage archéologique de Bactres," *Compte-Rendus de l'Académie des Inscriptions et Belles-Lettres*, 2013 (sic)-3 (juillet-octobre): pp. 1155-71 (published 2015).

La Vaissière, Etienne de, Philippe Marquis, Julio Bendezu Sarmiento, "A Kushan military camp near Bactra", in Harry Falk (ed.) *Kushan Histories. Literary Sources and Selected Papers from a Symposium at Berlin, December 5 to 7, 2013*, (Bremen: Hempen, 2016): pp. 241-54.

La Vaissière, Etienne de, "The Monks and the control of the Irrigation networks", in Frantz Grenet, Étienne de la Vaissière, Simone Mantellini (eds) *L'irrigation en Asie centrale*, (Paris: Collège de France, forthcoming 2019).

Le Berre, Marc, Daniel Schlumberger, "Observations sur les remparts de Bactres", in Bruno Dagens, Marc Le Berre, and Daniel Schlumberger, *Monuments Préislamiques D'Afghanistan*, (Paris: Klincksieck, 1964).

Masʿūdī, *Murūj al-Ḏahab* trans. Charles Pellat *Les prairies d'or*, II, (Paris: Société asiatique, 1965).

Melikian-Chirvani, Assadullah Souren, "L'évocation littéraire du bouddhisme dans l'Iran musulman", in Jean Aubin (ed) *Le Monde Iranien et l'Islam. 2: Sociétés et Cultures*, (Paris, Genève: EPHE-Droz, 1974): pp. 1-72.

Pugačenkova, Galina, "K poznaniju antičnoj i rannesrednevekovoj arkhitektury severnogo Afganistana", in U. T. Kruglikova, V. I. Sarianidi (eds.), *Drevnjaja Baktrija*, (Moscow: Nauka, 1976): pp. 125-62.

Schefer, Charles Henri Auguste, *Chrestomathie Persane à l'usage des élèves de l'école spéciale des langues orientales vivantes*, (Paris: E. Leroux, 1883).
van Bladel, Kevin, "The Bactrian Background of the Barmakids", in Charles Burnett, Anna Akasoy, Ronit Yoeli-Tlalim (eds) *Islam and Tibet: Interactions Along the Musk Routes*, (Farnham: Routledge, 2010): pp. 45-87.
Wordsworth, Paul, "The hydrological networks of the Balkh Oasis after the arrival of Islam: a landscape archaeological perspective", *Afghanistan*, 1.1 (2018): pp. 182-208.
Xuanzang, *Da Tang Xiyu Ji*, ed. Taishô, LI, n° 2087; trans. Li Rongxi, *The Great Tang dynasty record of the Western regions*, (Berkeley: BDK America, 1996).
Ya'qūbī, *Kitāb al-Buldān*, ed M.J. de Goeje, (Leiden: Brill, 1892).

Biographical Note

Étienne de la Vaissière is a Professor of the Ecole des Hautes Etudes en Sciences Sociales, Paris, where he teaches Central Asian medieval history with a particular focus on economic and social history. His PhD dealt with the main (Sogdian) trading networks of Central Asia, (published as Sogdian Traders, Brill, 2005) and his Habilitation thesis was on the integration of central Asian elites in the Muslim World (Samarcande et Samarra, Louvain, 2007). He has also published extensively on nomads from the fourth to ninth century. He currently works with French teams operating in archaeological surveys in Afghanistan and Uzbekistan.

Among Saints and Poets: The Spiritual Topography of Medieval Shiraz

Denise Aigle
CNRS, UMR 8167 "Orient & Méditerranée"
denise.aigle@cnrs.fr

Abstract

Shiraz is distinguished from other cities due to its reputation as the city of saints and poets, as previously emphasised in the title of Arberry's book of 1960: "Shiraz, Persian City of Saints and Poets". In textual sources, the city is the "Fortress of saints" (*burj al-awliyā'*). Shiraz owes its sanctity to the many mausoleums dedicated to the descendants of ʿAlī b. Abī Ṭālib (and Šāh-i Čirāġ), as well as famous mystics such as Šayḫ Kabīr (d. 371/982) and Rūzbihān Baqlī (d. 606/1209). The poets Saʿdī (d. 691/1292) and Ḥāfiẓ (d. 792/1390) celebrated Shiraz as the city of roses and nightingales. Their sanctuaries, which still the object of pious visits, accentuate the "capital" of city's sacrality.

After reconstructing the urban space in which the sacred buildings are located, the purpose of this paper is to show how the specific sanctity of the city emerged from the textual sources. Two major texts addressing the sanctity of Shiraz date from the eighth/fourteenth century. In the *Šīrāz-nāma* (completed in 744/1343), Ibn Zarkūb unfolds the history of the city and speaks of its merits. In *Šadd al-īzār* (ca. 791/1389), a guide for pilgrimage to Shiraz's seven cemeteries, Junayd Šīrāzī describes the ritual geography of the city. He notes the places where the Shirazis are buried, thus establishing the symbolic presence of the deceased among the living. Alid shrines in particular thus contributed to the "capitalisation of the sacred" in Shiraz.

Keywords

Alid shrines – Saʿdī's tomb – Ḥāfiẓiyya – Rūzbihān Baqlī – Šayḫ Kabīr's sanctuary

 | DOI:10.1163/9789004434332_009

1 Introduction

Shiraz occupies a special place among the cities of medieval Iran, even though it shares many characteristics with others of the Persian plateau.[1] Similarly to Isfahan, Rayy and Hamadan, the city of Shiraz is situated at the crossroads of major trade routes, running north-south (Isfahan-Shiraz-Fīrūzābād-Sīrāf) and east-west (the former Achaemenid 'royal route' to Khuzistan via Bihbahān).[2] A second aspect is that, like other centres of the Iranian plateau, Shiraz is surrounded by mountains, providing the supply of water for the city and the agricultural development of its hinterland, via a network of small non-perennial rivers and *qanāt*.[3] Third, Shiraz and Isfahan share a comparable chronology of urbanisation: despite the longstanding occupation of the valley, attested by the nearby vestiges of Qaṣr-i Abū Naṣr,[4] it was not until the third/ninth to fourth/tenth centuries that the former Arab military camp was rapidly transformed as a major regional metropolis. Fourth, Shiraz resembles Isfahan in appearance: both are encircled by ramparts constructed during the Buyid age. Lastly, its role as a power centre ("*dār al-mulk*") resembles that of other regional centres such as Marw, Hamadan, Tabriz and Isfahan. In fact Shiraz served as the capital in successive periods: the Buyids in the fourth/tenth century; the Salghurids in the sixth/twelfth and seventh/thirteenth centuries, following the collapse of Saljuq authority, and the Injuids of the eighth/fourteenth century, subsequent to the Mongol Ilkhanate.

Nevertheless, as already emphasised in the title of Arberry's work (*Shiraz: Persian City of Saints and Poets*), Shiraz is different from other cities in terms of its reputation, at least as concerns the manner in which it is lauded in the sources. It is known as the city of poets (*šahr-i gul u bulbul*), renowned in particular thanks to the works of Saʿdī (d. 691/1292) and Ḥāfiẓ (d. 792/1390), but above all as the "Fortress of saints" (*burj al-awliyā'*), to use the expression of

1 My thanks go to J. Paul, D. Durand-Guédy, and R. Mottahedeh for their valuable comments, which greatly assisted in improving a first version of this article.

2 The road network of Fārs, centred around Shiraz, is described in detail in Schwarz, Paul, *Iran nach den arabischen Geographen des Mittelalters*, Leipzig, 1896 (reprint Hildesheim-New York: Georg Olms, 1969): pp. 183-99, 200-3, 232-234.

3 Ḥāfiẓ, *Dīwān*, ed. P. Khanlārī (Tehran: Ḫwārazmī, 1980): n.°274, bayt 2, celebrated the *qanāt* of Ruknābād with the words: "O Ruknābād, gushing fountain, may you be blessed, streaming Elixir!" This *qanāt*, which flows above ground at the Tang-i Allāh Akbar, was constructed during the Buyid age; Mustawfī Qazwīnī, Ḥamd Allāh, *Nuzhat al-qulūb*, ed. and trans. Le Strange, G., 2 vols (London & Leiden: Luzac & Brill, 1915, 1919): pp. 115, trans. 113. On the water supply of cities, see Rante's article in the present volume.

4 On Qaṣr-i Abū Naṣr and its transition from a Sasanian settlement to an Islamic city, see Whitcomb in this volume.

Junayd Šīrāzī.[5] The city is also noted for its sanctity owing to the many mausoleums dedicated to the descendants of ʿAlī b. Abī Ṭālib (and Šāh-i Čirāġ), and to famous mystics such as Šayḫ Kabīr (d. 371/982) and Rūzbihān Baqlī (d. 606/1209).

The intention of this current article is to study the manner in which this sacredness is inscribed in the urban landscape and defines the city's religious life and pilgrimages.[6] The analyses follow in a series of numbered sections, after a brief review of previous research on the topography of Shiraz.

2 Previous Research

Several modern works in Persian contain topological and architectural information on Shiraz. Ḥasan Imdād's monograph, published in 1960, was one of the first topographical studies. He first retraces the city's history, then provides notes on each edifice, with a description of the restorations undertaken by the various princes reigning over Fars.[7] Karāmat-Allāh Afsar, in the *Tārīḫ-i bāft-i qadīmī-yi Šīrāz*, published in 1974, studies the history of the city in chronological fashion, from the Buyids to the Qajars. He describes the transformation of the historic neighbourhoods and the evolution of the urban fabric, and reproduces a map of Shiraz from the Qajar period, drawn up by Furṣat al-Dawla Šīrāzī in his *Āṯār-i ʿAjam*.[8] More recently, in 2012, Razzāq Pārsa Amīrī dedicated numerous pages to Shiraz in his important study on the history of cities and villages of Fārs.[9] He describes the history of the edifices, with details of the successive stages of building. Information on the infrastructure of Shiraz (mosques, madrasas, shrines, gardens) can also be found in the dictionary of Kūruš Kāmalī-Sarwistānī.[10] The studies of Amīrī and Kamālī-Sarwistānī both constitute invaluable references on the topography of Shiraz. Finally, John

5 Junayd Šīrāzī, Muʿīn al-Dīn Abu'l-Qāsim, *Šadd al-izār*, ed. Qazwīnī, Muḥammad (Tehran: Čāpḫāna-yi Majlis, 1328š./1949): p. 3. Cf. Mustawfī Qazwīnī, *Nuzhat*: pp. 115, trans., 116.

6 I discuss the sanctity of Shiraz in greater detail in a monograph currently in preparation, titled *Saints hommes de Chiraz et du Fārs: Pouvoir, société et lieux de sacralité*.

7 Imdād, Ḥasan, *Šīrāz dar guḏašta u ḥāl* (Ṣīrāz: Mūsawī, 1339š./1960).

8 Afsar, Karāmat-Allāh, *Tārīḫ-i bāft-i qadīmī-yi Šīrāz* (Tehran: Našr-i Qatra, 1353š./1974, 1995^2).

9 Amīrī, Razzāq Pārsa, *Tārīḫ-i Fārs, Šahr-hā wa rusta-hā-yi ān*, 3 vols (Šīrāz: Nawīd-i Šīrāz, 1391š./2012): II, pp. 11-300.

10 Kamālī-Sarwistānī, Kūruš, *Dānišnāma-yi āṯār-i tārīḫī-yi Fārs* (Šīrāz-Tehran: Mu'assasa-yi farhangī wa pažūhišī-yi Dānišnāma-yi Fārs & Sāzimān-i mīrāṯ-i farhangī, 1384š./2006).

Limbert devotes a study to the political and social history of Shiraz at the time of Ḥāfiẓ, with several pages on the city's topography in the fourteenth century.[11]

These historical-topographic studies on Shiraz describe the mosques, the tombs of famous holy men and poets, and the Alid shrines. Nevertheless, they fail to discuss the city's 'sanctity', as recognised by Shirazi historians and travellers, who perceived Shiraz and its extramural areas as a territory bursting with "saintliness" and a special "spirituality."

The notion of "territory" has sparked the interest of anthropologists and specialists of sainthood insofar as sacred places are concerned.[12] Yet the question beckons: what are the criteria that characterise a particular place, space, or landscape deemed to be 'sacred'? The German theologian and scholar Rudolf Otto (1869-1937) was one of the first theorists of "the spirituality of the place."[13] In his view, sacred sites have a spiritual quality (*numen*) or divine power, which attributes a specific sacredness, *genius loci*, or *numen loci* to them.[14]

In Islam, a sacred site is above all a sacred territory, *ḥaram*, which can be a 'sanctuary' (*ḥaram*) or 'sacred pasture' (*ḥīma*). The meaning and definition of these terms derive from pre-Islamic Arabia,[15] however Islam invested these

---

11 Limbert, John, *Shiraz in the Age of Hafez. The Glory of a Medieval Persian City* (Seattle and London: University of Washington Press, 2004): pp. 52-66, includes useful tables on the city's pilgrimage sites (4.1), mosques (4.2), and madrasas (4.3), citing sources and, where possible, the founders and locations of the edifices. On Shiraz, see Ḫūbnaẓār, Ḥasan, *Tārīḫ-i Šīrāz* (Tehran: Suḫan, 1380š./2002), as well as the encyclopaedia entries in Lambton, Ann K. S., "Shīrāz", in *EI*², IX (1998): pp. 491-7, and Shahbazi, A. Shapour, "Shiraz i. History to 1940", in *Encyclopaedia Iranica online*, http://www.iranicaonline.org/articles/shiraz-i-history-to-1940.

12 Vincent, J. F., J. Dary and R. Verdier, *La Construction religieuse du territoire* (Paris: L'Harmattan, 1995); Vauchez, André, "Introduction", in *Lieux sacrés, lieux de culte, sanctuaires. Approches terminologiques, méthodologiques, historiques et monographiques*, Vauchez, André ed. (Rome: École Française de Rome, 2000): pp. 1-7; Vauchez, André, "Introduction", in La Genière, Juliette de, Vauchez, André, Leclant, Jean, (eds.), *Les sanctuaires et leur rayonnement dans le monde méditerranéen de l'Antiquité à l'époque moderne* (Paris: De Boccard, 2010): pp. 1-6.

13 Otto's best known work is Otto, Rudolf, *Das Heilige: über das Irrationale in der Idee des Göttlichen und sein Verhältnis zum Rationalen* (Breslau: Trewendt und Garnier, 1917), with numerous subsequent reprints.

14 On the notion of sacred space, see Azad, Arezou, *Landscape in Medieval Afghanistan: Revisiting the Faḍā'il-i Balkh* (Oxford: Oxford University Press, 2014): pp. 12-3.

15 Mayeur-Jaouen, Catherine, "Tombeau, mosquée et zâwiya: la polarité des lieux saints musulmans", in Vauchez (dir.), *Lieux sacrés, lieux de culte, sanctuaires. Approches terminologiques, méthodologiques, historiques et monographiques* (Rome: École Française de Rome, 2000): pp. 133-47, 133.

spaces with new values.[16] The quintessential *ḥaram* is the territory of Mecca. Its shrine – the Kaaba – existed prior to all other places of worship.[17] The shrines of the Muslim saints then multiplied the number of sacred places (*ḥaram*) by spreading the sacred fluid (*baraka*) of the corpse.

In his famous work *Du Sacré*, Alphonse Dupront established a typology of sacred places. He distinguishes "cosmic places" in the natural universe, meaning those known for their extraordinary character.[18] Isfizārī, for example, mentions two springs near Herat known for their curative virtues.[19] In Shiraz, people attributed medicinal qualities to the subterranean river Ruknābād, which watered the Muṣallā plain.[20] For Dupront, there are also "places of eschatological fulfilment", which are mythically situated at the centre of the world. The "sacred cities" of Jerusalem and Mecca, for example, are symbolic places serving to establish the unity of the world.[21] History is another source of sacred commemoration. "Historical places" can commemorate a significant event in the memory of the community,[22] but more often are locations rendered sacred by the presence of a venerated saintly body, or the remains of a group of personages endowed with special powers, as in the case of certain tombs of the anonymous saints of Shiraz. The pilgrimage to these shrines represents a unique experience of religious life, as part of an intense quest for sacralisation. In making their way to the 'sacred site', sometimes from afar, the pilgrim's attitudes and behaviours in seeking the remains become paramount. The presence of the venerated personage can then become manifest as a source of benediction or miraculous fluid.

---

16 The term *ḥarām* is literally translated as "prohibited" or "illicit", however it can also signify "saintly" or "consecrated": an example of the ambiguity of the sacred in Islam. See Mayeur-Jaouen, "Tombeau": pp. 133.

17 Quran: 3, 96-97.

18 Dupront, Alphonse, *Du Sacré. Croisades et pèlerinages. Images et langages* (Paris: Gallimard, 1987): pp. 379-80.

19 Isfizārī, Mu'īn al-Dīn Muḥammad, *Rawżāt al-Jannāt fī awṣāf-i madīnat Harāt*, ed. Sayyid Muḥammad Kāẓim Imām, 2 vols. (Tehran: Intišārāt-i Dānišgāh-i Tihrān, 1338-1339š./1959-60): I, pp. 101-2.

20 William Francklin, *Observations Made on a Tour from Bengal to Persia in the Years 1786-87* (London: T. Cadell, 1790): pp. 92.

21 Dupront, *Du Sacré*: pp. 386-389. Even before Islam, Mecca was the "mother of cities" (*umm al-qurā*), the metropolis of the world; ibid.: pp. 380.

22 A memorial at Yazd contains the remains of 30,000 people who perished from famine in the siege of the city by Pīr Muḥammad, son of the Timurid 'Umar-Šayḫ, in 797/1394-5. Miraculous powers emanated from the graves of victims; see Manz, Beatrice, *Power, Politics and Religion in Timurid Iran* (Cambridge: Cambridge University Press, 2007): p. 185.

Two major sources dealing with the sanctity of Shiraz date from the eighth/fourteenth century: the *Šīrāz-nāma* of Ibn Zarkūb[23] (completed in 744/1343), and the *Šadd al-īzār* (ca. 791/1389) by Junayd Šīrāzī. The founding narrative related by Ibn Zarkūb clearly conveys the sacredness of Shiraz: when the Umayyad governor Ḥajjāj b. Yūsuf sent his brother Muḥammad b. Yūsuf to the region, the latter had a dream vision of a celestial hermitage, whose members heralded "a land that would be the shrine (*qadamgāh*) of thousands of Sufis," furnished with "food for travellers who roam the world of the unknowable (*musāfirān-i ʿālam-i ġayb*)."[24] The next day, Muḥammad b. Yūsuf visited the place seen in his vision and contemplated the traces of a circle. Thanks to these visions, he ordered the construction of Shiraz on that site.[25]

This narrative establishes the city's sanctity by divine election (Muḥammad b. Yūsuf's dream) and also presages the ascetics, Sufis, and holy men who would later reside there. Here Ibn Zarkūb plays on the dual signification of the word "food" in both the terrestrial and spiritual senses. The term "traveller" likewise conveys both the literal meaning and the sense of spiritual itinerancy. The word "*ġayb*" also alludes to all the functions to be fulfilled by the pious men of Shiraz, in the sacred foundations of the future city. We should also note that for Ibn Zarkūb, the saintly status of Shiraz contributes to the sacredness of the entire territory of the province. As the author explains in his introductory chapter, Fars is the homeland of Salmān Fārsī: this famous Iranian companion of the Prophet was the archetype of the followers of ʿAlī, and his name is frequently cited in the hagiographical sources. Šahrbānū, the daughter of the last Sasanian sovereign Yazdegerd III (r. 632-651), is yet another "celebrity" (*faḍīla*) of the province. According to tradition, this Sasanian princess was the wife of Ḥusayn, the Prophet's grandson. She was a repository of the light of royal glory

23 The author's full name is Muʿīn al-Dīn Abu'l-ʿAbbās Aḥmad b. Abī l-Ḫayr Šīrāzī.

24 Ibn Zarkūb, *Šīrāz-nāma*, ed. Jawāḍī, I. Wāʿiẓ, (Tehran: Bungāh-i Farhang-i Vāža-javādī, 1350š./1972): p. 34.

25 Iṣṭaḫrī, *Kitāb al-Masālik wa-l-mamālik*, ed. de Goeje, (Leiden: Brill, 1967²): pp. 124-5; Ibn Ḥawqal, *Kitāb Ṣūrat al-arḍ*, ed. Kramers, J.H., 2 vols, (Leiden: Brill, 1938): p. 279; *Ḥudūd al-ʿālam*, Tr. and com. Minorsky, V., *The Regions of the World: A Persian Geography 372 AH-982 AD*, GMS N. S. XI, (London: Luzac, 1937): p. 126. After the taking of Iṣṭaḫr, Muḥammad b. Qāsim observed the auspiciousness of this place (*tabarraka Muḥammad b. Qāsim bi-hadā l-makān*) and determined to found the city; see Ibn Ḥawqal, *Ṣūrat*: p. 279. Other traditions indicate Muḥammad, a brother of Ḥajjāj b. Yūsuf's, as the founder. It is possible that the mention of Ḥajjāj b. Yūsuf is intended as a reference to the Umayyad dynasty, which was enhanced by the saintly figure of the caliph ʿUmar b. ʿAbd ʿAzīz; see Borrut, Antoine, *Entre mémoire et pouvoir. L'espace syrien sous les derniers Omeyyades et les premiers Abbassides (v. 72-193/692-809)* (Leiden: Brill, 2011): pp. 283-320. Unlike Ibn Zarkūb, the traditional geographic sources do not mention Sufis.

of ancient Iran, and is considered to be the 'mother of imams'. Her son, 'Alī b. al-Ḥusayn, is described as "noble on both sides," *karīm al-ṭarafayn*),[26] and the tombs of her descendants became sacred spaces of Shiraz.

Junayd Šīrāzī offers another important source describing a 'ritual topography'. In his book he identifies seven stages (*nawbat*) in the pilgrimage (*ziyārāt*) to the graves of holy men and local pious Muslims.[27] In naming the locations where the Shirazi saints are buried, he establishes the symbolic presence of a "society of the deceased" among the living. In other words, he inscribes the *memoria* of the dead within the urban space.[28] The shrines and individual graves situated in cemeteries were places of a kind of religious conviviality, frequently visited by the city's inhabitants, who came to pay homage at the graves of family members, and of those men recognised for piety and charismatic powers. Junayd Šīrāzī emphasises the sacred dimension of the city. Shiraz is a true "pole of saintliness" thanks to the benedictions (*barakāt*) conferred by its shrines, relics, cemeteries, and the rituals performed there.

2.1 *Urban Framework*

Before discussing the city's spiritual landscape, we must examine its topography. The works initiated by Karīm Ḫān Zand in the eighteenth century transformed the urban topography remarkably, meaning that any reconstruction can only be approximate. Although Fasā'ī described the neighbourhoods in his nineteenth century work, the accounts prepared at this time no longer reveal the topography of pre-modern Shiraz.[29] The descriptions below serve as a commentary to the map provided with this work, which combines Afsar's map with information gleaned from the textual sources.

The most remarkable feature of the city was its fortification, parts of which dated to the fourth/tenth century, when Shiraz was the capital of a Buyid emirate. The chronicles mention destruction of the fortification on several occasions. The damage was undoubtedly partial, and sources also note the efforts of city inhabitants and local princes to reinforce it against potential enemies.

---

26 Ibn Zarkūb, *Šīrāz-nāma*: p. 21.

27 'Īsā, son of Junayd Šīrāzī, translated the *Šadd al-izār* into Persian: 'Īsā b. Junayd Šīrāzī, *Hazār mazār*, ed. Wiṣāl, Nūrānī, (Šīrāz: Intišārāt-i Kitābḫāna-i Aḥmadī Šīrāz, 1364š./1975).

28 Van Renterghem describes the "funerary" and "living" geography of Baghdad as being equally important. The former was "less unstable and, as it were, definitive." Van Renterghem, Vanessa, *Les élites bagdadiennes au temps des Seldjoukides. Étude d'histoire sociale*, 2 vols (Beirut-Damascus, Institut Français du Proche-Orient, 2015): I, p. 431.

29 Ibn Zarkūb and Junayd Šīrāzī contain numerous topographic references. In addition to these we have: the chronicle of Waṣṣāf; the travel account of Ibn Baṭṭūṭā, from the late Ilkhanid period; two key texts from the Qajar period – the *Fārs-nāma-yi Nāṣirī* (1304/1887) of Mīrzā Ḥasan Fasā'ī, and the *Āṯār-i 'Ajam* of Furṣat al-Dawla Šīrāzī.

The city gates, recorded as a total of nine in the eighth/fourteenth century, are again frequently mentioned, and serve as important landmarks in localisation of infrastructure.[30] The urban fabric was divided into neighbourhoods (*maḥalla*). At the time of its peak under the Buyids, Shiraz counted 19 of these. In the eighth/fourteenth century, Ḥamd Allāh Mustawfī indicated 17, but provided no names.[31] The historical heart of the city (Sarrājān and Bālā Kaft) can be distinguished from the new neighbourhoods that developed during the prosperous period of the Salghurids (Bāġ-i Qutluġ, Bāġ-i Naw).[32]

The Bālā Kaft neighbourhood, mentioned in the *Šīrāz-nāma*, was in the southeast part of the city.[33] It was also in the southeast that Junayd Šīrāzī situated the *ḫānaqāh* of Faqīh Ḥusayn b. Muḥammad b. Salmānī (d. 664/1266), a well-known local jurist,[34] in the Saḫtūya lane of the Jaṣṣāṣīn neighbourhood ("neighbourhood of plasterers"), near Bāb-i Dawlat.[35] We can thus deduce that this neighbourhood was either a sub-district of Bālā Kaft, or in close proximity.[36] This was also the case with the Bāġ-i Qutluġ neighbourhood (now known as Āstāna) situated near Bāb-i Fasā. According to the *Šīrāz-nāma*, Maḥalla-yi Bāl Kat was another sub-district, with the famous Ribāṭ-i Abiš where the Salghurids were buried.[37]

Ibn Zarkūb mentions the neighbourhood of Maḥalla-yi Pīrāsta, situated in the city centre near the Jāmiʿ al-ʿAtīq. Two of the city's eminent religious personalities were interred there. The first was Jamāl al-Dīn Miṣrī (d. 653/1255), the *qāḍī al-quḍāt* of Fārs under the Atabek, Abū Bakr. This personage was buried in a madrasa that had been built under his sponsorship in the Bāzār-i Gīwa-dūzān

30 Mustawfī Qazwīnī, *Nuzhat*: p. 114, trans. p. 113 provides the names of the nine gates. Their names were later changed: Munḏir became Bāb-i Dawlat; Fahandar became Bāb-i Saʿādat; Šūštar became Kāzarūn; and Kawār, Fasā. The name "Bāb-i Naw" reveals that it is a new gate. The Āstāna and Ġassān gates became, respectively, Darāk-Mūsā and Bayḍā. See Afsar, *Tārīḫ*: p. 77.

31 Mustawfī Qazwīnī, *Nuzhat*: p. 114, trans. p. 113.

32 For descriptions of the neighbourhoods of Shiraz, see: Afsar, *Tārīḫ*: pp. 78-80; Ḫūbnaẓār, *Tārīḫ*: pp. 309-10; Amīrī, *Tārīḫ*: 11, p. 130.

33 Ibn Zarkūb, *Šīrāz-nāma*: p. 224.

34 Junayd Šīrāzī, *Šadd al-izār*: pp. 176-7.

35 Ḫūbnaẓār, *Tārīḫ*: pp. 310.

36 Junayd Šīrāzī, *Šadd al-izār*: p. 177. Ibn Zarkūb, *Šīrāz-nāma*: p. 159, mentions a "neighbourhood of plasterers" (*Maḥalla-yi Gačpazān*) in discussing the tomb of Abū Muḥammad Makkī. This is doubtless the same neighbourhood.

37 Ibn Zarkūb, *Šīrāz-nāma*: p. 78. The sources are in marked conflict concerning the location of this edifice. The Ribāṭ-i Abiš was situated in the Umm Kulṯūm cemetery, while the Madrasa-yi ʿAḍudiyya, the shrine where the Salghurid Atabeks were interred, was found in Maḥalla-yi Bāġ-i Naw. Qazwīnī supposes that both places were designated using the same name; see Junayd Šīrāzī, *Šadd al-izār*: pp. 217-8, n. 4.

("Bazaar of slipper makers").[38] The second was Dawūd b. Muḥammad b. Farīd al-Dīn Rūzbihān (d. 671/1272-1273), also highly respected by Abū Bakr, who was laid to rest in the *ḫānaqāh* that he had constructed in this neighbourhood.[39]

In the eighth/fourteenth century, the Baġ-i Naw neighbourhood, developed by the Salghurids, was the place of residence of one of the city's most eminent families, the ʿUmarī, whose ancestry is traced to the caliph ʿUmar.[40] The family had their own mosque here, and the sources also infer that the area had a bazaar.[41] Šayḫ Aḥmad b. ʿAbd Allāh (d. 683/1284-85) was buried in the neighbourhood cemetery, with Junayd Šīrāzī stating that his interment was near Bāb-i Mundir (*i.e.*, Dawlat).[42] This information allows us to situate Maḥalla-yi Bāġ-i Naw east of Maḥalla-yi Bālā Kaft. In this neighbourhood there was also the tomb of Rūzbihān Baqlī. Maḥalla-yi Bāġ, not far from Bāb-i Fasā, located in the southeast of the city, in what would appear to have been a sub-district of Maḥalla-yi Bāġ-i Naw.[43]

The shrine of an important Alid family, the Ḥusaynī-Daštakī, is located in Maḥalla-yi Daštak[44] in the southern part of the city near Bāb Salm.[45] Amīr Ṣadr al-Dīn Muḥammad Ḥusaynī-Daštakī (d. 764/1363) had a madrasa built on the tomb of his ancestors,[46] which, according to Fasāʾī, was called Madrasa-yi Manṣūriyya and founded in the name of his children.[47] Maḥalla-yi Sarrājān ("neighborhood of saddlers"), situated in the north, close to the Jāmiʿ al-ʿAtīq, was considered to be a sub-district of Maḥalla-yi Daštak.[48]

---

38 Ibn Zarkūb *Šīrāz-nāma*: p. 171. Sayyid Niẓām al-ʿAlawī (d. 530/1135-1136), an Alid *naqīb* of Shiraz, was buried in his madrasa, which is also situated in the "Bazaar of slipper makers"; *ibid*: pp. 202-3.

39 Ibn Zarkūb, *Šīrāz-nāma* pp. 180-181.

40 On the *nisba* of this family and the name of the neighborhood, see Junayd Šīrāzī, *Šadd al-izār*: p. 183, n. 2. Note that Ibn Zarkūb and Junayd Šīrāzī were members of this family.

41 In his note on Muḥammad b. Abī l-Fawāris (d. 667/1268-69), Junayd Šīrāzī states that this person's tomb was situated in Maḥalla-yi Bāġ-i Naw "between the market and the cemetery." The *Hazār Mazār* situates it in the middle of Bāzār-i Naw in the vicinity of the Jāmiʿ al-ʿAtīq. See Junayd Šīrāzī 1949, p. 267; ʿĪsā b. Junayd Šīrazī, *Hazār mazār*: p. 312.

42 Junayd Šīrāzī, *Šadd al-izār*: p. 269, where it is named as Bāb-i Jadīd (*i.e.*, Bāb-i Dawlat).

43 Afsar, *Tārīḫ*: p. 79.

44 Fasāʾī is a distant descendant of this family.

45 Junayd Šīrāzī, *Šadd al-izār*: pp. 320-1; Fasāʾī, Mīrzā Ḥasan Ḥusaynī, *Fārs-nāma-yi Nāṣirī*, ed. Muḥammad Rastigār Fasāʾī, 2 vols, (Tehran: Amīr Kabīr, 1367š./1988): pp. 1220-21.

46 Junayd Šīrāzī, *Šadd al-izār*: p. 321.

47 Fasāʾī, *Fārs-nāma*: p. 1220.

48 ʿĪsā b. Junayd Šīrazī, *Hazār mazār*: p. 348. Ibn Zarkūb, *Šīrāz-nāma*: p. 229, mentions "Maḥalla-yi Pālāgarān" as a sub-district of Daštak, but this does not appear to be identical with Maḥalla-yi Sarrājān.

Maḥalla-yi Bāhaliyya took its name from the cemetery that developed around the grave of Šayḫ Bāhalī.[49] Junayd Šīrāzī found no information on him in the sources at his disposal, but assumes that he was a virtuous sheikh who lived before Šayḫ Kabīr. The tomb of the famous grammarian of the Arabic language, Sībawayh (d. ca. 180/796), was also situated in this neighbourhood.[50] As the Bāhaliyya mosque was located near the cemetery at the Bāb-i Kāzarūn gate,[51] Maḥalla-yi Bāhaliyya would have been situated in the west of the city. Maḥalla-yi Maqārīḍī, the locality of the *ḫānaqāh* and tomb of sheikh Abū ʿAbd Allāh Muḥammad b. ʿAbd al-Raḥman Maqārīḍī (d. 411/1020-1021),[52] was probably a sub-district of Maḥalla-yi Bāhaliyya, as it was located between the Bāb-i Kāzarūn gate and the Bāhaliyya cemetery.[53]

The seven cemeteries of Shiraz, which structure the ritual geography in the *Šadd al-īzār* were: Rawḍat al-Kabīriyya, in the north; Maqbarat al-Bāhaliyya,[54] in the west; Maqbara Salm,[55] outside the city walls to the south; Mašhad Umm Kulṯūm, outside the walls to the southeast; Maqbarat Bāġ-i Naw, in the southeast of the city; Maqbarat al-ʿAtīq,[56] in the centre, near the Jāmiʿ al-ʿAtīq and grand bazaar; Maqābirat al-Muṣallā, outside the walls to the northeast.

2.2 *Mosques and Alid Shrines*

2.2.1 Old and New Mosques

The Jāmiʿ al-ʿAtīq ("Old Friday Mosque") was located in the heart of the city in the Sarrajān neighbourhood.[57] The *Šadd al-izār* provides a detailed account of the founding. The author recounts that the Saffarid emir ʿAmr b. al-Layṯ (d. 288/901) had heard of a pious woman from Sarwistān (a town east of

49 Junayd Šīrāzī, *Šadd al-izār*: p. 95; ʿĪsā b. Junayd Šīrazī, *Hazār mazār*: p. 137.

50 Junayd Šīrāzī, *Šadd al-izār*: pp. 45-9; ʿĪsā b. Junayd Šīrazī, *Hazār mazār*: pp. 137-41.

51 Ibn Zarkūb, *Šīrāz-nāma*: p. 154.

52 Junayd Šīrāzī, *Šadd al-izār*: p. 99; ʿĪsā b. Junayd Šīrāzī, *Hazār mazār*: p. 141.

53 It seems that another important person of Shiraz, frequently cited by Junayd Šīrāzī, is also buried in this neighbourhood: this would be Abū Šujāʿ Muḥammad b. Saʿdān Maqārīḍī (d. 509/1115), author of a *Taḏkirat al-mašā'iḫ*. See Junayd Šīrāzī, *Šadd al-izār*: p. 101; ʿĪsā b. Junayd Šīrazī, *Hazār mazār*: p. 142.

54 This cemetery is one of the oldest in Shiraz; Ḫūbnaẓār, *Tārīḫ*: p. 62.

55 The name of this cemetery derives from Salm b. ʿAbd Allāh al-Ṣūfī. He was related to Šayḫ Kabīr; Junayd Šīrāzī, *Šadd al-izār*: p. 132.

56 This cemetery probably developed due to the presence of the shrines of Aḥmad b. Mūsā al-Kāẓim and his brother Muḥammad; see above.

57 The mosque was in ruins by 1935. The only surviving part was the storage area used for copies of the Quran and a few other elements, essentially from the Safavid era. An excavation campaign was carried out: see Wilber, Donald N., *The Masjid-i ʿAtiq of Shiraz* (Šīrāz: Pahlavi University, 1972): pp. 9-31.

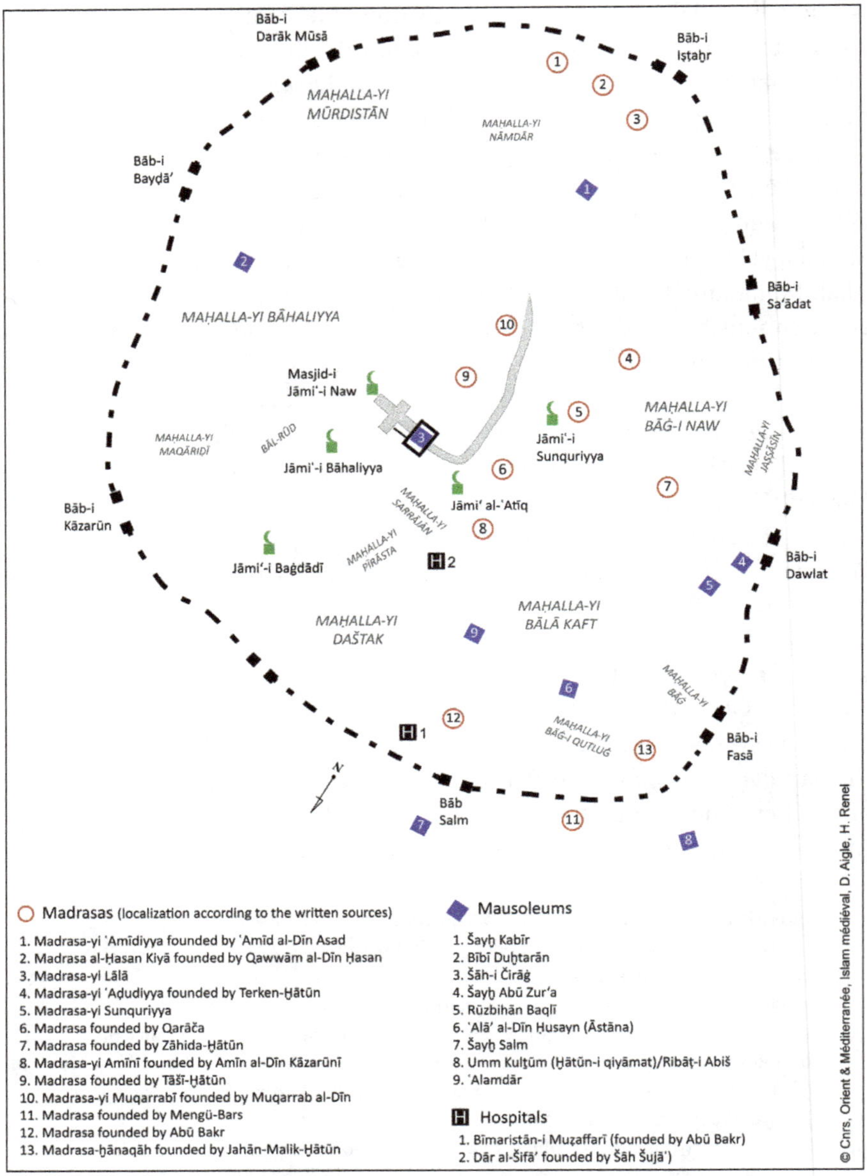

MAP 1 Map of Medieval Shiraz based on textual sources

Shiraz), who had many palm trees, all lawfully gained (*ḥalāl*). The emir sent an emissary to buy them from her for the construction of the mosque, but the woman refused any money, saying that the act of her gift was "*fī sabīl-illāh*". The account informs us that ʿAmr b. al-Layṯ eventually accepted the trees and completed the mosque.[58] Here, we should note that numerous passages of the Quran convey that a future reward is reserved for those who make a gift "in the way of God".[59] While the sources do not allow us to assess the role of the Saffarids in the development of Shiraz, this account also alludes to traditions concerning the construction of the mosque of Medina: in that case the land identified for construction belonged to two orphans who were under the tutelage of Muʿāḏ b. ʿAfrāʾ, but it was later acquired by Abū Bakr, because the Prophet refused to accept it free of charge.[60]

More than two centuries later, the Jāmiʿ al-ʿAtīq was embellished by an unnamed Salghurid Atabek. He had made two silver rings, which were sent to Mecca to be blessed by "the hands of the friends of God" in the holy city.[61] The sacredness of this mosque was reinforced by the presence of three versions of the Quran, hand-copied by the three most important Shi'ite imams, ʿAlī, Ḥusayn, and Jaʿfar al-Ṣādiq. The mosque also possessed a copy of the holy book penned by the caliph ʿUṯmān. According to popular belief, the caliph was reading this Quran at the time of his assassination, and traces of blood could still be seen on it.[62] These Qurans were stored in a room called "Bayt al-maṣāḥif", in the centre of the mosque courtyard.

No religious edifice other than the Jāmiʿ al-ʿAtīq is afforded highly detailed description in the local historiography.[63] While the origin of the account given above is difficult to determine, there was clearly an intention to attain an important Islamic identity for Shiraz, beginning in the fourth/tenth century. This was despite the strong presence of Zoroastrian, Christian, and Jewish communities at the time.[64] The account of the foundation of the Jāmiʿ al-ʿAtīq recalls

58 Junayd Šīrāzī, *Šadd al-izār*: p. 288.

59 Amri, Nelly, *Les saints en Islam, les messagers de l'espérance. Sainteté et eschatologie au Maghreb aux XIVe et XVe siècles* (Paris: Le Cerf, 2008): p. 86.

60 On these different traditions, see Pedersen, J., "Masdjid", in *EI*², 6 (1991): pp. 630-1.

61 Junayd Šīrāzī, *Šadd al-izār*: p. 288; ʿĪsā b. Junayd Šīrāzī, *Hazār mazār*: p. 332. These rings were most certainly intended for a door of the Jāmiʿ al-ʿAtīq.

62 Junayd Šīrāzī, *Šadd al-izār*: p. 288; ʿĪsā b. Junayd Šīrāzī, *Hazār mazār*: p. 332.

63 Junayd Šīrāzī, *Šadd al-izār*: pp. 287-8; Fasāʾī, *Fārs-nāma*: pp. 1206-12. Junayd Šīrāzī claims to draw from the *Niẓām al-tawārīḫ* of Nāṣir al-Dīn Bayḍawī, but the latter text does not provide any description of the Jāmiʿ al-ʿAtīq; Bayḍāwī, Nāṣir al-Dīn Abū ʿAbd Allāh, *Niẓām al-tawārīḫ*, ed. B. Karīmī, B., (Tehran: ʿIlmī, 1313š./1934): p. 87.

64 Muqaddasī, *Aḥsan al-taqāsīm fī maʿrifat al-aqālīm*, ed. de Goeje, *Descriptio imperii moslemici*, BGA VI, (Leiden: Brill, 1877): p. 429.

that of the mosque in Medina. The mosque is also described as a "reliquary" housing prestigious copies of the Quran.[65] It is also a place of holiness because of the *baraka* of the friends of God of Mecca. The Jāmiʿ al-ʿAtīq of Shiraz thus refers to the sacredness of the two holiest cities in Arabia since the time of the Prophet. The act of building a mosque as the foundation of this Islamic city can be conceived as replicating the construction of the Medina mosque. Such traditions symbolically link the Jāmiʿ al-ʿAtīq to the Prophet, source of the light of revelation. From this perspective, it is not surprising that the prayers offered near the minbar were fulfilled.[66]

In front of the shrine of Aḥmad b. Mūsā al-Kāẓim, and to the west of the original mosque, a new mosque, the Masjid-i Jāmiʿ-i Naw in the Bāhaliyya neighborhood, was constructed on the orders of the Atabeg Saʿd b. Zangī (r. 591-628/1196-1231).[67] Ibn Zarkūb suggests that by building this mosque, the Atabeg sought to imitate ʿAlī, whose mosque in Kūfa became the most important in the "Arabic country" (*bilād-i ʿarab*).[68] According to a popular belief reported in the nineteenth century, but probably dating to the Middle Ages, the cypresses planted in the courtyard of the mosque were more than 600 years old; one was called the "Lover" (*ʿāšiq*) and the other the "Beloved" (*maʿšūq*). Dervishes had spent the night under these trees, and it was said that even after death their presence continued.[69] This legend recalls one of the emblematic themes of Ḥāfiẓ's ghazals: that of the Lover's intense feeling (*ʿišq*) for the Beloved, whose slender waist is symbolised by the cypress. The names attributed to the cypresses in the courtyard of the Jāmiʿ al-ʿAtīq thus express the mystical love of the Sufis.

65 On the Qurans of ʿUṯmān, cf. the "*Muṣḥaf* ʿUṯmān", deposited in the mosque of the Umayyads in Damascus in the sixth/twelfth century, see Mouton, J.M., "De quelques reliques conservées à Damas au Moyen Âge: stratégie politique et religiosité populaire sous les Bourides", *Annales Islamologiques*, XXVII (1993): pp. 245-54, here. See also the supposed Quran of ʿUṯmān, in Kufic script, at Tashkent.

66 Junayd Šīrāzī, *Šadd al-izār*: p. 287; ʿĪsā b. Junayd Šīrāzī, *Hazār mazār*: p. 331; Ḥamd Allāh Mustawfī Qazwīnī, *Nuzhat*: p. 115, trans.: p. 116.

67 Waṣṣāf, *Tajziyat al-amṣār wa tazjiyat al-aʿṣār*, ed. M. M. Iṣfahānī, lithograph (Bombay, 1852): p. 155; Mustawfī Qazwīnī, *Nuzhat*: p. 115, trans.: p. 116; ʿĪsā b. Junayd Šīrāzī, p. 126. Begun in 598/1104-05, it was completed in 615/1121-2; see Kamālī-Sarwistānī, *Dāniš-nāma*: pp. 373-375; Amīrī, *Tārīḫ*: II, p. 136.

68 Ibn Zarkūb, *Šīrāz-nāma*: p. 78.

69 William Francklin, *Observations*: pp. 64-5.

2.2.2 Šāh-i Čirāġ

The descendants of the Prophet (the Alids) play a fundamental role in the sanctity of Shiraz. According to the local sources, after the death of the eighth imam, ʿAlī al-Riḍā b. Mūsā al-Kāẓim, in 203/818, three of his brothers, Ḥusayn,[70] Aḥmad[71] and Muḥammad,[72] sought refuge in Shiraz after the 'betrayal' of the Abbasid caliph al-Ma'mūn who had promised to make ʿAlī al-Riḍā his successor. Beginning in the seventh/thirteenth century, the shrine of Aḥmad b. Mūsā al-Kāẓim, known then as now as the *Šāh-i Čirāġ* ("King of light"), became the most venerated holy site for the people of Shiraz.[73]

The "discovery" of his grave dates from the Salghurid period, when Muqarrab al-Dīn Masʿūd, the vizier of the Atabeg Abū Bakr (r. 1231-1260), identified it between the two *masjid-i jāmiʿ*. Junayd Šīrāzī tells of a light emanating from a small mound. This led the faithful to dig into the earth, thereby revealing an intact body holding a copy of the Quran in one hand, and a seal with the engraving "al-ʿIzzat Allāh Aḥmad b. Mūsā" in the other.[74] Abū Bakr ordered the construction of a shrine on the site.[75] This narrative of the discovery of the grave of Aḥmad b. Mūsā al-Kāẓim corresponds to the traditional model describing the identification of graves that shelter the real or supposed remains of the descendants of ʿAlī. Such shrines were often built (or discovered) following a dream or miraculous sign, as in the case of the grave of Aḥmad b. Mūsā al-Kāẓim, or that of ʿAlī in Mazār-i Šarīf, found in a village near Balkh in the Saljuq period.[76] The phenomena illuminating the site indicate the sacredness of the place, while the preservation of the intact body attests to the saintliness of the individual. These signs serve as part of the mechanism of authentication for

70 Ibn Zarkūb, *Šīrāz-nāma*: p. 199; Junayd Šīrāzī, *Šadd al-izār*: pp. 260-1.

71 Ibn Zarkūb, *Šīrāz-nāma*: pp. 196-8; Junayd Shīrāzī, *Šadd al-izār*: pp. 289-92.

72 Ibn Zarkūb, *Šīrāz-nāma*: p. 199; Junayd Šīrāzī, *Šadd al-izār*: p. 292.

73 The shrine in its current state is described by Taqī Muṣṭafawī, Muḥammad, *Iqlīm-i Fārs*, (Tehran: Taban Press, 1964): pp. 63-5. On the history of this shrine and the description of the different modifications made since medieval times, see Kamālī-Sarwistānī, *Dāniš-nāma*: pp. 371-4; Amīrī, *Tārīḫ*: II, pp. 224-51.

74 Junayd Šīrāzī, *Šadd al-izār*: 290. Normally, ʿIzzat Allāh would be used in the *status constructus*; an almost identical account is found in Ibn Zarkūb, *Šīrāz-nāma*: pp. 196-9.

75 Ibn Zarkūb, *Šīrāz-nāma*: p. 198; Fasā'ī, *Fārs-nāma*: pp. 1186-7; Furṣat al-Dawla Šīrāzī, *Āṯār-i ʿAjam*, ed. ʿAlī Dibāšī, 3 vols (Tehran: Bāmdād, 1968): pp. 445-8.

76 On the discovery of ʿAlī's grave, see McChesney, R. D., *Waqf in Central Asia. Four Hundred Years in the History of a Muslim Shrine, 1480-1889* (Princeton: Princeton University Press, 1991): pp. 26-36; Subtelny, Maria, *Timurids in Transition. Turko-Persian Politics and Acculturation in Medieval Iran* (Leiden: Brill, 2007): pp. 208-12.

sacred centres,[77] and indeed the Alid shrines were discovered using the same process in Egypt and Syria.[78] The foundation narrative for these Alid shrines and for holy places in general share common features of visions, dreams, and miraculous events. The "discovery" of the graves of the illustrious dead is indeed a *Sitz im Leben* of the hagiographical and edification literature.

The shrine of Aḥmad b. Mūsā al-Kāẓim became a sacred space renowned as a "place of benedictions" (*maḥall al-barakāt*).[79] It attracted many pious visitors, as well as Shirazis, engaged in continuous recitation of the Quran.[80] In his early eighth/fourteenth century description of the shrines of the city, Ibn Baṭṭūṭa notes that the most important was that of Aḥmad b. Mūsā al-Kāzim. He writes that "his mausoleum (*mašhad*) is venerated by the inhabitants of the city. They seek God's favour through the sanctity of this monument."[81] He describes the ritual performed there every Sunday evening, attended by the mother of Šayḫ Abū Isḥāq Īnjū, the pious Tāšī-Ḫātūn.[82] The Quran reciters read from the holy book, and then dishes, fruits, and sweets were shared. After the meal the audience listened to the preacher's sermon, while Tāšī-Ḫātūn stood behind a grilled window, in an upper chamber over the hall. Then, Ibn Baṭṭūṭa recounts, timbals were sounded, and a bugle and trumpet were played at the door of the tomb, as was also done at the doors of kings.[83]

In the medieval sources, this tomb is never designated by the name of Šāh-i Čirāġ. This appellation only appear later in the works of Fasāʾī and Furṣat al-Dawla Šīrāzī. The latter describes it as the "blessed tomb Šāh-i Čirāġ, sepulcher of Ḥaḍrat Sayyid Amīr Aḥmad, known as the King of light and Sayyid of Sayyids."[84] This reference to light in the name of Aḥmad b. Mūsā al-Kāẓim's

77 See Meri, Josef W., *The Cults of Saints among the Muslims and Jews in Medieval Syria* (Oxford: Oxford University Press, 2002): pp. 43-7.

78 See Rāġib, Yūsuf, "Al-Sayyida Nafīsa, sa légende, son culte et son cimetière", *StIs* 44 (1976): pp. 61-86 and 45 (1977): pp. 27-55; Mulder, Stephennie, *The Shrines of the ʿAlids in Medieval Syria: Sunnis, Shiʿis and the Architecture of Coexistence* (Edinburgh: Edinburgh University Press, 2014); Talmon-Heller, Daniella, "Graves, Relics and Sanctuaries: the Evolution of Syrian Sacred Topography (eleventh-thirteenth centuries)", *ARAM Periodical* 18-9 (2006-7): pp. 601-20.

79 Expression borrowed from Amri, Nelly, *Sîdî Abû Saʿîd al-Bâjî (1156-1231)* (Sousse: Contrastes-Éditions, 2015): p. 159.

80 On the recitation of the Quran, see Nelson, Kristina, *The Art of Reciting the Qurʾān* (Austin: University of Texas Press, 1985).

81 Ibn Baṭṭūṭa, *The Travels of Ibn Baṭṭūṭa*, trans. Gibb, H.A.R., 2 vols (Cambridge: Hakluyt Society, 1959): II, p. 313.

82 Tašī-Ḫātūn had her tomb covered by a dome (*qubba*); Junayd Šīrāzī, *Šadd al-izār*: p. 292, n. 1.

83 Ibn Baṭṭūṭā, *Travels*: II, p. 314.

84 Furṣat al-Dawla Šīrāzī, *Āṯār*: p. 443.

tomb alludes to Shiite Imamite theories on the twelve imams. According to several traditions, as soon as an imam is born, God makes a "column of light" appear for him; the imam is thus a "lamp of light."[85] As descendants of Muḥammad, the Alids are bearers of the "*nūr al-walāya*" ("light of the Imamite mission") transmitted from imam to imam.[86]

2.2.3 Other Alid Shrines

The tomb of Muḥammad b. Mūsā al-Kāẓim, the brother of Aḥmad b. Mūsā al-Kāẓim, was also situated in the city centre, near the grand bazaar.[87] Junayd Šīrāzī calls it the "blessed tomb" where the "noblest Sayyids" and the "most pious of men" are laid to rest. Similarly to the case of the Jāmiʿ al-ʿAtīq, any vows uttered here held the promise of fulfilment.[88] And, like the Šāh-i Čirāġ, luminous phenomena attested to the sacredness of the place: on Friday nights, lights could be seen rising from a nearby hillock.[89]

The mausoleum of the third brother, Ḥusayn b. Mūsā al-Kāẓim, was situated in the Bāġ-i Qutluġ neighbourhood.[90] According to Junayd Šīrāzī it was surrounded by pleasant gardens, with fruit trees of all kinds.[91] This characteristic is typical of shrines, which often demonstrate privileged relationships with nature, such as in the growth of a specific tree or flowers, and often situated near a hillock or source of pure water.

The mausoleum of one of the brothers' nephews, ʿAlī b. Ḥamza b. Mūsā al-Kāẓim, was also discovered in Shiraz. He and a few friends had sought refuge in a cave near the city,[92] but were killed by Abbasid agents. His "blessed tomb" is located in the Muṣallā cemetery, outside the city and Bāb-i Iṣṭaḫr.[93] The author

85 Amir-Moezzi, Mohammad Ali, *Le guide divin dans le shîʿisme originel*, (Paris: Verdier, 1992): pp. 151-2.

86 On the role of light in Imamite Shiism, see Rubin, Uri, "Pre-existence and Light. Aspects of the Concept of Nūr Muḥammad", *Israel Oriental Studies*, v (1975): pp. 62-119; Amir-Moezzi, *Le guide*: pp. 73-154.

87 Ibn Zarkūb, *Šīrāz-nāma*: p. 199; Junayd Šīrāzī, *Šadd al-izār*: p. 292; Fasāʾī, *Fārs-nāma*: pp. 1187-8; Kamālī-Sawistānī, *Dāniš-nāma*: pp. 271-4.

88 Ibn Zarkūb, *Šīrāz-nāma*: p. 199; Junayd Šīrāzī, *Šadd al-izār*: p. 292.

89 Junayd Šīrāzī, *Šadd al-izār*: pp. 260-1.

90 Junayd Šīrāzī, *Šadd al-izār*: pp. 260-1; Fasāʾī, *Fārs-nāma*: pp. 1188-9; Furṣat al-Dawla Šīrāzī, *Āṯār*: pp. 453-4; Kamālī-Sawistānī, *Dāniš-nāma*: p. 232. Several manuscripts attribute the *laqab* ʿAlāʾ al-Dīn to Ḥusayn b. Mūsā; Junayd Šīrāzī, *Šadd al-izār*: p. 260, n. 4.

91 Ibn Zarkūb, *Šīrāz-nāma*: p. 199.

92 Ibn Zarkūb, *Šīrāz-nāma*: pp. 200-1; Junayd Šīrāzī, *Šadd al-izār*: pp. 366-8; Fasāʾī, *Fārs-nāma*: pp. 1189-90; Furṣat al-Dawla Šīrāzī, *Āṯār*: p. 457.

93 Ibn Zarkūb, *Šīrāz-nāma*: pp. 200-1; Junayd Šīrāzī, *Šadd al-izār*: pp. 366-368; Ibn ʿInaba, *ʿUmdat*: p. 217. Another descendant of Mūsā al-Kāzim, a certain Abū Jaʿfar Muḥammad Ṣūrānī, was also buried near Bāb-i Iṣṭaḫr; Ibn ʿInaba, *ʿUmdat*: p. 221. According to Junayd

of the *Šadd al-izār* describes his execution in hagiographical style: decapitated, with his head sent to Damascus.[94] The account given is evidently an allusion to the martyrdom of Ḥusayn by the Umayyads, in 680, however it involves a hagiographical commonplace, since the (Abbasid) caliph resided in Baghdad during the period in question. After his execution, the voice of ʿAlī b. Ḥamza was forever heard proclaiming: "There is no God but God."[95]

The city also had several shrines for Alid women venerated by the Shirazis. According to Ibn Zarkūb, Sayyida Umm Kulṯūm, a woman of Ḥasanī ancestry, came to Shiraz after the murder of her family.[96] She remained there in disguise, dedicating her life to the praise of God. Šayḫ Aḥmad b. al-Ḥusayn (d. 367/977-978), a relative of Šayḫ Kabīr, reportedly hid her and placed himself at her service.[97] But eventually, pursued by Abbasid agents who had come to Shiraz intending to execute her, she fell into a well and died. According to the author of the *Fārs-nāma-yi Nāṣirī*, her tomb was known at the time as "Ḫātūn-i qiyāmat,"[98] and there was even a cemetery bearing her name (Mašhad Umm Kulṯūm). However, the source information on the precise location of her tomb is conflicting.

The tomb of another Alid woman, Sayyida Umm ʿAbd Allāh Ḥusniyya, known by the name of Bībī Duḫtarān, was situated in the middle of a large garden.[99] The sanctity of this site is illustrated in an anecdote recounted by Junayd Šīrāzī. In this narrative, a woman, condemned to be stoned, begged God for pardon. After the event, her body was buried near that of Bībī Duḫtarān. A few days later a poor woman settled nearby. She built a small enclosure in which she spent her lifetime serving the poor in the name of God. She asked to be interred near the graves of Bībī Duḫtarān and the woman who had been stoned. According to popular rumour, lights radiated from the shrine where the three women are buried as a sign of their sainthood. Once again, prayers made there are fulfilled. Bībī Duḫtarān's esteemed Alid status extended to the

---

Šīrāzī, "out of respect" for the *Ahl al-bayt*, ʿAḍud al-Dawla had a wall erected around his *turba*. See also Fasāʾī, *Fārs-nāma*: p. 1191.

94 Junayd Šīrāzī, *Šadd al-izār*: p. 367.

95 Junayd Šīrāzī, *Šadd al-izār*: p. 367.

96 Ibn Zarkūb, *Šīrāz-nāma*: p. 199; Junayd Šīrāzī, *Šadd al-izār*: pp. 161-162; Fasāʾī, *Fārs-nāma*: p. 1192. Mustawfī Qazwīnī (in Junayd Šīrāzī, *Šadd al-izār*: p. 162, n. 1) supposes that she fled to Shiraz during the reign of the Buyid ʿImād al-Dawla (r. 322-338/934-949).

97 Junayd Šīrāzī, *Šadd al-izār*: pp. 162-163. He lived in a *ḫānaqāh* and copied versions of the Quran to support himself.

98 Fasāʾī, *Fārs-nāma*: p. 1193; Furṣat al-Dawla Šīrāzī, *Āṯār*: p. 458.

99 Junayd Šīrāzī, *Šadd al-izār*: pp. 113-4; ʿĪsā b. Junayd Šīrāzī, *Hazār mazār*: pp. 154-5. Fasāʾī, *Fārs-nāma*: p. 1192; Furṣat al-Dawla Šīrāzī, *Āṯār*: p. 457-8. According to Amīrī, *Tārīḫ*, II, p. 206, Bībī Duḫtarān was a Timurid princess, whose tomb was erected during this period.

two other women, who benefited from her spiritual impulse (*baraka*) after death. This feminine shrine, situated in the Bāhaliyya cemetery near Bāb-i Bayḍā and the ramparts,[100] was also considered sacred at the time of Mubāriz al-Dīn Muẓaffar (d. 765/1363), a follower of rigorous Sunni Islam. His wife, Ḫātūn-Qutluġ-Bek,[101] had a high dome constructed above the shrine of these saintly women, as well as a nearby madrasa, which was very popular. The positioning of a madrasas near a shrine was in fact common, both representing important institutions of religious life.

2.2.4 Haft tanān and Čihil tanān

Pilgrimage guides mention numerous shrines in which anonymous individuals with miraculous powers were interred: for example, the shrine of the "Forty young girls" (Čihil duḫtarān) near Herat. People considered it to be a "blessed tomb" (*mazār-īst mutabarrak*) and went there in search of grace. Such shrines do not commemorate the memory of a specific individual, but rather a type of persona endowed with special virtues in the eyes of the community.[102]

One such case at Shiraz was the tomb of the "Seven Men" (*al-rijāl al-sabaʿa*, pers. *haft tanān*), in the Muṣallā cemetery beyond the city walls.[103] The *Šadd al-izār* recounts the traditions of this place in detail.[104] One of these was that a pious man, a washer of the dead at the Iṣṭaḫr gate, was called by a young man whose face bore traces of devotion and the light of *walāya*. The two men departed together for an enclosure (*muḥawwaṭat*) near the Muṣallā cemetery. The youth asked his companion to wait for a voice, which would allow them entry. But when the man entered, he saw that the youth was dead, with his body facing the *qibla*. Six other people appeared, carrying a shroud and fragrances, and these helped him wash the body of the young man. After together wrapping the body in the shroud, the six individuals carried it outside the enclosure. The washer of the dead then left, but the walls and other people had disappeared. He prayed throughout the night, and the next morning he saw a tomb covered with fresh water. Soon afterward so it is said, a new tomb covered with fresh water appeared, and this continued until there were seven tombs in total. The tombs are said to be the seven "pillars" (*awtād*), which guide people

100 Junayd Šīrāzī, *Šadd al-izār*: p. 114.

101 She was the daughter of Quṭb al-Dīn Šāh-Jahān b. Jalāl al-Dīn Suyurġatmiš, the Qara Khitai sovereign of Kirmān. She died in 785/1383; see Mustawfī Qazwīnī, in Junayd Šīrāzī, *Šadd al-izār*: p. 114, n. 1 and 2.

102 Isfizārī, *Rawżāt*, I, p. 145.

103 Amīrī, *Tārīḫ*, II, pp. 203-5.

104 Junayd Šīrāzī, *Šadd al-izār*: pp. 409-411; ʿĪsā b. Junayd Šīrāzī, *Hazār mazār*: pp. 433-4.

away from misfortune and bring rain. However, in truth there are only four *awtād* in the spiritual hierarchy of saints. Junayd Šīrāzī claims to draw from oral sources in providing this narrative, however his informers had evidently confused the *awtād* with the "substitutes" (*abdāl*), which are indeed seven in number. Regardless of the confusion, this tradition seeks to place the city of Shiraz under the protection of the hierarchy of invisible saints, who support the entire world.[105]

Located close to this tomb is another shrine sheltering the remains of anonymous individuals. Known by the name of the "Forty people" (*čihil tanān*), it was first mentioned in the *Fārs-nāma-yi Nāṣirī*[106] and *Āṯār-i ʿAjam*.[107] Karīm Ḫān Zand had this small *takiyya* with iwans around its courtyard constructed for dervishes. Situated in a garden planted with trees, to the northeast of Shiraz, it is a short distance away from the tomb of Ḥāfiẓ and 200 metres from the 'Haft tanān' mausoleum. According to Furṣat al-Dawla Šīrāzī, it shelters the remains of a certain Jamāl al-Dīn Maḥmūd b. Naṣīr al-Dīn Muḥammad Afzarī who lived in the eighth-ninth/fourteenth century.[108] The building of this shrine, in proximity to the grave of the Haft tanān, signifies that the location was considered to be a place filled with sacredness, stemming from the miraculous fluid (*baraka*) that emanated from the graves of the city's patron saints.

2.3 *Mausoleums of Sufis and poets*

The spiritual landscape of the medieval city, in addition to the Friday mosques and the shrines with origins marked by supernatural and indeed miraculous features, was also integrated with mausoleums of other illustrious Shirazis.

2.3.1 Šayḫ Kabīr

The major spiritual figure of Shiraz was Abū ʿAbd Allāh Muḥammad b. Ḫafīf, otherwise known as "al-Šayḫ al-Kabīr" ("the Great Sheikh") (d. 371/982). Sulamī, who probably met this person during his visit to Shiraz around 370/981, attested that he was the "supreme master, the memory of his time, the oldest of Sufis, with the most perfect spiritual state." Indeed, Sulamī records this Great

105 A life story of Rūzbihān Baqlī refers to this legend; Šaraf al-Dīn Ibrāhīm b. Rūzbihān II, *Tuḥfat ahl al-ʿirfān fī ḏikr sayyid al-aqṭāb Rūzbihān*, in *Rūzbihān-nāma*, ed. M. Dāniš-Pažūh (Tehran, Silsila-i Intišārāt-i Anjuman-i Āṯār-i Millī, 1969): pp. 1-149, 55-57.

106 Fasā'ī, *Fārs-nāma*: pp. 1202-03.

107 Furṣat al-Dawla Šīrāzī, *Āṯār*: p. 443. Concerning this shrine, see Afsar, *Tārīḫ*: p. 117; Amīrī, *Tārīḫ*: II, pp. 202-3. "Čihil tanān" indicates an interment place of saints unknown in the spiritual hierarchy; see Dihḫudā, *Luġat-nāma* (Tehran: Tehran University Publications, 1988), VI, p. 8385.

108 Furṣat al-Dawla Šīrāzī, *Āṯār*: p. 474.

Sheikh as the thousandth and final master in his *Ta'rīḫ al-ṣūfiyya*.[109] Much later, Ibn Baṭṭūṭa reported that his tomb attracted many pilgrims:

> Another of the shrines of Šīrāz is the mausoleum of the imām and pillar (*al-quṭb*), the saint (*al-walī*) Abū 'Abd Allāh b. Ḫafīf, who is known by them [simply] as "the šayḫ." They come to it morning and evening, and rub their hands on it [for blessing]. I myself saw the Qāḍī Majd al-Dīn [Ismā'īl Fālī][110] come to it on visitation and kiss it.[111]

The Moroccan traveller was impressed by the charisma of Šayḫ Kabīr, evident in popular belief. He narrates a miracle accomplished by the holy man in the island of Ceylon.[112] The account is clearly apocryphal, since Šayḫ Kabīr never visited the Indian Subcontinent, however it emphasises the links between Shiraz and the ports of the Indian Ocean, from Zanzibar to Ceylon. The mausoleum of Šayḫ Kabīr, situated in the north of Shiraz and south of Bāb-i Iṣṭaḫr,[113] also lent its name to the al-Rawḍat al-Kabīriyya cemetery.[114] Through time, according to Junayd Šīrāzī, many people were buried alongside the blessed tomb of Šayḫ Kabīr, beginning with his close servants Abū Aḥmad al-Kabīr[115] and Abū Aḥmad al-Ṣaġīr.[116]

The hagiographical sources recount numerous anecdotes contributing to the holiness of the family, from both genealogical and spiritual perspectives. The accounts surrounding the sheikh's birth and family are indeed legendary. His father was a commandant in the army of the Saffarid 'Amr b. al-Layṯ, and his mother, Umm Muḥammad, was the daughter of a *karrāmiyya* sheikh from Nishapur. She gave birth after Shiraz was captured by the Saffarids.[117] This

109 Thibon, Jean-Jacques, *L'oeuvre d'Abū 'Abd al-Raḥmām al-Sulamī 325/937-412/1021 et la formation du soufisme* (Damas: Institut Français du Proche-Orient, 2009): p. 113. The *Ta'rīḫ al-ṣūfiyya* did not survive, but the sources included numerous references to this work; ibid.: 568, n. 28.

110 Majd al-Dīn Ismā'īl Fālī (d. 756/1335) was the *qāḍī al-quḍāt* of Fārs who opposed Ilkhan Öljeitü's will to impose Shiism; note in Ibn Zarkūb, *Šīrāz-nāma*: pp. 127-9; Junayd Šīrāzī, *Šadd al-izār*: pp. 423-6; 'Īsā b. Junayd Šīrāzī, *Hazār mazār*: pp. 144-6; Subkī, Tāj al-Dīn, *Ṭabaqāt al-šāfi'iyya al-kubrā*, ed. Nūr al-Dīn Šarība (Cairo: al-Maṭba'a al-Ḥusainīya, 1323-1324/1906), IX: pp. 400-3.

111 Ibn Baṭṭūṭā, *Travels*, II: p. 314.

112 Ibn Baṭṭūṭa, *Travels*, II: p. 315.

113 Afsar, *Tārīḫ*: p. 83.

114 See the description of the shrine in Amīrī, *Tārīḫ*, II: pp. 187-8.

115 Junayd Šīrāzī, *Šadd al-izār*: pp. 46-7.

116 Junayd Šīrāzī, *Šadd al-izār*: pp. 47-8.

117 Daylamī, *Sīrat al-Šayḫ al-Kabīr Abū 'Abd Allāh b. Ḫafīf al-Šīrāzī*, ed. Schimmel, A. (Ankara: Türk Tarih Kurumu Basımevi, 1955): 9; Junayd Šīrāzī, *Šadd al-izār*: pp. 368-71.

particularly pious woman accompanied her son to Mecca to contemplate God.[118] In the family home, she also received the grace of the vision during the "night of destiny" (*laylat al-qadr*), when celestial light filled the world. She was buried in the "arcade of saints" (*qanṭarat al-awliyā'*), in a small enclosure near the Iṣṭaḫr gate of Musallā cemetery, which is also the site of the tomb of the "Haft tanān".[119] The selection of this location, near the *awliyā'*, is another sign of the recognition of the sainthood of Umm Muḥammad for the people of Shiraz.

A curious anecdote describes the circumstances of Šayḫ Kabīr's marriage. One night he dreamed of a child, who made his parents cross "*al-ṣirāṭ*" on the Day of Resurrection. On awakening, he said to his servant: "Find me a girl to marry!" The servant, unable to find a spouse in the middle of the night, offered his own daughter. The sheikh accepted and married her that very night. After seven months, the young woman brought a son into the world, however this child died in youth. In the *Šadd al-izār*, Junayd Šīrāzī provides a note on a son of Šayḫ Kabīr,[120] calling him "al-Šayḫ 'Abd al-Salām" and saying that he had acquired *'ilm* and *ma'rifa*. 'Abd al-Salām was buried near the sheikh's favorite servant, Abū Aḥmad al-Ṣaġīr, opposite the "face of his mother"(*rūy-i mādar-i ḫud*).[121] Given 'Abd al-Salām's interment near Abū Aḥmad al-Ṣaġīr's tomb, it may well be that his mother was the daughter of this faithful servant. As with other anecdotes, authenticity could be questionable, however the account of this son, born from the union of Šayḫ Kabīr and his servant's daughter, seeks to establish both carnal and spiritual parentage. Junayd Šīrāzī also presents Abū Aḥmad al-Ṣaġīr as a full member of Šayḫ Kabīr's household. One day, the latter said: "How can I not love Abū Aḥmad? For thirty years, he covered me and my family with blankets." Abū Aḥmad al-Ṣaġīr was buried near Abū Aḥmad al-Kabīr, the second faithful servant of the sheikh. The two tombs were located in the blessed shrine of Šayḫ Kabīr (*bi-ḥaẓīrati-hi al-muqaddasat*), behind that of the holy man, with their bodies having the same *qibla*.[122] The burials of Šayḫ Kabīr, his son, and perhaps also the latter's mother, as well as the two servants who shared for decades in the life of their master, thus probably served as the central core of the Rawḍat al-Kabīriyya cemetery.

The sacredness of this holy man's grave was recorded only a few decades after his death, in the form of a meaningful anecdote. Šayḫ 'Imād al-Dīn Abū

118 Daylamī, *Sīrat*: pp. 5, 220, 368.

119 Junayd Šīrāzī, *Šadd al-izār*: p. 371.

120 Junayd Šīrāzī, *Šadd al-izār*: pp. 48-9.

121 Junayd Šīrāzī, *Šadd al-izār*: p. 48.

122 Junayd Šīrāzī, *Šadd al-izār*. p. 48. Concerning the position of the bodies in the tombs, see Halevi, Leor, *Muhammad's Grave. Death Rites and the Making of Islamic Society* (New York: Columbia University Press, 2007): pp. 182, 188-9.

Ṭāhir ʿAbd al-Salām (d. 661/1263) was buried opposite the gate (*šabaka*) of Šayḫ Kabīr's tomb. He had taken a stone from the tomb of Šayḫ Kabīr, and had placed it on the head of his own intended tomb to benefit from the *baraka* of the holy man.[123] In a dream he saw Šayḫ Kabīr, who said: "Who made you more deserving of this stone than the saints (*awliyāʾ*) and the poor (*fuqarāʾ*)?" ʿImād al-Dīn Abū Ṭāhir thus realised his duty toward the holy man's shrine and donated a valuable mill to it. The significance of the anecdote was that, if one wished to benefit from the miraculous *baraka* of Šayḫ Kabīr, then their duty was to contribute to the continuation of worship at the tomb.

Similarly to the Šāh-i Čirāġ (see above), the shrine of Šayḫ Kabīr became an important site of religious conviviality. Ibn Baṭṭūṭa described the ritual performed at the holy man's grave, and noted that Tāšī-Ḫātūn visited the mausoleum every Thursday evening.[124] A *ḫānaqāh* and madrasa were built near the shrine of the patron saint of Shiraz, where qadis and jurists would congregate.[125] Although Ibn Baṭṭūṭa does not identify who erected these buildings, the person responsible may have been the Salghurid Atabek Zangī b. Mawdūd. In the *Nuzhat al-qulūb*, Mustawfī Qazwīnī writes that Zangī b. Mawdūd had constructed a building with endowments, near the tomb of Šayḫ Kabīr.[126] The entire group of buildings constituted an important complex for practice of religious rituals and teaching. At the *ḫānaqāh*, food was distributed to the destitute, in keeping with a well-established custom of Shiraz.

2.3.2 The Shrine of the Rūzbihāniyya

Rūzbihān Baqlī is the second emblematic saint of Shiraz, often referred to as the "sheikh of ecstatic sayings" (*šayḫ-i šaṭṭāḥ*). His grandson Šaraf al-Dīn Ibrāhīm, the author of a long hagiography composed 94 years after his grandfather's death, describes his fame. He was "the guide of the masters of the path" (*muqtadā-yi arbāb-i ṭarīqat*), renowned for his miraculous gifts and spiritual states (*karāmāt u aḥwāl*).[127] After Rūzbihān Baqlī's death in 606/1209, his descendants perpetuated the "order" of Rūzbihāniyya at the shrine of the holy man.[128]

---

123 Junayd Šīrāzī, *Šadd al-izār*: pp. 56-7; ʿĪsā b. Junayd Šīrazī, *Hazār mazār*: pp. 98-99.

124 The tomb of Maḥmūd-Šāh Īnjū was adjacent to Šayḫ Kabīr's shrine; see Ibn Baṭṭūṭā, *Travels*, II: p. 314.

125 Ibn Baṭṭūṭā, *Travels*, II: p. 314.

126 Mustawfī Qazwīnī, *Nuzhat*: p. 116.

127 Šaraf al-Dīn Ibrāhīm b. Rūzbihān II, *Tuḥfat*: p. 10.

128 On the Rūzbihāniyya, see Ernst, Carl, *Ruzbihan Baqli. Mysticism and the Rhetoric of Sainthood in Persian Sufism* (Richmond: Curzon, 1996).

Rūzbihān Baqlī's grave is located in a *ḫānaqāh*, built for him in 560/1165, next to the Ḥaddāš b. Manṣūr gate[129] in the Bāġ-i Naw neighbourhood in the southeast of Shiraz.[130] His son Faḫr al-Dīn Aḥmad (d. 645/1247), and grandson Ṣadr al-Dīn Ibrāhīm (d. 685/1286), were buried next to him. His great-grandson ʿIzz al-Dīn Masʿūd expanded and restored the *ribāṭ*, which had become a spiritual centre and place of pilgrimage. The section added to the shrine was called the "new *ribāṭ*" (*al-jadīd*), while the burial of Rūzbihān was in the "old *ribāṭ*" (*al-qadīm*).[131]

The first miracle accomplished by the sheikh occurred during the construction of this very *ribāṭ*. The beam intended to cover the audience hall (*bārgāh*) was too short, so the sheikh threw his prayer mat over the end of the wood. When he removed the mat, the beam had become long enough to reach the wall. According to tradition, a piece of this wood remained. Whenever someone was ill in Shiraz, he or she could come to the *ribāṭ*, where a piece of the wood would be burned and their illness would disappear. These stories on the origin of the order instituted by Rūzbihān have a supernatural dimension. Šaraf al-Dīn Ibrāhīm, author of the noted hagiography, recounts that at the time when his grandfather chose the practices of sobriety (*ṣahw*) over inebriation (*sukr*), he had a *ribāṭ* built for the *awliyā'*.[132] It was also at this time that the sheikh defined the principles of the Rūzbihāniyya practice, founded on knowledge rather than the mystical inebriation of his initial spiritual orientation. Ibn Baṭṭūṭa observed the practices of this order in 725/1325, at the time of his visit to the shrine of Rūzbihān: "his grave is a mosque where Friday prayers are held, and the *ḫuṭba* is delivered." Ibn Baṭṭūṭa heard the great qadi Majd al-Dīn Ismāʿīl Fālī read the *Musnad* of Abū ʿAbd Allāh Muḥammad Šāfiʿī and the

129 Junayd Šīrāzī, *Šadd al-izār*: p. 240; ʿĪsā b. Junayd Šīrazī, *Hazār mazār*: pp. 286-7. The *laqab* signifies he who "cut the head of enemies with his arm." For a description of the tomb and its improvements, see Amīrī, *Tārīḫ*, II: pp. 189-93.

130 Šaraf al-Dīn Ibrāhīm b. Rūzbihān II, *Tuḥfat*: p. 14; Šams al-Dīn ʿAbd al-Laṭīf b. Rūzbihān II, *Rūḥ al-jinān fī sīrat al-Šayḫ Rūzbihān*, in *Rūzbihān-nāma*, ed. M. Dāniš-Pažūh (Tehran: Silsila-i Intišārāt-i Anjuman-i Āṯār-i Millī, 1969): pp. 152-370, 217.

131 Junayd Šīrāzī, *Šadd al-izār*: p. 247. Excavations at the site revealed the graves of seven individuals, six of which identified by inscriptions: Rūzbihān and his son, Faḫr al-Dīn Aḥmad; a grandson, Ṣadr al-Dīn Ibrāhim Rūzbihān al-Ṯānī; two great-grandsons, Jalāl al-Dīn Yaḥyā, Šaraf al-Dīn Ibrāhīm; a great-great-grandson, ʿIzz al-Dīn Masʿūd. The seventh tomb bore an illegible inscription, apart from the date 701/1301-13; cf. Corbin, Henry and Moʿīn, Mohammed, *Le jasmin des fidèles d'amour* (Téhéran: Éditions Manoutcheri, 1987): p. 72; Amīrī, *Tārīḫ*, II: p. 193.

132 Šaraf al-Dīn Ibrāhīm b. Rūzbihān II, *Tuḥfat*: p. 14.

Mašāriq al-anwār of Raḍī al-Dīn Abu'l-Faḍā'il al-Ḥasan al-Ṣaġānī (d. 650/1252).[133] Rūzbihān was buried in an enclosure within an open-air area to the south of the main building of the *ribāṭ*. As he was dying, he ordered: "Put me there, as that is where I often saw the messenger of God."[134] Rūzbihān himself indicated the ritual to be performed by pilgrims at his tomb, to obtain his intercession. Speaking to the disciples and followers gathered at his deathbed, he said:

> Whoever needs something should draw water from the well of this *ribāṭ* and use it for his ablutions. He should then prostrate twice at the head of my grave (*bālīn-i man*) and twice at the foot of my grave (*pāyīn-i man*), so that I will intercede for him and his prayer will be granted. This promise was made to Rūzbihān (by God): "Whoever visits your grave after your death or comes there on pilgrimage with a sincere heart [this] will be granted for you."[135]

The author of the *Šadd al-izār* records anecdotes that attest to the practice of worship at the tomb of Rūzbihān, in a manner similar to that indicated in the *manāqib*. His descendants organised activities for the promotion of the shrine. However, the order seems to have experienced a sharp decline in popularity by the time of the final quarter of the ninth/fifteenth century. The famous description of the military parade organised by prince Sulṭān Ḫalīl Aq-Qoyunlu at Band-i Amīr in 881/1476 mentions several *ṭarīqa* – the orders of Šayḫ Kabīr (*kabīriyya*) and Šayḫ Abū Isḥāq Kāzarūnī (*muršidiyya*), as well as the dervish followers of Aḥmad al-Rifāʿī – but not the Rūzbihāniyya.[136]

According to the testimonies of Junayd Šīrāzī, Mustawfī Qazwīnī, and Ibn Baṭṭūṭa, the graves of Rūzbihān and his descendants were well kept, as suited an important holy place. It was only later that the shrine would be abandoned. The neglect of Rūzbihān's tomb was perhaps the result of anti-Sufi feeling during the Safavid period. Henry Corbin has also proposed the possibility of a homonymous confusion, which would have contributed to the disappearance of Rūzbihān from popular memory.[137]

---

133 Ibn Baṭṭūṭa, *Travels*, II: pp. 316-7. The *Mašāriq al-anwār* is a work on the hadith; Baalbaki, Ramzi, "al-Ṣaghānī", in *EI*² 8 (1995): pp. 848-9.

134 Šaraf al-Dīn Ibrāhīm b. Rūzbihān II, *Tuḥfat*: p. 42.

135 Šaraf al-Dīn Ibrāhīm b. Rūzbihān II, *Tuḥfat*: pp. 141-2.

136 Dawānī, Jalāl al-Dīn, *ʿArż-i sipāh-i Uzun Ḥasan*, ed. I. Afšar in *Majallat-i dāniškada-yi adabiyāt-i Tihrān*, 1335š. (3) (1956): pp. 1-41, 22.

137 Corbin, *Le jasmin*: pp. 73-74.

2.3.3 The "Sanctuary Gardens" of Poets

In Shiraz, city of the poetic arts, there were numerous tombs of poets which attracted popular devotion. These were situated outside the city, in gardens where families went to relax. Regrettably, the medieval sources offer little on the popular practices associated with these tombs.

Junayd Šīrāzī opens the *Šadd al-izār* with the biography of Šayḫ Kabīr and closes with that of Saʿdī.[138] The two figures seem to incarnate the essence of Shiraz, in both saintliness and poetry.[139] Šīrāzī presents "al-Šayḫ Mušarraf al-Dīn Muṣliḥ b. ʿAbd Allāh al-Saʿdī al-Šīrāzī" as the most virtuous among Sufis and sojourners at the shrine of Šayḫ Kabīr.[140] The tomb of Saʿdī was situated in a *ḫānaqāh* which he had built in the Muṣallā area to the northeast of the city, at the foot of the mountains bounding the territory of Shiraz in this direction. There, during his lifetime, he distributed food to the poor and saints. Ibn Baṭṭūṭa described it as:

> ... an attractive building with a beautiful interior garden, situated near a source of the Ruknābād stream and in the vicinity of the ancient fortress of Fahandar. The sheikh had built small marble pools to wash clothes.

As noted above, the shrines of holy men were often connected with nature. The Arabic traveller continues: "Today, people come from the city to visit his grave, eat the food that is offered, wash their clothes, and go home,"[141] and finally adds: "That is just what I did." For Shirazis, the shrine of Saʿdī was evidently a site of both pious visitation and conviviality.[142]

The tomb of Ḥāfiẓ, named in nineteenth century sources as "Takiyya-yi Ḥāfiẓ" or "Ḥāfiẓiyya,"[143] was likewise situated northeast of the city in the garden

138 Junayd Šīrāzī, *Šadd al-izār*: pp. 462-3; ʿĪsā b. Junayd Šīrazī, *Hazār mazār*: pp. 477-8; Fasāʾī, *Fārs-nāma*: pp. 1200-01; Furṣat al-Dawla Šīrāzī, *Āṯār*: pp. 467-9.

139 In the same vein, Arberry considers the saints and poets of Shiraz as part of the legacy of Persia to the world: "Shiraz has little to teach of the art of the wise, just and stable government; it has bequeathed to us no pattern for better society. Its legacy is in no way material, but rather exclusively spiritual and artistic; and have we the right to say that it is less precious on that account?" Arberry, Arthur J., *Shiraz. Persian City of Saints and Poets* (Norman: University of Oklahoma Press, 1960): p. vii.

140 Junayd Šīrāzī, *Šadd al-izār*: p. 462; ʿĪsā b. Junayd Šīrazī, *Hazār mazār*: p. 477.

141 Ibn Baṭṭūṭa, *Travels*, II: p. 318.

142 On the description of the shrine, see Amīrī, *Tārīḫ*: II: pp. 177-9; on the history of its construction, see ibid.: 182-3.

143 Fasāʾī, *Fārs-nāma*: pp. 1201-02; Furṣat al-Dawla Šīrāzī, *Āṯār*: pp. 469-73. See Sarvestani, Kuros Kamali, "Hafez's tomb (Ḥāfiẓiya)", in *EIr* 11 (2003): pp. 505-507; Amīrī, *Tārīḫ*, II: pp. 171-176.

of Muṣallā. This place was in fact often cited in the poems of Ḥāfiẓ, for example: "Pour, cupbearer, this wine that abides: paradise will offer you neither the fountain of Ruknābād nor our gardens of Muṣallā".[144] In 856/1452, some sixty years after the death of Ḥāfiẓ, the Timurid governor of Fārs, Abu'l-Qāsim Bābūr b. Bāysunqur, ordered the construction of a dome over his tomb.[145] A man of wit wrote on the wall of the building: "Although he plundered all the pious endowments of the city, may God have mercy on [Abu'l-Qāsim Bābūr] for having erected this building."[146]

William Francklin (d. 1839) recorded the religious practices of the people of Shiraz in regards to the shrines of the city's emblematic poets. At the tomb of Saʿdī, visitors would lay flowers and leave prayer beads and other objects of piety, or place a copy of his *Dīwān* on the marble tomb.[147] He noted similar practices at the shrine of Ḥāfiẓ. Shirazis loved to visit the tomb of this poet in spring and summer, where they could smoke and play chess or other games in a spirit of conviviality. A copy of his *Dīwān* could be placed on the tomb, and his works were read here in homage. Francklin also recorded that "the most beautiful youths of the city would gather there", and suggested their profound respect for this favourite poet, as in his honour they made "numerous libations of excellent Shiraz wine, renowned even in Europe."[148] Although Ibn Baṭṭūṭa's testimony only concerns Saʿdī, the shrine of Ḥāfiẓ would have been continuously frequented since his death, by Shirazis gathering to read his poems and enjoy themselves.[149]

As for the shrines of these two emblematic poets of Shiraz, the sources also describe that of Abu'l-ʿAṭā' Kamāl al-Dīn Maḥmūd (689-753/1290-1352) as a popular place of conviviality. This poet was also known as Ḫwājū Kirmānī,[150] indicating that he was originally from Kirman. He had come to Shiraz to serve "scholars" (*ʿulamā'*) and virtuous men (*fuḍalā'*)",[151] and then visited Kāzarūn to seek spiritual education from Amīn al-Dīn Balyānī, referred to in one of his poems as his "guide" (*pīr-i ḫud*).[152] The body of Ḫwājū Kirmānī was interred

144 Ḥāfiẓ, *Dīwān*: n.°3, bayt 2.
145 Sarvestani, "Hafez's tomb": p. 505. Curiously, Fasā'ī and Furṣat al-Dawla do not mention that the tomb was constructed at this time. The work of ʿAlī Sāmī, referenced by Sarvestani, was unavailable to me.
146 Ḫwāndmīr, *Ma'āṯir al-muluk*: pp. 170-1, in Subtelny, *Timurids*: p. 164.
147 William Francklin, *Observations*: p. 101.
148 William Francklin, *Observations*: pp. 90-2.
149 See Fig. 1 for a photograph of a man visiting the tomb of Ḥāfiẓ to pay his respects, in 2005.
150 Ḫwānsārī, Suhaylī, "Zandaganī-i Ḫwājū", in *Dīwān-i aš'ār*, ed. Ḫwānsārī, S. (Tehran, 1336š./1956): pp. 2-85, 68-71.
151 Ḫwānsārī, "Zandaganī": pp. 4-5.
152 Ḫwājū Kirmānī, *Dīwān-i aš'ār*, ed. Ḫwānsārī, Suhaylī (Tehran: Bārānī, 1336š./1957): p. 75.

along the road leading from Isfahan to Shiraz, on a hill of the Tang-i Allāh Akbar pass facing the Gunbad-i ʿAḍud.[153] The tomb was situated in a verdant garden filled with roses.[154]

This area of the Tang-i Allāh Akbar pass, north of Shiraz, became the burial place of many poets and scholars. Ahlī Šīrāzī (d. 942/1535), author of panegyrics in honour of Šāh Ismāʿīl and some of his dignitaries, is interred near the mausoleum of Ḥāfiẓ.[155] Just ahead and on the right along the road from Isfahan, there is the grave of the Ilkhanid historiographer Waṣṣāf. The tomb is situated in a small garden called "Siyāfiyya", and identified by an inscription.[156] To the left, or northeast of Ḥāfiẓ's tomb, there is the burial of the Muẓaffarid prince Šāh-Šujāʿ, who was recognised for his poetic talents.[157] The tomb is also close to the shrine of the "Haft tanān" or "forty people" (see above).[158] Prior to his death, Šāh-Šujāʿ had asked that his body be placed to rest in the "lands of Muṣallā" (*zamīn-hā-yi Muṣallā*). His tomb was further embellished during the Zand period (eighteenth century). Four columns are surmounted by a ceramic-covered dome (*gunbad*), and the poems of both Ḥāfiẓ and Šāh-Šujāʿ are inscribed in the interior.[159] Furṣat al-Dawla, who scattered poems throughout his *Āṯār-i ʿAjam*, also lies near the Takiyya-yi Ḥāfiẓ.

Recently, Setrag Manoukian has described the alternation of prose and verse in Furṣat al-Dawla's as "making poetry a component of the landscape". Fārs, and more particularly Shiraz, are "poetic territories".[160] The Muṣallā plain and its gardens, irrigated by the water of the Ruknābād, so lauded by poets, are indeed a "place of memory" for the poetic arts of Shiraz, which have brought the city renown throughout the Muslim world and beyond.

153 The Gunbad-i ʿAḍud was erected by ʿAḍud al-Dawla as housing for his combat elephants; see Kamālī-Sarwistānī, *Dāniš-nāma*: 331-2.

154 Furṣat al-Dawla Šīrāzī, *Āṯār*: pp. 482-3; Ḫʷānsārī, "Zandagānī": pp. 79-80. See the detailed description and study of the shrine by Kamālī-Sarwistānī, *Dāniš-nāma*: pp. 29-30.

155 Thackston, Wheeler, "Ahlī Šīrāzī", in *EIr*², I (1985): pp. 637-8.

156 Kamālī-Sarwistānī, *Dāniš-nāma*: p. 40.

157 After taking back Shiraz from his brother Šāh-Maḥmūd, Šāh-Šujāʿ composed a *maṯnawī* entitled *Rūḥ al-ʿuššāq* in which the fratricidal conflict is presented as a quarrel between two lovers; see Wing, "Mozaffarids," *EIr*, online edition.

158 Amīrī, *Tārīḫ*, II: pp. 201-2.

159 Kamālī-Sarwistānī, *Dāniš-nāma*: pp. 35-6, reproduces the poems featuring in the shrine.

160 Manoukian, Setrag, *City of Knowledge in Twentieth Century Iran. Shiraz History and Poetry* (London and New York: Routledge, 2012): p. 19.

3 Conclusion

The shrines of the most celebrated spiritual figures of Islam, and the tombs of the holy men recognised for exemplary conduct, are suffused with sacredness. In these places there can be supernatural phenomena, such as illuminations, apparitions and diverse forms of inspiration. They are sources of wonders, with powers of attraction and influence over local, regional, and even international areas.[161] The burial shrine is first and foremost a place of mediation between profane and sacred. It is indeed a place of presence where people enter into contact with God or with whoever is laid to rest there and protects. Invocations made at the grave of a saint are granted. His *baraka* infuses the entire environment, including the earth, flowers and trees.[162] More so than mosques, shrines are places where the people can express emotions, shed tears, and ease the tensions of worldly life.

Shiraz had numerous renowned shrines and smaller graves: distinctive sites integrated with the natural landscape, both within and beyond the city walls. The "*ḫānaqāh*-shrines" of its many holy men served as meeting places, where people gathered to recall the episodes of their illustrious lives, and to experience the manifestations of their powers after death. Many miracles occurred in the necropolis (*rawḍa*) of Shiraz, which gradually developed in the *ḫānaqāh* of Šayḫ Kabīr and Rūzbihān Baqlī, originally as the site of the tombs of saintly figures, and then for their biological and spiritual descendants. These shrines attracted pious pilgrims, including from faraway lands.

The city of Shiraz was also the site of shrines to anonymous personages, such as the "Seven Men" (Haft tanān) and "Forty People" (Čihil tanān). These saints, who entered into the spiritual hierarchy of Sufi Islam, were considered the "guardians of the territory" (*ṣāḥib al-balad*). Such "territorialisation" of the authority of the men of God is one of the central themes in the hagiographical literature.[163] Jean-Claude Garcin notes that in Mamluk Egypt: "the invisible hierarchy [of saints], as the secret spiritual framework of the world, is distributed by geographic region."[164] In Shiraz, the two shrines of these anonymous

161 Vauchez, "Introduction" 2010: p. 2.

162 Amri, Nelly, "Zâwiyya et territoire en Ifriqiya du VIIe/XIIIe siècle à la fin du IXe/XVe siècle", in La Genière, Juliette de, Vauchez, André and Leclant, Jean (eds.), *Les sanctuaires et leur rayonnement dans le monde méditerranéen de l'Antiquité à l'époque moderne* (Paris: De Boccard, 2010): pp. 243-94, 278.

163 Amri, "Zâwiyya et territoire": p. 254, uses this term in discussing Ifriqiyya of the fourteenth and more so the fifteenth century.

164 Garcin, J.-Cl., "Le sultan et le Pharaon (le politique et le religieux dans l'Égypte mamelouke)", in *Hommages à François Daumas* (Montpellier: Université Paul-Valéry, 1986):

patron saints were situated in the Muṣallā gardens, a place with sacred functions, shared with the city's other illustrious sepulchres.

However, it was the important Alid shrines that were most important in earning Shiraz its lasting reputation of sanctity. As shown above, the "discovery" of the tombs involved a transcendent aspect. Their traditions of founding sacralisation identified links with the devotion to the family of the Prophet, the "People of the House" (*ahl al-bayt*). The origin narratives confer a sacred quality on these Alid shrines, associating them with the saintly history of the descendants of ʿAlī, persecuted by the Umayyad and then Abbasid caliphs. Some of these individuals had come to Shiraz seeking refuge from repression, where they nevertheless met their tragic death.

In the thirteenth century, luminous phenomena emanating from a mound signalled the presence of the saintly body of Aḥmad b. Mūsā al-Kāẓīm, near the first cathedral mosque of Shiraz. His tomb, known as the "King of Light" (*Šāh-i čirāġ*), became the site of numerous commemorative rituals. Luminous phenomena could also be seen every Friday night, near the tomb or "threshold" (*āstāna*) of Muḥammad, a brother of Aḥmad, also situated near the Jāmiʿ al-ʿAtīq. The many Alid shrines thus contributed to the "capitalisation of the sacred" for Shiraz.[165]

What happened after the conquest of Iran by the Safavids and the establishment of Shiism as the state religion? Was devotion encouraged at the Alid shrines? Topographical studies have not revealed significant maintenance of the shrines by the Safavids. However the Safavid sovereigns did travel in pilgrimage to the tomb of the imam ʿAlī al-Riḍa, in Mashhad.[166] Indeed in the autumn of 1010/1601, Šāh ʿAbbās I (r. 1571-1629) left Isfahan on foot, travelling to Mashhad in fulfilment of a vow, of uncertain nature.[167] Toward the end of Safavid rule, the shahs provided important patronage for the tomb of Fāṭima, the sister of ʿAlī al-Riḍā, in Qom.[168] Charles Melville noted that the emergence of Qom, to the detriment of Mashhad, could be linked to the fact that the former city was much closer to Isfahan. We could suppose that for similar

---

pp. 261-71 [Reprint in *Espaces, pouvoirs et idéologies de l'Égypte médiévale* (London: Variorum Reprints, 1987)]: p. 271.

165 Expression used by Chabbi, Jacqueline, *Le seigneur des tribus. L'islam de Mahomet* (Paris: CNRS, 1997): p. 50.

166 Savory, Roger M., "The Safavid State and Polity", *IrSt*, VII/1-2 (1974): pp. 179-212, illustrates the interaction between religion and politics at this time.

167 Melville, Charles, "Shah 'Abbas and the Pilgrimage to Mashhad", in Melville, Ch. (ed.), *Safavid Persia, Pembroke Papers*, IV (1996): pp. 191-229. Šāh Ismāʿīl also visited Mashhad before important battles; ibid.: p. 193.

168 Melville, "Shah Abbas": pp. 218-20.

reasons of proximity, the Safavids would have encouraged pilgrimages to the Alid shrines of Shiraz. However, this hypothesis would require confirmation based on textual sources. It should be noted that in his study on religion under the Safavids, Rasūl Jaʿfariyān does not mention worship at the Alid tombs of Shiraz.[169]

It is well known that the Safavids were hostile to Sufi masters.[170] According to Ḥusayn Karbalāʾī, author of the *Rawḍāt al-jinān wa jannāt al-janān*:

> Šāh Ismāʿīl crushed all the *silsila*. The graves of their ancestors were destroyed, not to mention what befell their successors.[171]

The tombs (*biqāʿ*) of the Sufi sheikhs of Fārs were desacralised.[172] Subsequent Sunni sheikhs managed the pilgrimage sites of this devout province, where votive offerings were made.[173] In 909/1503, two years after the conquest of Tabriz, the *kāzarūnī* order founded by Šayḫ Abū Isḥāq was eradicated, with the massacre of four-thousand preachers of the *ṭarīqa*. Shah Ismāʿīl's great fear was that the wealth and strong organisation of the dervishes might encourage them to incite revolts against the Shiite Safavid authorities.[174] In Shiraz, the tombs of Šayḫ Kabīr and Rūzbihān Baqlī were spared, however it seems that the interests in the associated spiritual movements gradually faded, leading to abandonment of the shrines. No charismatic leader had emerged, capable of diffusing the teachings of Šayḫ Kabīr beyond Fārs, within an organised *ṭarīqa*. Likewise, the family order of the Rūzbihāniyya had never extended beyond the province. The Rūzbihāniyya thus never attained the strong social force as the *kāzarūnī*, and perhaps for these reasons, the Safavids refrained from suppressing the order.

Local poets and visitors alike described the "garden cemetery" of Muṣallā, with its rose plantings and orange groves, as an enchanting place fed by the curative and nourishing waters of the Ruknābād. The most famous poets of

---

169 Jaʿfariyān, Rasūl, *Ṣafawiyya dar ʿarṣa-i dīn, farhang wa siyāsat* (*The Safavids in the Arenas of Religion, Culture ans Politics*), 3 vols (Qum: Pazūhiškada-yi Ḥawza wa Dānišgāh, 1379š).

170 Arjomand, Said Amir, "Religious Extremism (*ghuluww*), Ṣūfism and Sunnism in Safavid Iran: 1501-1722", *Journal of Asian History*, XV/1 (1981): pp. 1-35.

171 Karbalāʾī, Ḥāfiẓ Ḥusayn, *Rawżāt al-jinān*, ed. Jaʿfar Sulṭān Qurrāʾī (Tehran: Bungāh-i Tarjuma va-Našr-i Kitāb, 1344-1349š./1965-70): I, p. 490.

172 Aubin, Jean, "Shāh Ismāʿīl et les notables de l'Iraq persan. Études safavides 1", *JESHO*, II/1 (1959): pp. 37-81, 58; Arjomand, "Religious Extremism": p. 10.

173 Aubin, Jean, "L'avènement des Safavides reconsidéré. Études safavides 3", *Moyen Orient et Océan indien*, V (1988): pp. 1-130, 95.

174 Aubin, "Shāh Ismāʿīl": p. 58. According to later sources, the tombs of the most venerable sheikhs (Abū Isḥāq and Amīn al-Dīn Balyānī) were spared; ibid.: p. 58, n. 1.

FIGURE 1
Man in Sufi garb in prayer at Ḥāfiẓ's mausoleum
PHOTO B. SALESSE 2005

Shiraz were interred there. Their tombs and the surrounding gardens were sites of pious visitation, and above all, highly popular places of family conviviality. These tombs, completely restored since the eighteenth century, still preserve the fame of Shiraz as the city of the poetic arts.

Bibliography

Afsar, Karāmat-Allāh, *Tārīḫ-i bāft-i qadīmī-yi Šīrāz* (Tehran: Našr-i Qatra, 1353š./1974, 2nd edition 1995).

Aigle, Denise (forthcoming), *Saints hommes de Chiraz et du Fārs médiéval. Pouvoir, société et lieux de sacralité, Xe-XVe siècles.*

Amir-Moezzi, Mohammad Ali, *Le guide divin dans le shî'isme originel* (Paris: Verdier, 1992).

Amīrī, Razzāq Pārsa, *Tārīḫ-i Fārs, Šahr-hā wa rusta-hā-yi ān*, 3 vols (Šīrāz: Nawīd-i Šīrāz, 1391š./2012).

Amri, Nelly, *Les saints en Islam, les messagers de l'espérance. Sainteté et eschatologie au Maghreb aux XIVe et XVe siècles* (Paris: Le Cerf, 2008).

Amri, Nelly, "Zâwiyya et territoire en Ifriqiya du VIIe/XIIIe siècle à la fin du IXe/XVe siècle", in La Genière, Juliette de, Vauchez, André and Leclant, Jean (eds.), *Les sanctuaires et leur rayonnement dans le monde méditerranéen de l'Antiquité à l'époque moderne* (Paris: De Boccard, 2010): pp. 243-4.

Amri, Nelly, *Sîdî Abû Sa'îd al-Bâjî (1156-1231)* (Sousse: Contrastes-Éditions, 2015).

Arberry, Arthur J., *Shiraz. Persian City of Saints and Poets* (Norman: University of Oklahoma Press, 1960).

Arjomand, Said Amir, "Religious Extremism (*ghuluww*), Ṣūfism and Sunnism in Safavid Iran: 1501-1722", *Journal of Asian History*, XV/1 (1981): pp. 1-35.

Aubin, Jean, "Shāh Ismāʿīl et les notables de l'Iraq persan. Études safavides 1", *JESHO*, II/1 (1959): pp. 37-81.
Aubin, Jean, "L'avènement des Safavides reconsidéré. Études safavides 3", *Moyen Orient et Océan indien*, V (1988): pp. 1-130.
Azad, Arezou, *Landscape in Medieval Afghanistan: Revisiting the Faḍāʾil-i Balkh* (Oxford: Oxford University Press, 2014).
Baalbaki, Ramzi, "al-Ṣaghānī", in *EI*² 8 (1995): pp. 848-9.
Bayḍāwī, Nāṣir al-Dīn Abū ʿAbd Allāh, *Niẓām al-tawārīḫ*, ed. B. Karīmī, B. (Tehran: ʿIlmī, 1313š./1934).
Borrut, Antoine, *Entre mémoire et pouvoir. L'espace syrien sous les derniers Omeyyades et les premiers Abbassides (v. 72-193/692-809)* (Leiden: Brill, 2011).
Chabbi, Jacqueline, *Le seigneur des tribus. L'islam de Mahomet* (Paris: CNRS, 1997).
Corbin, Henry and Moʿīn, Mohammed, *Le jasmin des fidèles d'amour* (Tehran: Éditions Manoutcheri, 1987).
Dawānī, Jalāl al-Dīn, *ʿArż-i sipāh-i Uzun Ḥasan*, ed. I. Afšar in *Majallat-i dāniškada-yi adabiyāt-i Tihrān*, 1335š., III (1956): p. 1-41.
Daylamī, *Sīrat al-Šayḫ al-Kabīr Abū ʿAbd Allāh b. Ḫafīf al-Šīrāzī*, ed. Schimmel, A. (Ankara: Türk Tarih Kurumu Basımevi, 1955).
Dihḫudā, *Luġat-nāma* (Tehran: Tehran University Publications, 1988).
Dupront, Alphonse, *Du Sacré. Croisades et pèlerinages. Images et langages* (Paris: Gallimard, 1987).
Ernst, Carl, *Ruzbihan Baqli. Mysticism and the Rhetoric of Sainthood in Persian Sufism* (Richmond: Curzon, 1996).
Fasāʾī, Mīrzā Ḥasan Ḥusaynī, *Fārs-nāma-yi Nāṣirī*, ed. Muḥammad Rastigār Fasāʾī, 2 vols (Tehran: Amīr Kabīr, 1367š./1988).
Furṣat al-Dawla Šīrāzī, *Āṯār-i ʿAjam*, ed. ʿAlī Dibāšī, 3 vols (Tehran: Bāmdād, 1968).
Garcin, J.-Cl., "Le sultan et le Pharaon (le politique et le religieux dans l'Égypte mamelouke)", in *Hommages à François Daumas* (Montpellier: Université Paul-Valéry, 1986): 261-271 [Reprint in *Espaces, pouvoirs et idéologies de l'Égypte médiévale* (London: Variorum Reprints, 1987)].
Ḥāfiẓ, *Dīwān*, ed. P. Khanlārī (Tehran: Ḫʷārazmī, 1980).
Halevi, Leor, *Muhammad's Grave. Death Rites and the Making of Islamic Society* (New York: Columbia University Press, 2007).
Ḫūbnaẓār, Ḥasan, *Tārīḫ-i Šīrāz* (Tehran: Suḫan, 1380š./2002).
Ḥudūd al-ʿālam, tr. and com. Minorsky, V., *The Regions of the World: A Persian Geography 372 AH-982 AD*, (London: Luzac, 1937 [GMS N. S. XI]).
Ḫʷājū Kirmānī, *Dīwān-i ašʿār*, ed. Ḫʷānsārī, Suhaylī (Tehran: Bārānī, 1336š./1957).
Ḫʷānsārī, Suhaylī, "Zandagānī-i Ḫʷājū", in *Dīwān-i ašʿār*, ed. Ḫʷānsārī, S. (Tehran, 1336š./1957): pp. 2-85.

Ibn Baṭṭūṭa, *The Travels of Ibn Baṭṭūṭa*, trans. Gibb, H.A.R., 2 vols (Cambridge: Hakluyt Society, 1959).

Ibn Ḥawqal, *Kitāb Ṣūrat al-arḍ*, ed. Kramers, J.H., 2 vols (Leiden: Brill, 1938).

Ibn ʿInaba, *ʿUmdat al-ṭālib* (Najaf: al-Maṭbaʿa al-ḥaydarīya, 1380/1918).

Ibn Zarkūb, *Šīrāz-nāma*, ed. Jawāḏī, I. Wāʿiẓ (Tehran: Bungāh-i Farhang-i Vāža-javādī, 1350š./1972).

Imdād, Ḥasan, *Šīrāz dar guḏašta u ḥāl* (Ṣīrāz: Mūsawī, 1339š./1960).

ʿĪsā b. Junayd Šīrāzī, *Hazār mazār*, ed. Wiṣāl, Nūrānī (Šīrāz: Intišārāt-i Kitābḫāna-i Aḥmadī Šīrāz, 1364š./1975).

Isfizārī, Muʿīn al-Dīn Muḥammad, *Rawżāt al-Jannāt fī awṣāf-i madīnat Harāt*, ed. Sayyid Muḥammad Kāẓim Imām, 2 vols. (Tehran: Intišārāt-i Dānišgāh-i Tihrān, 1338-1339š./1959-60).

Iṣṭaḫrī, *Kitāb al-Masālik wa-l-mamālik*, ed. de Goeje, 2nd edition (Leiden: Brill, 1967).

Jaʿfariyān, Rasūl, *Ṣafawiyya dar ʿarṣa-i dīn, farhang wa siyāsat* (*The Safavids in the Arenas of Religion, Culture ans Politics*), 3 vols (Qum: Pazūhiškada-yi Ḥawza wa Dānišgāh, 1379š).

Junayd Šīrāzī, Muʿīn al-Dīn Abu'l-Qāsim, *Šadd al-izār*, ed. Qazwīnī, Muḥammad (Tehran: Čāpḫāna-yi Majlis, 1328š./1949).

Kamālī-Sarwistānī, Kūruš, *Dānišnāma-yi āṯār-i tārīḫī-yi Fārs* (Šīrāz-Tehran: Mu'assasa-yi farhangī wa pažūhišī-yi Dānišnāma-yi Fārs & Sāzimān-i mīrāṯ-i farhangī, 1384š./2006).

Karbalāʾī, Ḥāfiẓ Ḥusayn, *Rawżāt al-jinān*, ed. Jaʿfar Sulṭān Qurrāʾī (Tehran: Bungāh-i Tarjuma va-Našr-i Kitāb, 1344-1349š./1965-70).

Lambton, Ann K. S., "Shīrāz", in *EI*², IX (1998): pp. 491-7.

Limbert, John, *Shiraz in the Age of Hafez. The Glory of a Medieval Persian City* (Seattle and London: University of Washington Press, 2004).

Manoukian, Setrag, *City of Knowledge in Twentieth Century Iran. Shiraz History and Poetry* (London and New York: Routledge, 2012).

Manz, Beatrice, *Power, Politics and Religion in Timurid Iran* (Cambridge: Cambridge University Press, 2007).

Mayeur-Jaouen, Catherine, "Tombeau, mosquée et zâwiya: la polarité des lieux saints musulmans", in Vauchez (dir.), *Lieux sacrés, lieux de culte, sanctuaires. Approches terminologiques, méthodologiques, historiques et monographiques* (Rome: École Française de Rome, 2000): pp. 133-47.

McChesney, R. D., *Waqf in Central Asia. Four Hundred Years in the History of a Muslim Shrine, 1480-1889* (Princeton: Princeton University Press, 1991).

Melville, Charles, "Shah ʿAbbas and the Pilgrimage to Mashhad", in Melville, Ch. (ed.), *Safavid Persia, Pembroke Papers*, IV (1996): pp. 191-229.

Meri, Josef W., *The Cults of Saints among the Muslims and Jews in Medieval Syria* (Oxford: Oxford University Press, 2002).

Mouton, Jean-Michel. "De quelques reliques conservées à Damas au Moyen-Âge. Stratégie politique et religiosité populaire sous les Bourides", *Annales Islamologiques* 27 (1993): pp. 245-54.

Mulder, Stephennie, *The Shrines of the 'Alids in Medieval Syria: Sunnis, Shi's and the Architecture of Coexistence* (Edinburgh: Edinburgh University Press, 2014).

Muqaddasī, *Aḥsan al-taqāsīm fī ma'rifat al-aqālīm*, ed. de Goeje, *Descriptio imperii moslemici*, BGA VI, (Leiden: Brill, 1877).

Mustawfī Qazwīnī, Ḥamd Allāh, *Nuzhat al-qulūb*, ed. and trans. Le Strange, G., 2 vols (London & Leiden: Luzac & Brill, 1915, 1919).

Nelson, Kristina, *The Art of Reciting the Qur'ān* (Austin: University of Texas Press, 1985).

Otto, Rudolf, *Das Heilige: über das Irrationale in der Idee des Göttlichen und sein Verhältnis zum Rationalen* (Breslau: Trewendt und Garnier, 1917).

Pedersen, J., "Masdjid", in *EI*², VI (1991): pp. 630-1.

Rāġib, Yūsuf, "Al-Sayyida Nafīsa, sa légende, son culte et son cimetière", *StIs*, XLIV (1976): pp. 61-86 and 45 (1977): pp. 27-55.

Rubin, Uri, "Pre-existence and Light. Aspects of the Concept of Nūr Muḥammad", *Israel Oriental Studies*, V (1975): pp. 62-119.

Šams al-Dīn 'Abd al-Laṭīf b. Rūzbihān II, *Rūḥ al-jinān fī sīrat al-Šayḫ Rūzbihān*, in *Rūzbihān-nāma*, ed. M. Dāniš-Pažūh, (Tehran: Silsila-i Intišārāt-i Anjuman-i Āṯār-i Millī, 1969): pp. 152-370.

Šaraf al-Dīn Ibrāhīm b. Rūzbihān II, *Tuḥfat ahl al-'irfān fī ḏikr sayyid al-aqṭāb Rūzbihān*, in *Rūzbihān-nāma*, ed. M. Dāniš-Pažūh, (Tehran, Silsila-i Intišārāt-i Anjuman-i Āṯār-i Millī, 1969): pp. 1-149.

Sarvestani, Kuros Kamali, "Hafez's tomb (Ḥāfiẓiya)", in *EIr* 11 (2003): pp. 505-7.

Savory, Roger M., "The Safavid State and Polity", *IrSt*, VII/1-2 (1974): pp. 179-212.

Schwarz, Paul, *Iran nach den arabischen Geographen des Mittelalters*, Leipzig, 1896 (reprint Hildesheim-New York: Georg Olms, 1969).

Shahbazi, A. Shapour, "Shiraz i. History to 1940", in *Encyclopaedia Iranica online*, http://www.iranicaonline.org/articles/shiraz-i-history-to-1940.

Subkī, Tāj al-Dīn, *Ṭabaqāt al-šāfi'iyya al-kubrā*, ed. Nūr al-Dīn Šarība (Cairo: al-Maṭba'a al-Ḥusainīya, 1323-1324/1906).

Subtelny, Maria, *Timurids in Transition. Turko-Persian Politics and Acculturation in Medieval Iran* (Leiden: Brill, 2007).

Talmon-Heller, Daniella, "Graves, Relics and Sanctuaries: the Evolution of Syrian Sacred Topography (eleventh-thirteenth centuries)", *ARAM Periodical*, XVIII-XIX (2006-7): p. 601-20.

Taqī Muṣṭafawī, Muḥammad, *Iqlīm-i Fārs* (Tehran: Taban Press, 1964).

Thackston, Wheeler, "Ahlī Šīrāzī", in *EIr* 1 (1985): pp. 637-8.

Thibon, Jean-Jacques, *L'oeuvre d'Abū 'Abd al-Raḥmām al-Sulamī 325/937-412/1021 et la formation du soufisme* (Damas: Institut Français du Proche-Orient, 2009).

Van Renterghem, Vanessa, *Les élites bagdadiennes au temps des Seldjoukides. Étude d'histoire sociale*, 2 vols (Beirut-Damascus: Institut Français du Proche-Orient, 2015).

Vauchez, André, "Introduction", in *Lieux sacrés, lieux de culte, sanctuaires. Approches terminologiques, méthodologiques, historiques et monographiques*, Vauchez, André (dir.) (Rome: École Française de Rome, 2000): pp. 1-7.

Vauchez, André, "Introduction", in La Genière, Juliette de, Vauchez, André, Leclant, Jean, (eds.), *Les sanctuaires et leur rayonnement dans le monde méditerranéen de l'Antiquité à l'époque moderne* (Paris: De Boccard, 2010): pp. 1-6.

Vincent, J. F., J. Dary and R. Verdier, *La Construction religieuse du territoire* (Paris: L'Harmattan, 1995).

Waṣṣāf, *Tajziyat al-amṣār wa tazjiyat al-aʿṣār*, ed. M. M. Iṣfahānī, lithograph (Bombay, 1852).

Wilber, Donald N., *The Masjid-i ʿAtiq of Shiraz* (Šīrāz: Pahlavi University, 1972).

William Francklin, *Observations Made on a Tour from Bengal to Persia in the Years 1786-87* (London: T. Cadell, 1790).

Wing, Patrick, "Mozaffarids", http://www.iranicaonline.org/articles/mozaffarids.

Biographical Note

Denise Aigle is Emeritus Research Professor at the Ecole Pratique des Hautes Etudes, Paris. Her interests include Mongol rule in Iran and Syria, issues of reciprocal representations between the Muslim East and Latin West, and hagiographic literature studied from the perspective of historical anthropology. Her published monographs include *Le Fārs sous la domination mongole (XIIIe-XIVe s.). Politique et fiscalité* (Leuven, Peeters, 2005) and *The Mongol Empire Between Myth and Reality. Historic Anthropological Studies* (Leiden, Brill, 2015). She has edited several collective volumes, including *Miracle et karāma. Hagiographies médiévales comparées* (Turnhout: Brepols, 2000), *Le Bilād al-Šām face aux mondes extérieurs. La perception de l'Autre et la représentation du souverain* (Beirut: IFPO, 2012). She studied with Jean Aubin and has recently published a variorum volume of his articles: *Jean Aubin, Études sur l'Iran médiéval. Géographie historique et société* (Leuven: Peeters, 2018).

The Fortified Landscape of Isfahan

Ali Shojaee Esfahani
a.shojaei@aui.ac.ir

Abstract

This article presents an exposition of the fortifications around Isfahan based on fieldwork and a thorough analysis of written sources and the available satellite images, processed with new GIS technologies. The fortress of Šāhdiz, of Ismaili fame, is well known and documented in the written sources, however the major conclusion of the current research is that in fact a *network* of fortifications was developed, strategically located to afford observation and communication. The heights surrounding Isfahan were exploited to full potential, gaining security over roads and settlement processes, which in turn contributed to urban growth. The evolution of this network conformed with the trends of urbanisation: as long as the main towns remained Qih and Jayy, it was the mountains in the east of the *rustāq* of Isfahan that played the decisive role. But from the fourth/tenth century onward, with the growth of Yahūdiyya and the integration of the southern province of Fars and Khuzistan in the Buyid and Saljuq polities, it was the fortified of sites in the mountainous areas south of Zāyanda-rūd that were key to securing the access roads.

Keywords

Isfahan – Landscape Archaeology – Network of fortification – roads – Satellite imagery – GIS

1 Introduction

This article deals with the issue of access to urban centres in the early centuries of Islam. Access implies both roads and the fortifications designed to control them: to send relevant information to the centres of population, whether cities or smaller centres, and if necessary to protect them. The paper deals with the issue through the case study of the heartlands of the province of Isfahan, in western Iran. Here, settlement of the very fertile Zāyanda-rūd valley lands was progressively polarised between the Sasano-Islamic town of Jayy, and the

 | DOI:10.1163/9789004434332_010

subsequent growth of the nearby urban centre of Yahūdiyya (later known as Isfahan), under the Abbasids.[1] The growth of these twin cities of Jayy and Yahūdiyya has been dealt with through analysis of key sources, such as Abū Nu'aym's local history,[2] however the relationships between these urban centres and their hinterlands, and the outside world beyond, are much less known. The existence of Isfahan was a gift of the Zāyanda-rūd, and the city became one of the main emporia of the Iranian plateau, roughly centred between the provinces of Fars, Jibal, Khurasan and Khuzistan (Fig. 1).[3] Its growth relied on the roads which led to it, and here as elsewhere, the road network was strongly determined by the topography.

The road network around Isfahan was first examined in the seminal works of Wilhelm Tomaschek and Paul Schwarz, on the basis of data from historic geographies and the then available maps.[4] The French engineer Maxime Siroux was the first to carry out a true field survey of the road network, following World War II. His results, based on detailed explorations extending to 100 kilometres distant from Isfahan, were published in the 1971 folio *Anciennes voies et monuments routiers de la région d'Ispahan*. Although Siroux noted some of

1 This article synthesizes the results of my doctoral dissertation, defended in 2014 at the University of Tehran, Shojaee 2014a. I am grateful to Natasha Shahidi and David Durand-Guédy for translating my draft for this current article. Except Fig. 1, all the figures are adapted from Shojaee, Ali, *Barrisī-yi bāstān-šināḫtī-yi qilāʿ wa istaḥkāmāt-i Iṣfahān az manẓar-i juġrafiyā-yi idārī* (PhD, Univ. Tehran, 2014). Toponyms are vocalized according to *Farhang-i juġrafiyā-yi Īrān*. X. *Iṣfahān* (Tehran: Intišārāt-i dāyirah-yi juġrāfiyā'i-yi sitād-i ārtiš/ Publications de la section géographique de l'Etat-major de l'armée 1355š.[/1976]).

2 See Golombek, Lisa, "Urban Patterns in Pre-Safavid Isfahan", *IrSt*, VII/1 (1974): pp. 18-44.

3 Isfahan is described as the emporium (*furḍa*) of Fars, Jibal, Khurasan and Khuzistan by al-Iṣṭaḫrī, *K. al-Masālik wa l-mamālik*, ed. M. de Goeje (Leyden: Brill, 1870): p. 199; Ibn Ḥawqal, *K. Ṣūrat al-arḍ*, ed. Michael de Goeje revised by Joel H. Kramers (Leyden: Brill, 1939): pp. 362-3; al-Muqaddasī, *K. Aḥsan al-taqāsim fī maʿrifat al-aqālīm*, ed. M. de Goeje, second edition (Leyden: Brill, 1906): p. 389; Zakariyyā Qazwīnī. *K. Āṯār al-bilād wa aḫbār al-ʿibād*, Persian trans. ed. M. Šāhmurādī, (Tehran: Intišārāt-i Danišgāh-i Tihrān, 1371š.): 2, pp. 17-18. See Durand-Guédy, David, *Iranian Elites and Turkish Rulers: a History of Isfahan in the Saljuq Period* (London and New York, 2010): p. 24, 30. Both Ibn Rusta, *K. al-Aʿlāq al-nafīsa*, ed. Michael; de Goeje, second edition (Leyden: Brill, 1967): p. 152, and al-Muqaddasī, *Aḥsan al-taqāsim*, note that Isfahan was equidistant (80 *farsaḫ*) from the main urban centers of Jibal (Rayy and Hamadan): Khuzistan ('Askar Mukram) and Fars (Iṣṭaḫr, then Shiraz).

4 Tomaschek, Wilhelm, *Zur historischen Topographie von Persien* (1883, reprint Osnabrück 1972): pp. 24, 27-8; Schwarz, Paul, *Iran im Mittelalter nach den arabischen Geographen* (1896, reprint Hildesheim-New York, 1969): pp. 929-30 (road from Kashan to Isfahan): 936-9 (Isfahan to Khuzistan): 939-42 (Isfahan to Fars).

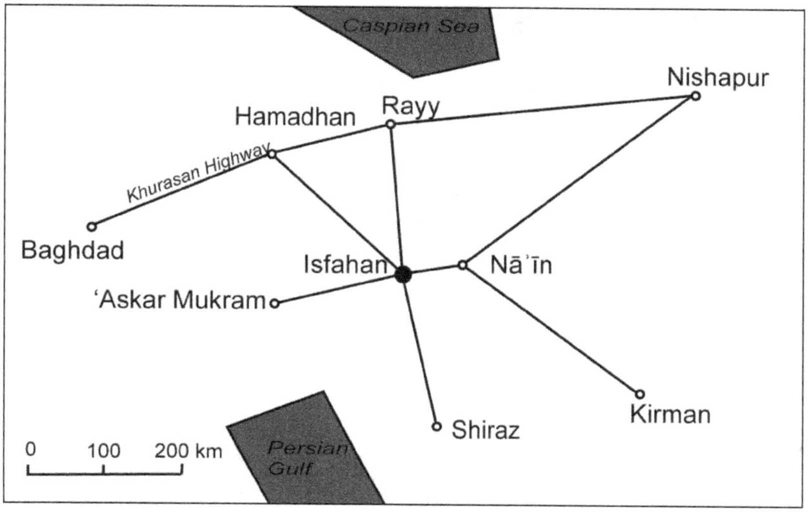

FIGURE 1 Road links between central Isfahan province and main cities of Iran, according to early Islamic geographers.

the visible fortifications, this was not his main concern.[5] Also in 1971, the local amateur historian, Caro Minasian, published *Shahdiz of Ismaili fame*, an in-depth study of the fortifications in a mountainous area immediately south of Isfahan, known as Kūh Ṣuffa.[6] Minasian was the first to correctly identify the site of Šāhdiz, the mighty stronghold taken by the Ismaili in the late fifth/eleventh century, and in this work also offered a precise map of the remains in the surrounding areas.[7] Again in 1971 (clearly an exceptional year for our field of interest), Samuel Stern, Elisabeth Beazley and Andrew Dobson published a very detailed article on the 'Fortress of Khān Lanjān', in the journal *Iran*. This is a site located 23 km south of Šāhdiz, above the lush valley of the Zāyanda-rūd.[8] Other fortifications have been noted in publications in Persian, most importantly by 'Alī Riḍa Ja'farī-Zand, but many in the area had never been recorded

5 See Siroux, Maxime, *Anciennes voies et monuments routiers de la région d'Ispahan* (Cairo: IFPO, 1971). Historical data regarding Šāhdiz had been first gathered by Schwarz, *Iran*: pp. 656-7.

6 Minasian, Caro O., *Shahdiz of Isma'ili Fame. Its Siege and Destruction* (London: Luzac & Company Ltd., 1971).

7 Willey, Peter, *Eagle's Nest. Ismaili Castles in Iran and Syria* (London: Tauris, 2005): pp. 206-12, on the fortress of Šāhdiz, does not add much to Minasian concerning this Ismaili stronghold.

8 Stern, Samuel M., Beazley, E. and Dobson, A., "The Fortress of Khān Lanjān", *Iran*, IX (1971): pp. 45-57.

prior to the surveys reported here,[9] and no comprehensive analysis has been attempted.

I will argue that the fortifications in question should be perceived as an integrated network serving in protection of both roads and settlements. The research reported takes the approach of landscape archaeology, in which the landscape is conceptualised not only as a backdrop for archaeological remains, but as "an entity that exists by virtue of its being perceived, experienced, and contextualized by people".[10] Methodologically, this work implies constant movement from desk to field and back again. The first step is to gather data from a broad array of textual sources and other materials: the 1971 mapping by Minasian, and to a lesser extent Siroux; the Corona satellite images captured in the years 1960-61, meaning prior to permanent territorial alteration from intense urbanisation and transport infrastructure;[11] oral accounts gathered from hunters, hikers and other locals familiar with the territory; historic maps from the nineteenth and twentieth centuries. While our focus is on the early period, roughly up to the Mongol era, we make full use of the classical texts of Islamic geography in Arabic and Persian, as well as the travelogues up to the early Pahlavi period.[12] Our investigations are assisted by the fact that the settlement geography and administrative boundaries remained remarkably stable over time. Indeed there are only slight differences between the descriptions of the *rustāq*s (districts) from the fourth/tenth century, and those

9 E.g. Jaʿfarī-Zand, ʿAlī-Riḍā. *Iṣfahān pīš az Islām* (Tehran: Intišārāt-i ān, 1381š./2002): pp. 181-7, on Taḫt-i Rustam, p. 188 on Kūh Dastgird and pp. 190-1 on Kūh Buzī.

10 See Knapp, Bernard and Ashmore, Wendy, "Archeological landscapes: Constructed, Conceptualized, Ideational", in W. Ashmore and B. Knapp, *Archaeologies of Landscape. Contemporaries Perspectives* (Malden, Mass., 1999): pp. 1-30, here p. 1. There had been no significant work on the landscape archaeology of Islamic-period Iran until some recent but still unpublished doctoral dissertations (e.g. Jaleh Kamalizad on the area of Ḫān Lanjān, see "The Landscape Archaeology in Lanjān Plain of Iṣfahān, Iran, during the Islamic Period (650-1722 AD) (Preliminary Report)", *Archaeologies*, XII/1 (2016): pp. 91-109). For the pre-Islamic period, the site http://www.archatlas.org/journal/journal_bydate.php (last consulted April 2018) offers some interesting studies of road networks, based on analysis of satellite images.

11 Corona images are provided free of charge by the Center for Ancient Middle Eastern Landscapes of the Oriental Institute, University of Chicago.

12 Indeed the most detailed data on the *rustāq* (district) of Jayy are found in the geographical opus written by Arbāb-Iṣfahānī, Muḥammad Mahdī, *Niṣf-i jahān fī taʿrīf al-Iṣfahān*, ed. Manūčihr Sutūda (Tehran: Amīr Kabīr, 1340š./1962): pp. 296-99, in the Qajar period.

in the reports surviving from later periods.[13] Even today, we can still identify large numbers of the villages mentioned in the earliest reports.[14] All of our data are then processed and analysed using geographical information system software (ArcGIS, IDRISI). The GIS analyses help reconstruct elements such as roads and fortifications, which may have left no visible remains, and assist us in understanding why a given fortification would have been built at its specific site. Any newly derived results then require confirmation through further field reconnaissance.

We draw on digital data from two sources, both made available by the US Geological Survey: the Shuttle Radar Topography Mission (SRTM) and Advanced Spaceborne Thermal Emission and Reflection Radiometer (ASTER).[15] These projects contribute to a Global Elevation Model, which is particularly useful for allowing checks of visibility from any point on the map. With these tools, supported by fieldwork survey, we can map the visual potential of any position, meaning the distances of sightlines in all potential directions. This has proven very useful in determining how a fortress could communicate with other fortresses or settlements over long distances, such as by using smoke signals, or fire by night.[16]

---

13 See Abū Nuʿaym al-Iṣfahānī, *Ḏikr Aḫbār Aṣbahān*, ed. Sven Dedering, *Geschichte Isbahans*, 2 vols (Leyden: Brill, 1931): I, p. 17, 113, II, p. 267; Yāqūt, *Muʿjam al-buldān*, ed. F. Wüstenfeld, *Jaqut's Geographisches Wörterbuch*, 6 vols (Leipzig: in comm. bei F.A. Brockhaus, 1866-73): 2, p. 415; Āvī, Ḥusayn, *Tarjumah-yi maḥāsin-i Iṣfahān*, ed. ʿAbbās Iqbāl, revised ed. by Muḥammad-Riḍā Zādhūš (Isfahan: Sāzmān farhangī va tafrīḥī-yi šahrdārī-yi Iṣfahān, 1385š./2006): p. 124. Cf. ʿImād al-Dīn al-Iṣfahānī, *Ḫarīdat al-qaṣr*, partial ed. ʿA. Āl Ṭu'ma, 3 vols (Tehran: Mīrāṯ-i maktūb, 1378š./1999): I, p. 326.

14 Cf. Ibn Rusta: p. 152; Ḥamza al-Iṣfahānī, *Taʾrīḫ sinī mulūk al-arḍ waʾl-anbiyāʾ*, ed. S.J. Īrānī-Tabrīzī (Berlin: Kaviani, 1340/1921-22): p. 28; *Mujmal al-tavārīḫ wa'l-qiṣaṣ*, ed. Malik al-Šuʿarā Bahār (Tehran: Intišārāt-i Ḫāvar, 1318š./1939): p. 54; Ḥamd-Allāh al-Mustawfī, *K. Nuzhat al-qulūb*, ed. M. Dabīrsiyāqī (Tehran: Ṭahah, 1378š.[/1999-2000]): pp. 54-6; Arbāb Iṣfahānī, *Niṣf-i jahān*: p. 296.

15 SRTM data (2003) offer maximum horizontal resolution of around 90 metres and global digital elevation of ± 16 metres. The more recent (2009) ASTER data provide better resolution: 30 metres, global digital elevation ± 18 metres.

16 See for example Kāšānī, Suhayl and Ḍarabī, Mirzā ʿAbd al-Raḥīm, *Mirʾāt-i Qāsān yā Tārīḫ-i Kāšān*, ed. Īraj Afšār, *Farhang-i Īrān-zamīn*, III (1334š.): pp. 105-262, here pp. 138-9, on the communication of three fortresses of Kashan. Yāqūt, *Buldān*: I, p 438, notes the use of smoke signals at the Caucasian city of Derbend. For Isfahan, see the account of Chardin, Jean, *Voyages du chevalier Chardin en Perse et autres lieux de l'Orient* (Paris: Le Normant, 1811): VIII, p. 119. Communication between fortresses could also have been by carrier pigeons, see Ragheb, Youssef, *Les messagers volants en terre d'Islam* (Paris: CNRS Editions, 2002).

The territory under investigation, represented in Fig. 2, is about 2300 km2. At the centre of this lay Isfahan, in the *rustāq* of Jayy, bordered to the north by the *rustāq* of Burḫuwār (now Burḫār), to the northeast by Quhāb (site of modern Isfahan International Airport), and to the east by Bara'ān. To the west lay the village of Dastgird (since incorporated in the city proper), and beyond this the area of Mārbīn. The Zāyanda-rūd crosses this territory from west to east. This territory is flanked by a part of the Zagros range of mountains (Persian '*kūh*'), trending northwest-southeast. Of these, the highest peak is Mount Kulāh Qāḍī (2523 m asl), rising some 1000 metres above the plain.[17]

We have identified and analysed 18 fortified sites, composing two groups: fortifications, or modern Persian *istiḥkāmāt*; and fortresses, referred to in the sources as *qal'a* and *diž* (rendered simply *diz* in Arabic script, as in 'Šāhdiz'). Both represent modifications of the natural setting. Fortresses are characterized by larger size and greater numbers of spaces for residential (rooms, water reservoirs) and defensive (towers, rampart) functions. A large number of persons could live for an extended period of time in a fortress. Fortifications are of much smaller dimension, generally consisting of only one or at most two rooms, sometimes with an added isolated tower and wall. All 18 sites have been indicated on the Global Elevation Model (Fig. 2) and Corona images (Fig. 3).

In this report we present the sites grouped in four geographic zones surrounding the metropolis: east, south, southwest and west. Each zone includes three to eight sites (Table 1). These groupings are solely for purposes of clarity in presentation. In fact we will show that the various sites were part of a single network, continually modified over the centuries.

No sites are observed in the crescent to the north of Jayy, a fact which could possibly be explained by the absence of major topographical relief until arrival at the first hill of the Karkās Range, some 50 km further north. The lack of relief in the northern reaches meant that the fortresses in the south, west and east of the *rustāq* of Jayy would still have maintained good visibility (see Fig. 9, 12 and 19) over the area.[18]

Before proceeding to the network of roads and fortifications, we begin with a few words about its centre, which is to say the urban agglomeration of Isfahan.

17 As the name suggests, the peak of Kulāh Qāḍī resembles the hat of an Islamic judge (see Fig. 17).

18 The most important fortress in the northerly direction could have been the formidable Wushāq Qal'a, a full 97 km north of Isfahan, in the Karkās Range, whose early history is unknown. The lack of fortified sites would fit well with observations by al-Iṣṭaḫrī (p. 228) and al-Muqaddasī (p. 487): in the fourth/tenth century on the lack of security over the northern road. However the absence of fortifications to the north of Isfahan (Table 1) could also be due to as yet insufficient research in the area. During the 2017 field season we observed remains in Sanja Fortress, northwest of Mayamah.

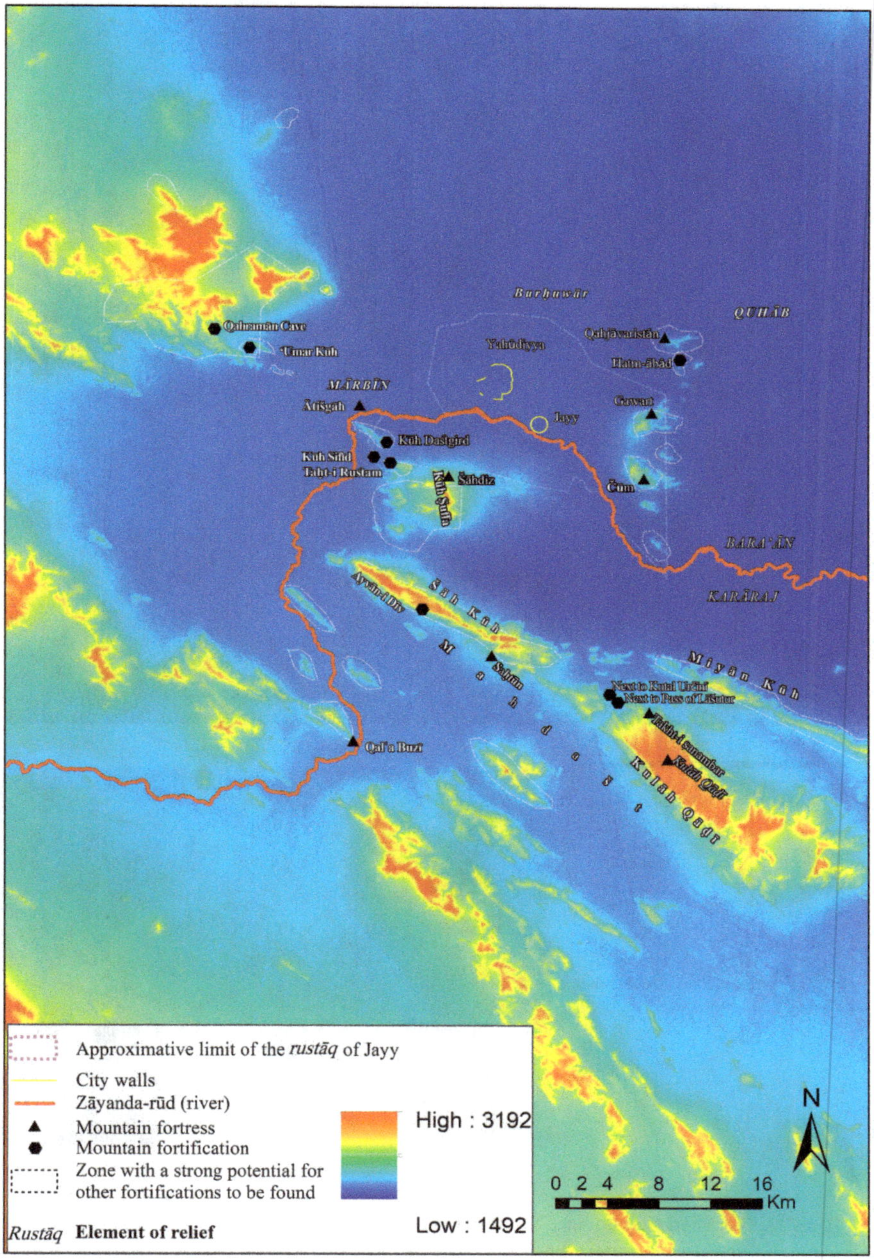

FIGURE 2 Topography of central Isfahan province, with fortified sites

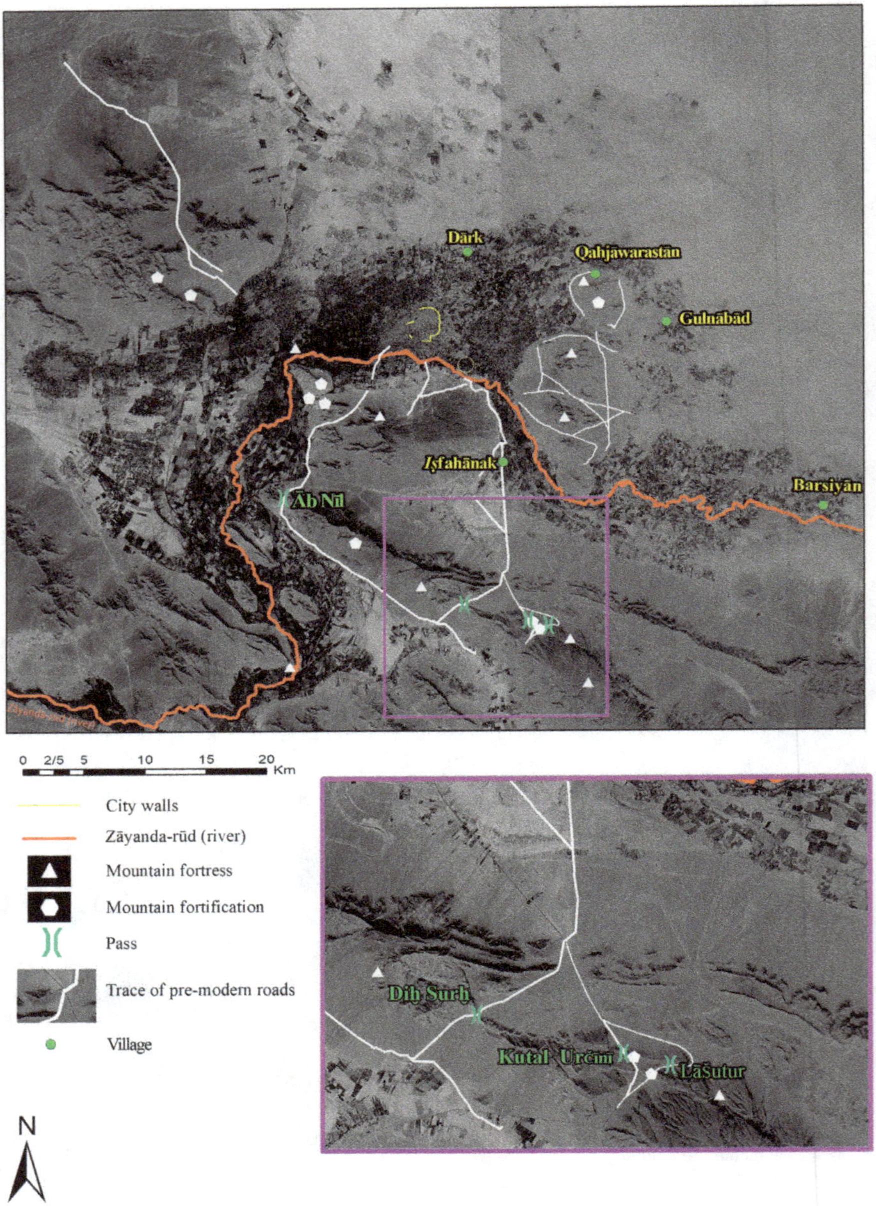

FIGURE 3A Fortified sites and pre-modern road network at the centre of Isfahan province
SOURCE: CORONA PHOTO, MODIFIED BY A. SHOJAEE ESFAHANI

TABLE 1 Geographical distribution of fortified sites in central part of Isfahan province

Location	Name of fortified site	Direction of nearby road(s)
East of the *rustāq* of Jayy	Fortress of Qahjāwarastān Fortress of Kūh Gavart Fortress of Kūh Čūm Fortification of Kūh Hatm-ābād	Kirman, Khurasan
South of the *rustāq* of Jayy	Fortress of Šāhdiz Fortress of Qalʿa Buzī (Ḫān Lanjān) Fortification of Saḫtūn Fortress of Kulāh Qāḍī Fortress of Taḫt-i Ṣanambar Fortification of Ayvān-i Dīv Tower by the Určīnī Road Tower by the Lāšutur Pass	Fars, Khuzistan
Southwest of the *rustāq* of Jayy	Fortification of Taḫt-i Rustam Fortification of Kūh Sifīd Fortification of Kūh Dastgird	Fars, Khuzistan
West of the *rustāq* of Jayy	Fortress of Ātišgāh (or Mārbīn) Fortification of ʿUmar Kūh Fortification of the Qahrimān Cave	Hamadan

2 The Cities of Jayy and Yahūdiyya

Jayy and then Yahūdiyya, the main urban centres of Isfahan province (Persian '*kurra*'), were situated on the left side of the Zāyanda-rūd, north of the Kūh Ṣuffa, within the *rustāq* of Jayy. The importance of this *rustāq* was stressed by early geographers, including al-Yaʿqūbī (d. 292/905) and the Isfahani Ibn Rusta (early fourth/tenth century). The latter, whose account of Isfahan is based on first hand observations, says that Jayy is the "*qaṣaba*" of the province. We can understand this polysemic term as 'the most essential part', also 'citadel'[19] or 'main town' of the province. The same author reports that the Isfahan's *madīna* (administrative capital) and markets are in Jayy, and that this was the place where all inhabitants meet (*majmaʿ ahlihā*). He explains:

19 In the current article, 'citadel' refers only to an urban fortress.

> Many good qualities (*ma'ānī*) contributed to give this place pre-eminence over the other *rustāq*-s. The first [most important?] is that the person who, despite possessing all the lands [of the province], made [the *rustāq* of Jayy] a *qaṣaba* and built a *madīna* there, only did it because it had the best soil, the purest air and the sweetest water and, in fact, that it possessed a set of attributes (*ḫilāl*) that suited it to the residence of the kings and the subjects (*maskan al-mulūk wa'l-ra'iyya*).[20]

Ibn Rusta explicitly describes the *madīna* of Isfahan as being a military centre (*qaṣaba*), since earliest times. This could even explain the origins of the name 'Isfahan' – a word that has been likened to Sipahān (the cavaliers, armies; the military camp).[21]

Over later centuries, an array of authors, both Isfahani and non, would continue to celebrate the quality of Isfahan's air.[22] Indeed, nature had afforded particular advantages to the site of Jayy. Considering morphology, the northward incline of the land in this part of the valley provided a natural assist in development of irrigation of the lands north of the Zāyanda-rūd, using a network of surface canals called *mādī*. The elevations to the south of the *rustāq* (the area of Kūh Ṣuffa south of the river, the nearby area of Hizār Darra, to the west) are on the other hand very dry and unfit for farming (Fig. 3b). Not surprisingly, most of the settlements during the Sasanid period were located north of the Zāyanda-rūd. This includes Jayy and Qih, as well as 15 villages later absorbed by the growth of Yahūdiyya, in the Abbasid period. The mountains to the south also protected the *rustāq* of Jayy from the supposedly unhealthy winds arriving from that direction, while leaving the location open to the good ones from north (the *šurṭa*) and east (the *ṣabā*).[23]

---

20 Ibn Rusta (p. 152, lines 16-21). In this passage, Ibn Rusta may have been thinking of the Iranian king Kay Ḫusraw, introduced later in the text as the first king to have been crowned in Isfahan (*ibid*: p., 196): and mentioned again in regards to his coronation (ibid.: p. 228).

21 Ḥamza al-Iṣfahānī (fl. ca. 350) is our oldest Islamic source on this etymology. See also Yāqūt, *Buldān*: 1, pp. 292-3. But previously, the Greeks had also linked the name of Isfahan with military activities, see Christensen, Peter, *The Decline of Iranshahr: Irrigation and Environments in the History of the Middle East, 500 BC to AD 1500* (Copenhagen: Museum Tusculanum Press, 1993): p. 131, 146 and Tafazzoli, Ahmad, *Sasanian Society I. Warriors II. Scribes III. Dehqâns* (New York: Center for Middle Eastern Studies, Harvard University, 2000): pp. 1-7.

22 al-Ya'qūbī Ibn Wāḍiḥ, *K. al-Buldān*, ed. M. de Goeje, second edition (Leyden: Brill, 1967): p. 275; al-Iṣṭaḫrī: p. 199; Ibn Ḥawqal: p. 363; al-Muqaddasī: p. 34, 396; Ibn al-Nadīm, *K. al-Fihrist*, ed. Riḍā Tajaddud (Tehran: Intišārāt-i dānišgāh-i Ṭihrān, 1350š./1971): p. 301; Nāṣir-i Ḫusraw, *Safar-nāma*, ed. M. Dabīrsiyāqī (Tehran: Zawār, 1375š.): p. 165, 167.

23 On the winds in general see al-Mas'ūdī, *K. al-Tanbīh wa l-išrāf*, ed. M. de Goeje (Leiden: Brill, 1894): p. 29. On the winds in Isfahan see Arbāb Iṣfahānī, *Niṣf-i jahān*: pp. 88-9, 335;

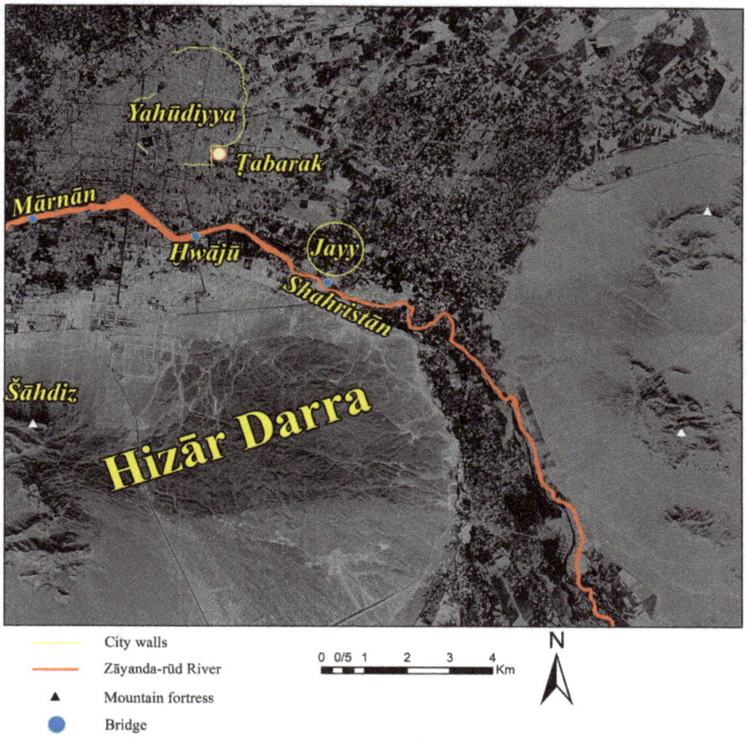

FIGURE 3B The city of Isfahan and its surroundings
SOURCE: CORONA PHOTO, MODIFIED BY A. SHOJAEE ESFAHANI

Jayy, the *madīnat Iṣfahān* mentioned by Ibn Rusta and other early authors, was situated near the Šahristān Bridge (still visible today), crossing the Zāyanda-rūd. The only other remnant of the Sasano-Islamic city is the very late mausoleum of the thirty-second Abbasid caliph, al-Rāšid, killed near the walls of Jayy/Šahristān in 532/1138. However, by analysing Ibn Rusta's highly precise first-hand descriptions in combination with the findings from archaeological investigations conducted prior to intense urbanisation, we have been able to reconstruct the probable path of the Sasanid walls together with the locations of the gates (Fig. 3b, 4).[24]

Taḥwīldār Iṣfahānī, Mīrzā Ḥusayn Ḫān, *Tārīḫ-i Iṣfahān. Juġrāfīyā-yi ṭabīʿī va insānī va āmār-i aṣnāf-i šahr*, ed. M. Sutūdah (Tehran: Intišārāt-i dānišgāh-i Ṭihrān, 1342š.): p. 13. On the knowledge of the Iranians on the wind, see Qarahčānlū, Ḥusayn, "Baḥs̱-i kutāh pīrāmūn-i bādhā az dīdgāh-i dānišmandān-i islāmī", *Majalla-yi dāniškadah-i ilāhiyyat*, XL (1369š.): pp. 119-39, here pp. 123-7.

24 See Ibn Rusta: pp. 160-2; Hunarfar, Luṭf-Allāh, *Ganjīnah-yi ās̱ār-i tārīḫī-yi Iṣfahān* (Isfahan: Imāmī, 1344š./1966): pp. 8-9; *Mīrfattāḥ*, ʿAlī-Aṣġar, "Ās̱ār-ī bāstānī-yi Jayy-i bāstān

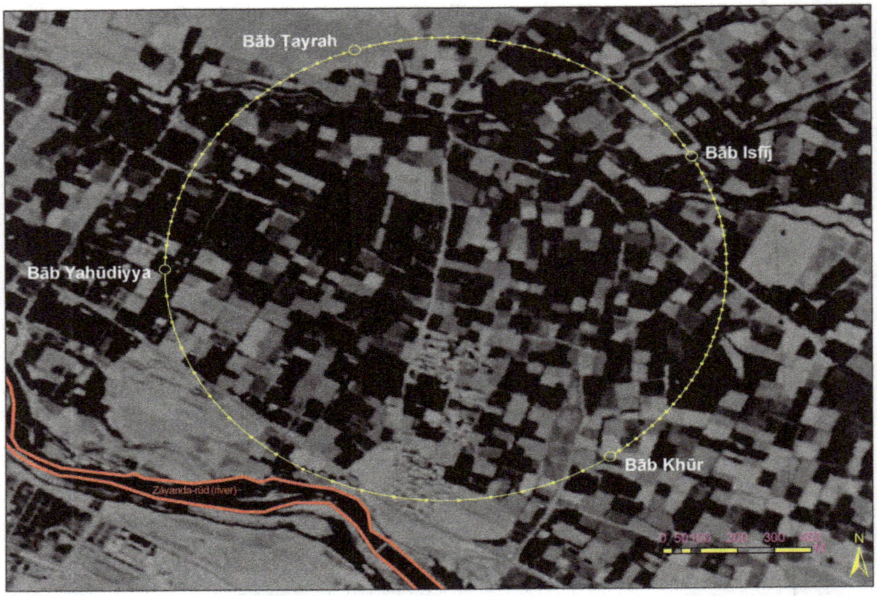

FIGURE 4 The walls of Sassanid Jayy
SOURCE: CORONA PHOTO, MODIFIED BY A. SHOJAEE ESFAHANI

The name, 'the Jewry', would suggest that the newer centre of 'Yahūdiyya', established northwest of Jayy, was initially a predominantly a Jewish settlement. After the Abbasid governor moved his residence here in the second/eight century, settlement grew denser around the area now occupied by the Friday Mosque, leading to the integration of neighbouring villages into an ever larger agglomeration. By the fourth/tenth century, Yahūdiyya had probably outgrown Jayy, although according to al-Muqaddasī the annual expenditures of their respective *masjid-i jāmiʿ* were still the same.[25] The Buyid rulers may have intentionally promoted this growth, and it was certainly them who surrounded Yahūdiyya with city walls (Fig. 5).[26] These played a major role in the political events of the Saljuq and Turko-Mongol period,[27] but lost their defensive functions during the Safavid period as the urban fabric of the new capital

(paygardī-yi masjid-i Jayy)", *Barrasīhā-yi tārīḫī*, LXVIII (1355š./1976): pp. 193-240, here 231-4. The location of the gate of Jayy can be deduced from the astronomical indications given by Ibn Rusta. The subject is investigated in detail in Shojaee, Ali, "Makānyābī-yi šahrhā-yi Jayy wa Yahūdiyya dar rūstāq-i Jayy-i Iṣfahān wa jābijā'ī-yi markaz-i qudrat tā bar-āmadan-i Saljūqiyyān', *Pažūhišhā-yi Īrānšināsī*, IV/1 (1393š.): pp. 57-76.

25 The growth process was studied by Golombek in her pioneering 1974 article.

26 See Shojaee, *Barrisī*: pp. 97-100.

27 See Durand-Guédy, in this volume.

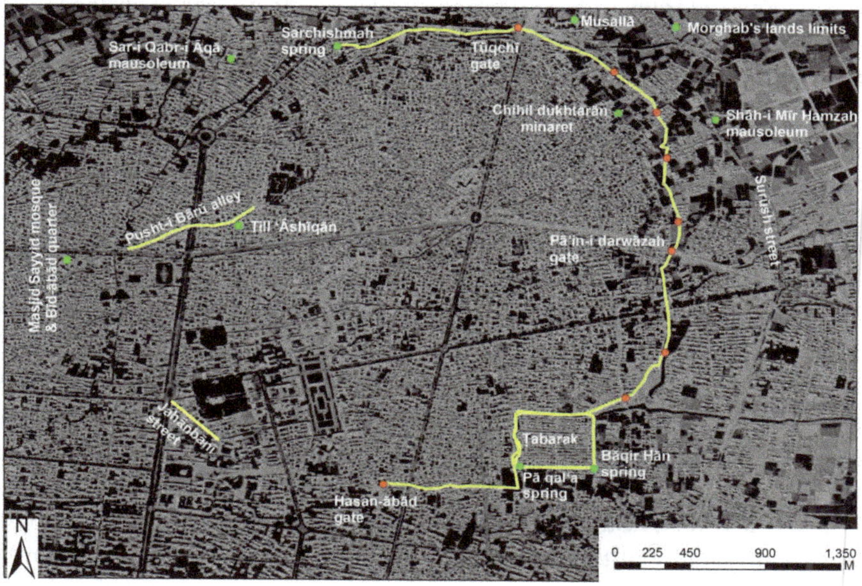

FIGURE 5 The walls of Yahūdiyya/Isfahan
SOURCE: CORONA PHOTO, MODIFIED BY A. SHOJAEE ESFAHANI

extended well beyond the medieval centre. Lack of maintenance and constant encroachments eventually led to the disappearance of the walls,[28] however it has been possible to reconstruct their outlines by analysis of the combined evidence from medieval sources, geo-referencing of Cherikov's map of 1851, and field observation and excavations.[29]

---

28 Chardin (*Voyages*: 7, p. 284) mentions six pathways through the walls and describes them as gradually disappearing. Tavernier does not mention the outline of the walls but says that some towers are in very poor condition, and that the inhabitants have opened several passages in the walls, see Tavernier, Jean-Baptiste, *The Six Voyages of John Baptiste Tavernier* (London, 1678): p. 389. See also Olearius, Adam, *The Voyages and Travels of the Ambassadors sent by Frederick, Duke of Holstein, to the Great Duke of Muscovy and the King of Persia* ..., trans. John Davies (London, 1669): p. 218.

29 Shojaee, *Barrisī*, deals extensively with the medieval walls of Yahūdiyya, building on al-Muqaddasī: p. 389; al-Māfarrūḫī, *K. Maḥāsin Iṣfahān*, ed. J. Ṭihrānī (Tehran, Maṭbaʿa majlis, 1933): p. 47, 91; Ḥamd-Allāh al-Mustawfī, *Nuzhat*: pp. 89-90; and the maps included in Mihryār, Muḥammad et al, *Asnād-i taṣwīrī-i šahrhā-yi īrānī-yi dawrah-yi qājār* (Tehran: Dānišgāh-i Šahīd Bihishtī; Sāzmān-i Mīrāṯ-i Farhangī-i Kišvar/Pažūhišgāh, 1378š./1999) pp. 170-173. My recent excavations on the site of Safavid palace of Jahān-namā, in the western part of the city, may have brought to light a portion of the walls; see Shojaee, Ali, "Excavations at the Jahānnamā Complex: urban archaeology at Isfahan, Iran", *Antiquity*, 91/355 (online publication last accessed 18 June 2018).

At about the same time as the walls were erected, a citadel called Ṭabarak was built in the southeast corner of the city (Figs. 3b, 5). This citadel was destroyed in the early twentieth century, leaving only the moat still visible.[30] However, maps and photos produced in the Qajar and early Pahlavi periods provide evidence of the buildings and external walls of the citadel.[31]

The location chosen for the Ṭabarak citadel can be considered from various perspectives. The placement at the margins of the town might be interpreted as a proof of the distance between the Shiʿi Buyid rulers, credited with the citadel's founding, and the Sunni population.[32] The specific choice of southeast corner of the city would then be due to two factors (Fig. 5). First, the springs of Bāqir Ḫān and Pā Qalʿa (literally 'Foot of the Citadel'), in the west and east of the location, as still seen today, would have allowed for the opening of a moat.[33] Second, this side was closer to Jayy, making control over communications in the entire eastern sector easier. The citadel fortress of Ṭabarak was crucial to the history of Isfahan.[34] As late as the Safavid period, it still served as the location for the royal treasury and custody of prisoners.[35] From its high walls there was good visibility of the fortresses in the surrounding countryside. Even in ruins at the beginning of the twentieth century, the remains rose to heights of 15 to 20 metres.[36]

---

30 See Golombek, "Urban Patterns": p. 20, Mihryār, *Asnād*: p. 170, 173.

31 See pictures by the German photographer Hoeltzer, active in Isfahan in the Qajar period, Hoeltzer, Ernst, *Īrān dar yakṣad wa sīzdah sāl-i pīš bā šarḥ wa taṣvīr.* I, *Iṣfahān*, ed. M. ʿĀṣimī (Tehran: Intišārāt-i vizārat-i farhang u hunar, 1355š./1976-7): p. 129, 132, 141). From the Pahlavi period, see Godard, André, "Iṣfahān", *Athār-é Īrān* (*Annales du service archéologique de l'Īrān*): II/1 (Paris: Geuthner, 1937): p. 161.

32 Durand-Guédy, *Iranian Elites*: p. 32.

33 Arbāb Iṣfahānī, *Niṣf-i jahān*: p. 106. Cf. Durand-Guédy, *Iranian Elites*: p. 57 note 30. According to Šafaqī, the fact that water naturally descended to this part of the city made it more feasible to create a deep surrounding moat, and contributed to the citadel's importance over time. Šafaqī, Sīrūs, *Juġrāfiyā-yi Iṣfahān* (Isfahan: Intišārāt-i dānišgāh-i Iṣfahān, 1381š.): pp. 178-9. For the importance of water in understanding the morphology of premodern cities of Iran, see Rante in this volume.

34 History shows that whoever controlled the citadel controlled the city. See Kutubī, Maḥmūd, *Tārīḫ-i āl-i Muẓaffar*, ed. ʿA. Nawāʾī (Tehran: Intišārāt-i Amīr Kabīr, 1364š./1975): p. 123 (Tamerlane became ruler of Isfahan when he was given the keys of Ṭabarak.)

35 See first-hand accounts by ʿImād al-Dīn Iṣfahānī, abridged by Bundārī, *Zubdat al-nuṣra*, ed. Martijn Theodoor Houtsma, *Recueil de textes relatifs à l'histoire des Seljoucides*, II: Histoire des Seljoucides de l'Irâq (Leyden: Brill, 1889): p. 154 and 156. For the Safavid period, see Chardin, *Voyages*: IV, pp. 1504-9, VII, pp. 483-91; Della Valle, Pietro, *Viaggi di Pietro della Valle.... la Turchia, la Persia, e l'India* (Brighton: Gancia, 1843): II, p. 117; Olearius, *Voyages*: p. 220.

36 Beaudouin, Eugène, *Ispahan sous les Grands Chahs*, special issue of *Urbanisme*, X (1932): pp. 1-47, here p. 14. Hoeltzer, *Īrān*: p. 101.

3 East of the Rustāq of Jayy

To the east, Isfahan was connected to the ancient town of Nā'īn, and beyond that to Khurasan and Kirman. From the Corona satellite images we can detect two main historic paved roads, respectively from Gulnābād and from Barsiyān to Nā'īn, which would agree with Maxime Siroux's description of the routes in this area.[37] Apart from these roads, other routes can be identified in the mountainous fields, which would have been under full control from the fortresses and fortifications east of Jayy (Fig. 3,9).

The defences to the east of the *rustāq* of Jayy included the three fortresses of Qahjāwaristān, Gawart and Čūm, and the isolated watchtower of Hatm-ābād (Fig. 2), each of which is situated within two to five km of another of the defences.[38] The fortresses of Qahjāvaristān (1749 m asl) and Gavart (1856 m asl) each extend over areas between 2000 and 3000 m^2. The water reservoirs in the ruins of these fortresses show numerous layers of material used to seal against leaks. Field surveys have also identified the remains of ancillary chambers and walls created along the edge of the mountain, as protection for the entrances to the fortresses (Figs. 6, 7). From their high positions, the fortresses ensured protection of the settlements immediately below, with their surrounding fields and roads. It is noteworthy that for both fortresses, construction involved preparation of the tallest part of the mountain as a smooth platform. In addition, at Qahjāvaristān the west wall of the mountain, overlooking Jayy, had been cut through in a way that allowed visual communication with other fortresses and settlements.[39] The Hatm-ābād watchtower (1761 m asl; with ruins now just over two metres in height) was built in a location with far-reaching visibility, compensating for the lack of view between the towers of Qahjāwaristān and Gavart.

This complex of fortified sites was particularly important for defence of the rich lands of the *rustāqs* of Quhāb, Burẖuvār and Jayy. In the Sasanid period, the fortresses held total command over the cities of Jayy (distant 8 to 12 km) and Qih (var. 'Qahjāwaristān'), in the *rustāq* of Quhāb, considered the second city of Isfahan province (Fig. 8).[40] The structures communicated with each other as well as with much more distant fortresses, such as Kulāh Qāḍī, at 22 to 33 km (Fig. 9).

37 Siroux, *Anciennes Voies*: pp. 26-7 and map.

38 This was also briefly noted by Siroux, Maxime, "Atesh-gah près d'Ispahân", *Iranica Antiqua*, v (1965): pp. 39-82, here pp. 77-8. The fortress site of Čūm is within an inaccessible military area, but can be seen on Google Earth.

39 On Qahjāvaristān, see also Schwarz, *Iran*: pp. 654-5.

40 Abū Nu'aym, *Ḏikr Aẖbār Aṣbahān*: I, p. 28.

FIGURE 6 The site of the fortress of Kūh-i Gavart
PHOTO SHOJAEE ESFAHANI 2012

FIGURE 7 Water reservoir in the fortress of Qahjāvaristān
PHOTO SHOJAEE ESFAHANI 2012

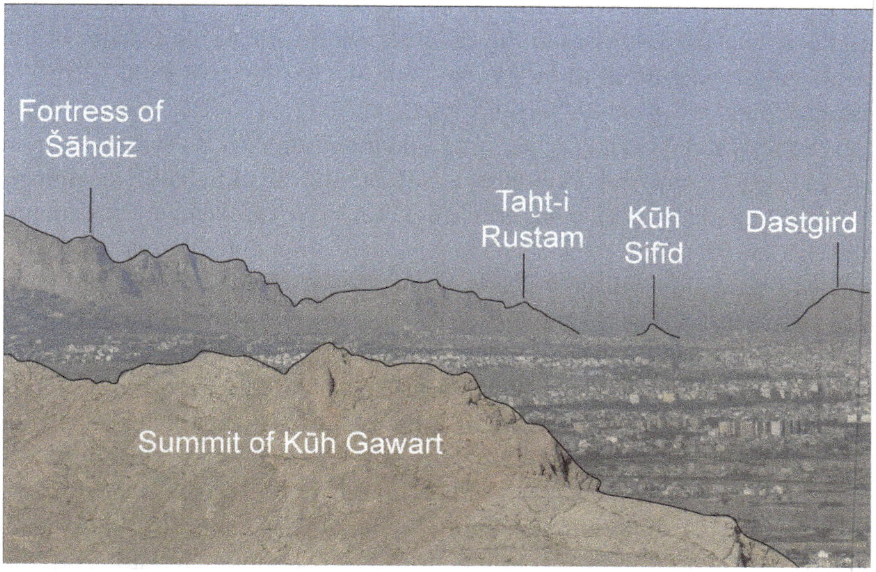

FIGURE 8 Fortified sites visible from the top of fortress of Kūh Gavart
PHOTO SHOJAEE ESFAHANI 2012

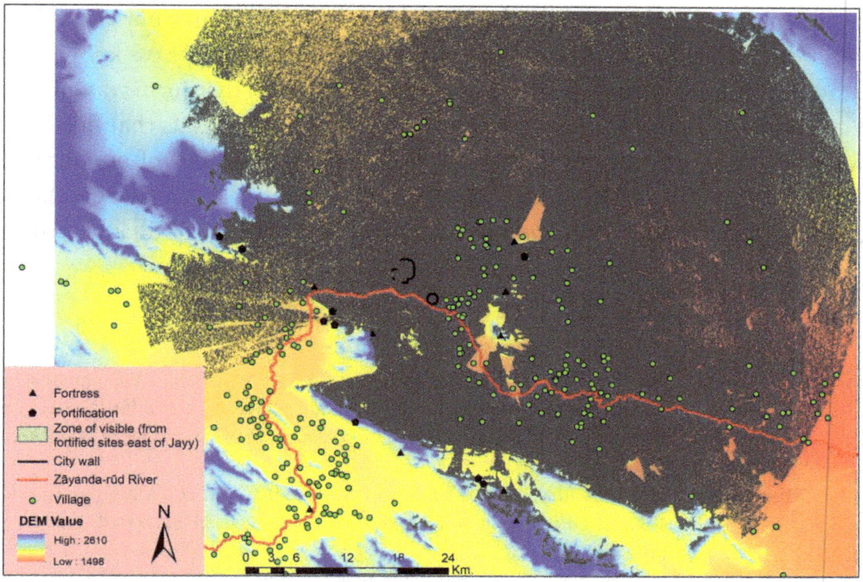

FIGURE 9 Visibility from fortified sites east of the *rustāq* of Jayy

The written sources describe events illustrating the importance of the fortifications and fortresses in the early Islamic period. In his account of the Arab conquest, al-Balāḏurī (d. 279/892) tells us that the aristocrats (*ašrāf*) of Isfahan found refuge in the "fortifications" (*maʿāqil*) of Qih. The local traditionist Abu'l-Šayḫ al-Iṣfahānī (d. 369/979), and after him Abū Nuʿaym al-Iṣfahānī (d. 430/1038), quote the account of a certain ʿAbd Allāh b. Badīl, recounting that the conqueror of Isfahan, Abū Mūsā al-Ašʿārī, descended on "the mountains of Jayy, the metropolis of Isfahan" (*jibāl Jayy madīnat Iṣfahān*), and then moved towards the village of Dārk, about one *farsaḫ* away.[41] Dārk has been precisely located (Fig. 3), and is seen to be much closer to the mountains of Qahjāwaristān and Gawart (9-13 km) than to other ones. This also makes it more accessible, and we can thus surmise that the conqueror's descent on '*jibāl Jayy madīnat Iṣfahān*' of al-Balāḏurī's refers to the mountains located in the east of Jayy, where ruins of fortresses and fortifications are still found.

An event taking place some three centuries later can also be located in this area. The Buyid historian Miskawayh recounts the Daylami leader Mardāwīč's death in Isfahan in 323/934-5, and describes the commemoration of the Zoroastrian festival of *sadah*, noting that "there was not a mountain (*jabal*) overlooking Jurīn of Isfahan nor any rising ground (*till ẓāhir*) but had arranged upon it firewood and brambles". Miskawayh informs us that Mardāwīč had his camp in Jurīn (Persian 'Gurīn'),[42] and continues that after three days, he would go to the palace (*qaṣr*) of Abū ʿAlī b. Rustam (*bi l-madīnat*).[43] We know from other sources (in particular a fourth/tenth *vaqf-nāmah*) that Jurīn was located East of Jayy.[44] In describing this same event, Ibn al-Aṯīr tells us that fireworks were set "from the bottom to the top" (*min asfalihi ilā aʿlāhi*) of a mountain called Karīm Kūh, in addition to other "mountains and hills which were there" (*jamīʿ al-jibāl wa'l-tilāl allatī hunāka*), and also that this Karīm Kūh looked over

---

41 al-Balāḏūrī, *Futūḥ al-buldān*, ed. M.J. de Goeje (reprint Beirut, 1988): p. 306; Abu'l-Šayḫ, *K. Ṭabaqāt al-muḥaddiṯīn bi-Iṣfahān*, ed. ʿAbd al-Ġafār Sulaymān al-Bindārī and Sayyid Ḥasan Kasrawī (Beirut: Dār al-ṯaqafat al-ʿilmiyya, 1988): I, p. 68; Abū Nuʿaym, *Ḏikr Aḫbār Aṣbahān*: I, p. 63.

42 Miskawayh, *Tajārib al-umam*, ed. and trans. in H.F. Amedroz and D.S. Margoliouth, *The Eclipse of the ʿAbbasid Caliphate*, 7 vols. (Oxford – London: Basil Blackwell, 1920-1): I, pp. 310-11, trans. vol. 4, p. 351.

43 Ibid.: I, p. 312, trans. IV, p. 353. This palace is also located by Ibn Ḥawqal (p. 363) "close to the city" (*bi l-madīnat Iṣfahān*). Cf. Schwarz, *Īrān*, p. 600.

44 See ʿĀbidī, Muḥammad-Amīn, "Matn-i vaqf-nāmah-yi Aḥmad ibn Rustam", *Vaqf. Mīrāṣ-i jāvīdān* XLVI (1383š./2004): pp. 63-7, here p. 65. The *vaqf-nāmah* mentions several buildings in the *rabaḍ* (suburb) of Jayy, including the bazaar of Garīn and its wall, Garīnah. Jurīn may be the Jizīn mentioned by Yāqūt and quoted by Schwarz, *Īrān*, p. 653.

(*al-mušrif ʿalā*) Isfahan.[45] Other authors have identified Karīm Kūh as Kūh Ṣuffa,[46] however if Jurīn was located east of the walls of Jayy,[47] it would be better identified with one of the mountains to the east of Isfahan, in particular Gavart or Čūm, both of which were closer to the outskirts of Jayy. We could also see a resemblance in the word forms 'Gūrīn' (> Jurīn) and 'Gavart'. The site of Gavart is still more probable considering that al-Yahūdiyya had not yet become the undisputed urban centre of the territory, a status which it only attained towards the end of the fourth/tenth century.[48]

4 South and Southwest of the rustāq of Jayy

Remains of fortresses have been identified on a number of mountains south of the *rustāq* of Jayy, among these: on Kūh Ṣuffa (fortress of Šāhdiz) and Kūh Qalʿa Buzī; on the eastern part of the mountainous range of Mahdašt, referred to as Kūh Kulāh Qāḍī (fortresses of Kulāh Qāḍī and Ṣanambar); on the western part of this range, referred to as Šāh Kūh (fortress of Saḫtūn, fortifications of Ayvān-i Dīv). Remains of lesser fortifications can also be seen at the lower levels of the Mahdašt range (next to Kutal Určīnī and the Pass of Lāšutur) and on the west side of Kūh Ṣuffa (see Table 1, Fig. 2).

4.1 *Šāhdiz and Qalʿa Buzī*

Kūh Ṣuffa stands only eight kilometres from Jayy and Yahūdiyya and offers a position of total command over the area (Figs. 3b, 10).[49] The strategic importance of this location grew as the urban configuration of the *rustāq* of Jayy developed:

---

45 Ibn al-Aṯīr, *al-Kāmil fi'l-ta'rīḫ*, ed.ʿU. Tadmurī, 13 vols (Beirut: Dār al-kutub al-ʿarabī, 1422/2001): VII, p. 33.

46 Ṣadr-Hāšimī, Muḥammad, "Baʿd az ḥamlah-yi ʿArab bi Īrān", *Farhang-i Iṣfahān*, XVII (1336š.): pp. 21-3, here p. 22; Minassian, *Shahdiz*: p. 47; Genito, Bruno and Saiedi Anaraki, Fariba, *ADAMJI PROJECT. From the Excavation (1972-1978) to the Archive (2003-2010) in the Masjed-e Jom'e, Isfahan* (Tehran: Embassy of Italy, 2011): p. 25.

47 See Maẓāhirī, Jamšīd, "Āfarījgān-Kažīn", *Yaġmā*, 318 (1353š./1974): pp. 703-4.

48 In Ibn Rusta (p. 160) and Ibn Faqīh (p. 530), the expression "*madīnat Iṣfahān*" still refers to Jayy, not to Yahūdiyya/Isfahan. Thermoluminescence dating of surface samples from Gavart points to a period from the sixth/twelfth to ninth/fifteenth centuries. All thermoluminescence analysis for research reported in this paper was conducted in 2015 at the Research Center for Conservation of Cultural Relics of the Iranian Cultural Heritage and Tourism Organization (Tehran). I thank Mrs. Bahrolulumi and Mrs. Azimi for their assistance in processing the samples.

49 A sixth/twelfth century source indicates that the then fortress of Šāhdiz was "immediately in front" of the city of Isfahan (*muqābila li-baldat Iṣfahān*), see Ibn al-Qalānisī, *Ta'rīḫ Dimašq*, ed. Henry Frédérick Amedroz, *History of Damascus* (Leyden: Brill, 1908): p. 153.

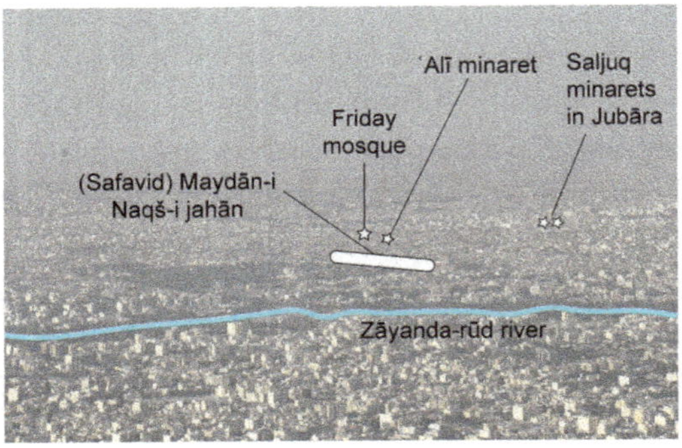

FIGURE 10
Panorama of modern Isfahan from the site of Šāhdiz/KūhṢuffa
PHOTO SHOJAEE ESFAHANI 2013

with the destruction of the town of Qih, the population began to concentrate in the town of Jayy, then in the *rabaḍ* of Jayy, eventually leading to formation of the town of Yahūdiyya.[50] The mountain offers substantial height (540 metres above the plain) and a vast, smooth platform, both of which were important conditions for settlement and construction. In the early period, when the economic centre of the *rustāq* was more to the east, Kūh Ṣuffa would likely have had a fortification or small fortress connected to the larger fortresses east of the *rustāq* (Gawart and Čūm, both eight kilometres from Jayy).[51] The site was in any case totally transformed during the Saljuq period. According to the sources, it was here that the Saljūq sultan Malik-Šāh (d. 485/1092) spent huge sums of money to build a fortress called Šāhdiz ('Royal Fortress', 'Fortress of the King'). The account of Ibn al-Aṯīr describes the events of construction at length.[52] The sultan intended this fortress to serve as the keep for his treasury and other

50 On the ruin of Qih see Abū Nu'aym, *Ḏikr Aḫbār Aṣbahān*: I, p. 15, 28. On the growth of Yahūdiyya see Golombek, "Urban Patterns"; Kamaly, Hossein, *Four Moments in the Early Islamic History of Isfahan* (PhD dissertation, Columbia University, 2004): p. 3; Bulliet, Richard W., *Islam. The View from the Edge* (New York: Columbia University Press, 1994): p. 73.

51 This may explain why there is no written evidence of Šāhdiz before the fifth/eleventh century, whereas the sources do report on the fortresses of Ḫān Lanjān and Ātišgāh. The identification of the Šāhdiz mentioned in the written sources with the archaeological remains on Kūh Ṣuffa was not established until the 1960s. The paternity of this discovery is controversial, although Minasian's claim seems more established; see Durand-Guédy, *Iranian Elites*: p. 91.

52 On the construction of Šāhdiz by Malik-Šāh, see Minasian, *Shahdiz*: pp. 17-19. The scholar includes the dubious story of the vizier Niẓām al-Mulk advising the sultan not to build a fortress in such a strategic place, for fear it would be taken by opponents of the state – events which duly occurred soon after the deaths of the two protagonists in 485/1092. The Šāhdiz was first taken by Terken Khatun (see Ibn al-Aṯīr, *Kāmil*: VIII, p. 360), then a few years later by the Ismailis of Aḥmad Ibn 'Aṭṭāš (ibid.: VIII, p. 452), see Durand-Guédy, *Iranian Elites*: pp. 154 and 158-9.

valuables when he was not present in Isfahan.[53] Thermoluminescence dating of ceramics recovered during the 2004 excavations indeed points to the early Islamic period for the deepest layers at the site.[54] The remains of Šāhdiz and its dependencies are distributed over an area of some 50 hectares. The inner citadel (*arg*) was constructed on a platform 150 metres in length and 15 to 20 metres in width, situated at the tallest point of the mountain, and therefore with optimum visibility, as well as protection to the north and east from an impenetrable precipice.

The fortress of Šāhdiz and the citadel of Ṭabarak, with its walls 15 to 20 metres in height, maintained mutual vision and command. With these fortifications on Kūh Ṣuffa and the plain of Isfahan, the Daylami and the subsequent Turkish rulers achieved communication between city and mountain, and increased their control over the town of Yahūdiyya, the *rustāq* of Jayy and its roads.

The sources refer to Qalʿa Buzī, also to the south of Jayy, as the "fortress of Ḫān Lanjān".[55] This fortification is situated within the *rustāq* of the same name, one of the most fertile of the region (Fig. 11), and unlike Šāhdiz was recorded as early as the Buyid period. Although the precise location of the city of Ḫān Lanjān is unknown, there is no doubt that the wording "fortress of Ḫān Lanjān" refers to Qalʿa Buzī, rather than some other military site of the *rustāq*. Indeed, Ibn Ḥawqal says that "[the town of Ḫān Lanjān] has a large fortress which served as the treasury (*ḫizāna*) of the [local] rulers (*umarāʾ*); it overlooks Ḫān Lanjān and its districts up to the outskirts of Isfahan (*ilā qurb-i Iṣfahān*)".[56] The only possible match with this description is the position of Qalʿa Buzī (see Fig. 12, showing visibility from the fortress looking south and north). The nearby large fortress of Saḫtūn can be excluded, since from this

53 This is noted for the reign of Malik-Šāh by Ẓahīr al-Dīn Nīšāpūrī, ed. A. Morton, *The Saljūqnāma of Ẓāhir al-Dīn Nīshāpūrī. A critical text making use of the unique manuscript in the Library of the Royal Asiatic Society* (Gibb Memorial Trust, 2004): p. 46.

54 Excavations at the site were directed by Mrs. Saʿīdī, who graciously provided the ceramics samples used for thermoluminescence dating.

55 Ibn Ḥawqal: p. 365, lines 15-17; Miskawayh: v, p. 315; Yāqūt, *Buldān*: II, p. 394. See Stern, Beazley and Dobson, "Fortress of Khān Lanjān", for a detailed description of the site. Classical geographers (Ibn Rusta: p. 153; Ibn Ḥawqal: p. 365) report the site of the city of Ḫān Lanjān as some seven to nine *farsaḫ* from the Isfahan centre. The position has still not been identified with precision, but would be somewhere between Ušturjān and Mubāraka. Ḥamd-Allāh Mustawfī (*Nuzhat*, p. 94) only mentions Fīrūzān (probably around Suhrufīruzān, at 2.5 km southwest of Pīr Bakrān). Ibn Baṭṭūṭa only mentions Ušturjān and Fīrūzān, Ibn Baṭṭūṭa. *Tuḥfat al-nuẓẓār*, ed. ʿAbd al-Hādī al-Tāzī (Rabat: Publications de l'académie royale du Maroc, 1407/1997): II, p. 29.

56 Ibn Ḥawqal: p. 365.

FIGURE 11 The fortress of Qalʿa Buzī (Ḫān Lanjān) above the Zāyanda-rūd
PHOTO COURTESY OF SĀZMĀN-I NAQŠA-BARDĀRĪ-YI KIŠWĀR, 1387Š./2008-9

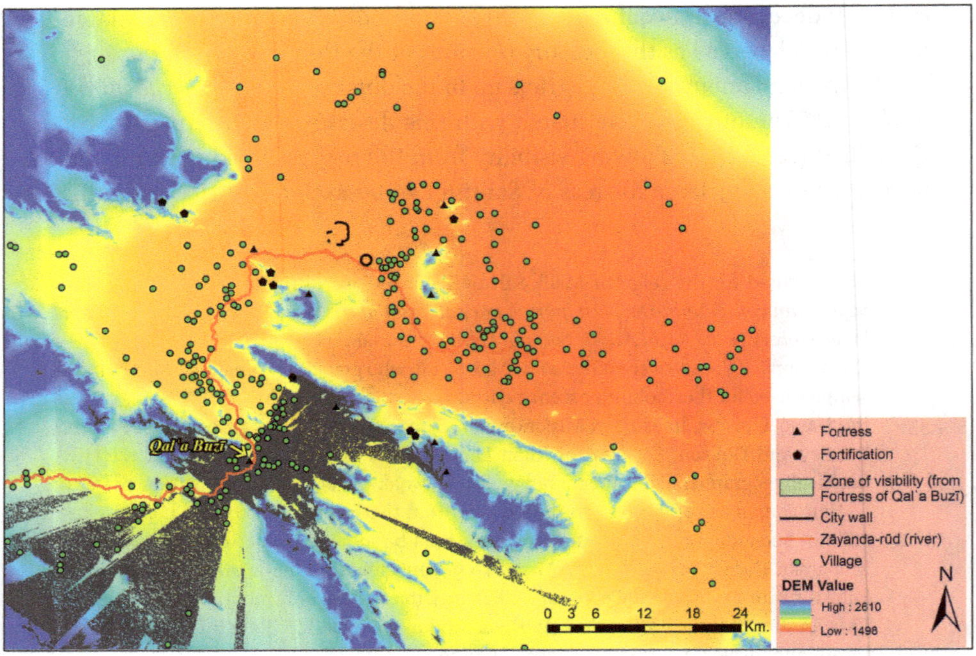

FIGURE 12 Visibility from the fortress of Qalʿa Buzī (Ḫān Lanjān)

location only the plain of Lanjān can be seen. Several mentions in the sources suggest the importance of Qalʿa Buzī in the events of the mid-fourth/tenth century, when the Samanids attacked Isfahan and the local Buyid ruler took refuge in the fortress, and again a century later, when it appears to be in the hands of the Niẓām al-Mulk's family.[57]

Significantly, Šāhdiz and Qalʿa Buzī are the two fortresses explicitly mentioned as taken by the Ismaili leader Aḥmad b. ʿAṭṭāš at the end of the fifth/eleventh century. From these fortresses the Ismailis were able to control the countryside and levy taxes, until Muḥammad b. Malik-Šāh took them back in 500/1107.[58] Control of Ḫān Lanjān would certainly have helped strengthen the Ismaili network to the south, particularly as concerns their positions around Arrajān (near modern Bihbihān), about 250 km in that direction.[59] Both Šāhdiz and Qalʿa Buzī were destroyed after the Saljuqs took them back.[60] Although these two fortresses had particular importance, the Ismailis must clearly have gained possession of other fortified sites.[61] Given the control they exerted over the countryside, they probably held secondary positions such as those of Taḫt-i Rustam and Kūh Sifīd, with Dastgird, and the fortress of Saḫtūn and fortifications of Ayvān-i Dīv (Fig. 2). In other words, one cannot imagine that Ismaili forces held only the great fortresses of Šāhdiz and Ḫān Lanjān, while all the less important ones remained in possession of pro-Saljuq forces in Isfahan.

4.2 *Between Qalʿa Buzī and Šāhdiz*

Qalʿa Saḫtūn and the fortifications of Ayvān-i Dīv were located in the western part of the Mahdašt range, on the southern skirts of Šāh Kūh, directly between Qalʿa Buzī and Šāhdiz, (Fig. 2). Both locations are about 12 km north of Qalʿa Buzī, with clear views to the latter fortress (Fig. 12).

The southward route from Isfahan to Fars and Khuzistan passed through a plain, approximately 11 km wide, between the mountains of Qalʿa Buzī and the Šāh Kūh. Given the topographic situation, and the locations of mountains and fields referenced in the geographic texts, it would appear that this was the only possible route for travel (Fig. 3). This further supports the hypothesis that the

57 Miskawayh: V, p. 315; Ibn al-Aṯīr, *Kāmil*: VIII, p. 453. On the role of Niẓām al-Mulk's family in Isfahan, see Durand-Guédy, *Iranian Elites*: pp. 112-129.

58 Ibn al-Aṯīr, *Kāmil*, VIII, p. 53. Durand-Guédy, *Iranian Elites*: p. 162.

59 Peter Willey, *Eagle's Nest*: p. 96, emphasises the necessity of communication between the various Ismaili fortresses.

60 Ḫān Lanjān was destroyed first, Šāhdiz was soon after. See Ibn al-Aṯīr, *Kāmil*: VIII, pp. 543-4; detailed account in Durand-Guédy *Iranian Elites*: pp. 174-181.

61 See ʿImād al-Dīn, *Ḫarīdat al-qaṣr*: I, p. 242 (speaking of the *qilāʿ al-malāḥida*, 'fortresses of the heretics').

Ismailis controlled more than the two fortresses of Ḫān Lanjān (Qalʿa Buzī) and Šāhdiz. Qalʿa Buzī could also have communicated with the towns of Jayy and Yahūdiyya through Qalʿa Kulāh Qāḍī. Command from Qalʿa Buzī over the Pass of Dih Surḫ on the one hand, and over the Isfahan-Ḫān Lanjān route on the other hand (Figs. 3, 12), would have achieved good control over communications.

The remains of Qalʿa Saḫtūn lie 1840-1900 metres above sea level and cover an area of 2.5 hectares – parameters closely resembling those of the fortress of Qalʿa Buzī. Considerable remains of stone and mortar constructions are found at the northwestern extremity of the fortress, on the highest cliff of the mountain (Fig. 13). Given the sightlines, it appears that these remains would pertain to a tower communicating with the plain of Lanjān and the two fortresses of Qalʿa Buzī and Qalʿa Kulāh Qāḍī. Inside the fortress are the remains of more than 70 rooms of different dimensions, executed in dry-stone masonry technique (Persian '*ḫuškī-čīn*'). There is also evidence of a rubble stone (Persian '*lāšah-sang*') and mortar wall surrounding the platform on which the fortress was built. The Darra Nāy spring, less than one kilometre distant, was probably the main source of water.

The fortification of Ayvān-i Dīv is situated 11 km north of Qalʿa Buzī, in a natural recess of a steep cliff (1980 m. asl) on the part of Šāh Kūh situated exactly opposite Kūh Qalʿa Buzī (Figs. 2, 14, 14b). The site appears as a platform about 20 metres long and 5.5 metres in height. The cliff-side location offers obstacle-free communication with Kūh Qalʿa Buzī. The position and type of construction leave no doubt of the exclusive function of this site as a further component of the network of fortresses, strengthening the position of Qalʿa Buzī in particular (Figs. 2, 12). This example clearly shows that a single fortress, whatever the excellence of its site, was never considered sufficient for protection of a territory, but was instead supported and informed by a network of other fortresses and fortifications.

4.3 *Mountain of Kulāh Qāḍī*

The fortresses of Kulāh Qāḍī and Taḫt-i Ṣanambar were situated in the eastern range of Mahdašt (Figs. 2, 3).

Qalʿa Kulāh Qāḍī was positioned near the mountain peak (2523 m asl), making it the highest fortress of all those currently in examination. The ruins cover an area of about 1000 m². Construction was both in hand-carved dry stone and in masonry using mortar. The remains of one of the northern walls indicate the presence of a two-story structure, however the rest is now destroyed, leaving substantial surface debris. The limited availability of flat surface did not allow for much ancillary construction or population, and the heavily mountainous location would have made access difficult and time-consuming. Had

FIGURE 13 Remains of fortifications at the highest point of the Qalʿa Saḫtūn
PHOTO SHOJAEE ESFAHANI 2013

FIGURE 14A Fortifications of Ayvān-i Dīv on the eastern slope of Šāh-kūh
PHOTO SHOJAEE ESFAHANI 2014

FIGURE 14B Fortifications of Ayvān-i Dīv, detail: wall built inside a natural gap in the mountain
PHOTO SHOJAEE ESFAHANI 2014

construction been carried out some 200 metres below the current location, at 2370-2390 metres altitude, there would have been greater possibilities for further structures, travel and trade. The fact that this fortress was built at the very top of the mountain indicates that its main purpose was to complete the communication network amongst the fortified sites. Indeed, the surrounding plain and routes to the towns of Isfahan province are clearly visible from the chosen location (Fig. 15).

Qalʿa Taḫt-i Ṣanambar is situated four kilometres northwest of Kulāh Qāḍī, at an altitude of 1800 metres. The fortress was installed on a descent from the heights of Miyān Kūh, to the north of the mountain of Kulāh Qāḍī, a location chosen for its visibility over a number of areas: the fortresses of Šāhdiz and Kūh Čūm (in the southwest and east of the *rustāq-i Jayy*); fortified sites of the Mahdašt range (e.g. Ayvān-i Dīv, Saḫtūn); the plain leading to the cities of Isfahan (Yahūdiyya and Jayy), as well as the southern route. The strategy of this positioning is further confirmed by the identification of five towers to the north and east, 20 to 25 metres distant from one another, making further communication possible towards the mountains of Čūm and Kūh Ṣuffa. The fortress site is about 10,000 m^2, and includes a wall and chambers made of stone, mortar and mud-brick (Persian '*ḫišt*'). The higher, main part of Ṣanambar is completely covered with mud-brick debris and stones (Fig. 16). The principal water was probably the spring of Čišm-i Bīdah, east of the fortress.

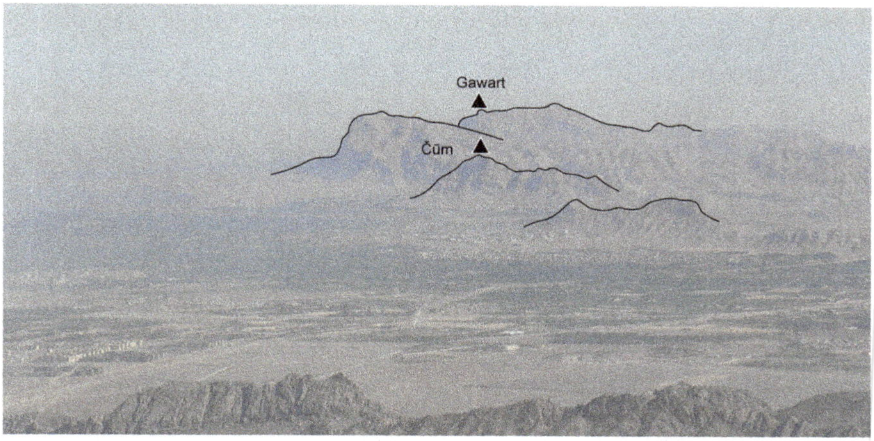

FIGURE 15 Some fortresses visible from the top of the fortress of Qalʿa Kulāh Qāḍī
PHOTO SHOJAEE ESFAHANI 2012

FIGURE 16 The core of the fortress of Ṣanambar
PHOTO SHOJAEE ESFAHANI 2014

In addition to the fortresses of Ṣanambar and Kulāh Qāḍī, situated on the peaks of the Mahdašt range, circular fortifications in stone and mortar were built at lower levels, where travel and communication were easier. The ruins of two towers, at present five to seven metres in height and 10 metres in width, can be seen on the road now known as Kutal Určīnī and Lāšutur (Figs. 2, 17). These towers would have supported control over the travel, trade and communications along the southern and southwestern routes.

FIGURE 17 The Určīnī road and its watchtower, close to the modern road opened through the mountain
PHOTO SHOJAEE ESFAHANI 2014

4.4 *Northwest of Kūh Ṣuffa*

The topography of low mountains in the southwestern part of the *rustāq* of Jayy permitted construction of communicating fortresses and fortifications, giving command over the inhabited zone and the southwestern route. Remains are still present at three sites (Fig. 2): Taḫt-i Rustam, Kūh Sifīd, and Kūh Dastgird.

Taḫt-i Rustam is the best known of the three sites, due to mention in itineraries of the Safavid era. Here, the remains include the stone foundation of a square tower, four metres on each side.[62]

The remains on Kūh Sifīd (also 'Kūh Šīrī'; sometimes 'Kūh-i Abu'l-Hawl, in reference to its leonine shape) are the most extensive, with rubble from a former building or buildings covering an area of 1000 square metres. The only intact structural remains are stone steps leading to a platform on the mountain.

62 Tavernier, *Six Voyages*: p. 378; Yuzukčiyān, Virṭāns, trans. L. Mīnāsīyān, *Vaṣf-i bināhā-yi mašhūr-i Iṣfahān* (Isfahan: Intišārāt-i Ġazal, 1377š.): p. 73; Gemelli Careri, Giovanni, *Giro del mondo del dottor d. Gio. Francesco Gemelli Careri* (Naples: Roselli, 1699): II, p. 88; Kaempfer, Engelbert, *Amoenitatum exoticarum politico-physico-medicarum*, V (Bonn: Wolfenbüttel, 1712): V, p. 196; De Bruyn, Cornelis, *Voyage au Levant* (Rouen, 1725): IV, pp. 125-6; Bell, John L.L., *Travels from St. Petersburg, in Russia, to diverse parts of Asia* (Glasgow: R. and A. Foulis, 1763): I, p. 125.

FIGURE 18
Remains of a paved road, two metres wide, in the Pass of Dih Surḫ
PHOTO SHOJAEE ESFAHANI 2012

Kūh Dastgird is the northernmost of the three areas. Here it was possible to see levelled platforms on the cliff side, buried under masses of rubble, and the remains of a road some 1.55 metres wide. However recent construction has destroyed this evidence, and no further study or identification of the remains appears possible.[63]

4.5 *Roads towards Fars and Khuzistan*

The earliest itineraries recorded in Islamic geography speak of two routes from Isfahan to Khuzistan and Fars, but with both of these via Ḫān Lanjān, meaning that there would have been a fork in the road after passing through that town.[64] Given the topographic constraints, the route can be identified with some precision: after crossing the Zāyanda-rūd, it would have passed west of Kūh Ṣuffa and then bordered the rich agricultural plain of Lanjān. Portions of this road can be identified on the Corona photo (Fig. 3). As late as the fifth/eleventh and eighth/fourteenth centuries respectively, the travellers Nāṣir-i Ḫusraw and Ibn Baṭṭūṭa both took this route from Khuzistan to Isfahan.[65] At the end of the

63 Ja'farī-Zand, *Iṣfahān*, p. 188.

64 Ibn Ḫurdāḏbih, *K. al-Masālik wa l-mamālik*, ed. M.J. de Goeje, second edition (Leyden: Brill, 1967): pp. 57-8; Qudāma b. Ja'far. *Kitāb al-ḫarāj*, ed. M.J. de Goeje (Leyden: Brill, 1889): p. 197; Ibn Rusta: p. 152.

65 Nāṣir-i Ḫusraw, *Safar-nāmah*: p. 165 and Ibn Baṭṭūṭa: II, p. 29.

fourth/tenth century, however, we hear of another road to Fars, not through Ḫān Lanjān but now through the village of Iṣfahānak and Qumša (Šāhriḍā).[66] In later periods this was the only route indicated in the itineraries as passing through Qumša.[67]

The difficulties of the travelling through the mountainous area from Kūh Ṣuffa to Hizār Darra, and further south from Miyān Kūh to Mahdašt, could explain why the southern road was not as popular in the early period, even though from Sasanid Jayy it would have been much shorter to go to Fars through Iṣfahānak (Figs. 2, 3, 3b). The later popularity of the Iṣfahānak road is probably connected to the rise of Yahūdiyya and to the fortifications constructed during the Buyid and Great Saljuq period, making the mountainous areas more secure. The opening of passes in the Mahdašt range, such as Kutal Určīnī and the Pass of Dih Surḫ, can also be attributed to these periods, given that without them the southern route would have remained relatively impractical (see Fig. 3).

On the southwest route, Ḫān Lanjān would have been connected to Ahwaz in Khuzistan by one road, and to Shiraz in Fars by another.[68] The description left by Ibn Ḥawqal is key to clarifying the route between Ḫān Lanjān and Yahūdiyya. He relates that a person coming from Fars and standing atop the Pass of Sarfarāz (literally: 'eminent', 'high') will see the *madīna*s of Jayy and Yahūdiyya and the gardens surrounding the urban centre.[69] Such a location could only be near the three mountains of Dastgird, Taḫt-i Rustam and Kūh Sifīd, which would be higher than other areas close to Isfahan, without intervening mountains in the northeastern direction leading to Jayy and Yahūdiyya, and so allow clear visibility. Having identified the location of the Pass of Sarfarāz, this further clarifies the route between Isfahan and Ḫān Lanjān. We can also better understand the functions of the fortifications in the heights of this area, commanding control over the southern and southwestern routes. Ibn Baṭṭūṭa's description of his arrival in Isfahan by the southwestern route makes clear that he travelled the same road as Ibn Ḥawqal, via the Pass of Sarfarāz.[70]

Ibn Baṭṭūṭa also mentions a village with a mosque, a half day's ride from Yahūdiyya, which was the last station before Isfahan. The presence of this mosque to the north of the Zāyanda-rūd proves that there was also a bridge

66 al-Muqaddasī (pp. 453, 457-8) is the first to mention this route. See Schwarz, *Iran*: pp. 653-4.

67 In the account of Ḥamd-Allāh al-Mustawfī, the road through Ḫān Lanjān is no longer mentioned.

68 Ibn al-Aṯīr, *Kāmil*: VII, p. 192; Nāṣir-i Ḫusraw, *Safar-nāmah*: p. 165.

69 Ibn Ḥawqal: p. 363, line 7.

70 Ibn Baṭṭūṭa: II, p. 29.

in this locality. Safavid sources mention the Mārnān bridge, and an inscription present as late as the nineteenth century recorded the later repairs by Shāh 'Abbās (Fig. 3b).[71] Travellers had a choice between crossing the river here or at the old bridge of Šahristān, just south of Jayy. Corona satellite images show two routes heading southwest after the Mārnān Bridge. One goes through the Pass of Sarfarāz and the lush plain of Lanjān, and the other runs between Kūh Ṣuffa and Taḫt-i Rustam. After a further 16 kilometres, both routes reach the Pass of Āb Nīl (Fig. 3). After the Pass of Āb Nīl, the remains of the two roads continue in two branches; one towards Qal'a Ḫān Lanjān and the other a further 20 kilometres to the Pass of Dih Surḫ, from which one could ultimately reach Khuzistan and Fars.

The itineraries and maps of the Safavid and Qajar eras provide information on the southern route towards Fars and Khuzistan.[72] The topography of the southern part of Zāyanda-rūd, with the Kūh Ṣuffa and hills of Hizār Darra, left travellers with little choice but to continue on their path towards the east, crossing the river via the Šahristān Bridge. This is why the sources mention this bridge and the village of Iṣfahānak as a possible route south, even though it was also possible to cross the Zāyanda-rūd at the Ḫwājū Bridge.[73] While passage through Hizār Darra would have been possible, the route would then have veered east, due to the heights of Miyān Kūh and Mahdašt. For this reason we can say that the main function of the Ḫwājū Bridge was not to facilitate passage on the southern route, but simply to allow communication with the southern side of the river.[74]

From the Corona satellite images we can make out a route crossing the easternmost part of Hizār Darra, and continuing to different locations in the range of Mahdašt: alternatively to Kutal Určīnī and the Pass of Lāšutur, or to the Pass of Dih Surḫ (Fig. 3). All these passes allowed travellers to reach Mihyār and then Qumšah in the south. The route to the Pass of Dih Surḫ and Kutal Určīnī is

---

71 Arbāb Iṣfahānī, *Niṣf-i jahān*: pp. 22, 23, 107-8; Yuzukčiyān, *Vaṣf*: p. 52.

72 These include Tavernier, *Six Voyages*: p. 245; Gemelli Careri, *Giro*: II, p. 229; Stocqueler, Joachim Hayward, *Fifteen months' Pilgrimage through Untrodden Tracts of Khuzistan and Persia, in a Journey from India to England* (London, 1832): I, p. 133, II, pp. 211-15; De Bode, Clement A., *Travels in Luristan and Arabistan* (London: J. Madden & co., 1845): II, p. 39; maps reproduced in Alai, Cyrus, *Special Maps of Persia 1477-1925* (Leiden and Boston: Brill, 2010): p. 114, 230, 237.

73 Tavernier, *Six Voyages*: p. 399; Della Valle, *Viaggi*: II: p. 54; Chardin, *Voyages*: vol. 4, p. 130; Dieulafoy, Jane, *La Perse, la Chaldée et la Susiane* (Paris: Hachette, 1887): pp. 335-341 and 354-9. Alai, Cyrus, *General Maps of Persia 1477-1925* (Leiden and Boston: Brill, 2005): pp. 149, 209, 250.

74 Luschey, Heinz, "The Pul-i Khwaju in Isfahan: a Combination of Bridge, Dam and Water Art", *Iran*, XXIII (1985): pp. 143-51, here p. 148.

particularly clear, while that to the Pass of Lāšutur is more difficult to make out due to the early asphalting of the gentle slope. None of these traces are visible on today's satellite images. The Pass of Dih Surḫ offered a route through one of the lower parts of the Šāh Kūh Mountain, convenient for travel south and southwest. Today a 200-metre long section of the route remains, designed to facilitate the passage of humans and animals on these steep slopes: 1 to 1.5 metres wide, partially carved in rock and partially executed in cobbles, including steps (Fig. 18). We can observe that the network of fortresses in this particular area had clear communication with the fortress of Ḫān Lanjān, which helps us to further understand the function of the overall system (Fig. 12). Another route for crossing the Mahdašt range was known as Kutal Určīnī or Určīnah (*kutal*: 'high hill', 'skirts of a hill'). Safavid and later sources frequently mention "Určīnī" as the southern entryway to Isfahan.[75] Here the natural relief was modified and parts of the road were paved with stone, to facilitate travel along the slopes. These works have since been destroyed or buried by modern road construction. A controlling watchtower was constructed less than 500 metres from the route, with sightlines to the surrounding fortresses (Fig. 17). However the Určīnī road was unsuited to camels, requiring the caravans took another road somewhat further east, along the gentler slopes of the Pass of Lāšutur (Figs. 3, 17). Here there was another watchtower, similar in appearance and function to that on the Určīnī route.

5 West of the *Rustāq* of Jayy

The fortified sites to the west of the *rustāq* of Jayy were situated along the road leading to the Baghdad-Khurasan Road, passing through Hamadan (Fig. 1). The best known is the so-called Qalʿa Ātišgāh, or Qalʿa Mārbīn, mentioned in the earliest Islamic texts.[76] Ibn Rusta, who locates it atop a mountain overlooking the Zāyanda-rūd, says it had command over "all the *rustāq*-s". Indeed, although Qalʿa Ātišgāh location was only 100 metres above the lands below, this was sufficient to provide clear sightlines to the heights in the east, stretching from Qahjāwārastān to Čūm, to the southern heights including Kūh Ṣuffa, Taḫt-i Rustam and Dastgird, and to the mountains north and northwest of Mārbīn,

75 Chardin, *Voyages*: IV, p. 1360. Floor, Willem, "The Bandar Abbas-Isfahan Route in the Late Safavid Era (1617-1717)", *Iran*, XXXVII (1999): pp. 67-94, here p. 81.

76 Ibn Ḫurdāḏbih: p. 20; Ibn Rusta: p. 153; Ḥamza al-Iṣfahānī, *Taʾrīḫ*: p. 28; al-Masʿūdī, *Murūj al-ḏahab*, ed. ʿAbd al-Ḥamīd (Beirut: Intišārāt Dār al-fikr, 1973): II, p. 238; Ibn Ḥawqal: p. 365; Āvī, *Tarjumah*: p. 95; Ḥamd-Allāh al-Mustawfī, *Nuzhat*: p. 92.

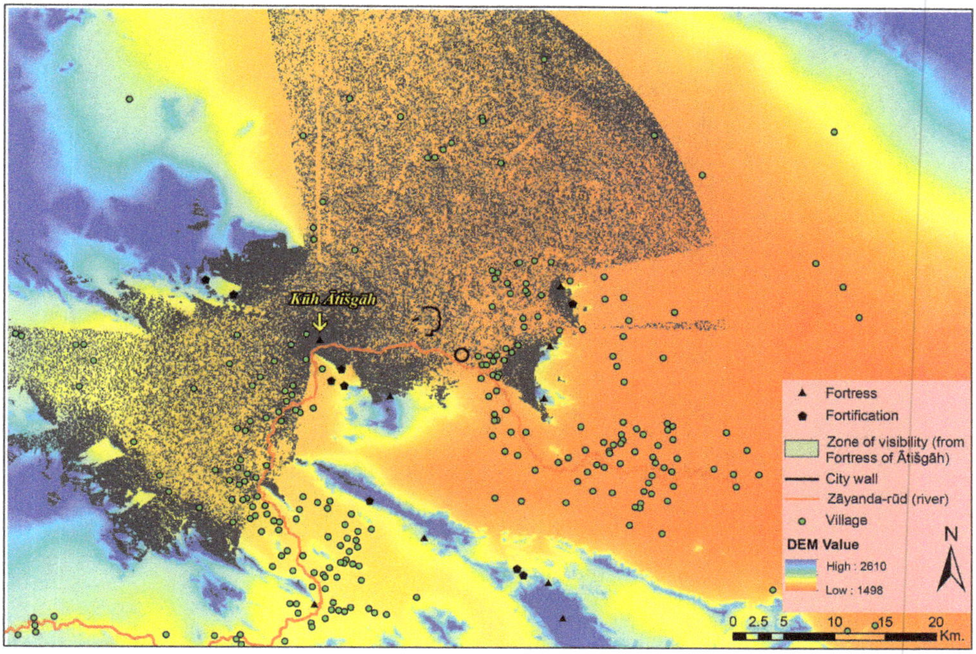

FIGURE 19 Visibility from the fortress of Kūh Ātišgāh

including the heights of Kūh ʿUmar (Fig. 19).[77] The earliest references to the "*qalʿat Mārbīn*" date back to the Islamic conquest, indicating it as one of the shelters for the people of Isfahan. The references continue in sources of the fourth/tenth and fifth/eleventh centuries, and in the Timurid and Safavid periods.[78]

About 9.6 km northwest of the Kūh-i Ātišgāh stands the isolated mountain Kūh ʿUmar, offering views of Kūh Ṣuffa, Ātišgāh, Āb Nīl Pass, Kūh Taḫt-i Rustam, Qalʿa Kulāh Qāḍī and also the Qahramān Cave. When necessary, it would have been possible to communicate between these mountains. Ascending towards the peak, we encounter the remains of a dry-stone masonry wall, 300 metres long, one metre wide, and still as much as two metres high (Fig. 20).

77 Siroux, Atesh-gah: p. 168, noted that the mud-brick tower at the height of this small mountain could have used smoke signals for communication up to Nāʾīn, via Kūh Čūm and Kūh-i Mīl (see full bibliographical references ibid.: pp. 77-78).

78 al-Balāḏurī, *Futūḥ*: p. 306; Āvī, *Tarjumah*: p. 113; Ḥāfiẓ-i Abrū, *Juġrāfiyā*, ed. Ṣādiq Sajjādī, 3 vols (Tehran: Mīrāṯ-i maktūb, 1997-9): II, p. 338; *ʿĀlam ārā-i ṣafavī*, ed. Yad-allāh Šukrī (Tehran: Intišārāt-i bunyād-i farhang-i Īrān, 1350š./1971): pp. 164, 170, 476. It is for this reason that the fortress of Šāhdiz, mentioned in Saljuq-period sources, was first identified with this site. It was only later that Minasian identified Šāhdiz and located it on Kūh Ṣuffa.

FIGURE 20 The fortified wall in the Kūh ʿUmar Kūh
PHOTO SHOJAEE ESFAHANI 2013

The Qahramān Cave is situated on the side of a small mountain, in a position 3.2 km northwest of the fortifications of Kūh ʿUmar, and is reached by climbing a six-metre escarpment. The remains of chambers, stone walls and a water reservoir with several layers of insulation can be seen. Two holes were opened in the corner of the largest room, enabling watch over the Mārbīn *rustāq* as far as the Āb Nīl Pass and the peak of Kulāh Qāḍī – yet another example of construction enabling communication between the lesser fortifications and the regional network of fortresses. Indeed, although the fortifications of Kūh ʿUmar and Qahramān Cave in the northwest of Kūh-i Ātišgāh appear to have required little construction work, they were effective in controlling communication amongst other fortified sites and the west road.

The west road connected Isfahan with the Khurasan Highway, via Karaj (Abū Dulaf) and Nihāwand. It is mentioned at the very outset of the Islamic period as the escape road taken by the Sasanid king Yazdegerd III, after his defeat at Nihāwand in the so-called 'Victory of Victories' and subsequently as the route taken by the Muslims for the conquest of Isfahan.[79] The road is then further described in the itineraries of the third/ninth (Qudāma) and fourth/tenth (al-Muqaddasī) centuries. From a very comprehensive description provided by

79 al-Ṭabarī, *Taʾrīḫ al-rusul wa'l-mulūk*, ed. M.A. Ibrāhīm (Cairo: Dār al-maʿārif, 1387/1967): IV, p. 140, 295; Abū Nuʿaym, *Ḏikr Aḫbār Aṣbahān*, I, p. 24, 63, 68.

Ḥamd-Allāh al-Mustawfī,[80] we learn that the first way station was four *farsaḫs* (around 26 km) from Isfahan, and was called Jūy-i Kūšk. Siroux placed this stop five kilometres from the centre of modern Najafābād,[81] however, considering that there is a locality named Kūšk situated five to six kilometres west of Kūh Ātišgāh, and that the distance between the Ātišgāh and Isfahan has been described as three *farsaḫs* (around 20 km), the Jūy-i Kūšk of Ḥamd-Allāh al-Mustawfī could instead be identified as the village named Kūšk, rather than the location proposed by Siroux, 25 km from Kūh Ātišgāh. However the precise location of Jūy-i Kūšk may be resolved, it is clear that the route continued from here to Gulpāyigān and then to Hamadan. This was the road taken by Della Valle in 1617, and still later by Flandin and Brugsch.[82] The Corona satellite images show that besides this itinerary, described in the travelogues, there was also a route running 26 km northward between the elevations of this area, up to the village of Čāl-i Siyāh. From here the travellers could continue north towards Kashan, or east towards Hamadan (Fig. 3). The fortifications of ʿUmar Kūh were situated exactly opposite the pass for this road. Together with the watch tower at Qahramān Cave, this gave the necessary control over the route and the means for relaying information through the entire area network of fortresses.

6 Conclusion

The waters of the Zāyanda-rūd and the fertility of lands north of Kūh Ṣuffa were the main factors permitting development on the plain of Isfahan. The surrounding heights were exploited to full potential for security over communications and settlement, which in turn contributed to urban growth. Although archaeological surveys have not yet achieved precise dating of the relative occupations, it is clear that each stage of urban growth in the *rustāq* of Jayy went hand in hand with new developments of fortresses and fortifications. As long as the main towns were still Qih and Jayy, it was the mountains in the east of the *rustāq* that played a decisive role. But in the next period, with the growth of Yahūdiyya and the integration of the southern province of Fars

80 Ibn Ḫurdāḏbih: pp. 18-19; Ibn Rusta: p. 166; Qudāma: pp. 198-9; al-Muqaddasī, p. 401; Ḥamd-Allāh al-Mustawfī, *Nuzhat* (ed. Le Starange 1915): pp. 171-2.

81 Siroux, *Anciennes voies*: p. 8; Tomaschek, *Topographie*: p. 24, confuses Jūy-i Kūšk with Bāb-i Kūšk, which Ḥamd-Allāh al-Mustawfī describes as a district of Isfahan.

82 Flandin, Eugène and Coste, Pascal, *Voyage en Perse* (Paris: Gide et Baudry, 1851): pp. 373-5; Brugsch, Heinrich Karl, *Reise der K. Preussischen Gesandtschaft nach Persien* (Leipzig: J.C. Hinrichs, 1863): II, pp. 19-48. Della Valle, *Viaggi*: I, pp. 449-51.

and Khuzistan in the Buyid and Saljuq polities, it was the fortified sites of the mountainous areas south of Zāyanda-rūd that were key to securing the access roads. Through fieldwork and application of new GIS technologies, we have demonstrated that the individual sites around Isfahan must be considered as elements in a network, in visual communication with the surrounding areas and with other fortress and fortifications.

These results are significant in understanding the modalities of a city's control over its hinterlands. We can see that when a number of these sites fell into the hand of hostile forces, as was the case during the Ismaili uprising, the city's control over the hinterland was placed in doubt, and even the survival of the city could be at stake. In other words, attention to the hinterland is very useful in understanding what happened inside the city in the different periods. In particular, the events of 485-500/1092-1107 provide an excellent example.

Research into the role of fortified sites in urban development can proceed on two levels. First, around Isfahan, we can continue the search for evidence of other fortresses or fortifications, which we expect would have composed part of the system (zones indicated in Fig. 1). Second, comprehensive surveys similar to the one reported here can be conducted around other Iranian cities, for crucial evidence of their history.

Bibliography

ʿĀbidī, Muḥammad-Amīn, "Matn-i vaqf-nāmah-yi Aḥmad ibn Rustam", *Vaqf. Mīrāṯ-i jāvīdān*, XLVI (1383š.): pp. 63-7.

Abu'l-Šayḫ, *K. Ṭabaqāt al-muḥaddiṯīn bi-Iṣfahān*, ed. ʿAbd al-Ġafār Sulaymān al-Bindārī and Sayyid Ḥasan Kasrawī (Beirut: Dār al-ṯaqafat al-ʿilmiyya, 1988).

Abū Nuʿaym al-Iṣfahānī, *Ḏikr Aḫbār Aṣbahān*, ed. Sven Dedering, *Geschichte Isbahans*, 2 vols (Leyden: Brill, 1931).

ʿĀlam ārā-i ṣafavī, ed. Yad-allāh Šukrī (Tehran: Intišārāt-i bunyād-i farhang-i Īrān, 1350š).

Alai, Cyrus, *General Maps of Persia 1477-1925* (Leiden and Boston: Brill, 2005).

Alai, Cyrus, *Special Maps of Persia 1477-1925* (Leiden and Boston: Brill, 2010).

Arbāb-Iṣfahānī, Muḥammad Mahdī, *Niṣf-i jahān fī taʿrīf al-Iṣfahān*, ed. Manūčihr Sutūdah (Tehran: Amīr Kabīr, 1340š./1962).

Āvī, Ḥusayn, *Tarjumah-yi maḥāsin-i Iṣfahān*, ed. ʿAbbās Iqbāl, revised ed. by Muḥammad-Riḍā Zādhūš (Isfahan: Sāzmān farhangī va tafrīḥī-yi šahrdārī-yi Iṣfahān, 1385š./2006).

al-Balāḏūrī, *Futūḥ al-buldān*, ed. M.J. de Goeje (reprint Beirut, 1988).

Beaudouin, Eugène, *Ispahan sous les Grands Chahs*, special issue of *Urbanisme*, X (1932): pp. 1-47.

Bell, John L.L., *Travels from St. Petersburg, in Russia, to diverse parts of Asia* (Glasgow: R. and A. Foulis, 1763).

Brugsch, Heinrich Karl, *Reise der K. Preussischen Gesandtschaft nach Persien* (Leipzig: J.C. Hinrichs, 1863).

Bulliet, Richard W., *Islam. The View from the Edge* (New York: Columbia University Press, 1994).

Bundārī, *Zubdat al-nuṣra*, ed. Martijn Theodoor Houtsma, *Recueil de textes relatifs à l'histoire des Seljoucides*, II: Histoire des Seljoucides de l'Irâq (Leyden: Brill, 1889).

Chardin, Jean, *Voyages du chevalier Chardin en Perse et autres lieux de l'Orient* (Paris: Le Normant, 1811).

Christensen, Peter, *The Decline of Iranshahr: Irrigation and Environments in the History of the Middle East, 500 BC to AD 1500* (Copenhagen: Museum Tusculanum Press, 1993).

De Bode, Clement A., *Travels in Luristan and Arabistan* (London: J. Madden & co., 1845).

De Bruyn, Cornelis, *Voyage au Levant* (Rouen, 1725).

Della Valle, Pietro, *Viaggi di Pietro della Valle.... la Turchia, la Persia, e l'India* (Brighton: Gancia, 1843).

Dieulafoy, Jane, *La Perse, la Chaldée et la Susiane* (Paris: Hachette, 1887).

Durand-Guédy, David, *Iranian Elites and Turkish Rulers: a History of Isfahan in the Saljuq Period* (London and New York, 2010).

Farhang-i juġrafiyā-yi Īrān. X. Iṣfahān (Tehran: Intišārāt-i dāyirah-yi juġrāfiyā'i-yi sitād-i ārtiš/Publications de la section géographique de l'Etat-major de l'armée 1355š. [/1976]).

Flandin, Eugène and Coste, Pascal, *Voyage en Perse* (Paris: Gide et Baudry, 1851).

Floor, Willem, "The Bandar Abbas-Isfahan Route in the Late Safavid Era (1617-1717)", *Iran*, XXXVII (1999): pp. 67-94.

Gemelli Careri, Giovanni, *Giro del mondo del dottor d. Gio. Francesco Gemelli Careri* (Naples: Roselli, 1699).

Genito, Bruno and Saiedi Anaraki, Fariba, *ADAMJI PROJECT. From the Excavation (1972-1978) to the Archive (2003-2010) in the Masjed-e Jom'e, Isfahan* (Tehran: Embassy of Italy, 2011).

Godard, André, "Iṣfahān", *Athār-é Īrān (Annales du service archéologique de l'Īrān)*, II(1) (Paris: Geuthner, 1937).

Golombek, Lisa, "Urban Patterns in Pre-Safavid Isfahan", *IrSt*, VII/1 (1974): pp. 18-44.

Ḥāfiẓ-i Abrū, *Juġrāfiyā*, ed. Ṣādiq Sajjādī, 3 vols (Tehran: Mīrāṣ-i maktūb, 1997-9).

Ḥamd-Allāh al-Mustawfī, *K. Nuzhat al-qulūb*, ed. M. Dabīrsiyāqī (Tehran: Ṭahah, 1378š. [/1999-2000]).

Ḥamza al-Iṣfahānī, *Ta'rīḫ sinī mulūk al-arḍ wa'l-anbiyā'*, ed. S.J. Īrānī-Tabrīzī (Berlin: Kaviani, 1340/1921-22).

Hoeltzer, Ernst, *Īrān dar yakṣad wa sīzdah sāl-i pīš bā šarḥ wa taṣvīr. 1, Iṣfahān*, ed. M. 'Āṣimī (Tehran: Intišārāt-i vizārat-i farhang u hunar, 1355š./1976-7).

Hunarfar, Luṭf-Allāh, *Ganjīnah-yi āṯār-i tārīḫī-yi Iṣfahān* (Isfahan: Imāmī, 1344š./1966).
Ibn al-Aṯīr, *al-Kāmil fi'l-ta'rīḫ*, ed.ʿU. Tadmurī, 13 vols (Beirut: Dār al-kutub al-ʿarabī, 1422/2001).
Ibn Baṭṭūṭa. *Tuḥfat al-nuẓẓār*, ed. ʿAbd al-Hādī al-Tāzī (Rabat: Publications de l'académie royale du Maroc, 1407/1997).
Ibn Ḥawqal, *K. Ṣūrat al-arḍ*, ed. Michael de Goeje revised by Joel H. Kramers (Leyden: Brill, 1939).
Ibn Ḫurdāḏbih, *K. al-Masālik wa l-mamālik*, ed. M.J. de Goeje, second edition (Leyden: Brill, 1967).
Ibn al-Nadīm, *K. al-Fihrist*, ed. Riḍā Tajaddud (Tehran: Intišārāt-i dānišgāh-i Ṭihrān, 1350š./1971).
Ibn Rusta, *K. al-Aʿlāq al-nafīsa*, ed. Michael de Goeje, second edition (Leyden: Brill, 1967).
Ibn al-Qalānisī, *Ta'rīḫ Dimašq*, ed. Henry Frédérick Amedroz, *History of Damascus* (Leyden: Brill, 1908).
ʿImād al-Dīn al-Iṣfahānī, *Ḫarīdat al-qaṣr*, partial ed. ʿA. Āl Ṭu'ma, 3 vols (Tehran: Mīrāṯ-i maktūb, 1378š./1999).
al-Iṣṭaḫrī, *K. al-Masālik wa l-mamālik*, ed. M. de Goeje (Leyden: Brill, 1870).
Jaʿfarī-Zand, ʿAlī-Riḍā. *Iṣfahān pīš az Islām* (Tehran: Intišārāt-i ān, 1381š./2002).
Kaempfer, Engelbert, *Amoenitatum exoticarum politico-physico-medicarum*, V (Bonn: Wolfenbüttel, 1712).
Kamalizad, Jaleh, "The Landscape Archaeology in Lanjān Plain of Iṣfahān, Iran, during the Islamic Period (650-1722 AD) (Preliminary Report)", *Archaeologies*, XII/1 (2016): pp. 91-109.
Kamaly, Hossein, *Four Moments in the Early Islamic History of Isfahan* (PhD dissertation, Columbia University, 2004).
Kāšānī, Suhayl and Ḍarabī, Mirzā ʿAbd al-Raḥīm, *Mir'āt-i Qāsān yā Tārīḫ-i Kāšān*, ed. Īraj Afšār, *Farhang-i Īrān-zamīn*, III (1334š.): pp. 105-262.
Kutubī, Maḥmūd, *Tārīḫ-i āl-i Muẓaffar*, ed. ʿA. Nawā'ī (Tehran: Intišārāt-i Amīr Kabīr, 1364š./1975).
Knapp, Bernard and Ashmore, Wendy, "Archeological landscapes: Constructed, Conceptualized, Ideational", in W. Ashmore and B. Knapp, *Archaeologies of Landscape. Contemporaries Perspectives* (Malden, Mass., 1999): pp. 1-30.
Luschey, Heinz, "The Pul-i Khwaju in Isfahan: a Combination of Bridge, Dam and Water Art", *Iran*, XXIII (1985): pp. 143-51.
al-Māfarrūḫī, *K. Maḥāsin Iṣfahān*, ed. J. Ṭihrānī (Tehran, Maṭbaʿa majlis, 1933).
al-Masʿūdī, *K. al-Tanbīh wa l-išrāf*, ed. M. de Goeje (Leiden: Brill, 1894).
al-Masʿūdī, *Murūj al-ḏahab*, ed. ʿAbd al-Ḥamīd (Beirut: Intišārāt-i Dār al-fikr, 1973).
Maẓāhirī, Jamšīd, "Āfarījgān-Kažīn", *Yaġmā*, CCCXVIII (1353š.): pp. 703-4.

Mihryār, Muḥammad et al, *Asnād-i taṣwīrī-i šahrhā-yi īrānī-yi dawrah-yi qājār* (Tehran: Dānišgāh-i Šahīd Bihishtī; Sāzmān-i Mīrāṯ-i Farhangī-i Kišvar/Pažūhišgāh, 1378š.).
Minasian, Caro O., *Shahdiz of Isma'ili Fame. Its Siege and Destruction* (London: Luzac & Company Ltd. 1971).
Mīrfattāḥ, ʿAlī-Aṣġar, "Āṯār-ī bāstānī-yi Jayy-i bāstān (paygardī-yi masjid-i Jayy)", *Barrasīhā-yi tārīḫī*, LXVIII (1355š./1976): pp. 193-240.
Miskawayh, *Tajārib al-umam*, ed. and trans. in H.F. Amedroz and D.S. Margoliouth, *The Eclipse of the 'Abbasid Caliphate*, 7 vols. (Oxford – London: Basil Blackwell, 1920-1).
Mujmal al-tavārīḫ wa'l-qiṣaṣ, ed. Malik al-Šuʿarā Bahār (Tehran: Intišārāt-i Ḫāvar, 1318š./1939).
al-Muqaddasī, *K. Aḥsan al-taqāsim fī maʿrifat al-aqālim*, ed. M. de Goeje, second edition (Leyden: Brill, 1906).
Nāṣir-i Ḫusraw, *Safar-nāmah*, ed. M. Dabīrsiyāqī (Tehran: Zawār, 1375š).
Olearius, Adam, *The Voyages and Travels of the Ambassadors sent by Frederick, Duke of Holstein, to the Great Duke of Muscovy and the King of Persia ...*, trans. John Davies (London, 1669).
Qarahčānlū, Ḥusayn, "Baḥṯ-i kutāh pīrāmūn-i bādhā az dīdgāh-i dānišmandān-i islāmī", *Majalla-yi dāniškadah-i ilāhiyyat*, XL (1369š.): pp. 119-39.
Qudāma b. Jaʿfar, *Kitāb al-ḫarāj*, ed. M.J. de Goeje (Leyden: Brill, 1889).
Ragheb, Youssef, *Les messagers volants en terre d'Islam* (Paris: CNRS Editions, 2002).
Ṣadr-Hāšimī, Muḥammad, " Baʿd az ḥamlah-yi ʿArab bi Īrān ", *Farhang-i Iṣfahān*, XVII (1336š.): pp. 21-3.
Šafaqī, Sīrūs, *Juġrāfiyā-yi Iṣfahān* (Isfahan: Intišārāt-i dānišgāh-i Iṣfahān, 1381š./2003).
Schwarz, Paul, *Iran im Mittelalter nach den arabischen Geographen* (1896, reprint Hildesheim-New York, 1969).
Shojaee, Ali, *Barrisī-yi bāstān-šināḫtī-yi qilāʿ wa istaḥkāmāt-i Iṣfahān az manẓar-i juġrafiyā-yi idārī* (PhD, Univ. Tehran, 2014).
Shojaee, Ali, "Makānyābī-yi šahrhā-yi Jayy wa Yahūdiyyah dar rūstāq-i Jayy-i Iṣfahān wa jābijā'ī-yi markaz-i qudrat tā bar-āmadan-i Saljūqiyyān", *Pažūhišhā-yi Īrānšināsī*, IV/1 (1393š.): pp. 57-76.
Shojaee, Ali, "Excavations at the Jahānnamā Complex: urban archaeology at Isfahan, Iran", *Antiquity*, 91/355 (online publication).
Siroux, Maxime, "Atesh-gah près d'Ispahân", *Iranica Antiqua*, V (1965): pp. 39-82.
Siroux, Maxime, *Anciennes voies et monuments routiers de la région d'Ispahan* (Cairo: IFPO, 1971).
Stern, Samuel M., Beazley, E. and Dobson, A., "The Fortress of Khān Lanjān", *Iran*, IX (1971): pp. 45-57.
Stocqueler, Joachim Hayward, *Fifteen months' Pilgrimage through Untrodden Tracts of Khuzistan and Persia, in a Journey from India to England* (London, 1832).
al-Ṭabarī, *Ta'rīḫ al-rusul wa'l-mulūk*, ed. M.A. Ibrāhīm (Cairo: Dār al-maʿārif, 1387/1967).

Taḥwīldār Iṣfahānī, Mīrzā Ḥusayn Ḫān, *Tārīḫ-i Iṣfahān. Juġrāfīyā-yi ṭabīʿī va insānī va āmār-i aṣnāf-i šahr*, ed. M. Sutūdah (Tehran: Intišārāt-i dānišgāh-i Ṭihrān, 1342š.).

Tafazzoli, Ahmad, *Sasanian Society I. Warriors II. Scribes III. Dehqâns* (New York: Center for Middle Eastern Studies, Harvard University, 2000).

Tavernier, Jean-Baptiste, *The Six Voyages of John Baptiste Tavernier* (London, 1678).

Tomaschek, Wilhelm, *Zur historischen Topographie von Persien* (1883, reprint Osnabrück 1972).

Willey, Peter, *Eagle's Nest. Ismaili Castles in Iran and Syria* (London: Tauris, 2005).

al-Yaʿqūbī Ibn Wāḍiḥ, *K. al-Buldān*, ed. M. de Goeje, second edition (Leyden: Brill, 1967).

Yāqūt, *Muʿjam al-buldān*, ed. F. Wüstenfeld, *Jaqut's Geographisches Wörterbuch*, 6 vols (Leipzig: in comm. bei F.A. Brockhaus, 1866-73).

Yuzukčiyān, Virṭāns, trans. L. Mīnāsīyān, *Vaṣf-i bināhā-yi mašhūr-i Iṣfahān* (Isfahan: Intišārāt-i Ġazal, 1377š).

Ẓahīr al-Dīn Nīšāpūrī, ed. A. Morton, *The Saljūqnāma of Ẓāhir al-Dīn Nīshāpūrī. A critical text making use of the unique manuscript in the Library of the Royal Asiatic Society* (Gibb Memorial Trust, 2004).

Zakariyyā Qazwīnī. *K. Āṯār al-bilād wa aḫbār al-ʿibād*, Persian trans. ed. M. Šāhmurādī, (Tehran: Intišārāt-i Danišgāh-i Tihrān, 1371š.).

www.archatlas.org.

Biographical Note

Ali Shojaee Esfahani (PhD 2014, University of Tehran) is Chairman of the Department of Archaeology at the University of Isfahan. His doctoral research dealt with military architecture of the pre-Mongol period in and around Isfahan, taking a landscape archaeology approach. He has participated in excavation projects and archaeological surveys in Nishapur and other parts of Iran. His publications (in English) include: "Archaeological Survey of Kamar-e Marj Castle in Kerman Province", in *Studies on the Iranian World II Medieval and Modern*, eds. A. Krasnowolska and R. Rusek-Kowalska (Krakow, Jagiellonian University); and "Excavations at the Jahānnamā Complex: urban archaeology at Isfahan, Iran", in the online journal *Antiquity* (2017). His current projects deal with a number of sites (excavation and survey) in the valley of the Zāyanda-rūd, extending through various periods (Varzana was a particular focus for 2018).

Yazd: a "Good and Noble City" and an "Abode of Worship"

Jamsheed K. Choksy
Indiana University
jchoksy@indiana.edu

Abstract

This article explores what is known and not known about Yazd and its nearby settlements from antiquity through the ninth/fifteenth century. The article begins with deliberation on the location of the city in pre-Islamic and Islamic times. The investigation expands to include examination of the urban center, its polity and faith-based communities, the arrival and flourish of Islam, plus the survival of Zoroastrianism and Judaism there. Reconstruction and new construction, administrative and economic activities, and roads and waterways within the wider Dašt-i Yazd ('Yazd plain/valley') are discussed in relation to the city.

Keywords

Yazd – Zoroastrians – Muslims – trade – shrines

1 Identifying Antecedents and Developments

This article briefly revisits previous scholarship on the central Iranian city of Yazd and seeks to contribute by adding to, modifying and clarifying aspects of what is known and not known about the city and its *aḥvāl*, or 'regional conditions and situations.' It explores the data on the location of pre-Islamic Yazd and continues with discussion of the Islamic city and its surrounding settlements. The focus is on continuity and change in settlement patterns, construction and reconstruction of physical structures, administrative and economic activities, roads and waterways, faiths and communities, and the spread of

 DOI:10.1163/9789004434332_011

Islam and survival of Zoroastrianism and Judaism to the ninth/fifteenth century in the Dašt-i Yazd ('Yazd plain/valley') and its immediate surroundings (Fig. 1).

The situation at Yazd is complicated when compared with that of some other major cities of Islamic Iran, which can be directly connected, textually and/or archeologically, with their pre-Islamic predecessors, in a manner that evidences settlement at the same location or at several locations that eventually became contiguous or at least semi-contiguous as Muslims assumed demographic majority. *Hangmatāna, at Mount Alvand, known to the Greeks as Ecbatana, together with its *paridaida* (Median: **paridaiza*; Avestan: *pairidaēza*;' enclosed area' or walled garden), is one such composite city that persisted as Islamic Hamadan. Medieval Hamadan essentially developed from three localities: Till (or Tell) Hagmatāna, Sang-i Šīr, and Muṣallā.[1] Till Hagmatāna began when a citadel was erected during the Median kingdom (ca. 673-550 BCE) and persisted through the Abbasid caliphate (132/749-656/1258). Sang-i Šir developed under Hellenistic rule (Alexander III reached the city in 324 BCE). Muṣallā dates back to the Parthian or Arsacid kingdom, which controlled Media from ca. 155 BCE-224 CE.

At other urban locales, not all the formerly discreet pre-Islamic areas contributed to formation of the Muslim habitat. Achaemenian (550-330 BCE) Ragā served as the foundation for Sasanian (224-651 CE) and subsequently Islamic Ray (Rayy). As with Hamadan, the city was not a single unified locale in ancient times, instead having three main centers. The site now known as Čašma 'Alī was occupied from the Neolithic through the Achaemenid and Parthian periods, and habitation continued under Muslim rule. The Šahristān ('citadel') further south, saw settlement during the Parthian and Sasanian regimes, and continued occupation as a Muslim site. The location called Tepe Mil (Tapa-yi Mīl), with its Zoroastrian fire temple and magian administrative complex, began during Sasanian times. As the Muslim population expanded with the spread of Umayyad (41/661-132/749) through to Seljuq (428/1037-590/1194) settlement, over the previously semi-distinct Sasanian areas, Čašma 'Alī and the Šahristān lost their individual boundaries. Tapa-yi Mīl, however, still lay outside the second/eighth century city wall, possibly serving as a suburb where the Zoroastrian minority tried to isolate itself and thus avoid intermingling with Muslims, as also evidenced at Bukhara for example. However, ceramics

1 Briefly overviewed by Brown, Stuart C., "Ecbatana", in *EIr*, VIII (1997): pp. 80-4 (updated at http://www.iranicaonline.org/articles/ecbatana); and Planhol, Xavier de, "Hamadān iii. Historical Geography", in *EIr*, XI (2003): pp. 605-7 (updated at http://www.iranicaonline.org/articles/hamadan-iii).

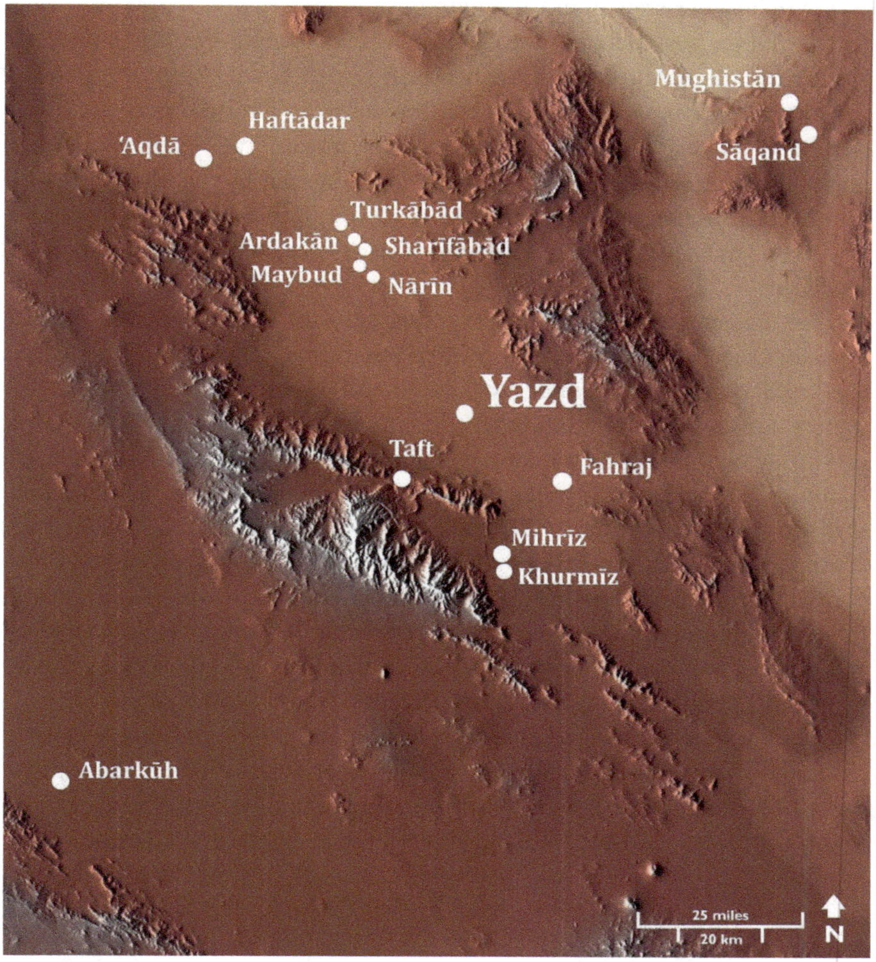

FIGURE 1 Yazd and its surroundings
CARTOGRAPHY: J.K. CHOKSY

and stucco remains indicate that the locality eventually came to include some Muslim inhabitants, before being abandoned and falling into ruin as all the residents gradually relocated to the main Muslim urban area.[2]

2 Rante, Rocco, *Rayy: From Its Origins to the Mongol Invasion, an Archaeological and Historiographical Study* (Leiden: Brill, 2014): pp. 12, 68, 73-4, 102, 121, 123, 140-1; more concisely Rante, Rocco (2010) "Ray i, Archeology", in *EIr*, at http://www.iranicaonline.org/articles/ray-i-archeo; both with references to previous archeological publications. Matheson, Sylvia A., *Persia: An Archaeological Guide*, 3rd edn, (London: Faber, 2001): pp. 23-4, sketched out the paths that connected the individual locations. On Zoroastrians' attempts to sequester their

A similar coexistence and ultimate decline occurred at Qaṣr-i Abū Naṣr, or Old Shiraz, a settlement dating to the Achaemenian Empire. Under the Sasanians, Qaṣr-i Abū Naṣr flourished as a walled citadel, mint and magian administrative center. These roles continued briefly after the Arab conquest, as indicated by data from buildings, coins and seals. However, the population gradually transferred to the city of Shiraz, established by the Muslims as the new administrative center during the second/eighth century, about 10 kilometers) to the southeast.[3] At that time, Qaṣr-i Abū Naṣr entered permanent decline.

The Achaemenian administrative site of Gaba (Greek: Gabae; Middle Persian: Gay; New Persian: Jayy) and the Jewish neighborhood of Yahūdiyya, were said to be still existent and separated by a distance of around three kilometers until as late as the fourth/tenth century, despite the fact that an Islamic center lay between the two centers and likely bridged them. Both of these localities can be textually and archeologically correlated with their Sasanian through Abbasid counterparts. As at Tepe Mīl and Old Shiraz, and perhaps through similar processes, the Sasanian administrative and Zoroastrian residential stronghold at Jayy eventually declined. Instead, it was Yahūdiyya and its surroundings that would become the important Islamic urban area and assumed the toponym of its province, Isfahan (from the Middle Persian provincial name Spahān), due to that term's prior association with Jayy.[4]

However, classical textual sources and Iranian written and archaeological data often fail to reveal precise information regarding the locational origin of many other cities.[5] Yazd, with its surrounding urban centers, offers a case in point. The Dašt-i Yazd is situated in central Iran at an elevation of around 1261

community in distinct neighborhoods and suburbs see Choksy, Jamsheed K., *Conflict and Cooperation: Zoroastrian Subalterns and Muslim Elites in Medieval Iranian Society* (New York: Columbia University Press, 1997): p. 29.

3 Frye, Richard N., "Introduction", in Richard N. Frye (ed.) *Sasanian Remains from Qasr-i Abu Nasr: Seals, Sealings and Coins* (Cambridge, Massachusetts: Harvard University Press, 1973): pp. 1-5, here pp. 2-4; Upton, Joseph M., "The Site of Qasr-i Abu Nasr and a Description of the Sealings", in ibid.: pp. 6-25, here pp. 8-9, 19, 23-5; Miles, George, "The Coins", in ibid.: pp. 26-36, here pp. 26-35.

4 Hansman, John, "Isfahan iv. Pre-Islamic Period", in *EIr*, XIII (2006): pp. 635-8 (updated at http://www.iranicaonline.org/articles/isfahan-iv-pre-islamic-period), here pp. 635-7; Planhol, Xavier de, "Isfahan ii. Historical Geography", in *EIr*, XIII (2006): pp. 617-22 (updated at http://www.iranicaonline.org/articles/isfahan-ii-historical-geography), here pp. 617-20; and especially Golombek, Lisa, "Urban Patterns in Pre-Safavid Isfahan", *IrSt*, VII/1 (1974): pp. 18-44.

5 See the observation and critique by Christensen, Peter, *The Decline of Iranshahr* (Copenhagen: Museum Tusculanum Press, 1993): pp. 25-7.

meters, and extends around 260 kilometers along a northwest-southeast axis.[6] To the north is the Dašt-i Kavīr, or 'barren desert', and to the east and southeast is Dašt-i Lūṭ, or 'bare desert'. The Dašt-i Yazd is essentially a peripheral and inter-desert region where the presence of water permits habitation. However, there is no mention of a city named Yazd or of other nearby towns within the Dašt-i Yazd in extant Achaemenian royal inscriptions and transactional tablets, and Yazd does not appear in any surviving official or private records of the Parthian kingdom. Nor is there a clear archaeological *till* or *tapa* ('mound') marking a site of ancient occupation, as for example can be seen at Hamadan.[7] Consequently, compared to cases such as Hamadan, Ray, Shiraz, and Isfahan, we face a more difficult task in pinpointing the origins and progression of from ancient settlement or settlements to the well-known, multi-confessional, mercantile, medieval Islamic city of Yazd, connected to other nearby habitations.

2 Pre-Islamic Yazd

The ancient Persian Asagarta or Sagartian tribe (rendered in Greek as Istachae) were said to have lived in the region where Yazd is located. In his fifth century BCE *History*, Herodotus (d. 425 BCE) referred to the Sagartians as a Persian-language speaking tribe of *νομάδες*, 'nomads', who fought alongside other Iranians in the Achaemenid army.[8] They were mentioned by King Darius I (r. 522-485 BCE) in his *Behistun Inscription*, regarding a failed uprising led by a self-proclaimed local ruler named Ciçantakhma, who was defeated, captured,

6 A description of the Yazd plain is given by Bonine, Michael E., *Yazd and Its Hinterland* (Marburg: Geographisches Institut der Universität, 1980): pp. 7, 10; the same author also provides maps of the topography and main settlements, p. 8 fig. 1, p. 9 fig. 2 and p. 198 fig. 72. Aerial photographic maps are available in Murtaḍā'ī, Muḥammad and Mīr-Dahqān Aškaḏarī, Sayyid Faḍl-Allāh, "Bāznigarī dar jāygāh-i šahr-i Yazd dar dawrān-i ṣadr-i Islām bā tikiya bar asnād-i maktūb va dāda-hā-ya bāstān-šināsī", *Pažūhiš-hā-yi bāstān-šināsī Īrān*, VI (1393š./2014): pp. 159-74 (with an English abstract pp. 20-21), here pp. 164-6; Mīr-Dihqān Aškaḏarī, Sayyid Faḍl-Allāh and Raḍvānī, Ḥasan and Kāẓimnažand, Ibrāhīm "Historical and Cultural Sites [and] Relative Chronology of Yazd-Ardakan Plain according to Archaeological Data", in M.Ḥ. ʿA. Ḫarānaqī, M. Ḫānīpūr, and R. Nāṣirī (eds), *Proceedings of the International Conference of Young Archaeologists* (Tehran, 1393sh./2014): pp. 358-68, here p. 366.

7 Fischer, Michael M.J., *Zoroastrian Iran between Myth and Praxis* (PhD dissertation, University of Chicago, 1973): pp. 119-20, based on comments by officials at the Yazd branch of the Ministry of Culture, attributes the lack of mound formation to the shifting of wind-blown sand, which would also have worn away ancient structures.

8 Herodotus, *History*, ed. and trans. Alfred D. Godley, 4 vols, (London: Heinemann, 1926-30): I, p. 125, and VII, p. 85.

disfigured, and impaled.[9] The accompanying relief carving depicts and names "Ciçantakhma, a Sagartian" as the sixth of ten main rebels.[10] More important in geographical terms is Darius' *Persepolis Inscription E*. Here, after naming each southwestern, western, and northwestern administrative province, or '*dahyu*', including "Elam [Fars], Media, Babylonia, Arabia, Assyria, Egypt, Armenia, Cappadocia, Sardis, Ionia", Sagartia is indicated as the first in a series of *dahyu*- progressing eastward, northward, and southward from central Iran: "Sagartia, Parthia, Drangiana, Aria, Bactria, Sogdia, Chorasmia, Sattagydia, Arachosia, Sind, Gandhara, Scythia, Maka", thus essentially placing Sagartia in the region subsequently known as the Dašt-i Yazd.[11]

In his *Geography*, Claudius Ptolemy (d. 168 CE) placed the Sagartian pasturelands along the southern part of the Dašt-i Kavīr, which is continuous with the north of Dašt-i Yazd, and also along the Pušt-i Kūh, immediately southwest of the Dašt-i Yazd,[12] in this way situating the Sagartians within the north and west of the Dašt-i Yazd. Both regions still receive the most rainfall in the Dašt-i Yazd region, and were likely best suited for grazing even in ancient times.[13] Indeed, potsherds and other materials at a site about 12 kilometers north of present-day Yazd city, over an area of around eight square kilometers, indicate settlement going back to the Achaemenid period – though those items have not been linked to the Sagartians or any other specific group.[14]

Gaius Plinius Secundus (Pliny the Elder, d. 79 CE), in his *Natural History*, mentioned a town of the Parthian period called "Issatis", which would seem

9 *Behistun Inscription* 2: 78-91, abridged at 4: 22-3. Inscription in Kent, Roland G., *Old Persian: Grammar, Texts, Lexicon*, 2nd edn (New Haven: The American Oriental Society, 1953): p. 122, 124, 128 and 131.

10 *Behistun Minor Inscription* G 8-9. Inscription in Kent, *Old Persian*: p. 135, with image on pl. 1.

11 *Persepolis Inscription E* line 15. Inscription in Kent, *Old Persian*: p. 136. For the Achaemenid provinces see Cook, John M., "The Rise of the Achaemenids and Establishment of Their Empire", in Ilya Gershevitch (ed.) *The Cambridge History of Iran*, II (Cambridge: Cambridge University Press, 1985): pp. 200-91, here p. 283 map 11; Frye, Richard N., *The History of Ancient Iran*, (Munich: Beck, 1984): p. 415 map 2.

12 Ptolemy, Claudius, *Geography*, ed. and trans. Edward L. Stevenson (New York: New York Public Library, 1932): 6, 2.6.

13 On rainfall and grazing see Bonine, *Yazd*: p. 7; and Lambton, Ann K. S., *Landlord and Peasant in Persia: A Study of Land Tenure and Land Revenue Administration*, rev. ed. (Oxford: Oxford University Press, 1969): p. 353.

14 The survey field report has not been published. The director is quoted in a brief public announcement: Khademzadeh, Mohammad H., "Archaeologists Discover Achaemenid Predecessor of Yazd City", *Paywand*, 16 May 2008, http://www.payvand.com/news/08/may/1154.html. More useful data is provided by Murtaḍā'ī and Aškaḏarī, "Bāznigarī".

to correspond with the tribal name of the Asagarta/Istachae/Sagartians.[15] Some scholars have proposed that Yazd thus began as Issatis (rendered also Isatichae, Ysatis, Yasatis) in Median or Achaemenian times.[16] Pliny, however, asserted that Issatis was situated in a *chora* (also *choara, choarena*: 'district') near the southern foothills of the Alburz Mountains – this would perhaps be at or close to Simnān (Semnān), meaning with the Alburz range to the north and the Dašt-i Kavīr to the south. That location would fit well with Pliny's statement that Issatis was built on a clifftop, rather than in the long inter-desert oasis valley where Yazd arose.[17] However, because the Sagartians were nomads, their range of habitation may very well have extended from Simnān down to Yazd by the time of the Parthian kings. Their territory would then skirt the Dašt-i Kavīr, lying between those two pastoral zones, and permit the single toponym 'Issatis' for both locations in pre-Islamic times.[18]

Information from the *Tabula Peutingeriana*, which survives as a thirteenth century CE copy of a fourth or fifth century CE illustrated *itinerarium*,[19] and is ascribed to a Roman archetype of the first century BCE, is also valuable for locating ancient Yazd. The *Tabula* notes a Parthian way station called Cetrora (perhaps derived from *chora*), along a road heading eastward from Hamadan. The place name Cetrora has been compared with an Islamic era toponym *Kaṯa/Kata/Kasa*, meaning 'hollow' and hence 'valley' from Old Iranian and Avestan *kata-*, 'enclosure', for Yazd.[20] Indeed, much later, the fourth/tenth

15 Pliny the Elder (Gaius Plinius Secundus), *Natural History*, ed. and trans. Horace Rackham and others, 10 vols (London: Heinemann, 1938-63): VI, pp. 44, 113.

16 Tomaschek, Wilhelm, "Zur Historischen Topographie von Persien I: Die Strassenzüge der tabula Peutingeriana", *Sitzungsberichte der Philologisch-Historischen Classe der Kaiserlichen Akademie der Wissenschaften* 102 (Vienna: Gerold, 1883): pp. 145-231, here p. 165; Markwart (Marquart), Josef, "Beiträge zur Geschichte und Sage von Erān", *ZDMG*, XLIX (1895): pp. 628-72, here p. 669.

17 As observed by Christensen, *Decline*: pp. 288-9 note 21.

18 For additional discussion on the tribe's possible locations and range see also Frye, Richard N., *The History of Ancient Iran* (Munich: Beck, 1984): here p. 90; Eilers, Wilhelm, "Asagarta", in *EIr*, II (1987): p. 701 (updated version at http://www.iranicaonline.org/articles/asagarta-sagartia-asagartiya-sagartian-old-persian-elamite-as-s-kar-ti-ia-babylonian-kursa-ga-ar-ta-a-a-greek-s.); and Potts, Daniel T., *Nomadism in Iran: From Antiquity to the Modern Era* (Oxford: Oxford University Press, 2014): pp. 102-8.

19 A map of roads; the *Tabula Peutingeriana* correctly notes the importance of Iṣṭaḫr as a Persian town.

20 Tomaschek, "Topographie": p. 16. Tomaschek's etymological suggestion, rather than his incorrect geographical locator, was followed by Williams Jackson, Abraham V., *Persia Past and Present* (London: Macmillan, 1906): pp. 351-2, but erroneously rejected by Christensen, *Decline*: p. 288, note 21. The toponym is also pronounced *Kuththa*, according to Barthold, Wilhelm, *An Historical Geography of Iran*, trans. Svat Soucek (Princeton: Princeton University Press, 1984): p. 167. Miller, Konrad, *Itineraria romana*, (Stuttgart:

century Muslim geographer al-Iṣṭaḫrī mentioned Kaṯa as the *ḥawma* ('main urban area') of Yazd in his *Kitāb al-Masālik wa 'l-mamālik*, "Book of Routes and Countries", while placing the city and its hinterland within the Kirman administrative province.[21] But it is unclear whether the Yazd area fell within the southwestern-most part of Parthia (Parthua, Paraetacene, Parthyene) *šahr* or rather the northwestern-most part of Kirman (Karmania) *šahr*, during the reigns of the Parthian kings. It could have been reassigned from Parthia to Kirman, an administrative shift which would explain the divergence in provincial allocation between the Strabo's (d. post 21 CE) account in his first century *Geography*, and Ptolemy's subsequent account in the second century *Geography*.[22]

During the Sasanian period of the third to seventh centuries CE, the locality of Yazd may have been encompassed within the northeastern end of Stakhr/Istakhr *kura* ('administrative district', likely assumed into Middle Persian from the Greek *chora*, also rendered *kūra/kust*), and within this, the 'province' or *šahr* of Fars (Pars).[23] But pre-Islamic terminology lacked clarity for the Middle Persian designation *šahr*, which was also used as part of the official name of the Sasanian Empire, i.e., Ērān Šahr, 'Land of Iran', due to its original meaning of 'territory'. The term was further applied to administrative districts like Stakhr, as attested on seals and sealings, and even in denoting cities.[24] In all likelihood, Yazd, in its pre-Islamic urban form, was categorized by Sasanian administrators as a *rōstāg*, or 'subdistrict/division', and based on the earlier positioning by Ptolemy and that subsequently provided by al-Iṣṭaḫrī, it may have

---

Strecker & Schröder, 1916): p. 797, linked Cetrora to the village of Ḫabīs (present-day Šādāb in Kirman province) which is much too far to the east.

21 al-Iṣṭaḫrī, Ibrāhīm b. Muḥammad, *Kitāb al-masālik wa'l-mamālik*, anonymous Persian translation, ed. Īraj Afšār (Tehran: Bungāh-i tarjama va-našr-i kitāb, 1340š./1961): pp. 101, 112 and 125.

22 Strabo, *Geography*, ed. and trans. Horace L. Jones, 8 vols. (London: Heinemann, 1917-32): 15: 2.14 and 15: 3.12; Ptolemy, *Geography*: 6: 6.

23 Another administrative term used for 'province' was *ōstān*. The location of Yazd within Stakhr *kūra* was noted by Frye, "Introduction": p. 4 and *History*: p. 333. On the administrative divisions see Frye, "Introduction": pp. 4-5, who cites Yāqūt al-Ḥamawī, *Muʿjam al-buldān*, ed. Ferdinand Wüstenfeld, 6 vols. (Leipzig: Brockhaus, 1866-73): I, p. 40; and Daryaee, Touraj, "The Effect of the Arab Muslim Conquest on the Administrative Division of Sasanian Persis/Fars", *Iran* XLI (2003): pp. 193-204, here pp. 195-6.

24 The interchange of the designations/names for administrative area and city, due to sharing of regional contexts, was noted broadly by Aubin, Jean, "Eléments pour l'étude des agglomérations urbaines dans l'Iran médiéval", in Albert H. Hourani and Samuel M. Stern (eds.), *The Islamic City: A Colloquium* (Oxford: Cassirer, 1970): pp. 65-75 (translated in this volume). This occurred during both the pre-Islamic and Islamic period. Aubin includes Isfahan as an example.

been governed within Kirman rather than Stakhr, or been transferred from one to the other.

Nonetheless the location of Yazd either as a valley or even as a *šahristān* (also *šahr*, by then meaning 'main city' and even 'provincial capital') was not mentioned in the Sasanian dynasty's founding tale, the *Kār-nāmag ī Ardešīr ī Pāpakān* or "Book of Feats of Ardešīr the son of Pāpak", which lists districts and cities included in the dominion of the dynasty's first kings of kings, of the third century CE (namely Ardešīr I, 224-40 CE; Šāpūr I, 240-70 CE; Hormīzd I, 270-1 CE). Neither is there any description of the area in the chapters describing provinces, well-known areas, and cities of Iran in the Sasanian-era Pahlavi *Bundahišn* ("[Book of] Primal Creation", chapters 8, 29, 31, 31), which was updated in Abbasid times and last revised in the late fifth/eleventh century. Nor can any indication of the Yazd area's importance during the early seventh century CE be gleaned from the Middle Persian or Pahlavi *Mādayān ī Hazār Dādestān*, "Book of a Thousand Judgments", which provides administrative details for Fars province. It likewise went unmentioned in the Middle Persian *Šahrestānīhā ī Ērānšahr*, "The Provincial Capitals of the Land of Iran", even though that sixth century CE text was revised under the Abbasid caliphate during the second/eighth century.

Omission of Yazd from this *Šahrestānīhā ī Ērānšahr* is especially noteworthy, since the text attributes construction of a number of cities, or most likely renovations, to Yazdegird (also 'Yazdegerd'; r. 399-420 CE, son of Šāpūr III i.e., Yazdegird I), in the early fifth century CE. These include Qūmis, Hamadan, Susa, Šuštar, and Jayy (Isfahan).[25] Only later, in the *Tārīḫ-i Yazd*, "History of Yazd", from the year 845/1441, do we find the notion that Yazd was reconstructed under the Sasanians, and hence named for Yazdegird, at whose command this restoration may have taken place in connection with the installation of a military garrison in the city. According to an account presented in the *Tārīḫ-i Yazd*, the city-founding king was Yazdegird son of Bahrām (Wahrām, Varahrān), i.e., Yazdegird II (r. 438-57 CE), son of Bahrām V.[26] Whether linked merely by etymology or substantially by building activity, that founding narrative for Yazd caught on and was reproduced in subsequent local histories, including in the *Tārīḫ-i jadīd-i Yazd*, "New History of Yazd".

---

25 *Shahrestānīhā ī Ērānshahr*, ed. and trans. Josef Markwart (Marquart), *A Catalogue of the Provincial Capitals of Ērānshahr* (Rome: Pontificio Istituto Biblico, 1931): lines 18, 26, 47, 53.

26 Jaʿfarī, Jaʿfar b. Muḥammad b. Ḥasan, *Tārīḫ-i Yazd*, ed. Īraj Afšār (Tehran: Bungāh-i tarjama va-našr-i kitāb, 1338š./1960): p. 29. Bonine, *Yazd*: p. 12, 21 note 26, incorrectly attributed Yazdegird I's patrimony to Bahrām IV.

Despite the failure to appear in extant pre-Islamic Iranian documents, the area of Yazd did not remain a backwater. One clue to the Yazd plain's increasing importance in early Sasanian times, prior to the reigns of Yazdegird I and Yazdegird II, is preserved in the Middle Persian and Parthian title Yazdbed/Yazdbad (Middle Persian: *yzdpt*, Parthian: *y'zdpt*), attested from the mid-third century CE in Šāpūr I's *Kaʿba-yi Zardošt Inscription*. Usually misread as the proper name Yazdbād, 'protégé of god', the title more likely meant 'chief/head of Yazd', with the suffix *bed* (from *pati*) paralleling that of *herberd*, 'chief of theologians', *mowbed*, 'chief of magi', *spāhbed*, 'chief of military/military commander', *handarz bed*, 'chief advisor', and similar official titles. Šāpūr's inscription indicates that regional officials and their family members were included as part of the royal court, and that this particular individual functioned as the *handarzbed bāmbišnān*, 'advisor to queens'.[27] Subsequently the title was also etched on Sasanian seals and bullae, rendered as *yzdpt*.[28]

Another indication of the late-antiquity administrative prominence of Yazd was the establishment of a mint in this area. In part on the basis of the sigillographic evidence,[29] the Sasanian mint mark YZ on *drahms* ('silver coins') is now generally regarded as designating Yazd, from at least the end of the fourth century CE.[30] Indeed, silver coins were struck at Yazd during the reign of Yazdegird I.[31] The mint became a regular source of coinage during the fourth regnal year of Xusrō I (r. 531-79 CE).[32] It continued under his son and successor

27 *Kaʿba-yi Zardošt Inscription*, Middle Persian version line 33 reconstructed, Parthian line 27, Greek line 64. Back, Michael, *Die sassanidischen Staatsinschriften* (Leiden: Brill, 1978): pp. 275, 361; Nyberg, Henrik S., *A Manual of Pahlavi*, pt. 2 (Wiesbaden: Harrassowitz, 1974): p. 153.

28 See Gignoux, Philippe, *Iranisches Personennamenbuch*, II: *Mitteliranische Personennamen*, fasc. 3: *Noms propres sassanides en moyen-perse épigraphique, Supplément* [1986-2001] (Vienna: Österreichische Akademie der Wissenschaften, 2003): p. 69, who also misread the suffix.

29 See above footnote 3.

30 Göbl, Robert, *Sasanian Numismatics*, trans. Paul Severin (Braunschweig: Klinkhardt & Biermann, 1971): pp. 80, 83, and pl. 12, no. 200, reverse of *drahm* of Hormizd IV, where the mint is incorrectly attributed as ZR or Zarang. The mint mark is correctly identified as denoting YZ rather than ZR by Sellwood, David, Whitting, Philip, and Williams, Richard, *An Introduction to Sasanian Coins* (London: Spink & Son, 1985): pp. 44, 45, 49; and by Schindel, Nikolaus (2005) "Sasanian Coinage", in *EIr*, at http://www.iranicaonline.org/articles/sasanian-coinage.

31 For example, one dating to Yazdegird I's regnal year 8 AY/407 CE for which see www.vcoins.com/en/stores/zurqieh/171/product/sasanian_yazdgird_i_ad_399420_ar_drachm_28mm_412g_zr_mint_struck_in_regnal_year_8_ad_407/545563/Default.aspx., where its minting is incorrectly attributed to ZR (Zarang).

32 As demonstrated by Schindel, Nikolaus, "The Mint System under Husraw I: Some Preliminary Remarks", in Christelle Jullien (ed.) *Husraw Ier Reconstructions d'un règne.*

FIGURE 2 Hormizd IV, drahm, RY 4 or 582 CE, with Yazd mint mark
J. K. CHOKSY COLLECTION

Hormīzd IV (r. 579-90 CE) (see Fig. 2), was especially active through the reign of Ardešīr III (r. 628-30 CE), and was still issuing currency in the years in which Yazdegird III (r. 632-51 CE) held the Sasanian throne (see "Islamic Yazd", below).[33] However, we do not have evidence serving to pinpoint the exact location of the mint within the Yazd valley.

Local histories of the Islamic period, which claimed to preserve regional memories, attributed the founding of specific towns and infrastructure within the Yazd plain to specific Sasanian dignitaries, even though archeological surveys suggest settlements in earlier periods. A particular reference in this regard is the *Tārīḫ-i jadīd-i Yazd*, "New History of Yazd", written in 862/1458 and drawing on the *Tārīḫ-i Yazd*, or "History of Yazd", written 17 years earlier.[34] The towns of 'Aqdā and Maybud were linked to military commanders serving Yazdegird I. In addition, a *diz* ('fortress') at Ḫurmīz (Hormiz/Hormizd/Ḫormiz), situated on a route leading southeast of Mihrīz (Mehriz), was attributed to Hormīzd

Sources et documents (Paris: Association pour l'avancement des études iraniennes, 2015): pp. 15-40, here pp. 120, 129-131, 136, 140 fig. 14.

33 Regarding the Yazd mint attestation from RY 1 of Yazdegird III see C11 in K Kaff, Jonathan K., "Sasanian and Arab-Sasanian Silver Coins from Turfan: Their Relationship to International Trade and the Local Economy", *Asia Major*, 3rd series, XI/2 (1998): pp. 67-115, here p. 112.

34 These local histories of Yazd were succinctly mentioned by Rypka, Jan, *History of Iranian Literature* (Dordrecht: Reidel, 1968): p. 447. They were selectively utilized, particularly for the Islamic period, by Āyatī, 'Abd al-Ḥusayn, *Tārīḫ-i Yazd* (Yazd: Gulbahār, 1317š./1938), especially pp. 18-19; Minorsky, Vladimir, "Abarḳūh", in *EI*², I (1960): p. 3; Frye, Richard N., "Ardakan", in *EI*², I (1960): p. 626; Fischer, *Zoroastrian Iran*: p. 119, note 1; Bonine, *Yazd*: pp. 12-13, 21; Bosworth, Clifford E., "Ardakan", in *EIr*, 2 (1987): pp. 370-1 (updated at http://www.iranicaonline.org/articles/ardakan-e-yazd); Lambton, Ann K. S., "Yazd", in *EI*², XI (2002): pp. 302-9; Miller, Isabel, "Local History in Ninth/Fifteenth Century Yazd: The Tārīkh-i Jadīd-i Yazd", *Iran*, XXVII (1989): pp. 75-9.

IV.[35] A *qanāt*, 'underground water channel', at Mihrīz was supposedly commissioned by Princess Mehrnigar, a daughter of Xusrō I.[36] Other settlements in the Yazd region, such as Fahraj and Haftādar, were also ascribed to the Sasanian era.[37] Ardakān, however, would first be mentioned only after it had become a predominantly Muslim town, with a Sufi *ḫānigāh* ('gathering place') erected there in the seventh/thirteenth century. As for the economic basis of these settlements, in addition to agriculture sustained by *qanāt*-based irrigation (some *qanāt* are still active, others endure as ruins), medieval Muslim authors traced the local production of silk textiles to the Sasanian era.[38] From all these reports, irrespective of the veracity of the founding narratives, we can clearly see that local tradition held the Dašt-i Yazd had been the scene of several economically sound urban settlements prior to the Arab conquest, which then persisted as Islamic ones.

Muslim sources also provide data on how Yazd's pre-Islamic (and Islamic) urban centers were connected with one another and to the rest of Iran. One main trade route led from the areas near Kashan (whose Sasanian origins remain unclear) and Nā'īn (probably of Sasanian origin), northwest of the Yazd plain, through the Yazdi settlements of 'Aqdā, Maybud, Yazd city, and Mihrīz, then heading southeastward to the Kirman region. Another major trade route, connecting the southwestern cities of Shiraz and Stakhr (Istakhr) to northeastern Turbat and Ṭūs, passed through the Yazdi settlements of Abarkūh, Taft, Yazd city, Sāqand, and Muġistān.[39] These urban centers of the Yazd valley also seem to have been linked by waterways flowing from the southwest and south, such as the now seasonal (due to rising aridity and desertification) Taft River – features that persisted into Islamic times, and in some cases still

35 Ja'farī, *Tārīḫ-i Yazd*: pp. 14, 42. See further Bāfqī, Muḥammad Mufīd Mustawfī, *Jāmiʿ-i mufīdī*, ed. Īraj Afšār, 3 vols. (Tehran: Kitābfurūšī-yi Asadī, 1339-41š./1961-3) (writing second half of eleventh/seventeenth century): III, p. 704, who utilized information from the two earlier local historians of Yazd and most frequently from al-Kātib, Aḥmad b. Ḥusayn b. 'Alī, *Tārīḫ-i jadīd-i Yazd*, ed. Īraj Afšār (Tehran: Ibn Sīnā, 1345š./1966), on which see Afšār, Īraj, *Yādgārhā-yi Yazd*, 3 vols. (Tehran: Ḫāna-yi kitāb, 1348-54š./1969-75): I, pp. 212, 229-30.

36 Bāfqī, *Jāmiʿ-i mufīdī*: III, p. 704.

37 Ja'farī, *Tārīḫ-i Yazd*: pp. 29-33; Kātib, *Tārīḫ-i jadīd*: p. 44; Bāfqī, *Jāmiʿ-i mufīdī*: I, p. 24.

38 Ja'farī, *Tārīḫ-i Yazd*: pp. 26-29, 50; Kātib, *Tārīḫ-i jadīd*: p. 38; Bāfqī, *Jāmiʿ-i mufīdī*: I, p. 14-15. Additional data is available in Ibn al-Balḫī (writing before 510/1116), *Fārs nāma*, edited by Guy Le Strange and Reynold A. Nicholson (London: Luzac, 1921): p., 20; and Ḥamd-Allāh al-Mustawfī, *Nuzhat al-qulūb*, ed. and trans. Guy Le Strange, *The Geographical Part of the Nuzhat-al-Qulūb* (London: Luzac, 1919): p. 77.

39 On these routes see Le Strange, Guy, *The Lands of the Eastern Caliphate* (London: Cambridge University Press, 1905, reprint 1966): p. 297, 326-7. On the attribution of Nā'īn to Sasanian times see Matheson, *Persia*: p. 96.

endure.[40] Consequently, as elsewhere in Iran, *dihān*, 'towns/villages' within the Yazd plain were likely well inhabited by Zoroastrians and smaller Jewish and Christian communities, before Islam reached the residents.[41]

Sasanian military garrisons at locales like Ḫurmīz and Nārīn provided security along the roads and waterways, and one or more of these could also have served as mint sites.[42] The Nārīn *qalʿa*, whose construction was later piously attributed to the biblical Solomon in the *Tārīḫ-i jadīd-i Yazd*, would be another possibility for the site of the Sasanian mint.[43] The Nārīn *diz* or *qalʿa* (see Fig. 3) overlooks the town of Maybud, approximately 51 kilometers northwest of present-day Yazd city. A Sasanian defensive wall with four gates encompassed both Maybud and the Nārīn *qalʿa*.[44] These pre-Islamic sites all have names and founding stories independent of the place-name Yazd, but none of the local narratives recount them as serving as the valley's royal mint. Yet the numismatic and sigillographic data are unequivocal in attesting the toponym Yazd. Consequently, the best candidate for the pre-Islamic city of Yazd and its Sasanian mint site is the location at which the Islamic urban center of the same name subsequently flourished. That Yazd city had a pre-Islamic physical origin and housed a vibrant Zoroastrian polity by Sasanian times is further demonstrated by the presence there of a major *ātash kada*, 'fire temple', which eventually was transformed as the city's and valley's most important *masjid al-jāmiʿ* 'congregational mosque' (as discussed in "Islamic Yazd", below).

40 Ibn Ḥawqal (lived fourth/tenth century), *Kitāb ṣūrat al-arḍ*, ed. M.J. de Goeje (Leiden: Brill, 1873): pp. 263, 287; *Ḥudūd al-ʿālam* (written fourth/tenth century), trans. Vladimir F. Minorsky, *The Regions of the World: A Persian Geography 372 AH-982 AD*, 2nd ed., (London: Luzac, 1970): p. 29 See further Khademzadeh, Mohammad H., "Continuity and Change in the City of Yazd", in Philip Jodidio (ed.), *Iran: Architecture for Changing Societies, An International Seminar* (Turin: Allemandi, 2002): pp. 27-30, here pp. 27-8. Also consult maps in Bonine 1980, *Yazd*: p. 198, fig. 72; and Fischer, *Zoroastrian Iran*: p. 128 map 1, and p. 156 map 3.

41 The Jewish presence at Yazd was also noted by Lambton, "Yazd".

42 Mīr-Dahqān, Raḍvānī, and Kāẓimnažand, "Historical and Cultural Sites": pp. 359-60.

43 For the tale about Solomon see Kātib, *Tārīḫ-i jadīd*: p. 7.

44 Kātib, *Tārīḫ-i jadīd*: pp. 32, 38-40. The urban structures of Maybud are documented by Afšār, *Yādgārhā-yi Yazd*: 1, pp. 67-106; the function of the wall is described by Pūyā, Sayyid ʿAbd al-ʿAẓīm, *Sīmā-yi bāstānī-yi šahr-i Maybud: Barrisī-yi tārīḫī-yi sāzvāra va sāḫt-i yak šahr-i kavīrī* (Maybud: Dānišgāh-i āzād-i islāmī, 1371š./1992): pp. 15, 37-50.

FIGURE 3 Sasanian Fortress, Nārīn Qalʿa, Maybud
PHOTO J. K. CHOKSY

3 Islamic Yazd

The earliest Muslim authors, such as the historians like al-Balāḏurī (d. 279/892) and al-Ṭabarī (d. 311/923), and geographers like al-Iṣṭaḫrī (d. 346/957) and Ibn Ḥawqal (d. 367/978), were clearly familiar with Yazd and its surroundings as an identifiable location in their times, however they were writing well after the events to which they refer. Similarly, even later accounts of Yazd draw significantly from local traditions. Like the fourth/tenth century histories of Bukhara and Isfahan, the histories of Yazd – particularly the *Tārīḫ-i Yazd* and its revised version, the *Tārīḫ-i jadīd-i Yazd* – would have been products of the latter city's economically prosperous and intellectually vibrant ninth/fifteenth century society. Infused by pride in their area's history and driven by desires to preserve the records, the various authors sought to convey the uniqueness and distinguishing features of the surrounding valley and the city of Yazd, in terms of its munificent rulers, learned residents, religious institutions (Muslim ones, but also Zoroastrian), public spaces, and its long history – including the pre-Islamic past.[45] Just as in universal histories, local histories do not distinguish between divine will and human action as causation of events and determination of outcomes, and so not surprisingly, we see secular and theological historiography combined in these histories of Yazd.[46] Moreover, ninth/fifteenth century local Yazdi historians, such as Jaʿfarī and Kātib, strengthened the trend of viewing Muslims as the deserved successors to earlier communities, and of rationalizing the tribulations of their own times by comparing to the grand events of the pre-Islamic era.[47]

The histories of Yazd are more than simple accounts of a city through time. They offer compilations of data, traditions, and lore on the entire area covered by the Dašt-i Yazd, and could be more accurately termed 'regional histories'.[48] The historians' sources were not only previous written accounts whose information they sought to verify and improve, but also narratives, which they included after favorable conclusion concerning the reliability or at least plausibility of the sources.[49] The authors sought to elucidate patterns of meaning from the events they documented. Similar to the compilations by Muslim universal historians and geographers, the accounts by local historians can therefore be used extensively, but cautiously, in reconstructing and analyzing Yazd's early Islamic past.

---

45 Miller, "Local History": pp. 73, 76, 79; Meisami, Julie S., *Persian Historiography: To the End of the Twelfth Century* (Edinburgh: Edinburgh University Press, 1999): pp. 12-3; Lambton, "Yazd"; Melville, Charles, "The Mongol and Timurid Periods", in Charles Melville (ed.) *Persian Historiography* (London, New York: Tauris, 2012): pp. 155-210, here pp. 180-1. More generally on local histories see Rosenthal, Franz, *A History of Muslim Historiography* (Leiden: Brill, 1968): p. 150; Humphreys, R. Stephen, *Islamic History: A Framework for Inquiry*, rev. ed., (Princeton: Princeton University Press, 1991): pp. 131-2; Robinson, Chase F., *Islamic Historiography* (Cambridge: Cambridge University Press, 2003): pp. 41, 139-40.

46 Miller, "Local History": p. 77; also Melville, "Mongol and Timurid Periods": p. 172, drawing upon Miller. More generally Rosenthal, *Muslim Historiography*: p. 151.

47 Compare the astute general observations on this point by Mottahedeh, Roy P., "Some Islamic Views of the Pre-Islamic Past", *Harvard Middle Eastern and Islamic Review*, I (1994): pp. 17-26, here p. 19, 20, 24.

48 Following Melville, Charles, "Persian Local Histories: Views from the Wings", *IrSt*, XXXIII/1 (2000): pp. 7-14, here p. 11; and Meisami, Julie S., "History as Literature", *IrSt*, XXXIII/1 (2000): pp. 15-30, here p. 15. The same could be said for other histories where writers expanded their purview to the hinterland, including Māfarrūḫī's fifth/eleventh century *Maḥāsin Iṣfahān*, which deals with that province, and the lost history of Hamadan (based on surviving information).

49 Kātib, *Tārīḫ-i jadīd*: p. 5.

According to the *Tārīḫ-i Yazd* and *Tārīḫ-i jadīd-i Yazd*, the region fell to Arab troops for the first time around the year 21/642. ʿUmar b. Muġīra is said to have then served as governor, under Caliph ʿUṯmān (reigned 23/644-35/656).[50] However this area, like much of Fars, Khurasan, and Kirman, was likely engulfed in periodic uprisings against foreign rule and so would have changed hands several times in ongoing battles between Iranians and Arabs, particularly in the period 27/648 to 32/653. Al-Ṭabarī, writing more two centuries after these events, in his *Taʾrīḫ al-rusul wa l-mulūk*, "History of the Prophets and Kings", commented on how very tumultuous the period had been, and on the uncertainty of the available historical data.[51] Al-Balāḏurī, in the third/ninth century, and Ibn al-Balḫī in the early sixth/twelfth century, wrote of similar impressions and recorded varied versions of the relevant events.[52] The *Tārīḫ-i jadīd-i Yazd* claims that in the year 30/651, the last Sasanian king of kings Yazdegird III (d. 30/651) camped at Yazd for two months before being forced to flee northeastward to Marw.[53] Even if he did not stop for long, Yazdegird was likely following a well-established military and mercantile route – one of many major roads passing through the Yazd valley from the four quadrants of Iran.[54] The king's presence also fits well with the location being a Sasanian mint, where he could have struck currency to fund the Iranian resistance to Arab invaders entering Yazd valley, as he did later at Marw. Local tales of treasure, allegedly found between the seventh/thirteenth and ninth/fifteenth centuries when Yazd city was ruled by the Atābaks, the Muzaffarids, and the Timurids, may recall memories of the bullion stored for production of Sasanian imperial coinage.[55] Al-Ṭabarī also recounts details of Yazdegird III's flight northeastward from western Iran in 30/651, while failing to mention Yazd specifically.[56]

50 Jaʿfarī, *Tārīḫ-i Yazd*: p. 16; Kātib, *Tārīḫ-i jadīd*: p. 53.

51 al-Ṭabarī, *Taʾrīḫ al-rusul waʾl-mulūk*, ed. M.J. de Goeje and others, 15 vols, (Leiden: Brill, 1879-1901): 1.2827-64.

52 al-Balāḏurī, *Kitāb futūḥ al-buldān*, ed. M.J. de Goeje (Leiden: Brill, 1866): p. 313; Ibn al-Balḫī, *Fārs-nāma*: pp. 114-17. See also Choksy, *Conflict and Cooperation*: pp. 20-2; Daryaee, Touraj, "The Collapse of the Sasanian Power in Fārs/Persis", *Nāma-yi Īrān-i bāstān* II/1 (2002): pp. 3-18, here pp. 8-12.

53 Kātib, *Tārīḫ-i jadīd*: pp. 46-8. Lambton, "Yazd", expressed skepticism, without giving reasons, about the last Sasanian king of kings having camped for a length of time in the Dašt-i Yazd, even though the time span described fits the chronology of his retreat.

54 For further details on the relevant Sasanian overland routes see Brunner, Christopher, "Geographical and Administrative Divisions: Settlements and Economy", in Ehsan Yarshater (ed.) *The Cambridge History of Iran*, III(2), (Cambridge: Cambridge University Press, 1983): pp. 747-77, here pp. 752-3.

55 Kātib, *Tārīḫ-i jadīd*: pp. 46-8.

56 al-Ṭabarī, *Taʾrīḫ*: 1.2863-4. al-Balāḏurī, *Futūḥ*: pp. 315-16, also did not mention Yazd specifically in connection with his report of Yazdegird's retreat to northeast, from Fars to Khurasan.

The first mention of Yazd in Islamic sources is by al-Ṭabarī himself, writing that ʿAbd-Allāh b. ʿĀmir b. Kurayz's Khurasan campaign moved through the locality (perhaps both valley and city) in the year 31/652.[57] It is also clear that Yazd continued as an administrative center, despite the tumult of military conquest and transfer of political power, since Arab-Sasanian *dirham* ('silver coins') bearing the Yazd mint mark (YZ) are attested between the years 62/682 and 85/704.[58] These indicate that the Sasanian currency operation at the Yazd mint had been incorporated into the Umayyad caliphate's administrative structure.

Kātib, in his *Tāriḫ-i jadīd-i Yazd*, following the then-established narrative of Islam being divinely decreed for conquered peoples, and therefore having been either swiftly accepted or enforced, claimed that the residents of the Yazd had converted to Islam during the first/seventh century.[59] However, contrary to this later stereotypical account, the *jizya* ('poll tax') required of non-Muslims would have been imposed as part of terms of capitulation, rather than conversion to Islam. As a consequence, tensions would have persisted between the Arab Muslim conquerors and their subject Iranian Zoroastrians, Jews, and perhaps also Christians – as is documentable for many other places in Fars, Kirman, and Khurasan.[60] Resistance from the followers of Iran's older faiths notwithstanding, a hypostyle *masjid* ('mosque') existed by the time of the late first/seventh century, demonstrating that Arab troops and perhaps immigrant families had begun to establish permanent roots on the Dašt-i Yazd. Another mosque, likewise among the earliest in Iran, was erected at Fahraj sometime between the second/eighth and fourth/tenth centuries.[61]

Al-Ṭabarī and al-Iṣṭaḫrī (cited above concerning the Arab conquest and the location of Kaṯa, respectively) provide the earliest textual mentions of the region in Islamic sources – suggesting that by the time of the early fourth/tenth century the city of Yazd was a well-known urban center. Al-Iṣṭaḫrī, both in his original Arabic text and as preserved in an anonymous Persian translation of the fifth/eleventh or sixth/twelfth century, noted Yazd as possessing a walled *ḥiṣār*, 'citadel', with two iron gates – the *bāb īzādī* and the *bāb al-masjid* – providing access to the older Zoroastrian fire temple and the newer Muslim

57 al-Ṭabarī, *Taʾrīḫ*: 1.2886.

58 Gaube, Heinz, *Arabosasanidische Numismatik* (Braunschweig: Klinkhardt & Biermann, 1973): pp. 105-7 with table 8; and Bates, Michael, "Arab-Sasanian Coins", in *EIr*, 2 (1986): pp. 225-9 (updated at http://www.iranicaonline.org/articles/arab-sasanian-coins), here p. 227; both revising the earlier conclusion by Walker that the mint mark should be interpreted as Sijistān/Sīstān (SK), see Walker, John, *A Catalogue of the Arab-Sassanian Coins* (London: The British Museum, 1941): p. xxxvi, cxxix,

59 Kātib, *Tāriḫ-i jadīd*: p. 16.

60 Choksy, *Conflict and Cooperation*: pp. 37-40. Lambton, "Yazd", concurs on the gradual conversion to Islam and the payment of the poll tax.

61 Choksy, *Conflict and Cooperation*: p. 95; Matheson, *Persia*: p. 97.

mosque.[62] Ibn Ḥawqal, writing around the year 366/977, elaborated on al-Iṣṭaḫrī's observations in asserting that "Yazd is the largest *rustāq* [division] in the *kūra* [district] of Istakhr ... and Kaṯa is the citadel."[63] Istakhr was considered the capital of Fars province even after its demographic eclipse by Shiraz, so in essence the *kūra* of Istakhr represented the province of Fars– thus placing the Dašt-i Yazd within Fars as a *nāḥiya*, or 'local administrative area'.[64] In roughly the same period the anonymous *Ḥudūd al-ʿālam*, written in Khurasan, likewise placed Yazdi urban centers "in the cold zone ... lying on the frontier of Fars with the desert".[65] The frontier location may explain why Iṣṭaḫrī had concluded that Yazd *rustāq* fell within the province of Kirman (see above, "pre-Islamic" section) instead of Fars and its Istakhr district.

Yāqūt (d. 626/1229) also placed Yazd, which he designated as a *kūra*, within Istakhr *kūra*. Drawing on al-Iṣṭaḫrī's previous report, he further commented that Kaṯa was the *ḥawma* or main urban area of Yazd, and also that Yazd's "*qalʿa* [citadel] is called Kaṯa".[66] So Kaṯa was the name of the neighborhood or quarter comprising the main urban area, within which was the citadel. Indeed, the *Tārīḫ-i jadīd-i Yazd* later claimed the toponym Kaṯa coincided with the *qalʿa*, 'citadel' (rather than its earlier and broader pre-Islamic sense of 'valley'), and also projected the construction back to Alexander III (r. 336-23 BCE), following yet another fanciful pattern of attribution.[67] Al-Iṣṭaḫrī, writing at around 339/950, said that "Kaṯa [is] the *ḥawma* of Yazd", and even noted road distances between Yazd and several destinations: Abarkūh, 153 kilometers to the southwest, a village called Kattū, 97 kilometers south-southwest, and Yazd and Maybud and Ardakān, respectively 56 and 61 kilometers to northwest. These measurements coincide well with ones subsequently noted by Ibn Ḥawqal. Additionally, al-Iṣṭaḫrī described Yazd as receiving its water not only through *qanāts*, but also via a river flowing from a *qalʿa* near a habitation with a lead mine. This would correspond to the village of Kattū, with its mountain-fed

62 al-Iṣṭaḫrī, *Kitāb al-masālik wa'l-mamālik*, ed. M.J. de Goeje, (Leiden: Brill, 1927): p. 77, 101, 112, 125; Iṣṭaḫrī (anonymous Persian translation): p. 111, 185.

63 Ibn Ḥawqal, *Ṣūrat al-arḍ*: p. 181, 187, 196.

64 Partially noted by Barthold, *Historical Geography*: pp. 152-3.

65 *Ḥudūd al-ʿālam*: p. 129, 380. As noted in Le Strange, *Lands*: p. 6, 249, 275, 255, 286, medieval geographers initially placed Yazd district within the 'cold' zone of Fars, then assigned it to Jibāl province. On the status of Yazd in terms of cold and hot zones and its geographical linkage to Kirman, but often placed in Fars, see also Barthold, *Historical Geography*: p. 148.

66 Yāqūt, *Būldān*: IV, p. 239, 1017. Relentless modern urbanization coupled with a disregard for cultural heritage has resulted in the *qalʿa* now functioning as a parking lot.

67 Kātib, *Tārīḫ-i jadīd*: p. 16.

stream and nearby mountain crag known as the Qalʿa-yi Zard.[68] The stream at Kattū may have been a tributary of the Taft River, which crossed the 23 kilometers of land between Taft and Yazd.

It should be noted that Taft itself has been proposed as the Kaṯa/Kata/Kasa of Yazd.[69] Yet the author of the *Tārīḫ-i jadīd-i Yazd* described Taft merely as an "ideal town" of the Dašt-i Yazd, rather than as an administrative center.[70] Moreover, Taft does not seem to have been enclosed by a pre-Islamic or early Islamic wall, nor does it have a ruined citadel or any evidence of minting activity. Indeed, the only nearby citadels and watchtowers date to the Qajar dynasty (i.e. from the late twelfth/eighteenth century).[71] Maybud, as already discussed, did have the architectural features necessary to be Kaṯa, including a *qalʿa*. Yet after the Arab conquest, this town and citadel experienced the same decline as did Qaṣr-i Abū Naṣr. Moreover there is no evidence that the Sasanian-Arab *dirham* of Yazd were minted there, and the extant medieval records do not attribute any administrative or economic importance to Muslim-controlled Maybud until after the eight/fourteenth century.[72]

Early Islamic monochrome, slashed, and sgraffiato pottery has been recovered during archeological investigations at Yazd city. The sherds have mainly been dated to the third/ninth to sixth/twelfth centuries, through comparison to ones found at sites such as Ġubayrā (approximately 70 kilometers south of Kirman city), and to earlier Parthian and Sasanian vessels from Šahr-i Qūmis (560 km north of Yazd) and the Qalʿa-yi Yazdegird (northwest of Kirmānšāh).[73] So the current locality of Yazd city was clearly occupied during those centuries.

---

68 Browne, Edward G., *A Year amongst the Persians* (London: Black, 1893): p. 392 specifically on Kattū, and pp. 391-5 on the road from Kattū via Taft to Yazd; and Williams Jackson, *Persia Past*: p. 352. Le Strange, *Lands*: p. 285, cited Ibn Ḥawqal on lead mining. The current location of Kattū village is unclear.

69 Murtaḍāʾī and Mīr-Dihqān, "Bāznigārī": pp. 164-5, 172 equate Taft with the original Kaṯa/Kata/Kasa while also making the more general observation that Yazd covered several urban areas.

70 Kātib, *Tārīḫ-i jadīd*: 214-19.

71 Afšār, *Yādgārhā-yi Yazd*: I, p. 423.

72 Modarres, Ali (2012) "Meybod", in *EIr*, at http://www.iranicaonline.org/articles/meybod.

73 For the pottery at Yazd see Murtaḍāʾī and Mīr-Dihqān, "Bāznigārī": pp. 169-70; and Mīr-Dihqān, Raḍvānī and Kāẓimnažand, "Historical and Cultural Sites": pp. 360, 368-9. Unfortunately, I do not have access to the unpublished data from surveys conducted at Yazd during 2005-7, during which pottery was collected and comparatively dated to sherds from sites such as Ġubayrā. For the ceramics from Ġubayrā see Bivar, A.D.H., *Excavations at Ghubayrā, Iran*, (London: School of Oriental and African Studies, 2000): pp., 130-42, 396 pl. 78, 402 pl. 84, 403 pl. 85; for those from Šahr-i Qūmis consult Hansman, John and Stronach, David, "Excavations at Shahr-i Qūmis, 1971", *Journal of the Royal Asiatic Society* CVI/1 (1974): pp. 8-22, especially p. 20; for those from Qalʿa-yi Yazdegird see Keall, Edward

Given that the written sources and archeological data suggest that the *ḥawma* encompassed several quarters, including Kaṯa with its *qalʿa*, then Islamic Yazd's *šahr/šahristān* likely extended over several former pre-Islamic neighborhoods.[74] The introduction to this paper noted the pattern in which larger Muslim cities aggregated proximate Achaemenian through Sasanian neighborhoods and settlements. This pattern is evidenced as well in northern Iran at Gurgān city, in eastern Iran at Kirman city, and at Mastūng town and the Panj Pāy (Panj Pai) archeological site in Baluchistan (the latter two now in Pakistan).[75]

Once Yazd city had become the unequivocal Islamic urban hub of the Dašt-i Yazd, it quickly rose in importance as a commercial and intellectual center. At the beginning of the fifth/eleventh century, Yazd became integrated in the Kākūyid state based in central Jibal. In particular it became one of the mints, as shown by a *dirham* struck in 421/1030 for ʿAlāʾ al-Dawla Muḥammad (d. ca. 433/1042), who was related to the weak Buyid ruler Majd al-Dawla at Rayy.[76] In the year 432/1041, meaning during the troubled period when the Türkmens and then Seljuqs arrived, the Kākūyid leader Abū Manṣūr Farāmurz and his

---

J. and Keall, Marguerite, "The Qal'eh-i Yazdigird Pottery: A Statistical Approach", *Iran* XIX (1981): pp. 33-80, here p. 51.

74 Additional discussion on Kaṯa/Kata in relation to the location of Yazd is found in Schwarz, Paul, *Iran im Mittelalter nach den arabischen Geographen*, 9 vols., (Stuttgart: Harrassowitz, later Wiegand, 1896-1936), specifically I, pp. 19-20. Drawing on Mustawfī and Ibn Baṭṭūṭa (d. 770/1369), he concluded that the name Yazd replaced Kata, and more generally III, p. 160, 192, 198, 263, V, 582. Barthold, *Historical Geography*: p. 167, concluded likewise.

75 For Gurgan city consult Kiani, M. Yussuf, *The Islamic City of Gurgān* (Berlin: Reimer, 1984): p. 16, 23-4, pls. 1-4, plans 1-3. The conclusion about Kirman is based on unpublished data from a survey I conducted during the summer of 2003. Likewise the conclusions for Mastūng and Panj Pāy are based on unpublished survey and excavation data which Clifford C. Lamberg-Karlovsky and I collected in Baluchistan, Pakistan, during the summers of 1999 and 2000. The Panj Pāy excavations were briefly noted by Khan, whose statement that settlement ended in the fifth century CE must be extended chronologically to include the early Islamic period, after which water for the citadel and surrounding fields dried up and the settlement shifted. See Khan, Makeen (2010) "A Brief Note on Archeological Site Panj Pai in Baluchistan", in Sita Pieris and Ellen M. Raven (eds) *ABIA: South and Southeast Asian Art and Archaeology Index*, III, sec. 2, (Leiden: Brill, 2010): p. 489.

76 Blair, Sheila S., *The Monumental Inscriptions from Early Islamic Iran and Transoxiana* (Leiden: Brill, 1992): p. 105, correctly notes that the numismatic evidence is not that of a temporary mint. This effectively discredits Bosworth's assertion that accounts in the *Tārīḫ-i Yazd* dating the integration of the city into the Kākūyid state before 443/1051 are inaccurate. Bosworth, Clifford E., "Dailamīs in Central Iran: The Kākūyids of Jibāl and Yazd", *Iran* VIII (1970): pp. 73-95, here p. 91, 93. Bosworth based his statement on the conclusion that there had only been a temporary mint. However Bosworth's inaccurate assessment was followed by Miller, "Local History": p. 78; and Lambton, "Yazd."

Daylamite followers were said to have built "the reinforced wall around the urban area, with the four gates having iron doors".[77] An iron plaque inscribed in Kufic, from the Ḫāṭir (Ḥaẓīr) Gate, names two of Abū Manṣūr Farāmurz's commanders, Abū al-Masʿūd al-Bihištī and Al-Muẓaffar Abū Yaʿqūb, as having overseen the construction of the gate, thereby confirming the written account in the *Tārīḫ-i Yazd* which names two more, Abū Jaʿfar and Abū Yūsuf. The inscription attributes the blacksmithing to "Muḥammad b. Abū Isḥāq, the ironworker of Isfahan" – suggesting the movement of skilled craftsmen from city to city in central Iran.[78] It remains unclear whether the gates and the wall (which is still visible today, see Fig. 4) were newly constructed or renovated from pre-Islamic fortifications. Much vaunted in the city's local histories, these defensive structures protected Yazd from being sacked by invaders in the fifth/eleventh and sixth/twelfth centuries. However, they could not safeguard the city from having to submit to the Seljuqs, apparently without siege, in 443/1051.

While the Kākūyids had been independent rulers over the period of around 443/1051-535/1141, under their Seljuk overlords, they became eventually vassals, referred to as Atābaks (Atabegs) (535/1141-719/1319). Over this arc of time, according to the *Tārīḫ-i Yazd* and the *Jāmiʿ-i mufīdī*, they became famed patrons of poets and other men of learning, luring them from cities in Iraq, Khurasan, and Kirman to the palaces, libraries, madrasas, and the locally celebrated Bihištī and Garšāspī gardens of Yazd city and the surrounding towns.[79] Certainly, the texts of these local writers are influenced by pride in Yazd. Yet royal patronage ensured that the Yazd area indeed did grow from its pre-Islamic commercial and administrative beginnings into a vibrant intellectual center, far from remaining a backwater or mere way station between other cities.

Yazd came to be called *dār al-ʿibāda*, or "abode of worship", in a charter supposedly given by the Seljuq ruler Ṭoġrıl Beg to the Kākūyid leader Abū Manṣūr

---

77 Jaʿfarī, *Tārīḫ-i Yazd*: p. 20. Another translation is provided by Bosworth, "Dailamīs": p. 89.

78 Ṣahba, Javād M. "Darvāzah-hā-yi nuhṣad va sī sāla", *Yādgār* I (1323š./1944): pp. 72-80: pp. 73-7 with illustrations, termed the structure *darvāzah-yi Ḥaẓīr* in Persian; Blair, *Monumental Inscriptions*: pp. 111-14 text, translation, commentary, 246 figs. 66-7, transliterated it as Ḫāṭir. Afšār, *Yādgārhā-yi Yazd*: II, pp. 681-94, also discussed the gate and its inscription. Compare the textual record in Jaʿfarī, *Tārīḫ-i Yazd*: p. 20, which indicates the local histories contain both core factual information and aggrandizements.

79 Jaʿfarī, *Tārīḫ-i Yazd*: pp. 21-3, 33, 35, 137, 148, 150-4; partially copied by Bāfqī, *Jāmiʿ-i mufīdī*: III, pp. 737-8. On the chronology of the Kākūyids and Atābaks see Bosworth, "Dailamīs": pp. 84-95; Fairbanks, S. C., "Atābakān-e Yazd", *EIr*, II (1987): pp. 900-2 (updated at http://www.iranicaonline.org/articles/atabakan-e-yazd-a-dynasty-which-governed-yazd-in-the-6th-12th-century). For the tenor of Atābak rule in Yazd see also Aigle, Denise, *Le Fārs sous la domination mongole: politique et fiscalité (XIIIe-XIVes)* (Paris: Association pour l'avancement des études iraniennes, 2005): p. 64, 113, 116, 130.

FIGURE 4 Yazd city wall
PHOTO J. K. CHOKSY

Farāmurz (d. post 455/1063), applying terminology that, according to local historians, reflected the zealousness of its occupants for Islam.[80] The Dašt-i Yazd had by this time begun to emerge as an important Muslim religious locality in central Iran, at the expense of its Zoroastrian and Jewish communities. The developmental timeline of sectarian buildings in medieval Yazd provides glimpses into the progress of Islamization.[81] As discussed earlier, al-Iṣṭaḫrī had mentioned a *bāb al-masjid*, or 'mosque gate', at Yazd during the early fourth century/tenth century. Yet the first minaret in Yazd city was erected only during the first part of the fifth/eleventh century, in the Durdā neighborhood, under the patronage of Arslān Ḫātūn, a daughter of the Seljuq Čaġrı Beg. That structure was said to have endured until the year 807/1405. By 832/1429 it lay in ruins, with sagging foundations and crumbling walls, perhaps due to limited resources and expertise at the time of construction, given the then small number of local Muslims.[82] A later congregational mosque in Durdā was attributed

80 Kātib, *Tārīḫ-i jadīd*: p. 59; reproduced by Bāfqī, *Jāmiʿ-i mufīdī*: III, p. 735. On the continuing zealousness of Yazd's Muslims see Fischer, *Zoroastrian Iran*: p. x.

81 For the monuments see also Afšār, *Yādgārhā-yi Yazd*.

82 Jaʿfarī, *Tārīḫ-i Yazd*: p. 19, 21; Kātib, *Tārīḫ-i jadīd*: p. 60; Bāfqī, *Jāmiʿ-i mufīdī*: I, p. 76, III, pp. 735-7.

to ʿAlāʾ al-Dawla Muḥammad's grandson, ʿAlāʾ al-Dawla Garšāsp (d. 536/1142), as was a "madrasa with two minarets".[83] Indeed the sixth/twelfth century, which corresponds to the period of Kākūyid rule as Atābaks, was said to have witnessed a proliferation of madrasas, which in turn ensured the spread of Muslim religious education among inhabitants of the Dašt-i Yazd.[84]

Yazd's main Zoroastrian *ātaš kada* was torn down to make way for construction of a *masjid al-jāmiʿ* (see Fig. 5), by the Kākūyid Abū Manṣūr Farāmurz, in service of the burgeoning Muslim community. According to the *Tārīḫ-i Yazd* in 513/1119: "They [Abū Manṣūr's amīrs Abū 'l-Masʿūd al-Bihištī, Al-Muẓaffar Abū Yaʿqūb, Abū Jaʿfar, and Abū Yūsuf] built the congregational mosque in the inner city."[85] Below the congregational mosque's courtyard flowed a *qanāt*, which would have provided water for the Muslim purification rites of *wuḍūʾ* and *ġusl*. This same *qanāt* would have originally served the water needs of the parallel Zoroastrian cleansing ceremonies of *padyāb* and *nāhn*, and of the *urwīs gāh*, "place of the ritual table", of the former *ātaš kada*. So at Yazd, as elsewhere across Iran, Zoroastrian converts to Islam began returning to their former places of worship under the auspices of Islam, often adapting rites of the older settings to suit the tenets and expectations of their new faith.[86]

The *mihrāb*, '*qibla* niche', and blue faience minarets of Yazd's main *masjid al-jāmiʿ* date to 766/1365 and are attributed to a rebuilding by Bībī Fāṭima Ḫātūn, wife of the Timurid governor Amīr Čaqmāq, followed by further restoration during the Safavid period. Likewise dating to the early to middle eighth/fourteenth century is the ʿAlīd inscription with the names of the twelve imams, suggesting a growing Twelver Shi'ite presence at Yazd (see Fig. 6).[87]

---

83 Jaʿfarī, *Tārīḫ-i Yazd*: p. 22.

84 Jaʿfarī, *Tārīḫ-i Yazd*: pp. 20-2. While the sources here deal specifically with Yazd, it should be noted that theological colleges flourished through central and western Iran in this period.

85 Jaʿfarī, *Tārīḫ-i Yazd*: p. 19.

86 See further Choksy, Jamsheed K., "Altars, Precincts, and Temples: Medieval and Modern Zoroastrian Praxis", *Iran* XLIV (2006): pp. 327-46, here p. 329, 331-3.

87 The seminal study of this congregational mosque was conducted by Siroux, Maxime, "La Masjid-e-Djum'a de Yezd", *Bulletin de l'Institut Français d'Archéologie orientale*, XLIV (1947): pp. 119-76, especially 122-48. Siroux's conclusions are followed and slightly expanded by Golshan, who similarly notes three mosques from prior to and after the Mongol invasion, combined into the current structure during the eleventh/seventeenth century. Golshan, Sedigheh, "A Comparative Study of Oral and Documentary History: The Great Grand Mosque of Yazd", *International Journal of Arts Theory and History*, VIII/4 (2014): pp. 1-27. A brief summary of the construction history of the congregational mosque is also provided by Matheson, *Persia*: p. 97.

FIGURE 5 Yazd congregational mosque, built on Zoroastrian fire temple site
PHOTO J. K. CHOKSY

The erection of the domed Dawāzdah Imām, "[Shrine to the] Twelve Spiritual Guides", situated east of Yazd city's congregational mosque, dates to the month of Ramaḍān in the year 429/1038. Two inscriptions painted in the interior, as well as citing passages from the Qur'an, record two Shi'ite Kākūyid military officers, Abu'l-Mas'ūd al-Bihištī and Al-Muẓaffar Abū Ya'qūb, previously involved in construction of the congregational mosque, who, "endeavoring for God's pleasure and seeking His recompense, ordered construction" of this shrine.[88]

88 Blair, "Monumental Inscription": p. 7, 103-7 text, translation, discussion, p. 245 fig. 64; and Qūčānī, 'Abd-Allāh, *Barrisī-yi katība-hā-yi banāhā-yi Yazd*, (Tehran: Sāzmān-i mīrāṯ-i

FIGURE 6
Alid inscription in the Yazd congregational mosque
PHOTO J. K. CHOKSY

This structure adds to evidence of ʿAlīd leanings among Yazd's Muslims, who were said to have publicly positioned themselves as Shafiʿis for the sake of avoidance of persecution from Sunnis.[89] By the time of constructing the shrine, Muslims would have comprised more than ninety percent of the local population.[90] Wealthy patrons and powerful commanders, such as those already mentioned, sponsored the mosques, theological colleges, and shrines that became the institutional basis of Islam in the Dašt-i Yazd.[91] Indeed, when writing in 670/1272 on the inhabitants of Yazd, Marco Polo asserted "*Il aosseut*

farhangī va-gardišgarī, 1383š./2004): pp. 13-16 discussion, pp. 50-6 figs. Further details on the structure, inscription, and sectarian affiliations are provided by Holod, Renata, "The Monument of Duvāzdah Imām in Yazd and Its Inscription of Foundation", in Dickran Kouymjian (ed.) *Near Eastern Numismatics, Iconography, Epigraphy and History: Studies in Honor of George C. Miles* (Beirut: American University of Beirut, 1974): pp. 285-8; Matheson, *Persia*: p. 97; and Anisi, Alireza, "The Davāzdah Imām Mausoleum at Yazd: A Re-Examination", *Iran* XLVII (2009): pp. 57-68, here p. 57, 65-6.

89 Āyatī, *Tārīḫ-i Yazd*: pp. 70-1. Other ʿAlīd shrines mentioned in Kātīb's local history are discussed by Miller, "Local History": p. 76.

90 Choksy, *Conflict and Cooperation*: p. 38, 95.

91 Holod provides detail in Holod, Renata, *Architecture, Patronage and Setting in Iran: A Case Study of Yazd 1300-1450* (PhD dissertation, Harvard University, Cambridge, Mass., 1972).

mahōmet",[92] and did not mention Zoroastrians at this location, although he referred to "fire worshippers" elsewhere "in Persia".[93]

But Zoroastrianism was not doomed at Yazd. The Sasanian royal family's fire temple of Ādur Anāhīd at Istakhr had been transformed into the hypostyle Masjid-i Sulaymān, shortly after that the Arab conquest of the city in 28/649,[94] but the Ādur Anāhīd holy fire was first relocated to a smaller temple, also at Istakhr, under protection by magi. The canonical fire temple of Ādur Farrōbay at Kāriyān (Kāryān), also in Fars province, was razed by ʿUbayd-Allāh b. Abī Bakra in the year 50/670, on orders of Ziyād b. Abīhi, the Umayyad governor of Iraq. However the magi had already divided the Ādur Farrōbay flame, regarded as the icon of Zoroastrian clergy and nobility, into two portions, to safeguard against extinguishing by zealous Muslims. One portion was reestablished at Kāriyān, while the other burned at Fasā.[95] Tradition has it that the Zoroastrian *dastūr dastūrān*, 'high priest of high priests', had moved to the village of Turkābād (Torkābād), 61 kilometers northwest of Yazd, in the sixth/twelfth century, and eventually to Yazd city in the eleventh/seventeenth century. Coinciding with the period of the high magus' initial arrival on the Yazd plain, in 569/1174, the Ādur Farrōbay and Ādur Anāhīd fires were relocated southeast to the village of Šarīfābād, adjacent to Ardakān, to burn inconspicuously within side chambers of a mudbrick *ātašgāh* ('fire temple').[96]

The reason for relocating the Zoroastrian high magus and two canonical fires to the Dašt-i Yazd would have been that this was a central locale, with direct

92 Written as *Il aożeut mahōmet* in Old French manuscripts such as M304 (Stephens Catalogue number XXXVII) dating from 1350 CE, in the Swedish National Library, Stockholm, on page 23 (page number on top right corner of the folio), ch. xxxiii (marked in red ink). This Old French phrase is perhaps best translated as "He is Muhammadan/They are Muhammadan." However, it has been translated by "The people are worshippers of Mahommet" in the English translation by Yule based on the Old French text published by the Geographical Society in Paris in 1824 and revised by Rusticiano Pauthier and Guillaume Pauthier in 1865, see Marco Polo, *The Book of Ser Marco Polo, The Venetian*, I, trans. Henry Yule, (London: Murray, 1871): p. 84, followed by Marco Polo, *The Travels*, trans. Ronald Latham, (London: Penguin, 1958): p. 62, as "The inhabitants worship Mahomet." Still other translations render it "Its inhabitants are of the Mahometan religion", see Marco Polo, *The Travels of Marco Polo*, trans. Thomas Wright, (London: Bohn, 1854): p. 52.

93 Marco Polo, trans. Yule: pp. 73-6; trans. Latham: pp. 58-60.

94 al-Masʿūdī, ʿAlī b. al-Ḥusayn, *Murūj al-ḏahab wa maʿādin al-jawhar*, rev. ed. C. Pellat, 7 vols., (Beirut: al-Jāmiʿa al-Lubnāniyya, 1965-79): IV, pp. 76-7. See additional information in Choksy, *Conflict and Cooperation*: p. 97, and "Altars": p. 331.

95 Ibn al-Faqīh (writing ca. 290/903), *Kitāb al-buldān*, ed. M.J. de Goeje (Leiden: Brill, 1885: pp. 246-7. Additional details in Choksy, *Conflict and Cooperation*: p. 97, and "Altars": p. 331.

96 See also Siroux, Maxime, "Le temple Zoroastrien de Sharīfābād", *Āṯār-é Īrān* III/1 (1938): pp. 83-9, here pp. 83-7. Images in Choksy, "Altars": p. 335, figs. 5a and 5b.

road links to western Iran, as well as to Kirman and from there to the western coastal towns of India, where other Zoroastrians had moved. This meant Yazd was also linked with these regions by pilgrims, merchants, and other travelers.[97] The centralization of Zoroastrian clerical and devotional activity on the Dašt-i Yazd corresponds to the period when Yazd city and its surrounding towns and villages were undergoing commercial, societal, and infrastructural revitalization under Kākūyids and Atābaks. Indeed, according to the local histories of Yazd, it was under the Atābaks that Turkābād and nearby Ardakān were founded.[98] The decision by the *dastūr dastūrān* to move Zoroastrian institutions and icons to the Yazd plain proved prudent for the faith's endurance: in the mid seventh/thirteenth century, Mongol invaders chose not to destroy the settlements because the Atābaks were in charge, although the tax officials of the Mongols allegedly extorted vast sums in place of collecting the amounts truly due.[99] Even the rise of Muḥammad b. Muẓaffar in 747/1346 had no special impact on the Yazdi Zoroastrians, given that this first Muzaffarid ruler had originated from Maybud. Instead, in villages such as Turkābād, both Muslims and Zoroastrians benefitted from greater protection when the Muzaffarids built surrounding defensive walls.[100]

Several subsequent *dastūr dastūrān* communicated with their Indian coreligionists the Parsis, via *rivāyat*s, 'treatises', written at Yazd, Turkābād, and Šarīfābād between the years 884/1479 and 1092/1681 (noted in those documents as Anno Yazdegird or 'AY' years 847 to 1049). The authors of each of these dated treatises locate the Dašt-i Yazd as being on the "edge of Fars province", and indicate Yazd as the principal city of these plains. Just as the author of the *Tārīḫ-i jadīd-i Yazd*, they also assert that the city is the "abode of worship", not just

97 Choksy, *Conflict and Cooperation*: p. 331. Contra Boyce, Mary, *A Persian Stronghold of Zoroastrianism* (Oxford: Clarendon, 1977): pp. 1-4, who incorrectly speculated the Yazd valley was remote and isolated: "The region was hardly one to attract Arab settlers, for the climate is harsh, with burning summers and searingly cold winters; and since the rainfall is scant, the plain is partly desert ... the villages of the inhospitable plain were long left more or less to themselves – which is presumably why the Dastur dasturān sought shelter there ... in Turkabad not later than the eleventh century." Also contra Boyce, Mary, "Ātaš", in *EIr*, III (1989): pp. 1-5 (updated at http://www.iranicaonline.org/articles/atas-fire), where relocation to the two flames to Šarīfābād is dated to the late seventh/thirteenth century. Āyatī, *Tārīḫ-i Yazd*: p. 16, conflated antiquity with this medieval relocation of holy fires when concluding that Yazd had long been a sacred site and pilgrimage place for Zoroastrians.

98 Jaʿfarī, *Tārīḫ-i Yazd*: pp. 35, 40-1; followed by Bāfqī, *Jāmiʿ-i mufīdī*: III, p. 705. Discussed by Bosworth "Dailamīs": p. 94; and Bonine, *Yazd*: p. 22, note 38.

99 Summarized by Lambton, "Yazd," citing the relevant sources. See also Aigle, *Le Fārs*: p. 93.

100 Jaʿfarī, *Tārīḫ-i Yazd*:52; Kātib, *Tārīḫ-i jadīd*: p. 85.

for Muslims but also for Zoroastrians. The *rivāyats* are clear in indicating that the Zoroastrians and Muslims resided within their own neighborhoods at Yazd city, the latter in the quarter known as Maḥmūdābād.[101] Local narratives later sought to explain the Jewish community's relocation from surrounding villages into the city of Yazd, specifically to the neighborhood around the *masjid al-jāmiʿ*, as the product of a Safavid royal decree during the reign of Šāh ʿAbbās I (996/1588-1038/1629), whereas such was not the case for the Zoroastrians.[102]

The *rivāyats* record names and locations – particularly in Turkābād and Šarīfābād – of *mubadān*, Zoroastrian 'magi/priests', and *hirbadān*, Zoroastrian 'theologians/religious teachers'. They mention anxiety among the regional Zoroastrians over periods of forced conversion to Islam, and document famine triggered by drought, which also kept merchants away from both Yazd and the Zoroastrian satellite towns, thus adding to the hardships.[103] The eventual growth in geographical extension of the towns of Turkābād and Šarīfābād brought them first to abut Ardakān and then became neighborhoods. Zoroastrians therefore increasingly lived alongside Muslims, reflecting the tendency of once distinct localities to agglomerate as parts of a larger urban area: a trend also observed by the author of the *Tārīḫ-i Yazd*.[104]

4 Yazd as a Nexus

Yazd and its surroundings appeared on a first century BCE road map and in classical geographies of the first and second centuries CE. It served as a mint from the late fourth century CE onward. The last Sasanian ruler fled through the Dašt-i Yazd on his way to Marw, in the first/seventh century. Muslim pilgrims began following a similar route to reach the eighth imam's shrine at Mashhad, stopping at the Yazdi *imāmzāda* and *ḫānaqāh*.[105] Zoroastrian pilgrims also began visiting six post-Arab-conquest *pīr* ('shrines') on the Dašt-i

101 *Persian Rivayats of Hormazyar Framarz and Others*, ed. and trans., Bamanji N. Dhabhar (Bombay: K.R. Cama Oriental Institute, 1932): pp. 614 and 579, respectively.

102 Recounted in Fischer, *Zoroastrian Iran*: pp. 312-26.

103 *Persian Rivayats*: pp. 579-80, 603, 614. On the droughts and famine see also Kātib, *Tārīḫ-i jadīd*: 2 p., 3, 10, 276. On the numerous Zoroastrian villages in the Dašt-i Yazd see further Fischer, *Zoroastrian Iran*: pp. 155-63.

104 Jaʿfarī, *Tārīḫ-i Yazd*: pp. 36-7. This trend continues today, with the areas from Ardakān southward to Maybud being built up contiguous with Yazd, and so gradually losing their distinct communal and administrative identities.

105 Kātib, *Tārīḫ-i jadīd*: pp. 7, 87-8, 129, 150-8.

Yazd, making it one of the focal points of this ancient Iranian faith.[106] Perhaps the presence of those shrines contributed to the relocation of the high magi and their holy fires on the Dašt-i Yazd. Yazd city's walls, gates, and citadel provided safety against brigands. In the fifth/eleventh century, the Seljuq vizier Niẓām al-Mulk (d. 485/1092) indicated Yazd's centrality as a transit point for travel and trade, by his comment on the importance of maintaining security "on the Yazd and Kirman road".[107]

Marco Polo's impression of seventh/thirteenth century Yazd provides further information on the city's function as an important transit point, particularly in connection with trade:

> Yasdi also is properly in Persia; it is a good and noble city, and has a great amount of trade. They weave there quantities of a certain silk tissue known as Yasdi, which merchants carry into many quarters to dispose of … When you leave this city to travel further, you ride for seven days over great plains, finding cover to receive you at three places only … At the end of those seven marches over the plain you come to a fine kingdom which is called Kirman.[108]

The first Parsi Zoroastrian recorded as visiting the Yazd area was Narīmān Hūšang of Navsari. Narīmān arrived in 883/1478 (846 AY), probably travelling via Kirman along the road mentioned by Marco Polo, but in the opposite direction. Initially speaking little Persian, Narīmān improved his skills through instruction from a local Zoroastrian tutor. He conducted business as a petty trader for one year before returning overland to India, bearing the first *rivāyat*. By then, according to the *rivāyat*s, the roadways had been made relatively safe for the merchants, priests, pilgrims, and emissaries travelling between Isfahan, Yazd, and Kirman, and to and from the Indian towns of Navsari, Surat, and Cambay.[109]

---

106 Choksy, *Conflict and Cooperation*: 336-7. For a map of the shrines within the Dašt-i Yazd see Fischer, *Zoroastrian Iran*: p. 207.

107 Niẓām al-Mulk, *Siyāsat nāma*, trans. Hubert Darke, *The Book of Government or Rules for Kings* (London: Routledge & Kegan Paul, 1960): p. 70.

108 Marco Polo: trans. Yule: p. 84. In Marco Polo, trans. Latham: p. 62, the passage is translated "Among the cities of Persia is one called Yazd, a very fine, splendid city and a center of commerce. A silken fabric called *Yazdi* is manufactured here in quantity and exported profitably to many markets … The traveler who leaves this city to proceed farther rides for seven days over a plain in which there are only three inhabited places where he can get shelter … And at the end of the seven days' journey lies a kingdom called Kirman."

109 *Persian Rivayats*: pp. li-xv, 605, 611, 616. On the routes see also Aigle, *Le Fārs*: p. 99, 182; and Fischer, *Zoroastrian Iran*: pp. 127-8.

The interconnectedness of other regions with pre-Islamic and Islamic Yazd and its nearby towns and villages created the administrative, economic, communal, and religious basis on which these developed, and endured as multi-communal centers.[110] After all, Yazd was and still is at the center of overland roads heading northward to Rayy and the Caspian Sea; northwestward to Tabriz and then to the Caucasus, the Black Sea and Anatolia (Turkey); westward via Isfahan to Baghdad; southwestward to Basra and the Persian Gulf; southward via Shiraz and Kāzirūn to Būshihr on the Persian Gulf; southeast toward Kirman, Bam, and the Indus River delta and onward to both the Arabian Sea and Gujarat; and finally northeastward to Mashhad and then on to Marw, Bukhara and Samarqand. Not surprisingly Iranians, Arabs, Turks and subsequently Europeans – adhering to Zoroastrianism, Judaism, Islam and Christianity – lived and worshiped in Yazd city and its surrounding settlements, and many more of them passed through the Dašt-i Yazd.

Bibliographical References

Afšār, Īraj, *Yādgārhā-yi Yazd*, 3 vols. (Tehran: Ḫāna-yi kitāb, 1348-54š./1969-75).

Aigle, Denise, *Le Fārs sous la domination mongole: politique et fiscalité (XIIIe-XIVe s)* (Paris: Association pour l'avancement des études iraniennes, 2005).

Anisi, Alireza, "The Davāzdah Imām Mausoleum at Yazd: A Re-Examination", *Iran* XLVII (2009): pp. 57-68.

Aubin, Jean, "Eléments pour l'étude des agglomérations urbaines dans l'Iran médiéval", in Albert H. Hourani and Samuel M. Stern (eds.), *The Islamic City: A Colloquium* (Oxford: Cassirer, 1970): pp. 65-75.

Āyatī, ʿAbd al-Ḥusayn, *Tārīḫ-i Yazd* (Yazd: Gulbahār, 1317š./1938).

Back, Michael, *Die sassanidischen Staatsinschriften* (Leiden: Brill, 1978).

Bāfqī, Muḥammad Mufīd Mustawfī, *Jāmiʿ-i mufīdī*, ed. Īraj Afšār, 3 vols. (Tehran: Kitābfurūšī-yi Asadī, 1339-41š./1961-3).

al-Balāḏurī, *Kitāb futūḥ al-buldān*, ed. M.J. de Goeje (Leiden: Brill, 1866).

Barthold, Wilhelm, *An Historical Geography of Iran*, trans. Svat Soucek (Princeton: Princeton University Press, 1984).

Bates, Michael, "Arab-Sasanian Coins", in *EIr*, 2 (1986): pp. 225-9; updated (2011) at http://www.iranicaonline.org/articles/arab-sasanian-coins.

110 Observed as well for modern Yazd by Fischer, Michael M. J., and Abedi, Mehdi, *Debating Muslims: Cultural Dialogues in Postmodernity and Tradition* (Madison: University of Wisconsin Press, 1990): pp. 225-44.

Bivar, A.D.H., *Excavations at Ghubayrā, Iran*, (London: School of Oriental and African Studies, 2000).

Blair, Sheila S., *The Monumental Inscriptions from Early Islamic Iran and Transoxiana* (Leiden: Brill, 1992).

Bonine, Michael E., *Yazd and Its Hinterland* (Marburg: Geographisches Institut der Universität, 1980).

Bosworth, Clifford E., "Dailamīs in Central Iran: The Kākūyids of Jibāl and Yazd", *Iran* VIII (1970): pp. 73-95.

Bosworth, Clifford E., "Ardakan", in *EIr*, 2 (1987): pp. 370-1 (updated at http://www.iranicaonline.org/articles/ardakan-e-yazd).

Bosworth, Clifford E. "Maybud", in *EI*², VI (1991): p. 912.

Boyce, Mary, *A Persian Stronghold of Zoroastrianism* (Oxford: Clarendon, 1977).

Boyce, Mary, "Ātaš", in *EIr*, III (1989): pp. 1-5 (updated at http://www.iranicaonline.org/articles/atas-fire).

Brown, Stuart C., "Ecbatana", in *EIr*, 8 (1997): pp. 80-4; updated (2011) at http://www.iranicaonline.org/articles/ecbatana.

Browne, Edward G., *A Year amongst the Persians* (London: Black, 1893).

Brunner, Christopher, "Geographical and Administrative Divisions: Settlements and Economy", in Ehsan Yarshater (ed.) *The Cambridge History of Iran*, III(2) (Cambridge: Cambridge University Press, 1983): pp. 747-77.

Choksy, Jamsheed K., *Conflict and Cooperation: Zoroastrian Subalterns and Muslim Elites in Medieval Iranian Society* (New York: Columbia University Press, 1997).

Choksy, Jamsheed K., "Altars, Precincts, and Temples: Medieval and Modern Zoroastrian Praxis", *Iran* XLIV (2006): pp. 327-46.

Christensen, Peter, *The Decline of Iranshahr* (Copenhagen: Museum Tusculanum Press, 1993).

Cook, John M., "The Rise of the Achaemenids and Establishment of Their Empire", in Ilya Gershevitch (ed.) *The Cambridge History of Iran*, II (Cambridge: Cambridge University Press, 1985): pp. 200-91.

Daryaee, Touraj, "The Collapse of the Sasanian Power in Fārs/Persis", *Nāma-yi Īrān-i bāstān* II/1 (2002): pp. 3-18.

Daryaee, Touraj, "The Effect of the Arab Muslim Conquest on the Administrative Division of Sasanian Persis/Fars", *Iran*, XLI (2003): pp. 193-204.

Eilers, Wilhelm, "Asagarta", in *EIr*, II (1987): p. 701 (updated version at http://www.iranicaonline.org/articles/asagarta-sagartia-asagartiya-sagartian-old-persian-elamite-as-s-kar-ti-ia-babylonian-kursa-ga-ar-ta-a-a-greek-s).

Fairbanks, S. C., "Atābakān-e Yazd", *EIr*, II (1987): pp. 900-2 (updated at http://www.iranicaonline.org/articles/atabakan-e-yazd-a-dynasty-which-governed-yazd-in-the-6th-12th-century).

Fischer, Michael M. J., *Zoroastrian Iran between Myth and Praxis* (PhD dissertation, University of Chicago, 1973).

Fischer, Michael M. J., and Abedi, Mehdi, *Debating Muslims: Cultural Dialogues in Postmodernity and Tradition* (Madison: University of Wisconsin Press, 1990).

Frye, Richard N., "Ardakan", in *EI*[2], I (1960): p. 626.

Frye, Richard N., "Introduction", in Richard N. Frye (ed.) *Sasanian Remains from Qasr-i Abu Nasr: Seals, Sealings and Coins* (Cambridge, Massachusetts: Harvard University Press, 1973): pp. 1-5.

Frye, Richard N., *The History of Ancient Iran* (Munich: Beck, 1984).

Gaube, Heinz, *Arabosasanidische Numismatik* (Braunschweig: Klinkhardt & Biermann, 1973).

Gignoux, Philippe, *Iranisches Personennamenbuch*, 2: *Mitteliranische Personennamen*, fasc. 3: *Noms propres sassanides en moyen-perse épigraphique, Supplément* [1986-2001] (Vienna: Österreichische Akademie der Wissenschaften, 2003).

Göbl, Robert, *Sasanian Numismatics*, trans. Paul Severin (Braunschweig: Klinkhardt & Biermann, 1971).

Golombek, Lisa, "Urban Patterns in Pre-Safavid Isfahan", *IrSt*, VII/1 (1974): pp. 18-44.

Golshan, Sedigheh, "A Comparative Study of Oral and Documentary History: The Great Grand Mosque of Yazd", *International Journal of Arts Theory and History*, VIII/4 (2014): pp. 1-27.

Hansman, John, "Isfahan iv. Pre-Islamic Period", in *EIr*, 13 (2006): pp. 635-8; updated (2012) at http://www.iranicaonline.org/articles/isfahan-iv-pre-islamic-period.

Hansman, John and Stronach, David, "Excavations at Shahr-i Qūmis, 1971", *JRAS*, CVI/1 (1974): pp. 8-22.

Herodotus, *History*, ed. and trans. Alfred D. Godley, 4 vols, (London: Heinemann, 1926-30).

Holod, Renata, *Architecture, Patronage and Setting in Iran: A Case Study of Yazd 1300-1450* (PhD dissertation, Harvard University, Cambridge, Mass., 1972).

Holod, Renata, "The Monument of Duvāzdah Imām in Yazd and Its Inscription of Foundation", in Dickran Kouymjian (ed.) *Near Eastern Numismatics, Iconography, Epigraphy and History: Studies in Honor of George C. Miles* (Beirut: American University of Beirut, 1974): pp. 285-8.

Ḥudūd al-ʿālam, trans. Vladimir F. Minorsky, *The Regions of the World: A Persian Geography 372 AH-982 AD*, 2nd ed., (London: Luzac, 1970).

Humphreys, R. Stephen, *Islamic History: A Framework for Inquiry*, rev. ed., (Princeton: Princeton University Press, 1991).

Ibn al-Balḫī, Abū Zayd, *Fārs nāma*, edited by Guy Le Strange and Reynold A. Nicholson (London: Luzac, 1921).

Ibn al-Faqīh, *Kitāb al-buldān*, ed. M.J. de Goeje (Leiden: Brill, 1885).

Ibn Ḥawqal, *Kitāb ṣūrat al-arḍ*, ed. M.J. de Goeje (Leiden: Brill, 1873).

al-Iṣṭaḫrī, Ibrāhīm b. Muḥammad, *Kitāb al-masālik wa'l-mamālik*, ed. M.J. de Goeje, (Leiden: Brill, 1927).

al-Iṣṭaḫrī, Ibrāhīm b. Muḥammad, *Kitāb al-masālik wa'l-mamālik*, anonymous Persian translation, ed. Īraj Afšār (Tehran: Bungāh-i tarjama va-našr-i kitāb, 1340š./1961).

Ja'farī, Ja'far b. Muḥammad b. Ḥasan, *Tārīḫ-i Yazd*, ed. Īraj Afšār (Tehran: Bungāh-i tarjama va-našr-i kitāb, 1338š./1960).

Kaff, Jonathan K., "Sasanian and Arab-Sasanian Silver Coins from Turfan: Their Relationship to International Trade and the Local Economy", *Asia Major*, 3rd series, XI/2 (1998): pp. 67-115.

al-Kātib, Aḥmad b. Ḥusayn b. 'Alī, *Tārīḫ-i jadīd-i Yazd*, ed. Īraj Afšār (Tehran: Ibn Sīnā, 1345š./1966).

Keall, Edward J. and Keall, Marguerite, "The Qal'eh-i Yazdigird Pottery: A Statistical Approach", *Iran*, XIX (1981): pp. 33-80.

Kent, Roland G., *Old Persian: Grammar, Texts, Lexicon*, 2nd edn (New Haven: The American Oriental Society, 1953).

Khademzadeh, Mohammad H., "Continuity and Change in the City of Yazd", in Philip Jodidio (ed.), *Iran: Architecture for Changing Societies, An International Seminar* (Turin: Allemandi, 2002): pp. 27-30.

Khademzadeh, Mohammad H., "Archaeologists Discover Achaemenid Predecessor of Yazd City", *Paywand*, 16 May 2008, http://www.payvand.com/news/08/may/1154.html.

Khan, Makeen, "A Brief Note on Archeological Site Panj Pai in Baluchistan", in Sita Pieris and Ellen M. Raven (eds) *ABIA: South and Southeast Asian Art and Archaeology Index*, III, sec. 2 (Leiden: Brill, 2010): p. 489.

Kiani, M. Yussuf, *The Islamic City of Gurgān* (Berlin: Reimer, 1984).

Lambton, Ann K. S., *Landlord and Peasant in Persia: A Study of Land Tenure and Land Revenue Administration*, rev. ed., (Oxford: Oxford University Press, 1969).

Lambton, Ann K. S., "Yazd", in *EI*², XI (2002): pp. 302-9.

Le Strange, Guy, *The Lands of the Eastern Caliphate* (London: Cambridge University Press, 1905, reprint 1966).

Marco Polo, *The Travels of Marco Polo*, trans. Thomas Wright (London: Bohn, 1854).

Marco Polo, *The Book of Ser Marco Polo, The Venetian*, I, trans. Henry Yule, (London: Murray, 1871).

Marco Polo, *The Travels*, trans. Ronald Latham, (London: Penguin, 1958).

Markwart (Marquart), Josef, "Beiträge zur Geschichte und Sage von Erān", *ZDMG*, XLIX (1895): pp. 628-72.

al-Mas'ūdī, 'Alī b. al-Ḥusayn, *Murūj al-ḏahab wa ma'ādin al-jawhar*, rev. ed. C. Pellat, 7 vols., (Beirut: al-Jāmi'a al-Lubnāniyya, 1965-79).

Matheson, Sylvia A., *Persia: An Archaeological Guide*, 3rd edn (London: Faber, 2001).

Meisami, Julie S., *Persian Historiography: To the End of the Twelfth Century* (Edinburgh: Edinburgh University Press, 1999).

Meisami, Julie S., "History as Literature", *IrSt*, XXXIII/1 (2000): pp. 15-30.

Melville, Charles, "Persian Local Histories: Views from the Wings", *IrSt*, XXXIII/1 (2000): pp. 7-14.

Melville, Charles, "The Mongol and Timurid Periods", in Charles Melville (ed.) *Persian Historiography* (London, New York: Tauris, 2012): pp. 155-210.

Miles, George, "The Coins", in Richard N. Frye (ed.) *Sasanian Remains from Qasr-i Abu Nasr: Seals, Sealings and Coins*, (Cambridge, Massachusetts: Harvard University Press, 1973): pp. 26-36.

Miller, Isabel, "Local History in Ninth/Fifteenth Century Yazd: The Tārīkh-i Jadīd-i Yazd", *Iran*, XXVII (1989): pp. 75-9.

Miller, Konrad, *Itineraria romana*, (Stuttgart: Strecker & Schröder, 1916).

Minorsky, Vladimir, "Abarḳūh", in *EI*², I (1954): p. 3.

Mīr-Dihqān Aškaḏarī, Sayyid Faḍl-Allāh and Raḍvānī, Ḥasan and Kāẓimnažand, Ibrāhīm "Historical and Cultural Sites [and] Relative Chronology of Yazd-Ardakan Plain according to Archaeological Data", in M.Ḥ. ʿA. Ḫarānaqī, M. Ḫānīpūr, and R. Nāṣirī (eds), *Proceedings of the International Conference of Young Archaeologists* (Tehran, 1393sh./2014): pp. 358-68.

Modarres, Ali (2012) "Meybod", in *EIr*, at http://www.iranicaonline.org/articles/meybod.

Mottahedeh, Roy P., "Some Islamic Views of the Pre-Islamic Past", *Harvard Middle Eastern and Islamic Review*, I (1994): pp. 17-26.

Murtaḍā'ī, Muḥammad and Mīr-Dahqān Aškaḏarī, Sayyid Faḍl-Allāh, "Bāznigārī dar jāygāh-i šahr-i Yazd dar dawrān-i ṣadr-i Islām bā tikiya bar asnād-i maktūb va dāda-hā-ya bāstān-šināsī", *Pažūhiš-hā-yi bāstān-šināsī Īrān*, VI (1393š./2014): pp. 159-74. (with an English abstract pp. 20-21).

Ḥamd-Allāh al-Mustawfī, *Nuzhat al-qulūb*, ed. and trans. Guy Le Strange, *The Geographical Part of the Nuzhat-al-Qulūb* (London: Luzac, 1919).

Niẓām al-Mulk, *Siyāsat nāma*, trans. Hubert Darke, *The Book of Government or Rules for Kings* (London: Routledge & Kegan Paul, 1960).

Nyberg, Henrik S., *A Manual of Pahlavi*, II (Wiesbaden: Harrassowitz, 1974).

Persian Rivayats of Hormazyar Framarz and Others, ed. and trans., Bamanji N. Dhabhar (Bombay: K.R. Cama Oriental Institute, 1932).

Planhol, Xavier de, "Hamadān iii. Historical Geography", in *EIr*, XI (2003): pp. 605-7; updated (2012) at http://www.iranicaonline.org/articles/hamadan-iii.

Planhol, Xavier de, "Isfahan ii. Historical Geography", in *EIr*, 13 (2006): pp. 617-22; updated (2012) at http://www.iranicaonline.org/articles/isfahan-ii-historical-geography.

Pliny the Elder (Gaius Plinius Secundus), *Natural History*, ed. and trans. Horace Rackham and others, 10 vols (London: Heinemann, 1938-63).

Potts, Daniel T., *Nomadism in Iran: From Antiquity to the Modern Era* (Oxford: Oxford University Press, 2014).

Ptolemy, Claudius, *Geography*, ed. and trans. Edward L. Stevenson, (New York: New York Public Library, 1932).

Pūyā, Sayyid 'Abd al-'Aẓīm, *Sīmā-yi bāstānī-yi šahr-i Maybud: Barrisī-yi tārīḫī-yi sāzvāra va sāḫt-i yak šahr-i kavīrī* (Maybud: Dānišgāh-i āzād-i islāmī, 1371š./1992).

Qūčānī, 'Abd-Allāh, *Barrisī-yi katība-hā-yi banāhā-yi Yazd*, (Tehran: Sāzmān-i mīrāṯ-i farhangī va-gardišgarī, 1383š./2004).

Rante, Rocco (2010) "Ray i, Archeology", in *EIr*, at http://www.iranicaonline.org/articles/ray-i-archeo.

Rante, Rocco, *Rayy: From Its Origins to the Mongol Invasion, an Archaeological and Historiographical Study* (Leiden: Brill, 2014).

Robinson, Chase F., *Islamic Historiography* (Cambridge: Cambridge University Press, 2003).

Rosenthal, Franz, *A History of Muslim Historiography* (Leiden: Brill, 1968).

Rypka, Jan, *History of Iranian Literature* (Dordrecht: Reidel, 1968).

Ṣahba, Javād M. (1944) "Darvāza-hā-yi nuhṣad va sī sāla", *Yādgār* I (1323š./1944): pp. 72-80.

Schindel, Nikolaus (2005) "Sasanian Coinage", in *EIr*, at http://www.iranicaonline.org/articles/sasanian-coinage.

Schindel, Nikolaus, "The Mint System under Husraw I: Some Preliminary Remarks", in Christelle Jullien (ed.) *Husraw Ier Reconstructions d'un règne. Sources et documents* (Paris: Association pour l'avancement des études iraniennes, 2015): pp. 15-40.

Schwarz, Paul, *Iran im Mittelalter nach den arabischen Geographen*, 9 vols., (Stuttgart: Harrassowitz, later Wiegand, 1896-1936).

Sellwood, David, Whitting, Philip, and Williams, Richard, *An Introduction to Sasanian Coins* (London: Spink & Son, 1985).

Shahrestānīhā ī Ērānshahr, ed. and trans. Josef Markwart (Marquart), *A Catalogue of the Provincial Capitals of Ērānshahr* (Rome: Pontificio Istituto Biblico, 1931).

Siroux, Maxime, "Le temple Zoroastrien de Sharīfābād", *Āṯār-é Īrān* III/1 (1938): pp. 83-9.

Siroux, Maxime, "La Masjid-e-Djum'a de Yezd", *Bulletin de l'Institut Français d'Archéologie orientale*, XLIV (1947): pp. 119-76.

Strabo, *Geography*, ed. and trans. Horace L. Jones, 8 vols. (London: Heinemann, 1917-32).

al-Ṭabarī, Abū Ja'far Muḥammad b. Jarīr, *Ta'rīḫ al-rusul wa'l-mulūk*, ed. M.J. de Goeje and others, 15 vols, (Leiden: Brill, 1879-1901).

Tomaschek, Wilhelm, “Zur Historischen Topographie von Persien I: Die Strassenzüge der tabula Peutingeriana”, *Sitzungsberichte der Philologisch-Historischen Classe der Kaiserlichen Akademie der Wissenschaften* 102 (Vienna: Gerold, 1883): pp. 145-231.

Upton, Joseph M., “The Site of Qasr-i Abu Nasr and a Description of the Sealings”, in Richard N. Frye (ed.) *Sasanian Remains from Qasr-i Abu Nasr: Seals, Sealings and Coins*, (Cambridge, Massachusetts: Harvard University Press, 1973): pp. 6-25.

Walker, John, *A Catalogue of the Arab-Sassanian Coins* (London: The British Museum, 1941).

Williams Jackson, Abraham V., *Persia Past and Present* (London: Macmillan, 1906).

Yāqūt al-Ḥamawī, *Muʿjam al-buldān*, ed. Ferdinand Wüstenfeld, 6 vols. (Leipzig: Brockhaus, 1866-73).

www.vcoins.com.

Biographical Note

Jamsheed K. Choksy (BA Columbia University; PhD Harvard University) is Distinguished Professor and Chairperson of the Department of Central Eurasian Studies at Indiana University, Bloomington. He was nominated by the US President and confirmed by the US Senate as a member of the Council overseeing the US National Endowment for the Humanities. Choksy has held fellowships from the Institute for Advanced Study (Princeton), National Endowment for the Humanities (Washington, DC), Guggenheim Foundation, the Center for Advanced Study in the Behavioral Sciences (Stanford), American Philosophical Society, and the American Academy of Religion. Choksy is a consulting editor for the *Encyclopedia Iranica*. He is the author of *Triumph over Evil: Purity and Pollution in Zoroastrianism* (1989), *Conflict and Cooperation: Zoroastrian Subalterns and Muslim Elites in Medieval Iranian Society* (1997), and *Evil, Good, and Gender: Facets of the Feminine in Zoroastrian Religious History* (2002).

Isfahan during the Turko-Mongol Period (11th-15th Centuries)

David Durand-Guédy
Independent scholar
david_durandguedy@yahoo.com

Abstract

This article deals with the history of Isfahan during the five centuries between the arrival of the Turks and the beginning of the Safavid period. It attempts to identify the continuities and ruptures from three different perspectives: To what extent can we speak of an urban decline? What was the relationship of the Isfahanis with the imperial rulers? How did the general context impact the makeup and the organisation of the society? It appears that Isfahan demonstrated remarkable resilience over the period, and that it is only beginning with the Timurid period that we can speak of a decline. Likewise, there is remarkable continuity in the ways in which the local elites collaborated with or resisted the imperial players. The role of the Ṣāʿid family, which held the cadial function over four centuries, is emblematic in this respect. Conversely, under special circumstances, the imperial players were able to exert great substantial influence on the local communities, and it is this political backing which explains the strengthening of the Hanafis in the sixth/twelfth century, and that of the Shiis in the following. One of the main accomplishments of this synthesis view of Isfahan history is that it assesses the effects of the Mongol domination relative to that of other periods.

Keywords

urban decline – agency – factionalism – guilds – Shiism – militia – *futuwwa*

1 Introduction

Forty-four years ago, a conference dedicated to Isfahan was convened at the University of Harvard. Most of the papers focused on the modern era, especially

 | DOI:10.1163/9789004434332_012

the period from 1597 to 1722, when Isfahan served as capital of the Safavid empire and was known as the half of the world. Two contributions addressed the pre-Safavid period, dealing with literature (Peter Chelkowski) and urbanism (Lisa Golombek).[1] As of that time (1974), there had been no rigorously critical study dedicated to the city itself.[2] Given the importance of Isfahan and its role in the history of Iran (the capital of the Great Saljuq state at the height of its power, and a major centre of Islamic culture throughout the centuries), such neglect would be surprising if we were to ignore the state of scholarship at the time. In his inspiring "Eléments for the study of urban agglomerations in medieval Iran" published in 1970, Jean Aubin sternly noted that the relevant entries of the *Encyclopaedia of Islam* (the only available encyclopaedic resource at the time) seemed to imitate the works of the nineteenth century geographer Malte-Brun. This was not a compliment.[3]

Since then, the situation has changed. Unlike some important medieval cities that have been excavated, or where this opportunity still exists (such as Rayy, Jūrjān, Nishapur, Āba, Abīvard or Baylaqān), the site of medieval Isfahan lies at the heart of the modern agglomeration, and so large excavations cannot be undertaken. In the early 2000s, the area south of the Friday Mosque was completely reshaped during construction of an underground interchange (Fig. 1-3), but these works did not produce the archeological results one would have hoped.[4] While major archaeological discoveries remain improbable, historians have been able to draw on a vast array of literary sources. Compared to 1974, we are much better informed about the history of the city before Shah

1 See Holod, Renata (ed.) *Studies on Isfahan. Proceedings of the Isfahan colloquium, sponsored by the Fogg Museum of Art, held at Harvard University, January 21-24, 1974*, special issue of *IrSt*, VII/1-2 (1974).

2 The voluminous compilations of Sayyid Muṣliḥ al-Dīn Mahdavī and Jalāl al-Dīn Humā'ī contain a vast amount of information on the history and the historical geography of Isfahan, but without a dynamic or problem-focussed angle. Mahdavī's biographical dictionaries have been recently reedited with supplements. The various volumes of Humā'ī's *Tārīḫ-i Iṣfahān* have been edited between 1375š. and 1395š. (1996-2016) by Māhduḫt Humā'ī.

3 Aubin, Jean, "Eléments pour l'étude des agglomérations urbaines dans l'Iran médiéval", in Hourani, Albert and Stern, Samuel (eds.), *The Islamic City: Papers on Islamic history* I (Carbondale: Southern Illinois University Press, 1970): pp. 65-75, here p. 65. See the translation in this volume.

4 Javeri, Mohsen, "The Excavations of Atiq square, Isfahan", in Genito, Bruno and Anaraki, Fariba Saiedi (eds.) *ADMAJI Project. From the Excavation (1972-1978) to the Archive (2003-2010) in the Masjed-e jome, Isfahan* (Tehran: Embassy of Italy, 2011): pp. 117-122. This being said, Ali Shojaee Isfahani may have found a portion of the city walls while excavating below the level of a Safavid building at the site of the Jahān Nāma. See Shojaee, Ali, "Excavations at the Jahānnamā Complex: urban archaeology at Isfahan, Iran", *Antiquity* 91/355 (2017), online publication last accessed 20/01/2017.

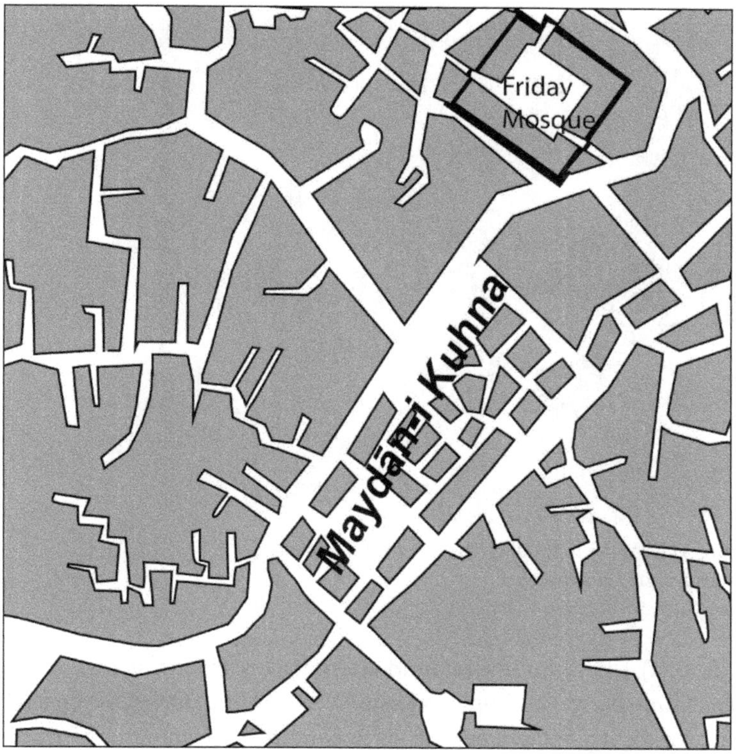

FIGURE 1 The Old Maydān in 1923 according to the map of Sayyid Riḍā-Khān, *Naqša-yi šahr-i Iṣfahān dar sāl-i 1302 hijrī-yi šamsī*
DRAWING: D. DURAND-GUÉDY

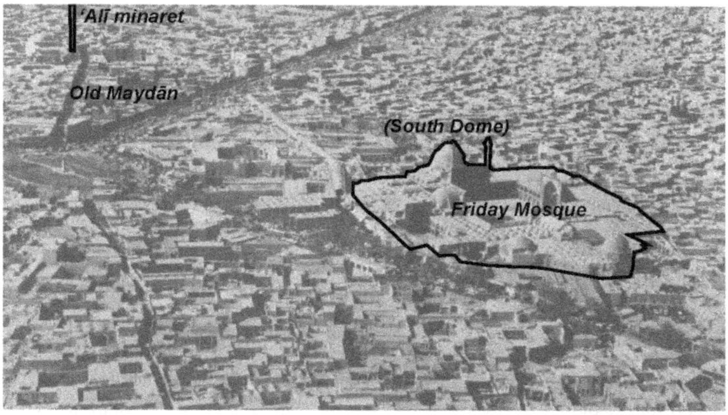

FIGURE 2 Aerial view of the Friday Mosque and the Old Maydān in the 1970s and situation of the main architectural features
DRAWING: D. DURAND-GUÉDY

FIGURE 3 Street view of the new "Old Maydān" built above an underground interchange
PHOTO D. DURAND-GUÉDY, 2015

ʿAbbās (d. 1038/1629), thanks to several fundamental works focusing on specific eras: the early Islamic period by Hosseyn Kamaly,[5] the Saljuq period by myself,[6] and the Timurid and early Safavid periods by Rosemarie Quiring-Zoche.[7] Apart from this we can also rely on many articles, most often in Persian, dealing with subjects ranging from architecture to medieval dialectology, as well as the very

5 Kamaly, Hosseyn, *Four Moments in the Early Islamic History of Isfahan* (unpublished PhD dissertation: Columbia University, 2004).

6 Durand-Guédy, David, *Iranian Elites and Turkish Rulers. A History of Iṣfahān in the Saljūq Period* (London & New York: Routledge, 2010), reviewed by Georges Lane (BSOAS), Robert Bulliet (*Iranian Studies*), Jürgen Paul (*Eurasian Studies*), A.C.S. Peacock (IJAS) and Elton Daniel (*American Historical Review*). The main findings are summed up in Durand-Guédy, David "What Does the History of Isfahan Tell us about Iranian Society during the Saljuq period?", in Edmund Herzig and Sarah Stewart (eds.), *The Idea of Iran, VI: the Age of the Seljuqs* (London and New York: Tauris, 2014): pp. 58-73. A book on Saljuq Isfahan has been published in Persian in 2011, but it merely offers a non-problematized accumulation of facts, through a non-critical approach of the sources, non-withstanding the many interpolations; see Kajbāf, ʿAlī Akbar, *Iṣfahān dar dawra-yi saljūqiyān* (Isfahan: Intišārāt-i šahrdārī-yi Iṣfahān, 1379š.).

7 Quiring-Zoche, Rosemarie, *Isfahan im 15 und 16 Jahrhundert. Ein Beitrag zur persischen Stadtgeschichte*, (Freiburg: Schwarz, 1980), reviewed by Jean Aubin (*Abstracta Iranica*), Robert D. Mcchesney (*The American Historical Review*) and Bert Fragner (WZKM). Fragner's enthousiasm contrasts with Aubin's reserve for an erudite tradition which only deals with social phenomena through institutional structures.

useful biographical notes concerning Isfahani individuals and families in the *Dāyirat al-maʿārif-i buzurg-i islāmī*.

Much remains to be done. Recent research has shown that landscape archaeology is invaluable in providing a clearer picture of the hinterland, and especially its interconnections with the city.[8] Important written sources still await detailed analysis. The discovery in Yazd of an *inšāʾ* collection, entitled *al-Muḫtārāt min al-rasāʾil*, has yielded invaluable glimpses of local society in the late sixth/twelfth century. This is a text (published as facsimile in 1976, in critical edition in 1999) which has served extensively in my research on pre-Mongol Isfahan.[9] Other recently published books on pre-modern Isfahan include: the Shiʿi treatise *Maṭlaʿ al-ṣabāḥtayn* of Abu'l-Saʿādāt Šufurwa (also: Šafrawayh) al-Iṣfahānī, written at the onset of the Mongol period;[10] the seventh/thirteenth century *Munāẓara-yi Baġdād wa Iṣfahān*, continuing a long tradition of literary debate about the respective virtues of Isfahan and its nemesis Baghdad;[11] and not least, a vast (11 volume) collection of endowment deeds (*waqf-nāma*), largely dating to the Safavid and Qajar periods, but also with documents relative to that of the Timurids.[12] Each of these works deserves detailed and specific study, but this is not the aim of the current paper.

Instead, I would like to attempt a synthesis on the lines suggested by Jean Aubin in the aforementioned article, which is "to move from the descriptive inventory [of events or spaces] to a dynamic approach and the formulation of complex problems."[13] I will deal with some of these problems for the fifth-ninth/eleventh-fifteenth centuries, when the main dynasties were of Turkic or Mongol ethnicity: the Saljuqs and their epigons (Atabegs of Azarbaijan, Khwarazm-Shahs); the Mongols and their epigons (Injuid, Chupanid); the

8 See Ali Shojaee Isfahani's contribution on the fortified landscape of Isfahan, in this volume. For a similar approach but focused on a specific *rustāq*, see Kamalizad, Jaleh, "The Landscape Archaeology in Lanjān Plain of Iṣfahān, Iran, during the Islamic Period (650-1722 AD) (preliminary report)", *Archaeologies*, XII/1 (2016): pp. 91-109.

9 *Al-Muḫtārāt min al-rasāʾil*, ed. facsimile Īraj Afšār (Tehran: Intišārāt-i anjūman-i āṯār-i millī, 2535 š.šahānšāhī), ed. Īraj Afšār & Ġulām-Riḍā Ṭāhir (Tehran: Bunyād-i mawqūfāt-i duktur Maḥmūd Afšār Yazdī, 1378š.).

10 Abu'l-Saʿādāt Šufurwa al-Iṣfahānī, *Maṭlaʿ al-ṣabāḥtayn wa majmaʿ al-faṣāḥtayn*, ed. Sayyid Ṣādiq Aškivarī (Qum: Pažūhišgāh-i ʿulūm-i insānī, 1385š.). See the presentation by Jaʿfariyān, Rasūl, "Abu'l-Saʿādāt Asʿad b. Qāhir-i Šufurwa, ʿālim-i shīʿa iṣfahānī dar qarn-i haftum-i hijrī wa yād-dāštanī dar barā-yi ḥamla-yi Muġūl bi Iṣfahān dar sāl-i 634", in R. Jaʿfariyān (ed.) *Maqālāt-i tārīḫī* (Qum: Intišārāt-i Dalīl-i mā), XVI (1386š.): pp. 171-87.

11 Kamāl Iṣfahānī, ʿAbd al-Salām b. Abi'l-Mujāhid, *Munāẓara-yi Baġdād wa Iṣfahān*, ed. Aḥmad-Riḍā Yalamih-hā (Dihāqān: Dānišgāh-i āzād-i islāmī vāḥid-i Dihāqān, 1391š.).

12 *Asnād-i mawqūfāt-i Iṣfahān*, 11 vols, ed. Sayyid Ṣādiq Ḥusaynī Aškivarī (Qum: Majmaʿ-i ḏaḫāʾir-i islāmī, 1388š./2009).

13 Aubin, "Elements": p. 67.

Muzaffarids (the only non-Turko-Mongol dynasty); the Timurids and lastly the Turkmen dynasties (Qara-qoyunlu and Aq-qoyunlu) (see Table 1 for a basic chronology).

I will not follow a chronological presentation, which is already available for the Saljuq and Timurid periods.[14] I will instead try to answer to several questions:

1. To what extent can we speak of an urban decline?
2. What was the relationship of the Isfahanis with the imperial rulers?
3. How did the general context impact the makeup and the organisation of the society?

Each of these questions will be the focus of one section of the text. Given the aims and the scope of the article, I will refrain from referring the primary sources where these have already been referenced in previous studies.

2 The Resilience of Isfahan in a Time of Urban Decline

A comparison between the account of Nāṣir-i Ḫusraw at the beginning of the Saljuq rule, and that of Giosafat Barbaro, the Venetian envoy at the Aq-Qoyonlu court, gives an impression of irremediable decline. The first saw no finer or more flourishing city than Isfahan in all the Persian-speaking lands. The second found the city miserable, but remembering its glorious past. What is the case? We should begin with a few words about the context.

2.1 *An Adverse Context*

The period is characterised by several phenomena. Usually the Turko-Mongol period refers to the three centuries from the Mongol conquest to the rise of the Safavids, and does not include the Saljuqs. True, the latter were still operating in the framework of the Abbasid caliphate, and moreover, it was thought that they had been swallowed up in the dominant Irano-Islamic culture, unlike the Mongols and their successors. However, as far as the location of the ruler is concerned, these dynasties all had much in common. From the fifth/eleventh

14 Durand-Guédy, *Iranian Elites*, and Quiring-Zoche, *Isfahan*. The main encyclopaedia articles on Isfahan also take a chronological approach. Lambton, Ann "Iṣfahān: history", in *EI*², IV, ed. by E. van Donzel *et al.* (Leiden: Brill, 1978): pp. 97-105 is the standard narrative. Saʿīdī, ʿAbbās "Iṣfahān. I. Juġrāfiyā wa juġrāfiyā-yi tārīḫī", *Dāyirat al-maʿārif-i buzurg-i islāmī*, IX, ed. by Muḥammad-Kāẓim Bujnūrdī (Tehran: Markaz-i Dāyirat al-maʿārif-i buzurg-i islāmī, 1379š.): pp. 156-81 is the most detailed. Kamaly, Hossein, "Iṣfahān – vi. Medieval period", in *EIr*, XIII, ed. by Ehsan Yarshater (New York: Bibliotheca Persica Press, 2006): pp. 641-50, is to be used with caution.

TABLE 1 The ruling dynasties of Isfahan

Date	Ruler of Isfahan (capital city indicated in brackets)	Main events in Isfahan
420/1029 to 427/1036	Ghaznavids (Ghazna)	421/1030: massacre of the population by the Ghaznavids
427/1036 to 443/1051	Kakuyids (Isfahan)	
443/1051 to 511/1118	Great Saljuqs (Isfahan) (443/1051 to 511/1118)	ca. 455/1063-485/1092: Niẓām al-Mulk based in Isfahan 490/1096-7 to 500/1107: *fitna* (Ismailis vs. 'Khurasani' networks)
511/1118 to 590/1194	Lesser Saljuqs (Hamadan), submitted to Sanjar (Khurasan) until 548/1153, then from 555/1160 to 588/1192 to the Atabegs of Azarbaijan (Tabriz)	ca. 521/1127: Isfahan ceases to be *dār al-mulk* of the Saljuqs 542/1147: *fitna* (Hanafis vs. Shafiis) 552/1157: *fitna* (Hanafis vs. Shafiis) 559-560/1164-5: *fitna* (Hanafis vs. Shafiis) 581/1186: *fitna* (Hanafis vs. Shafiis) 584-585/1188-9: *fitna* (Hanafis vs. Shafiis)
590/1194 to 596/1200	Khwarazm-Shahs (Urganch)	591-2/1195-6: Abbasid occupation 594/1197-8: occupation by rebel Khwarazmian general Mayājuq
596/1200 to 614/1217-8	various amirs (Hamadan)	successive rule of Gökche; Ay-Toghmïsh; Mengli; Ïghlamïsh
614/1217-8 to 633/1235-6	Khwarazm-Shahs (Urganch)	614/1217-8 Atabeg of Fars Sa'd b. Zangī, Atabeg of Azarbaijan Öz-beg b. Pahlawān, and Khwarazm-Shah Muḥammad b. Tekish occupy Isfahan 618/1221-2: *fitna* (Hanafis vs. Shafiis) 625/1228: Battle of Sīn between Khwarazm-Shah Jalāl al-Dīn Mengburnu and the Mongols

TABLE 1 The ruling dynasties of Isfahan (*cont.*)

Date	Ruler of Isfahan (capital city indicated in brackets)	Main events in Isfahan
633/1235 to 736/1335	Mongols/Ilkhanids (Tabriz)	633/1235: massacre by the Mongols bet. 663/1265-681/1282: Bahāʾ al-Dīn b. Muḥammad Juwaynī governor of Isfahan and ʿIrāq 710/1310: *fitna* after Öljeitü's attempt to declare Twelver Shiism as official religion
736/1335 to 742/1341-2	Chobanids (Tabriz)	
742/1341-2 to 758/1357	Injuids (Shiraz)	742/1341-2 Šayḫ Abū Isḥāq Īnjūʾ appointed governor of Isfahan by Amīr Pīr Ḥusayn
758/1357 to 789/1387	Muzaffarids (Yazd)	758/1357: Isfahan is taken by Mubāriz al-Dīn Muḥammad Muẓaffar (d. 765/1364), and after his demise it is ruled by his son Šāh Maḥmūd (d. 776/1374), and his grand-son Šāh Yaḥyā. 786/1384: the Isfahanis force Šāh Yaḥyā out of town; Muẓaffar-i Kāšī rules Isfahan
789/1387 to 856/1452	Timurids (Samarqand, then Herat)	790/1388 massacre of the population by Timur 817/1414: siege and looting of the city by Šāhruḫ
856/1452 to 874/1469	Qara-qoyunlu (Tabriz)	858/1454 massacre of the population and destruction of the city by Jahān-Šāh
874/1469 to 908/1503	Aq-qoyunlu (Tabriz)	
908/1503		Isfahan is taken by Shah Ismāʿīl

to the ninth/fifteenth centuries, imperial rulers did not live inside cities, but at the *ordo*, which is a moving tented encampment, set up outside the city-walls (the *ordo* is called *laškargāh* in pre-Mongol Persian sources). This is the 'Ruling from the Outside' paradigm, which I first introduced in my book *Iranian Elites and Turkish Rulers*, then in a series of articles dealing with the spatial context of the Saljuq courts.[15]

In the case of Isfahan, these extra-urban locations are relatively well-known for the Saljuq period. Close to the city walls, three of the four *bāġ* (walled gardens) mentioned by Ẓahīr al-Dīn Nīšāpūrī (fl. 573/1177-8) can be identified: Aḥmad-Siyāh, east of Jūbāra; Kārān, south of the citadel of Ṭabarak; finally Falāsān, on the southern side of the Zāyanda-rūd (probably at the site of the Safavid *bāġ* of Hazār Jarīb) (see Fig. 5). For the later periods, other *bāġ* are mentioned, such as the Bāġ-i Taḫt-i Qarāja where the Timurid prince Muḥammad Sulṭān b. Baysonqur put on a great feast for the local elites there, after they rallied in his support.[16] But many of the royal pastures were located much further from the city, such as Firaydūn, 140 kilometres to the West, on the pastoral road to Burūjird and Nihāvand, or in Gandumān, 100 kilometres to the South, near modern Burūjin. Both areas were used as royal summer resorts from the Saljuqs to the Qajars.[17]

Beyond the general contempt or distance of the rulers from city life, the second significant feature of the new context was the marginalisation of Isfahan on the political chessboard. In the fifth/eleventh century, the city had been the capital of a powerful regional state (the Kakuyids), and an even more formidable empire (the Great Saljuqs),[18] but in the second decade of the sixth/twelfth century, it lost its status to Hamadan, a city more strategically located in the 'Irāq-i 'ajam. Moreover, the province of 'Irāq-i 'ajam itself was increasingly sidelined by the rise of Azarbayjan and its new hub, Tabriz (Tabriz was the capital of the Eldigüzid Atabegs, then of the Mongols, and later of the

---

15 See my introduction, "Location of Rule in a Context of Turko-Mongol Domination," in D. Durand-Guédy (ed.) *Turko-Mongol Rulers, Cities and City Life* (Leiden-Boston: Brill, 2013): pp. 1-20, with references to my other publications.

16 Quiring-Zoche, *Isfahan*: pp. 35-6.

17 Ẓahīr al-Dīn Nīšāpūrī, edited by A. Morton, *The Saljūqnāma of Ẓāhir al-Dīn Nīshāpūrī* (Antony Rowe UK, 2004): pp. 30-1 (the *bāġ*), 116 (Gandumān and Balāsān). Balāsān, modern Čoġāḫūr, is also mentioned by the Ilkhanid author Ḥamd Allāh Mustawfī, edited by Guy Le Strange, *The Geographical Part of Nuzhat-al-qulūb* (Leyden and London: Brill & Luzac, 1915), p. 49. Le Strange mistakenly reads *bilā ša'an*, "unrivalled". For later mention of Gandumān, see Quiring-Zoche, *Iṣfahan*: p. 22.

18 Durand-Guédy, *Iranian Elites*: pp. 49-52 (Kakuyid capital), 75-101 (Saljuq capital), 207-10 (loss of rank).

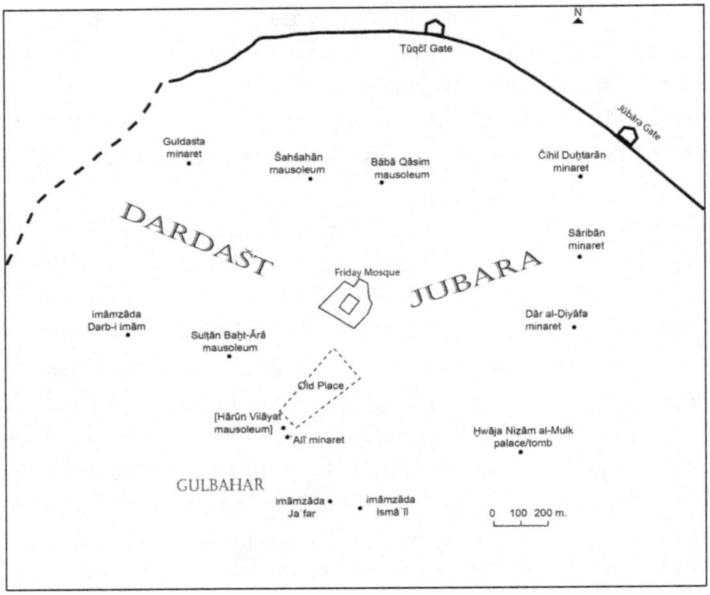

FIGURE 4 Main monuments of Isfahan during the Turko-Mongol period
DRAWING: D. DURAND-GUÉDY

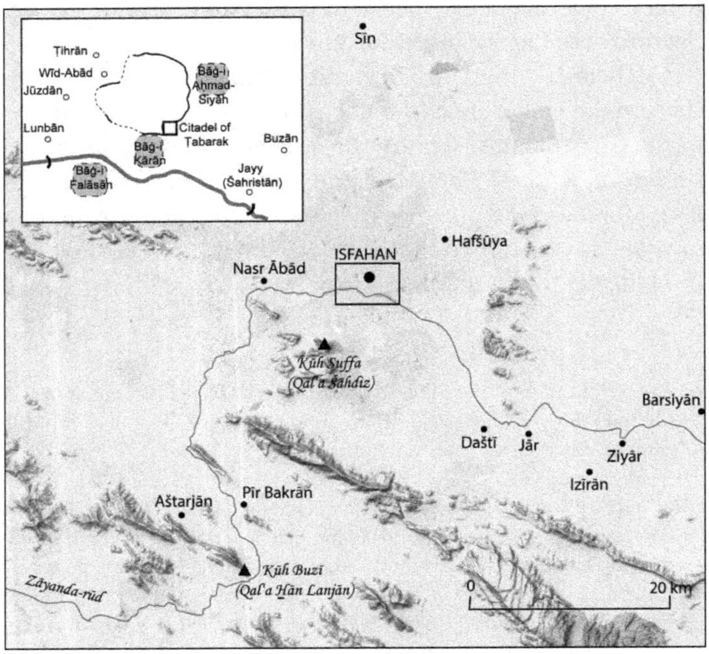

FIGURE 5 The hinterland of Isfahan
AUTHOR: D. DURAND-GUÉDY

Türkmen dynasties).[19] Until the Safavids succeeded in definitively reunifying the two reaches of the Iranian plateau, the major political centres were located in Azarbayjan on the one hand (Tabriz), and in Central Asia on the other (in Khwarazm, Khurasan or Transoxiana). Exceptionally, a prince with imperial ambitions would make Isfahan his capital, but this was always a short-lived arrangement. Shāh Maḥmūd, the Muzaffarid ruler of Isfahan, tried to challenge Jalayirid control over Azarbaijan and he occupied Tabriz, but he soon had to yield. During the Timurid period, Isfahan was twice made capital (first with Iskandar b. ʿUmar Šayḫ after he left Shiraz, in 815-7/1412-14, and then with Muḥammad b. Baysonqur in 850-5/1446-51), but again this did not last.

The rise to the role of imperial centre always brought a certain type of growth through diverse mechanisms. One of these was the influx of specialised personnel operating in divans and bazaars.[20] The arrival of Ibn Sīnā (Avicenna), the most exceptional scholar of his time, at the court of the Kakuyids is the best example of the attraction exerted by the new capital. Likewise, during his short rule, the Timurid prince Iskandar gathered a brilliant court, a glimpse of which can be seen in a document analysed in depth by Francis Richard.[21] After the fall of Iskandar, Šāhruḫ took several scholars and artists from the court of Isfahan with him, to Khurasan, as did his father Timur.[22] Indeed, Aubin says that when a city ceased to serve as a capital, it lost some of its inhabitants. (He quotes the example of Tabriz, which could have lost up to one-third of

19 The marginal situation of Isfahan in the later eighth/fourteenth century could be demonstrated by monetary evidence. For Stephen Album, Iran seems to have been divided into four monetary zones, each of which employed its own standards and saw its own sequence of weight reductions in the silver coinage. The four regions may be termed Baghdādī, Tabrīzī, Nīshāpūri, and Shīrāzī, after their chief cities. Album, Stephen, "The coinage of Nūr-Āward, Atabeg of Lur Buzurg, 751-57 H./AD 1350-56", *Museum Notes* 22 (1977): pp. 213-39, here p. 222, note 28.

20 For the post-Saljuq periods, among other evidence, this is perceptible through the rise of taxes on crafts and trade (*tamġa*), see Petrushevsky, Ilya Pavlovic, "The Socio-Economic Condition of Iran under the Īl-Khāns", *Cambridge History of Iran*, v, edited by John Boyle (Cambridge: Cambridge Univ. Press, 1968): pp. 483-537, here p. 508.

21 A letter sent from Iskandars chancery in Isfahan, summoning Sayyid Qiwām al-Dīn Yazdī to the court, gives the names of the many scholars and artists (theologians, historians, astronomers, physicians, calligraphers, poets) already present there. See Richard, Francis, "Un témoignage inexploité concernant le mécénat dEskandar Solṭān à Eṣfahān", *OM*, xv (1996): pp. 45-72.

22 It is only because he belonged to a family of scholars that Ṣāʾin al-Dīn ʿAlī Ibn Turka, who was 18 at the time of Timur's conquest, was deported to Samarqand, along with his father and brother, while the city inhabitants were slaughtered. See Melvin-Koushki, Matthew S. (2012) *The Quest for a Universal Science: the Occult Philosophy of Ṣāʾin al-Dīn Turka Iṣfahānī (1369-1432) and Intellectual Millenarianism in Early Timurid Iran* (PhD, Yale Univ.): p. 45.

its population after the capital was transferred to Qazwin).[23] Conversely, because of its marginal position between Fars and in Azarbaijan, which were the major centres of power in Western Iran from the sixth/twelfth to ninth/fifteenth century, competing polities often fought over Isfahan: atabegs of Fars versus atabegs of Azarbaijan in the sixth/twelfth century (the succession of three armies, provisioned from the local harvest in 532/1138, heralded the new era);[24] Chupanids versus Jalayirids versus Muzaffarids after the crumbling of the Ilkhanate; Timurid prince in Shiraz versus Türkmens in the ninth/fifteenth century.

Linked with the two previous phenomena, Isfahan witnessed during these five centuries a remarkably higher degree of violence than in the first centuries of Islam. The city was besieged, taken, and looted many times over the period (Table 1 and 2). It is important to note that here again, the Mongol period does not represent a clear watershed. At the beginning of the eighth/fourteenth century, Ḥusayn Āvī, an author based in Isfahan, insisted on describing the devastation of Isfahan *before* the Mongol conquest.[25] As we will see below, Āvī had an obvious interest in exonerating the Ilkhanids (or the Čingīzḫāniyān as he calls them) of wrongdoing. But still, long before the capture of Isfahan by the Mongols and especially after the demise of the Saljuq dynasty, the contemporary sources describe a miserable situation. This emerges clearly in the long qasida of Šaraf al-Dīn Šufurwa quoted by Āvī, in the writings of Rāwandī (fl. 631/1233-4), of the poet Kamāl al-Dīn Ismāʿīl (d. ca. 635/1237), but also in some desperate letters in the *al-Muḫtārāt*. The situation obviously then got worse with the Mongol conquest. Speaking about Timur's troops, at a century and half after the Mongol conquest, Jean Aubin noted that ancient beliefs and customs endured within the tribal nobility, who led these tough soldiers, horse-meat eaters, drinkers of wine, kumis and alcohol of millet. and he concludes: that world was naturally indifferent to Islamic solidarity.[26] (The same might be said about the Khwarazmians, who ruled Isfahan after the Saljuqs). The *Raubwirtschaft*, which was first experienced during the Ghaznavid oc-

23 Aubin, Jean, "Chiffres de population urbaine en Iran occidental autour de 1500", *Moyen Orient & Océan Indien*, IV (1976-7): pp. 37-54, here p. 42 (*apud* Barbaro). For a comparison with European cities, cf. Boucheron, Patrick and Menjot, Denis, "La ville medievale", in Jean-Luc Pinol (ed.), *Histoire de l'Europe urbaine, I. De l'Antiquité au XVIII*^e^ *siècle. Genèse des villes européennes* (Paris: Seuil, 2003): pp. 287-592, here p. 381.

24 See Durand-Guédy, *Iranian Elites*: p. 259. On the struggle between Fars and Azarbaijan for the control of Western Iran see Durand-Guédy, David, "1147. The Battle of Qara-Tegin and the Rise of Azarbayjan", *Der Islam*, XCII/1 (2015): pp. 161-96.

25 Ḥusayn b. Muḥammad Āvī, *Tarjuma-yi maḥāsin-i Iṣfahān*, edited by ʿAbbās Iqbāl (second edition, Isfahan: Sāzmān farhangī wa tafrīḥī-yi šahrdārī-yi Iṣfahān, 1385š.): pp. 150-1.

26 Aubin, Jean, "Comment Tamerlan prenait les villes?", *StIs*, XIX (1963): pp. 83-122, here p. 86.

cupation at the beginning of the fifth/eleventh century, reached new heights after the golden decades of the Great Saljuq rule, but it was probably nothing compared to the Mongol period, when the sedentary subjects were deemed to exist solely in order to be exploited.[27]

Above all, with the Mongols, the general massacre of the population, or *qatl-i ʿāmm*, became the new norm. This happened at least once every century, at the hands of Čormaġun, Timur, Jahān-Šāh Qara-qoyunlu and Ismāʿīl Safawī. We are "lucky" to have a very precise, first-hand account of the massacre ordered by Timur in 790/1388, for fully understanding of what a *qatl-i ʿāmm* really meant: Ḥāfiẓ-i Abrū, who was with the Timurids, counted between 1,000 to 2,000 skulls in each of the 28 skull minarets on Eastern side of Isfahan. After extrapolating for the other side, he concluded that the tally of 70,000 put forward by the Timurid bureaucracy was correct (the soldiers had to bring their quota of heads before they could pretend a reward).[28] The figure obtained, which is confirmed by Aubin, amounts to the number of victims by the atomic blast in Hiroshima.

Isfahan was also ravaged by *fitnas* (strife) so frequently that it surprised the non-Isfahanis. The struggle between the Ismailis and their local opponents after the double assassination of sultan Malik-Šāh and his vizier was particularly long and brutal (485-500/1092-1107). But it was only the first of a long series of *fitna*, which continued to erupt at irregular intervals, each with their share of looting, killing, and destroying (Table 1). Before and after the Mongol period, travellers such as Yāqūt (d. 626/1229)and Ibn Baṭṭūṭa (in Isfahan in 727/1326-7) noted that the *fitnas* had left much of Isfahan in ruins.[29] An Isfahani who had taken refuge in Shiraz remarked that for, this reason, the city was often referred to as a paradise where the pigs graze.[30]

As if this were not enough, beginning in the eighth/fourteenth century, Isfahan also suffered from the plague (*ṭāʿūn* in the sources), which was an

27 Morgan, David, *Medieval Persia 1040-1797* (London & New York: Longman, 1988): p. 53. On the Ghaznavid occupation, Durand-Guédy, *Iranian Elites*: pp. 62-5.

28 Aubin, "Comment Tamerlan": pp. 114-16. For the massacre by Jahānšāh, see Aubin, Chiffres: p. 50, note 23 (quoting Barbaro) and Aubin, Jean, *Deux sayyids de Bam au XVe siècle. Contribution à l'histoire de l'Iran timouride* (Wiesbaden: Franz Steiner Verlag, 1956): p. 78.

29 Yāqūt, *Muʿjam al-buldān*, ed. F. Wüstenfeld, *Jaquts Geographisches Wörterbuch*, 6 vols. (Leipzig: in comm. bei F.A. Brockhaus, 1866-73): 1, p. 296. H.A.R. Gibb, *The travels of Ibn Baṭṭūṭa* (Cambridge: Hakluyt Society, 5 vols, 1958-2000): p. 295.

30 Maḥmūd b. Muḥammad quoted by Durand-Guédy, *Iranian Elites*: pp. 228-9. The most famous text about Isfahan *fitnas* is Kamāl al-Dīn Ismāʿīls alleged quatrain, presciently calling for heathens (the Mongols) to solve the situation. See Durand-Guédy, "What Does the History of Isfahan".

TABLE 2 Sieges of Isfahan

Date	Within Isfahan	Besieging Isfahan	Outcome
423/1032	Kakuyids	Ghaznavid general ʿAlī b. ʿImrān	successful resistance
438/1046-7	Kakuyid Abū Manṣūr Farāmurz	Saljuq Ṭoġrïl Beg	successful resistance
442/1050-443/1051 (12 months)	Kakuyid Abū Manṣūr Farāmurz	Saljuq Ṭoġrïl Beg	city taken
486/1093 (1 month)	Saljuq Terken Khatun	Saljuq Berk Yaruq	siege left in exchange for money
495/1102	Saljuq Muḥammad b. Malik-Šāh	Saljuq Berk Yaruq	siege left after prince escaped the city
532/1138		Saljuq Dāʾūd b. Maḥmūd and former Abbasid caliph al-Rāšid	successful resistance
627/1229-1230	Khwārazmians	Mongols	successful resistance
633/1235-6	Khwārazmians	Mongols Chormaghun	city taken and massacre
755/1354	Injuid Abū Isḥāq	Muzaffarid Šāh Šujāʿ	successful resistance
756/1355	Mīr-Mirān, deputy of Injuid Abū Isḥāq	Muzaffarid Šāh Šujāʿ	siege left in exchange for money
757/1356	Mīr-Mirān	Muzaffarid Šāh Sulṭān	city taken
808/1406	Qadi Aḥmad Ṣāʿid & Timurid Rustam b. ʿUmar Šayḫ	Timurid Abū Bakr b. Mīrānšāh	successful resistance
814/1411	Timurid Iskandar	Timurid Šāhruḫ	city taken (with the help of the inhabitants)
817/1414		Timurid Šāhruḫ	city taken

important but poorly documented factor in urban decline throughout Persia, where the number of victims is estimated at 30% of total population.[31] As in Europe, rather than being a single event, the plague returned in waves and remained endemic for centuries, with intermittent and particularly lethal outbursts. In the case of Isfahan, for example, a reliable source indicates that in 810/1407, the plague wrought the death of 20,000 people, an event which also impaired the harvest.[32]

2.2 *The Advantages of the Setting*

From the figures given by Ḥamd-Allāh Mustawfī (fl. 740/1340), who relied on fiscal archives, the revenues of the province of ʿIrāq-i ʿajam were reduced to one seventh of former yields, between the first half of the sixth/twelfth and the first half of the eighth/fourteenth centuries.[33] This is a staggering decline. Nevertheless, Isfahan probably suffered less severely than the other cities of the province. It benefited from its unique position: in a desertic environment, but near a river – an uncommon combination of on the Iranian plateau. While most Iranian cities did develop near a river (Kirman being the exception), Isfahan is the only desert city that could do without *qanāt*. Water was drawn from the river Zāyanda-rūd and carried by a network of open air canals, called *mādī*, to the surrounding fields. These canals represented an important landscape feature, and crossing them was not very easy (a Timurid prince died after attempting to leap across one of them).[34] Only the rural districts furthest from the river relied on irrigation from *qanāt*. This is important, because in this period of history the destruction of the *qanāt*s, whether deliberate or by neglect, proved extremely damaging to the survival of urban centres in Iran. While dams could be destroyed during siege actions, as for example occurred in 809/1407, reconstruction was easier than for a collapsed underground channel.[35] The hinterland, although small, was very productive, and

31 The reference study on the plague in the Middle East, Dols, Michael W., *The Black Death in the Middle East* (Princeton: Princeton University Press, 1977), focused on Syria and Egypt.

32 Jaʿfarīs *Tārīḫ-i kabīr*: p. 47, cf. Manz, Beatrice Forbes, *Power, Politics and Religion in Timurid Iran* (Cambridge: Cambridge Univ. Press., 2007): p. 164. Quiring-Zoche, *Isfahan*: p. 267, speaks mistakenly of 200,000 victims.

33 Ḥamd-Allāh Mustawfī, *Geographical Part*: pp. 47-8, trans. pp. 54-5 (the two figures compared are in currency dinars, i.e. monies of account). See also Petrushevsky, "Socio-Economic Condition": pp. 498-9.

34 Quiring-Zoche, *Isfahan*, p. 24.

35 In 809/1407, the ruler of Shiraz Pīr Muḥammad b. ʿUmar Šayḫ occupied the province of Isfahan where his brother ruled. When he reached Fīrūzān, in the upper Zāyanda-rūd valley, he ordered a breach in the dam that served in regulating the waters. See Quiring Zoche, *Isfahan*: p. 21. Simon de Saint-Quentin comments that the waters of the Zāyanda-rūd were

this explains both the growth of Isfahan in the third-fourth/ninth-tenth centuries and its repopulation after the many catastrophes suffered from the sixth/twelfth century onward (especially after the general massacres). The local farmers were sufficiently mobile to fill the vacuum of the depopulated city (recall that in pre-industrial society, the rural population far outnumbered urban population, and in Islam, differently than in the European situation, no one was tied to the soil unless by debt).[36] If we are to believe the Safavid chronicler Faḍlī Ḫūzānī Iṣfahānī, who belonged to one of the major local families, Isfahan was again a city just six years after Timur's horrendous massacre.[37]

The geographical setting offered further advantages, as the arid environment was a natural deterrent in a context of nomadic empowerment. Beyond the limits of the oasis, the landscape of *dašt* (arid plain) was much too dry to serve useful in the kind of pastoral nomadism practiced by Türkmens, or by the Lurs who emulated them. Türkmens did settle in the 'Irāq-i 'ajam, but further to the northwest (along the Mazdaqān river, hence their presence in the Qum/Kashan area, as noted by Tenreiro) and to the south, in Fars, but not in Isfahan.[38] As for the Great Lurs, their migration routes did bring them to the extreme south of the province of Isfahan (the Zard-Kūh region) in spring, but later, as winter approached, they moved away to pastures in Khuzistan. Given such a context, it is not surprising that the sources only very rarely refer to nomads wintering in the Isfahan area.[39] This does not mean that the presence of nomads was incompatible with urban growth (the *marché-pâturage* dynamic,

---

also diverted by the Mongols when they besieged it, but this is probably a confusion with cases in Khurasan (Marw) and Khwarazm. See Boyle, John, "The Capture of Isfahan by the Mongols", in *The Mongol World Empire* (London: Variorum reprints, 1977): p. 334.

36 For the difference between theory and practice, see Petrushevsky, "Socio-Economic Condition": pp. 524-9.

37 Ḫūzānī, *Afżal al-tavārīḫ*, ms. Cambridge f. 125 r (quoted by Aubin). I was not able to check the edited text.

38 During the Saljuq period, the Türkmens mentioned in Isfahan were from outside the province (in 420/1029 they were from Khurasan; from Azarbayjan in 486/1093 and 474/1082). See Durand-Guédy, *Iranian Elites*: index (Türkmen). For the same ecological reason, Isfahan was not an *iqṭā'* land under the Mongols, see Petrushevsky, "Socio-Economic Condition": pp. 518-19. The Portuguese envoy Tenreiro, who travelled in Western Iran in 1515, notes a Türkmen population in Shiraz, even in Qum and Kashan, but not in Isfahan. See Aubin, "Chiffres": p. 46.

39 During Šāhruḫ's campaign against his nephew Iskandar in 817/1414, a group of nomads (*ḥašamiyān*) wintering beyond the walls of Isfahan found themselves caught between the belligerents. See Quiring-Zoche, *Isfahan*: p. 28. These nomads may have been Lurs, planning to spend the winter by the Gāvḫūnī marsh, East of Warzāna. However, I am not aware of such a migration route in modern times, and in any case the numbers involved must have been rather small.

identified by Aubin, explains the rise of many centres in the northeast, such as Baylaqān), but in general, the urban locations at some distance from the nomads were advantageous over the long term. In the Eastern provinces (Gurgan, Khurasan, Kirman), and even in ʿIrāq-i ʿajam, many towns disappeared due to nomadic pressure.[40] This does not necessarily mean that Isfahan remained free of occasional or specific pressures from the Lurs, which could especially have happened in the seventh/thirteenth, at the height of nomadic power.[41] But unlike Fars or more western parts of ʿIrāq-i ʿajam, not to speak of Khurasan, the region of Isfahan never experienced the "bedouinisation" discussed by Xavier de Planhol.[42] The comparatively high level of land tax (*ḫarāǧ*) from the country districts, as reported by Ḥamd-Allāh Mustawfī, is a good indicator: remarkably, these taxes were nearly twice as high as they were at Tabriz.[43]

The presence of sustained commercial activity, with long-range ramifications, also helped Isfahan to uphold its rank, and stimulated its recovery after the destructions. For Aubin, the growth of urban conurbations depends on the roads, because in cities (Islamic or not) the economic function was always paramount.[44] The overshadowing of the roads through Bayhaq, in Khurasan, by the road through Juwayn, is cited as a typical example of urban decline caused by commercial shifts. In our case, the sharp decline of Mesopotamia and of the Persian Gulf trade, with the shift of the commercial trends northwards through Azerbayjan and beyond Anatolia, was detrimental to Isfahan.[45] But the city

---

40 Petrushevsky, "Socio-Economic Condition": pp. 489-90.

41 Fīrūzān, one of the three cities of the Zāyanda-rūd valley, was given as *suyūrġāl* to the Atabeg of Luristan by Ilkhan Abaqa, see Lambton, Ann, "Eqṭāʿ", in *EIr*, VIII (1998): pp. 521-32, here p. 527.

42 *Bédouinisation* refers to the transformation of landscape, but also of the social structure, under the pressure of nomadic forces. This concept is at the core of Xavier de Planhol, *Les fondements géographiques de l'histoire de l'Islam* (Paris, 1968) and is taken up in his following publications. The typical case is Muġān, which the geographical literature describes as a prosperous little city, south of the Araxes river, in the fourth/tenth century; then as a region with some cities, but mostly pasturages, used by Türkmen in the early seventh/thirteenth century; finally a winter pasture for the Mongols, in the ninth/fifteenth century (see ibid.: p. 210, and Fig. 5 pp. 214-215).

43 Petrushevsky, "Socio-Economic Condition", p. 508. Aubin, "Chiffres": p. 52, note 37, is critical about the optimistic view of a peaceful cohabitation between agriculture and stock-raising.

44 See for example Aubin, Jean, "La survie de Shīlāu et la route du Khunj-ō-Fāl", *Iran* VII (1969) : pp. 21-37, here p. 32. Contrary to Aubin, research on European cities has concluded that international trade does not make the urban growth, see Boucheron and Menjot, "La ville médiévale": pp. 324, 372, 379.

45 This was already noted by André Godard, "Historique du Masdjid-é djuma d'Iṣfahān", *Athār-é Īrān* (Paris: Geuthner), I/2 (1936): pp. 213-82, here p. 230.

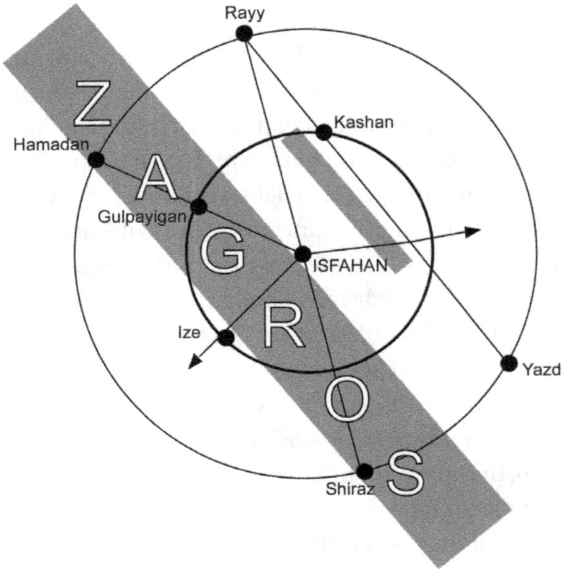

FIGURE 6 Position of Isfahan in Western Iran
AUTHOR: D. DURAND-GUÉDY

was too perfectly aligned with Rayy/Warāmīn, Kashan, Shiraz, and the emporia of Persian Gulf (Sīrāf, later Lār), to lose its role as a commercial hub on the North-South commercial axis (see Fig. 6).[46] Moreover, the only ready crossing of the Great Deserts was by a road passing through Ḫūr and Biyābānak (the oasis of Garma mentioned by Nāṣir-i Ḫusraw), and this was accessible via Nā'īn, east of Isfahan. The city also lies at the outlet of the mountainous road linking the plateau to Lower Iraq and Khuzistan, through Īḏa/Malāmīr.[47]

Nāṣir-i Ḫusraw marvelled at the wealth and activity of the bazaars of Isfahan, less than two years after the city had been besieged and famished by Ṭoġrïl Beg. This siege, which lasted for a year, had in turn followed on a decade of harsh Ghaznavid occupation. The account of Nāṣir-i Ḫusraw shows that the focus on military events in our main sources, the chronicles, could have communicated a distorted vision of the economic realities. The presence of a strong Jewish community in the city, dating to the Sassanid period, probably assisted the

46 Isfahan was also a centre of production, especially in the weaving of cotton and silk. See Petrushevsky, "Socio-Economic Condition": pp. 488, 508.

47 It may be that the road to the west (through Gulpāyigān, or through Gāpalā, further south) lost attractiveness over the period, due to the decline of central Iraq and of Hamadan, perhaps also due to the passage near Lur territory.

economy in times of both growth and hardship, particularly due to the key role of its members in credit activities.[48] Interestingly, Lambton writes that "the speed with which towns … recovered from sieges, disorders … is notable and presumes a widespread use of credit, which enabled society to maintain itself and survive disasters." By way of hypothesis, we can surmise a link between Isfahan's capacities for recovery and this strong community.

2.3 *Examples of Resilience*

The resilience of Isfahan over the Mongol period can be illustrated through distinctly different phenomena: the continuity of urban morphology, and the vibrancy of the literary tradition.

2.3.1 Urban Morphology

Under Turko-Mongol rule, the upheaval in urban hierarchies sometimes extended to the level of the 'displacement' of one city for another, for example Mashhad replacing Ṭūs, Warāmīn surpassing Rayy, or Isfārāʾīn outshining Sabziwār. In contrast, once the city was founded in the third/ninth century, the site of Isfahan remained stable. The main morphological features dated to before the Turko-Mongol period, and remained in place. The Kakuyid ruler ʿAlāʾ al-Dawla (d. 433/1041-2) erected the walls at great cost at the beginning of the fifth/eleventh century (Fig. 5). These served to protect the now fully grown agglomeration from new enemies coming from the east (the Ghaznavids and the Türkmens). The walls played a tremendous role in the following centuries, and as for cities of south-eastern France during in the High Middle Ages, we can say that "the defensive vocation of the city was a decisive factor of urban vitality."[49] The walls were constantly restored over the centuries, perhaps for the last time under Jahān-Šāh Qara-qoyunlu, when 50,000 men were mobilised for the task.[50]

Topographical studies inside the perimeter wall, enclosing an area of about 650 hectares,[51] are still insufficient to comprehend the organisation of the urban space. I would venture to say that there is no reason to suppose that the network of main streets indicated on the 1923 map of Riḍā Ḫān's had changed much since the medieval period. In any case, the main elements of the urban

48 As is well known, Isfahan developed in the second/eighth century around a Jewish neighbourhood, hence its early Islamic name, Yahūdiyya. Fishel, Walter Joseph, *Jews in the Economic and Political Life of Mediaeval Islam* (London: Franck Cass., 1937) and "Isfahan: the story of a Jewish community in Persia", in *Joshua Starr Memorial Volume* (New York: Conference on Jewish Relations, 1953): pp. 111-28. See also Quiring-Zoche, *Isfahan*: p. 207.

49 Georges Duby, quoted by Boucheron and Menjot, "La ville médiévale": p. 308.

50 Quiring-Zoche, *Isfahan* : pp. 44-5.

51 Durand-Guédy, *Iranian Elites*: p. 26, note 19.

landscape remained the fortress of Ṭabarak in the southeast corner, where the military representative of the ruler and his soldiers were garrisoned, and the Friday Mosque (*masjid-i jāmiʿ*) and nearby *maydān* at the centre of the city.[52]

The Friday mosque itself exemplifies the remarkable resilience of Isfahan. This building, which is one of the most famous of the Muslim word, serves as a textbook of Islamic art, precisely because it was continuously embellished by each of the imperial and local rulers controlling Isfahan.[53] The 'Ruling from the Outside' paradigm does not change anything, quite the opposite: the city was the place where the power of the state showed itself to its (sedentary) subjects, and this was still more true when the state had his power base partly or totally outside of cities. The difference between Isfahan and Ardistān is striking (Figs. 7 and 8). In the latter case, the Friday mosque is 'frozen' in its distinctive Saljuq style: four *īwān*-s courtyard, brick dome over the *šabistān*, stucco and brick decoration, Kufic inscriptions.[54] The stylistic unity of the Ardistān mosque demonstrates how the Mongol conquest crushed the urban dynamic of this city. On the contrary, the architecture and the decoration of the Isfahan mosque continued to evolve (Fig. 9). From the Buyids, the Saljuqs inherited the overall arrangement around the central courtyard. They made decisive changes: building a great domed chamber to the south, over the *maqṣūra* (where the ruler sat); adding a monumental entrance to the north (Robert Hillenbrand has recently delivered a brilliant analysis of this section);[55] as well, during the sixth/twelfth century, four *īwān*-s were built. However, unlike the case at Ardistān, the subsequent dynasties continued with further expansions. Under the Ilkhanids, there was an addition of a great chamber (with Öljeitü's famous *mihrāb*) to the northwest of the courtyard. Then, under the Muzaffarids, a prayer hall was constructed behind the narrow *īwān* (the Ṣuffa-yi Darwīsh) leading to the northern dome, as well as another prayer hall and a madrasa on

52 Apparently the Maydān-i naqš-i jahān (the heart of Safavid Isfahan) did not exist until the mid-ninth/fifteenth century. See Quiring-Zoche, *Isfahan*: p. 36, note 1.

53 The most recent synthesis on the Isfahan Friday Mosque remains Grabar, Oleg, *The Great Mosque of Isfahan* (New York: New York University Press, 1990), with Galdieri's postscript. Hunarfar, Luṭf-Allāh, *Ganjīna-yi āṯār-i tārīḫī-yi Iṣfahān* (Isfahan: Imāmī, 1344š.): pp. 67-128, includes the text of all the inscriptions.

54 Hunarfar, *Ganjīna*: pp. 210-27. The contrast is particularly clear in the two southern *īwān*, which have very similar structures for the mosques of both cities, but with completely different decorative programs. It was only in the Safavid period that some repairs were finally made to the Northern *īwān* of the Ardistān Friday mosque.

55 Hillenbrand, Robert, "Architecture and Politics: the North and South Dome Chambers of the Isfahan *jāmiʿ*" in Edmund Herzig and Sarah Stewart (eds), *The Idea of Iran, VI: the Age of the Seljuqs* (London and New York: IB Tauris, 2014): pp. 149-74.

FIGURE 7 Courtyard of the Friday Mosque of Ardistān
PHOTO D. DURAND-GUÉDY, 2015

FIGURE 8 Courtyard of the Friday Mosque of Isfahan, South *īwān*
PHOTO D. DURAND-GUÉDY, 2015

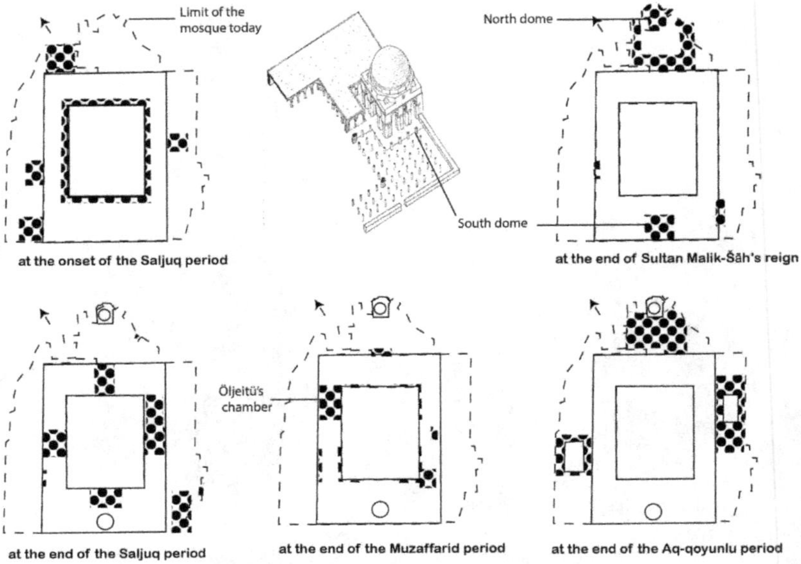

FIGURE 9 Evolution of the Friday Mosque of Isfahan
DRAWING: D. DURAND-GUÉDY, BASED ON GALDIERI, *IṢFAHĀN*

the eastern side. The Timurids erected a large winter hall on the western side, and the Aq-Quyunlu dynasty added the two minarets and blue tile works (as in Tabriz) on the southern *īwān*. Each dynasty added its personal touch in a particular direction, always endeavouring to surpass the previous ones: no less than 484 different vaulting techniques have been employed in the prayer halls.

This phenomenon can also be observed, on a less grandiose scale, in other buildings within and beyond Isfahan city. For example in the *imāmzāda* Jaʿfar, built at the end of the Ilkhanid period, the treatment of the zone of transition follows the earlier Saljuq type exactly (see Table 3).[56] And the famous *īwān* of the Pīr Bakrān mausoleum, South-West of Isfahan, is built next to an earlier chamber for a local saint. At Buzān and Jār, the Ilkhanid square chambers (and dome in the case of Buzān) are erected over a Saljuq structure.[57]

2.3.2 Local Histories

The tradition of local chronicles again provides illustrations of the resilience of Isfahan. At the beginning of the reign of Malik-Šāh, when the city was the

56 Wilber, Donald, *The Architecture of Islamic Iran: the Il-Khanid Period* (Princeton: Princeton Univ. Press, 1955), p. 162.

57 Ibid.: pp. 121 (Pīr Bakrān), 119 (Gar) and 149 (Buzān).

TABLE 3 Buildings with dated inscription in Isfahan and surroundings

Date of inscription	Location	Type of building
479-480/1086-7	Isfahan	Southern dome of the Friday mosque
481/1088-9	Isfahan	Northern dome of the Friday mosque
491/1097-8	Barsiyān	minaret
501/1007-8	Isfahan	minaret of Čihil Duẖtarān
515/1121-2	Jār	minaret
515/1121-2	Isfahan	entrance of the Friday mosque
526/1132	Sīn	minaret
528/1134	Barsiyān	sanctuary of the Friday mosque
528/1134	Buzān	*imāmzāda* Karrār
529/1134-5	Sīn	sanctuary of the mosque
ca. 6[90/1290]	Jār	chamber adjacent to the minaret
698/1299, 703/1303, 712/1312	Linjān	Pīr Bakrān mausoleum
708/1308	Aštarjān	*imāmzāda* Rabiʿa Ḫātūn
710/1310	Isfahan Friday Mosque	mihrāb of Öljeytü in the Friday mosque
713/1313?	Buzān	domed chamber adjacent to the Saljuq structure
715/1315-16	Aštarjān	Friday mosque
716/1317	Kārlādān	Minār-i junbān mosque
72-/132-	Isfahan	*imāmzāda* Jaʿfar
730/1330, 731/1330-1	Sīn	caravanserai
741/1340-1	Isfahan	Bābā Qāsim mausoleum
753/1352	Isfahan	Sulṭān Baẖt-Ārā mausoleum
755/1354	Isfahan	madrasa Imāmī (madrasa Bābā Qāsim)
768/1366, 759-76/1358-74	Isfahan	portal of the Friday Mosque
768/1366-7, 778/1376-7	Isfahan	madrasa in the Friday Mosque
851/1447-8	Isfahan	Bayt al-Šitāʾ of the Friday Mosque
854/1450-1, 855/1451-2	Mārbīn	*ḫānaqāh* Naṣr Ābād
861/1456	Kārlādān	the two "shaking minarets" of the Minār-i junbān mosque
8567/1453	Isfahan	*sardar* of the *imāmzāda* Darb-i imām
895/1489-90	Isfahan	*sardar* of the *ḫānaqāh* of Abū Masʿūd Rāzī
902/1496-7	Isfahan	*sardar* of the *zāwiya* of Darb-i Kušk

capital of the Saljuq empire, al-Mufaḍḍal al-Māfarrūḫī wrote his *K. Maḥāsin Iṣfahān*, or the "*Beauties of Isfahan*". He dedicated it to the governor of Isfahan, who is introduced, in excellent form, as the greatest of the beauties of Isfahan. This governor was the son of the great vizier Niẓām al-Mulk, and with his educated background, he would have appreciated the lengthy display of hyperbolic Arabic provided by al-Māfarrūḫī. In 729/1329, after one century of Mongol rule, this same work was adapted into Persian by Ḥusayn Āvī for the son of the famous vizier of his time, Rašīd al-Dīn Faḍl-Allāh (d. 718/1318).[58] This *Tarjuma* illustrates the shift occurring during the Saljuq period, when Persian replaced Arabic as the favoured language of the *belles lettres*. But such a work, based on a text then two centuries and a half old, also shows that Isfahan remained fundamentally unchanged. Ḥusayn Āvī merely reorganised the content and updated the poetic dimension, adding many pre and post-Mongol Persian qasidas.

Even more significant is the persistence of writings belonging to the genre of the *munāẓira* (Persian *munāẓara*), or 'verbal jousting', between Baghdad and Isfahan. It blossomed when the latter city emerged as a regional capital, at the onset of the fifth/eleventh century. A good example would be the *Ḥikāyat Abi'l-Qāsim* by Abū Muṭahhar al-Azdī. The main character is a well-spoken Isfahani living in Baghdad, who visits his native city. There, he invites himself to a reception and ridicules the boastful attempts of the local notables to emulate the refined Baghdadis: he compares the climate, the qualities of the river, the local products and habits to those of Baghdad, belittling them mercilessly. This vindication of Baghdad and prolix critique of Isfahan is clearly a reaction to the Isfahanis' ambitions of matching their model: the very violence of Azdī's prose could indicate how seriously that claim was taken by some in the caliphal capital.[59]

The aforementioned *K. Maḥāsin Iṣfahān* is also a *munāẓira* in disguise, since al-Māfarrūḫī endeavours to establish Isfahan's position as the permanent capital of the Saljuqs – and so prevent them from settling in Baghdad.[60] In the

58 The *Tarjuma* has been described in great detail by Browne, Edgar "Account of a Rare Manuscript History of Iṣfahán", *JRAS*, no volume mentioned/1-2 (1901): pp. 411-6 and 661-704.

59 Durand-Guédy, *Iranian Elites*: pp. 48-9. The work has been edited by Adam Mez, *Abulkasim, ein bagdâder Sittenbild* (Heidelberg: Carl Winter's Universitätsbuchhandlung, 1902) and translated by René Khawam, *24 heures de la vie d'une canaille* (Paris: Phébus, 1998). The most recent study is by Selove, Emily, *Ḥikāyat Abī al-Qāsim: A literary banquet* (Edinburgh: Edinburgh Uni. Press, 2016).

60 Durand-Guédy, David, "The political agenda of an Iranian *adīb* at the time of the Great Saljuqs: Māfarrūkhī's *K. Maḥāsin Iṣfahān* put into context", *Nouvelle Revue des Etudes Iraniennes*, 1 (2008): pp. 67-106.

following century, Ḫāqānī included a famous *munāẓira* between Baghdad and Isfahan in one of the qasidas of his celebrated *Tuḥfat al-ʿIrāqayn*.[61]

Here again, the genre continued after the Mongol conquest. A *Munāẓara-yi Baġdād wa Iṣfahān* was written for a local vizier called Mubārak-Šāh.[62] Differently from Azdī's *Ḥikāya* and Ḫāqānī's qasida, the merits and defaults of Isfahan and Baghdad are not portrayed through the words of fictional characters, but by the personification of the cities themselves (a trope known as *zabān-i ḥāl*). The themes dealt with are nonetheless identical to those of three centuries earlier: the virtues of the relative settings (soil, climate, river, fruits), and most importantly of the inhabitants. The text is structured around a series of declarations, which trigger detailed answer and new pretences, calling for further answers. For example, in the words of Isfahan, we hear:

> Since my foundation, I have been the residence of the successful sultans and the envy of the famous *ḫāqān*-s; not a single war has been fought between the kings of Iran and Turan without me being its prize.

To this, Baghdad can answer that it was the capital of the Chosroes (i.e. of the Sassanids), the caliphs, and so on.[63] The debate ultimately closes in favour of Isfahan, when the city plays its trump card: the presence of the dedicatee himself, who is a sun of overpowering brightness. At first glance, such literary contests seemed outdated, considering that the new centre of power was Tabriz, and the two former capitals of the Great Saljuq empire (Isfahan and Baghdad) had lost their political relevance. However, upon reflection, they fit well with what we have said about the resilience of Isfahan during the Turko-Mongol period. Just as Damascus, the Umayyad capital, did not at all vanish after the founding of Baghdad, Baghdad and Isfahan again lived on, thanks in part to their similar advantages of rivers and fertile hinterlands.

---

61 Ḫāqānī, *Dīvān,* edited by Ḍiyā Sajjādī (Tehran: Zavvār, 1357š.): pp. 353-8. Cf. Khanikof, N. de, "Mémoire sur Khâcâni, poëte persan du XIIe siècle. Seconde partie", *JA*, Sixième série, v (1865): pp. 295-367, here 332-40. The *munāẓira* stricto sensu is 15 verses long (v. 21-36).

62 Kamāl Iṣfahānī, ʿAbd al-Salām b. Abi'l-Mujāhid, *Munāẓara-yi Baġdād wa Iṣfahān*, ed. Aḥmad-Riḍā Yalamih-hā (Dihāqān: Dānišgāh-i āzād-i islāmī vāḥid-i Dihāqān, 1391š.). Nothing is known about the author, whose name is given in the first line of the text (Kamāl Iṣfahānī, *Munāẓarah*, p. 81), but from his *nisba* and the content of the text he was obviously an Isfahani. As for the dedicatee of the work, we know he lived in Isfahan (ibid.: pp. 159-60) and held the title of vizier (ibid.: p. 153). Pūrjavādī, Naṣr-Allāh, *Zabān-i ḥāl dar ʿirfān va adabiyāt-i pārsī* (Tehran: Hirmis, 1385š., p. 480), has identified him as the dedicatee of another *munāẓira*, dealing with chess and backgammon.

63 Kamāl Iṣfahānī, *Munāẓara*: pp. 86 and 117.

These features explained the rise of the cities, and could still be celebrated, even if there were no longer caliphs in Baghdad or sultans in Isfahan.

2.4 *Outline of an Evolution*

On the basis of what we have said, we could sketch the overall lines of a process of evolution. The apex was indubitably at the beginning of the period, when Isfahan was the capital of the Saljuq sultans Malik-Šāh (d. 485/1092) and his son Muḥammad (d. 511/1118). At the end of this golden age, the Hamadan based author of the *Mujmal al-tawārīḫ wa'l-qiṣaṣ* (520/1126-7) states that there was no greater city in ʿIrāq-iʿajam or Khurasan than Isfahan, which confirms Nāṣir-i Ḫusraw's perception of eighty years earlier.[64] The city never returned to such levels of prosperity until the Safavid Shah ʿAbbās I. The decline set in early in the sixth/twelfth century. The unfinished sanctuary room of the mosque of Barsiyān, 50 kilometres east of Isfahan, can be interpreted as a still visible sign of this process.[65] But the analysis of archaeological data remains difficult, given the lack of archives. Mary Crane remarks that the madrasa Imāmī was completed in 755/1354, at the very time of the siege of Isfahan by the Muzaffarid Mubāriz al-Dīn Muḥammad.[66] Pondering on dated documents in the development of architectural faïence in Isfahan and its surroundings, she concludes that "important building activity continued ... uninterrupted by the dissolution of Mongol authority and the turbulent rule of the Muzaffarids."[67] These contradicting conclusions show how little certitude we have here.

Similarly, for the first half of the eighth/fourteenth century, our written sources are contradictory: in 740/1340, Ḥamd-Allāh Mustawfī speaks of the destruction of the region, and Ibn Baṭṭūṭa of the city.[68] However, only slightly earlier, vizier Rašīd al-Dīn had noted that under Muḥammad Īḏajī (a governor obviously of Luri origin), efforts had been made to rebuild the city, in particular

64 Lambton, "Economy": p. 120.

65 Smith, Myron Bement, "Material for a corpus of early Iranian islamic architecture, II. Manār and masdjid, Barsīān (Iṣfahān)", *Ars Islamica*, IV (1937): pp. 7-41, here p. 8. Construction, begun in 528/1133-4, might have been interrupted because of the outbreak of the "wars of Fars". See Durand-Guédy, *Iranian Elites*: pp. 226-9.

66 Crane, Mary, "A Fourteenth-Century Mihrab from Isfahan", *Ars Islamica*, VII/1 (1940): pp. 96-100, here p. 99 (cf. the previous dating of 725 suggested by Godard is accepted by Hunarfar, *Ganjīna*, p. 303).

67 Ibid.: p. 100. Cf. Lambton, Ann, "Yazd. 2 History", *EI*[2]: pp. 302-9, here p. 306: "in spite of the prevailing turbulence and the internal warfare of the Muẓaffarids, the city apparently prospered under them and was extended".

68 Ḥamd-Allāh Mustawfī, *Geographical Part*, p. 48, trans. p. 55. Ibn Baṭṭūṭa, trans. p. 295 (*ḫuriba akṯaruhā*).

through *waqfs* (charitable foundations funded by endowed properties).[69] Writing in 729/1329, Ḥusayn Āvī also gives the impression of a flourishing economy.[70] This fits well with what we know about the situation in Qum.[71] The conclusion is that the breakdown of Ilkhanate and beginning of the troubled Chupanid rule ruined a region which had recovered from the destruction of the seventh/thirteenth century.[72]

Whatever the case, during the agitated sixth/twelfth and eighth/fourteenth centuries, as well as during the 70 years under the Mongol yoke, the situation of Isfahan would have been envied by many other cities in the region, thanks to its position and its walls. Perhaps we can interpret the stability (even the increase) in the number of villages of Isfahan hinterland, as reported by Yāqūt (360) just before the Mongol period, and one century later by Ḥamd-Allāh Mustawfī (400), as a sign of economic health.[73]

The majority of archaeological remains (domed chamber, mosque, minarets, dam) in the Zāyanda-rūd valley date from the Saljuq and the Ilkhanid period (table 3).[74] This indicates a lack of resources or of interest in the area for the two centuries between the Timurid and the Safavid conquest. Demographical considerations confirm this impression of decline for the eighth-ninth/fourteenth-fifteenth centuries. Applying a surface/density ratio, I have advanced the figure of 100,000 inhabitants at the time the walls were

69 Rashīd al-Dīn, ed. Karl Jahn, *Tāriḫ-i-Mubārak-i-Ġāzānī des Rašīd al-Dīn Faḍl Allāh Abī-l-Ḫair: Geschichte der Ilḫāne Abāǧā bis Gaiḫātū (1265-1295)* (Prague: Verlag der Deutschen Gesellschaft der Wissenschaften und Künste, 1941): pp. 40, 349.

70 Ḥusayn Āvī, *Tarjuma*: pp. 80-83.

71 Drechsler, Andreas, *Die Geschichte der Stadt Qom im Mittelalter (650-1350): politische und wirtschaftliche Aspekte* (Berlin, Klaus Schwarz Verlag 1999): pp. 338-41.

72 The same context could also explain the failure to finish the mosques of Izīrān, Daštī, and Jār, southeast of Isfahan, around 735/1334-5. Wilber, *Architecture*, p. 162-3, saw it as "the result of some local incident or of local economic conditions or of a sudden lack of interest on the part of the authorities". For Siroux, these buildings could show "a plan … for the establishment of a number of new communities in the vicinity of Isfahan"; Siroux, Maxime, *Anciennes voies et monuments routiers de la région d'Ispahan* (Cairo: IFAO, 1971), p. 264. Rosario Paone, "The Mongol Colonization of the Isfahan Region", *Quaderni del seminario di iranistica uralo-altaistica e caucasologia dell' universita' degli studi di Venezia*, X (1981): pp. 1-33, found this (fascinating) hypothesis unfounded, but does not give alternative explanation.

73 Cf. Yāqūt, *Muʿjam*, I, p. 294, line 7, and p. Ḥamdallāh, *Nuzhat*, p. 50. Ibn Rusta, *Aʿlāq*, 154, line 5, gives 2,800 villages (*ḍiyāʿ*), for the whole province. Petrushevsky, "Socio-Economic Condition", p. 496, notes that the number of villages in Herat hinterland fell from 400 to 167, but over five centuries.

74 Table 2 gathers data from Hunarfar, *Ganjīna*, Wilber, *Architecture* and Siroux, *Anciennes voies*. A handful of constructions can now be added, including the Warzāna dam, but they are not dated.

built, in the beginning of the fifth/eleventh century (Richard Bulliet proposes a population twice as large, but does not explain how he arrived at such a figure).[75] Relying on Ḥāfiẓ-i Abrū's "macabre calculation", Jean Aubin assessed that the population of Isfahan must have been in the range of 80,000 to 100,000 persons.[76] This is less than during the Saljuq period, but still represents very important numbers. Then, almost a century after Timur's devastating conquest, and the passage of more than two decades of recovery from Jahān-Šāh's massacre, Giosafat Barbaro assessed the population of Isfahan as only 25,000 people. (Interestingly, the inhabitants remembered a city of 150,000, which coincides with our own figures.)[77] Barbaro also reported that only one sixth of the Isfahan *intra muros* surface remained inhabited.[78] We know that Iranian cities were less densely built than their Arab or European counterparts, and the same Barbaro describes many cities of western Iran (such as Nā'īn or Sulṭāniyya) as "*mal habitata*", but still, this very low proportion is significant (he reported Marāġa as 'only' decreased to one-fourth of previous). The dated constructions are less frequent for this period (Table 3), and most of the city probably fell into ruin. The Portuguese Gil Simões reports that in 1515 Isfahan had only 3,000 hearths (*vezinhos*), which amounts to 20,000 inhabitants. This means that the 5,000 victims mentioned by Tenreiro for the Safavid conquest, thirteen years earlier, would have represented one quarter of the population.[79] (Proportionally, this nears the level of a *qatl-i 'amm*). We can compare the fate of Isfahan to that of Jurjān and Baghdad: in both cases, the destruction at the hand of Timur presented much greater difficulties in recovery than did the first destruction by the Mongols (it was irremediable in the case of Jurjān).

---

75 See Durand-Guédy, *Iranian Elites*: p. 26. I used the same coefficient applied by Richard Bulliet for Nishapur (100 to 200 person/ha.).

76 Aubin, "Chiffres" : pp. 40-41 and "Comment Tamerlan": pp. 114-16. By adding the corresponding numbers of women and infants to the 70,000 skulls of male victims, Aubin reaches a total number of 150,000 persons inside the city. But this figure includes the garrison as well as the inhabitants of the countryside, who had taken shelter within the walls (this was one of the functions a city: for example, when Azarbaijani troops approached in 808/1406, the peasants took refuge inside the city, with their provisions. See Quiring-Zoche, *Isfahan*, p. 19-20.).

77 Aubin, "Chiffres": p. 43. Petrushevsky ("Socio-Economic Condition": p. 507), Lambton ("Iṣfahān": p. 102), and Quiring-Zoche (*Isfahan*: p. 267), rely on the sixteenth century English translation of Barbaro's account and speak of 50,000 inhabitants).

78 Aubin, "Chiffres" : p. 39.

79 Aubin, "Chiffres": p. 40.

3 Relationship with Imperial Powers

As can be expected for a city of this size, the sources show a complex and stratified urban society. The upper level (the elites, notables, patricians, aristocrats – whatever name is chosen) are as always the best known, if only because many authors of our key sources were themselves members of this elite. (As well as al-Māfarrūḫī, other examples are the cases of ʿImād al-Dīn Iṣfahānī and Abū Bakr Tihrānī, the authors of respective chronicles of the Saljuq and Aq-Qoyonlu states). As Aubin says: "Members of the aristocracy, either as individuals or as families, at the same time held religious functions, administrative office and the management of large landed estates."[80]

Since Richard Bulliet's pioneering remarks on "the Patriciate and the state" in his study of Nishapur, many studies have focused on the politics of notables.[81] In the following sections we distinguish several types of the relationships between the elite of Isfahan and the imperial powers (Turko-Mongol sultans and their viziers).

3.1 *Situation 1: Submission*

A one-way relationship occurred when the imperial power is too powerful for the local elites to have a say. This case naturally includes the bloodsheds of 421/1030 (at the hand of the Ghaznavids), 633/1235-6 (Čormaġun), 790/1388 (Timur) and 858/1454 (Jahānšāh), but also during the brutal occupations of 497/1104 (by Berk Yaruq), 555-6/1160-1 (by Arslan b. Ṭoġrïl), 817/1414 (by Šāhruḫ). When Šāhruḫ took Isfahan, he first ordered it plundered, then he gave 100,000 *kebeki* dinars as alms to the mausoleums: demonstrated force was followed by a show of good will.[82] Conversely, the massacres ordered by Masʿūd of Ghazna and by Timur were triggered by the killing of tax collectors (*muḥaṣṣilān*) who had been sent to gather the tribute (*māl-i amān*).[83]

In some cases, this type of one-way relationship also applied over longer periods of time. For example, during the second half of the fifth/eleventh century, when Isfahan was the capital of the Great Saljuqs, the indigenous elites found their margin of manoeuvre reduced to nil, particularly when the all-powerful vizier Niẓām al-Mulk, a native of Khurasan, settled inside the city. Not only did he control the vizierate, but at that point he held the city under

80 Aubin, Jean, "Etudes safavides. I. Šāh Ismāʿīl et les notables de l'Iraq persan", *JEHSO*, II (1959): pp. 37-81, here p. 39.

81 Bulliet, Richard, *The Patricians of Nishapur. A Study in Medieval Islamic Social History* (Cambridge, Mass.: Harvard University Press): part II.

82 Manz, *Power*: p. 86.

83 Aubin, "Comment Tamerlan": pp. 101-2. Durand-Guédy, *Iranian Elites*: p. 64.

sway: he appointed his son (the aforementioned dedicatee of the *K. Maḥāsin Iṣfahān*) as governor of Isfahan; he used his immense wealth to build a local network (in particular through foundations of institutions such as a madrasa and a *ḫānaqāh*), which drew in the members of the old local elite (for example the Māfarrūḫī-s and the Kūšīḏī-s).[84] Niẓām al-Mulk's construction of a dome over the *maqṣūra* of the Friday mosque is a visual expression of this domination (see Fig. 9). I advanced the hypothesis that the impossibility of open opposition inside the city (unlike the Hanbali popular movement in Baghdad) fuelled the tensions unleashed after the death of Niẓām al-Mulk and the crisis of the Saljuq authority. An observation strongly in support of this thesis is that Isfahan appears to be the only major Iranian city in which Ismailism had a solid base.[85]

This period bears similarities to that of the governorate of Bahā' al-Dīn Muḥammad (d. 678/1279), the young son of the great Ilkhanid vizier Šams al-Dīn Ṣāḥib-i Dīwān Juwaynī (the brother of the historian). Like Niẓām al-Mulk, Bahā' al-Dīn came from a Khurasani family (which is no surprise, since during the Turko-Mongol period the historical dynamics came from the East). He was appointed governor at the beginning of Abaqa's reign (663/1265) and took his mission to restore law and order very seriously. In today's words, his policy was "zero tolerance". Waṣṣāf, who lived in nearby Shiraz at the beginning of the eighth/fourteenth century, left a grim portrait of him. A simple glance, a lack of respect, a negligence would bring terrible sufferings on the culprit, whatever his age. The population did enjoy security (the shopkeepers were asked to not lock their shops), but at the cost of living in continual fear of disproportionate punishment for a quibble. Interestingly, Waṣṣāf also tells us that Bahā' al-Dīn was fond of culture, generous with men of writing and ulamas, solicitous of the situation of the inhabitants (the history of the twentieth century teaches us not to be surprised by such a dysfunctional behaviour), and that his building policy (palaces, gardens and promenades) contributed to the embellishment of Isfahan. Two centuries later, the account of Ḫʷāndamīr omits the gory details about Bahā' al-Dīn's psychopathic behaviour, but notes that he was well educated, that the region of Isfahan thrived thanks to his policies, and that he patronised such brilliant man as the poet Badr Jājarmī (d. 686/1287).[86] Waṣṣāf's

---

84 Durand-Guédy, *Iranian Elites*: pp. 112-129.

85 Durand-Guédy, David, "Ismailism in Saljuq Isfahan. A Reassessment", in Eva Orthmann & Petra G. Schmidl (eds.), *Science in Context: The Dustūr al-munajjimīn and its World* (Berlin: EB-Verlag Dr. Brandt, 2017): pp. 223-61.

86 Lane, George E. *Early Mongol rule in thirteenth-century Iran: a Persian Renaissance* (London: Routledge, 2003): pp. 197-9. Āl-i Dāvūd, 'Alī, "Bahā' al-Dīn Juvaynī", *Dāyirat al-ma'ārif-i buzurg-i islāmī*, (Tehran, 1383š.): XIII: pp. 64-5.

contemporary, Ḥusayn Āvī, merely speaks of the governor's generosity and of his splendid palace.[87] What can be deduced from all these accounts is that Bahāʾ al-Dīn Juwaynī's policy in Isfahan resulted in increased security, even if, as Waṣṣāf said, he might have "gone too far" (*az ḥadd-i iʿtidāl tajavvuz kard*). However, as far as politics are concerned, Bahāʾ al-Dīn and Niẓām al-Mulk appear remarkably similar, in their grip on the city, their crushing of the opposition, and their cultural patronage.

3.2 *Situation 2: Forced Collaboration*

The second situation, much more frequent, happened when the urban elites were strong enough to collaborate with the central state in the control of local resources. Such collaboration arose from the structural weakness of the Iranian states during the pre-modern period.[88] Their representative(s) (*šiḥna, wālī, ḥākim, muqṭaʿ, dārūġah*) had neither the means nor mission to truly operate a city.[89] I have summed up this collaboration between the ruling sultan and the local notable(s) in terms of a quid pro quo: "I recognise your role in the city by granting you an official position, thereby making you my main interlocutor; in exchange for which you recognise my authority and you will ensure that the taxes are raised."[90] This applies to the period after 485/1092, and even more after Hamadan replaced Isfahan as the centre of imperial power (*dār al-mulk*) in ca. 521/1127. After Hamadan and Bagdad became the main centres of the Saljuqs, the latter had to acknowledge the Ḫujandī-s, a family installed by Niẓām al-Mulk at the head of the Shafii madrasa of Isfahan, which had played a leading role in fighting the Ismailis. The Saljuqs sultans recognised the power of the Ḫujandī-s by granting them the "*riyāsat al-balad*" on a hereditary basis.

But in the troubled environment of the sixth/twelfth century, each player also endeavoured to bypass the other. On the one hand, the Saljuqs, of Hanafi persuasion, strengthen the local Hanafis and their leading family, the Ṣāʿid-s. On the other hand, the Ḫujandī-s sought allies outside Isfahan (first and foremost the Abbasids caliphs), and backed rival (Saljuq) contenders at the

87 Ḥusayn Āvī, *Tarjuma*: pp. 85-6.

88 Since his *Herrscher, Gemeinwesen, Vermittler: Ostiran und Transoxanien in vormongolischer Zeit* (Stuttgart: Franz Steiner Verlag, 1996), the 'reach of the state' has been at the heart of Jürgen Paul's research.

89 Cf. Durand-Guédy, *Iranian Elites*: pp. 210-12 and Quiring-Zoche, *Isfahan*: pp. 114-56. Quiring-Zoche and myself have both endeavoured to present a table listing all the governors/ruling authorities in Isfahan, over the period studied (Quiring-Zoche identifies 77 governors, myself 19).

90 See Durand-Guédy, David, "An Emblematic Family of Seljuq Iran: the Khujandīs of Iṣfahān", in C. Lange and S. Mecit (eds), *The Seljuqs: Politics, Society and Culture* (Edinburgh: Edinburgh Univ. Press, 2011): pp. 182-202, here p. 191.

dynastic level.[91] In recognising this, instead of the standard presentation chosen by many historians, which focuses on the dynastic struggles and fails to deal with the urban strife (*fitna*) mentioned in chronicles, we can interconnect two levels of struggle, at the local and dynastic levels. The events of 541-2/1147, at the heights of the 'wars of Fars', are a perfect example of this. At that time, Bozabe, the powerful governor of Fars-Khuzistan and atabeg of a young Saljuq prince, challenged the authority of the sultan. In a move of defiance against both the sultan and his governor in Isfahan, the Ḫujandī-s sided with Bozabe and opened the gates to him. Taking control of Isfahan, the ancient capital of the Great Saljuqs, was an important step in their bid for power. Bozabe had his Saljuq prince enthroned as sultan in the palace of Malik-Šāh, and took a Ḫujandī as vizier. The sultan in Hamadan eventually prevailed on the battlefield of Qara-Tegin, and the Ḫujandī-s had to pay for their involvement. Their assets in Isfahan were plundered by rioters enjoying royal approval, and the heads of the family were sent into exile. After only one year the same sultan forgave them and reinstated their functions, titles and lands, which was a clear admission that the city could not be controlled without them.[92]

A similar pattern can be seen in the Timurid period – which has many features in common with the context of the sixth/twelfth century: the imperial ruler was far away in the east (Marw for Saljuq Sanjar; Herat for Timurid Šāhruḫ) while princes fought for power in Western Iran. In 850/1446, after four decades of Šāhruḫ's rule, some Isfahani notables decided to side with a grandson of Šāhruḫ, Muḥammad Sulṭān, who had ambitions of his own. These notables included the *naqīb* (the syndic of the Alids), the qadi as well as others *ḫwāja* and *amīr*. Šāhruḫ's representative in Isfahan was made prisoner, a local sayyid was killed, probably because he opposed the rebellion, and when Muḥammad Sulṭān arrived from his base in the north of ʿIrāq-i ʿajam, he sealed his alliance by taking two locals (one Turkish *amīr* and one Persian *ḫwāja*) as viziers. However, here again, the older sultan (Šāhruḫ) prevailed on the battlefield. Muḥammad Sulṭān fled to the mountains of Luristān.[93] Šāhruḫ granted the *amān* (protection) to the population, but had the leaders of the rebellion harshly punished: the *naqīb* Šāh 'Alā' al-Dīn, the qadi, and one *ḫwāja* were hung at the gate of Sāwa. The killings of sayyids by both camps was particularly revolting, given their relationships to the Prophet. When Šāhruḫ died shortly

91 Ibid.: pp. 192-3.

92 On these events, see Durand-Guédy, "1147": pp. 191-7.

93 'Alā' al-Dawla Ibn Kākūya in the fifth/eleventh century and Šāh Isḥāq Injū in the eighth/fourteenth also fled to Luristān (the first to escape the Ghaznavids, the second the Muzaffarids).

afterwards, Muḥammad Sulṭān returned from Luristān and endeavoured to make amends to the recent events: he took the same *ḫʷāja* as vizier, and a mausoleum was built to receive the remains of the murdered *naqīb* (the mausoleum, adorned with splendid *muqarnas*, is visible north of the old town, in a district named Šahšahān after him).[94]

In both cases (541-2/1147 as 850/1146-7) we see local elites changing sides to protect their interests, meaning their material situation. The *al-Muḫtārāt min al-rasā'il*, our best source on the functioning of the local society, contains numerous letters sent to community leaders, formulating a request that can be summed up as follows: "you are well connected, please help me to avoid taxation on such and such land and I will be your obedient servant."[95] Periods of crisis like these highlight the fiscal dimension in decision making. It is only after the Ismailis began to collect taxes in the countryside that the pro-Saljuq elites of Isfahan rallied Muḥammad b. Malik-Šāh to challenge his brother Berk Yaruq.[96] Likewise, the issue of taxes appears to have been an important factor in the decision of the elites of Isfahan to choose Muḥammad Sulṭān against Šāhruḫ. Isfahan apparently withheld taxes completely for some time, and after Šāhruḫ ordered the tax arrears of 'Irāq to be investigated, the Isfahanis were apparently unable to reach an agreement, unlike the notables of other cities.[97] Significantly, after Muḥammad Sulṭān entered Isfahan, he threw the tax registers in the Zāyanda-rūd and granted fiscal immunity (*suyūrġāl*) to the notables.[98] Four decades earlier, in 810/1407, the Timurid ruler of Shiraz Pīr Muḥammad entered Isfahan without meeting any resistance, obviously in part because of the plague, but possibly also because of a tax agreement with the locals – the taxes registers (*daftar-hā-yi mālī wa ḫarājī*) being similarly thrown in the river.[99]

---

94 On these events, see Quiring-Zoche, *Isfahan*: pp. 34-39. On the mausoleum, see Hunarfar, *Ganjīna*: pp. 333-41. *Šahšahān*, lit. "king of the kings" derives from the title *naqīb al-sādāt* (head of the sayyids), since the sayyids were called "*šāh*" or "*šāhzāda*", see Humā'ī, Jalāl al-Dīn, *Silsila-yi sādāt wa mušajarāt wa musaṭṭaḥāt. Insāb wa nasab-i imām-zāda-hā-yi-Iṣfahān* (1390š./2011): p. 351.

95 Durand-Guédy, *Iranian Elites* : pp. 249-55.

96 Durand-Guédy, *Iranian Elites*: pp. 162-7.

97 Manz, *Power*: pp. 150, 254-5.

98 Quiring-Zoche, *Isfahan*: p. 36 (especially for the new meaning taken by *suyūrġāl* in the ninth/fifteenth century).

99 See Quiring-Zoche, *Isfahan:* pp. 22-3. Throwing official documents in the Zāyanda-rūd seems to have been a well-established local practice, the most famous example being the destruction of the Safavid archives after Isfahan was taken by the Afghans.

3.3 *Situation 3: Alone, Looking for a Ruler or Trying to Stay Independent*

The *riyāsatal-balad*, granted by the sultan to a leading figure of Isfahan, cannot be compared in any way with the charters of Western medieval cities. In Isfahan, as elsewhere in Islamic Iran, self-rule was neither a legal nor practical option. In one anecdote told by al-Māfarrūḫī, Masʿūd of Ghazna, who had driven the Kakuyids out of Isfahan, asked a madman: "do you like me best, or Ibn Kākūya?" The madman answered, in Isfahani dialect, "May you leave and may he not return!"[100] Does the response express a desire for self-rule? A debatable question – the more likely point is that all rulers are likely to be unjust – but if such desires existed, it is significant that they are put into words by a madman, as if such a situation could not be envisioned by rational people.

Exceptionally, no ruler or military governor was present in the city. But in this case, the local elites immediately looked for an ally on the dynastic level. In 492/1099, for example, the Ismailis threatened the legitimacy, the political influence, the source of income, and even the life of the elites installed by the Saljuqs in Isfahan. In the absence of both Sultan Berk Yaruq and his governor Öner, these same elites (I call them the 'Khurasani networks' because of the eastern origin of their leaders) rallied to the rival of the ruling sultan, Muḥammad b. Malik-Šāh, hoping that he could better help them to overcome their enemies. The same situation repeated in 812/1409, when the Timurid ruler Pīr Muḥammad was killed in Kirman and after his son 'Umar Šayḫ had left Isfahan. The prominent local figure at that time in Isfahan was Qadi Aḥmad Ṣāʿid. He allied himself with a Muzaffarid prince, who hasted from Azarbayjan at the head of Qara-Qoyunlu troops.[101] In both cases, the elites longed for a restoration of golden age: Niẓām al-Mulk's in the first case, the Muzaffarid's in the second.

Can we say that no attempt for self-rule was ever made during the period? On two occasions, a century apart, we see a member of the Mīr-i Mīrān family in charge of the city. The first time (in 755/1354), Amīr Jalāl al-Dīn resisted a siege by the Muzaffarids and secured their withdrawal by paying a tribute (*naʿl-bahā*); the second time (in 856/1452), Amīr Zayn al-ʿĀbidīn seized power after the flight of the governor. In both cases, the 'independence' did not last long (in 757/1356 the Muzaffarids eventually prevailed and Jalāl al-Dīn took his family

100 al-Māfarrūḫī, *K. Maḥāsin Iṣfahān*, ed. J. Ṭihrānī (Tehran: Maṭbaʿa majlis, 1933): p. 47 (*to beši va-ān nīmeyād*). Cf. Ḥusayn Āvī, *Tarjuma*: pp. 145-6.

101 This prince was seeking revenge, as Timur had exterminated 70 members of the Muzaffarid family in the village of Mihyār, south of Isfahan, in 795/1395 (a striking contrast with the way Ṭoġrïl Beg treated the Kakuyids after he seized Isfahan from them).

to Kashan; as for Zayn al-ʿĀbidīn, he eventually had to yield power to Bābur's emir), but these cases deserve further investigation.[102]

3.4 *The Basis of a Political Culture*

As early as the fourteenth century, Barbaro noted that the inhabitants of Isfahan sometimes disobeyed their kings.[103] Jean Aubin speaks of the "*tendances particularistes*" of the notables, and Beatrice Manz remarks on a "city never fully quiescent".[104] What was the basis of this distinct political culture? In final analysis, the nature of the relationship between imperial and local levels mirrored the ratio of power between them. The mechanism of 'Ruling from the Outside' empowered the urban elites, but this applied to all cities, not just to Isfahan. The distance from the capital was also important. Weaker Saljuq sultans moving to Hamadan increased the margin of manoeuvre of the Isfahani elites, who became increasingly involved in imperial politics. Likewise, at the end of Timurid rule, the well-informed Abū Bakr Ṭihrānī writes that "when the Isfahanis grew aware of the weakness of Bābur Mīrzā [in Khurasan], they were inclined towards the Türkmens [of Azarbaijan] and they sent *ra'īs* Muḥammad to invite them."[105]

A key factor in the agency of the Isfahanis is the military potential of the locals themselves. This issue was first investigated by Claude Cahen in his ground-breaking article "Mouvements populaires et autonomisme urbain dans l'Asie musulmane du Moyen-Age," and after him by Jürgen Paul.[106] For Isfahan, the empowerment of the locals relied on the city walls and the capacity to fight. The walls were instrumental because no imperial player, either a ruler or his representative, could think of barring himself inside Isfahan without the collaboration of its inhabitants. In 527/1132, a Saljuq prince defeated on the battlefield wanted to resist from Isfahan, but as the source says "he learned that the inhabitants of Isfahan would not help him during the siege, so he left them and headed towards Fars."[107] This is also what happened to Timurid prince Iskandar after he barred himself within Isfahan in 814/1411

102 Quiring-Zoche, *Isfahan*: pp. 228-9.

103 Barbaro, *Travels*: p. 72.

104 Aubin, "Chiffres": p. 40. Manz, *Power*: p. 254.

105 Abū Bakr Ṭihrānī, edited by Lugal, Necati and Sümer, Faruk, *Kitab-i Diyarbakriyya. Ak-Koyunlular tarihi* (Ankara: Türk Tarih Kurumu Basımevi, 1964): pp. 326-7.

106 Cahen's article was published in *Arabica* V (1958) and VI (1959): pp. 1-91. Reference to further research on the subject can be found in the special issue of *IrSt* XXXVIII/4 (2005), "Private Warfare in Pre-Mongol Iran", edited by Jürgen Paul.

107 Durand-Guédy, *Iranian Elites*: p. 257. See also my "Iranians at War under Turkish Domination: the Example of Pre-Mongol Isfahan", *Iranian Studies* XXXVIII/4 (2005): pp. 587-606, here p. 592.

to resist his uncle Šāhruḫ. When the inhabitants provided help to Šāhruḫ's soldiers, the city was taken and Iskandar had to fall back on the citadel (this event shows well the distinction that should be made between the citadel and the city walls).[108]

These same walls gave the inhabitants a tactical advantage when they did not want to let in an imperial player (see Table 2). In 492/1099, the Saljuq sultan Berk Yaruq, beaten by his brother Muḥammad, headed for Isfahan but "its people would not open the gates of the city to him."[109] Conversely, after the Injuid Abū Isḥāq was forced by the Muzaffarid Mubāriz al-Dīn to leave his capital Shiraz, he took refuge in Isfahan. The city resisted for three years, until the commander of the citadel (*kwutvāl*) let the Muzaffarid in. The situation of 856/1452 is even more astonishing. At that time, the Qara-qoyunlu Türkmens were moving towards ʿIrāq-i ʿajam, but not only was the citadel of Isfahan still in the hands of the Chagatai (i.e. Timurid) garrison, there was also a pro-Timurid force of 10,000 men camped before the city. This did not prevent the Isfahanis, out of hate for Chagatai's rule, from barring themselves inside the city: they shut the gates and blocked "the streets and the access roads to the various districts" (*kūcha-hā wa sar-i maḥall-hā*).[110] Usually the citadel was out of reach of the townsfolk, but not always: in 855/1452, after the governor ordered the common people (in the source: "those who did not wear a turban") to shave their beards, the outburst of popular anger developed into open insurrection and a mob stormed the citadel, forcing the governor to flee through the donjon.[111]

The capacity and will of the locals to engage in fighting was well known. The secretary and chronicler of the last Khwarazmshah, Nasawī, writes that as far as fighting value is concerned, "the common people of Isfahan cannot be compared to that of any other city, because on the special days and on the New Year's day, they are used to parading outside the walls, wearing cuirasses of different colours."[112] In the late Ilkhanid period, Ḥusayn Āvī quotes a qasida by Saʿd al-Dīn Harawī extolling the military might of the Isfahanis, with the expected comparison, in the *iġrāq* (exaggeration) mode, to the pahlavans of the *Šāh-nāma*: "… if [Ḫʷāja Firdawsī] had seen the valorous men of Isfahan, in the manner of the *Šāh-nāma*, he would have told for their glory one hundred

108 Quiring-Zoche, *Isfahan*: p. 29.
109 Durand-Guédy, *Iranian Elites*: p. 164.
110 Quiring-Zoche, *Isfahan*: p. 43.
111 Quiring-Zoche, *Isfahan*: pp. 41-2.
112 Durand-Guédy, "Iranians at war": p. 594.

stories each night and for each one of them."[113] (The rest of the qasida is about the paradisal setting of Isfahan). At the same period, Ibn Baṭṭūṭa and Ḥamd-Allāh Mustawfī note that courage is the main characteristic of the inhabitants.[114] The *Munāẓarah-yi Baġdād va Iṣfahān* also insists heavily on this theme. During the debate, Baghdad is congratulating itself for being preserved from factional strife, unlike Isfahan: "*fitna* has taken its cloths from the inviolable limit of my territory to the plain of annihilation; in my lands, everyone, whatever his class, can enjoy tranquillity and security; the wind of savagery does not blow on my ruins…."[115] Isfahan deplores that Baghdad mistook a demonstration of courage for a *fitna*:

> What you said on the battles, the clashes, and the manner the way the people from here communicated with each other with swords, it is actually the very cradle of virtue, and you took it for a flaw, and you took for stupidity what is actually courage itself, because [these bellicose manifestations] derive from their formidable appearance (*mahābat*), their virility (*rujūliyat*), courage (*šujaʿāt*) and sense of honour (*ḥamiyat*) – all appreciable qualities found in only a very few people.[116]

The fact is that for each major period of political instability, the sources clearly shows the involvement of the Isfahani archers and foot-soldiers, protected by their famous raw silk armour (*qaz-āgand*). During the Ismaili crisis of the late fifth/eleventh century, the pro-Saljuq networks of Isfahan not only went all out to reclaim the control of the city, but also participated in the military operations around the fortress of Šāhdiz.[117] In 625/1228, after the great battle of Sīn between the Mongols and the Khwarazmshah, the Isfahanis under the leadership of the Ṣāʿid family prevented the Mongols from capturing Isfahan.

113 Ḥusayn Āvī, *Tarjuma*: p. 53, verses 3-10. Note that al-Māfarrūḫī, *Maḥāsin*: p. 41, refers to the military valour of the Isfahanis only through quotes and anecdotes about the Sasanid period, probably because this was a sensitive subject after the Saljuqs had been forced to resort to a long siege to take the city. But cf. Māfarrūḫī, *Maḥāsin*: pp. 37-40 (commented by Paul, Jürgen, "Histories of Isfahan", *IrSt*, XXXIII/1-2 (2000): pp. 117-32, here pp. 125-7) where self-defence is not mentioned as one of the options when confronted to tyranny (pp. 110-12 in Ḥusayn Āvī, *Tarjuma*).

114 Ibn Baṭṭūṭa: p. 295 (*al-ġālib ʿalayhim al-šujāʿa wa'l-najda*). Ḥamd-Allāh Mustawfī, *Geographical part*: p. 49 (*mardum … mardāna bāšand*), trans. p. 56.

115 Kamāl Iṣfahānī, *Munāẓara*: p. 118.

116 Ibid.: p. 132. This passage ends with a martial poem in Arabic. Significantly, Ḥusayn Āvī, *Tarjuma*: p. 54, commenting on the qasida of Saʿd al-Dīn Harawī, use the same lexicon as the *Munāẓara*: *rujūliyat, šujaʿāt, šawkat, ṣawlat.*

117 Durand-Guédy, *Iranian Elites*: pp. 162-181.

"They had an immense courage" remembers Nasawī. His contemporary, the Baghdadi scholar Ibn al-Ḥadīd, says that "Isfahan alone held out against the Tatars, who attacked it several times in 627/1229-30. The townsfolk resisted and many fell victim on both sides."[118] After the death of Timur, we see another Ṣā'id (the Qadi Aḥmad Ṣā'idī) who barred himself inside the city and victoriously resisted the Timurid ruler of Azarbayjan, Abū Bakr, who expected his victory near Hamadan to afford automatic control of 'Irāq-i 'ajam. Qadi Aḥmad Ṣā'idī was said to have commanded twenty thousand men during these events. It is he who is introduced as the soul of the resistance, although a Timurid ruler (Rustam b. 'Umar Šayḫ) was also present.[119] In 810/1407, the same Qadi Aḥmad fought alongside his men in the army of Rustam at Gandumān. This shows that the local troops did not content themselves with *intra-muros* missions.[120] Again in 812/1409, in the volatile context created by the assassination of Pīr Muḥammad, one source notes that Qadi Aḥmad, in coordination with the "people of Čahārdāngah and Dūdangāh", took control of the city.[121] And after Šāhruḫ's death, Abū Bakr Ṭihrānī tells us that the *javānān wa sipāhiyān* of Isfahan flocked to Muḥammad Sulṭān b. Baysonqur, who would head up of a strong army within just two months.[122] If *sipāhiyān* refers to professional soldiers (the Chagatay), the *javānān* were obviously the equivalent of the *aḥdāth* of the pre-Mongol sources (both *javānān* and *aḥdāth* mean "youths"). While we can debate whether the term of 'militia' should be use for the pre-Mongol period, it is no longer the case for the Timurid period: the authors of our sources speaks unambiguously of *čirīk*, which refers to back-up forces, opposed to the *yasaqī* (the imperial troops).[123]

118 Ibid.: p. 294.

119 Quiring-Zoche, *Isfahan*: p. 21; Manz, *Power*: pp. 157, 279.

120 Manz, *Power*: p. 158. In 625/1228 the Khwarazmshah sent the Isfahani fighters back into the city before the battle with the Mongols, which took place at the northern limit of the oasis. But this does not imply, as I had thought, that such troops were useless beyond the city walls. Actually Nasawī tells us that the Khwarazmshah acted like this to show his scorn of the Mongols, meaning that he did not see the need to use the wonderful Isfahani archers. A qasida by Sa'dī about one of these local fighters shows well that they were not afraid to fight the Mongols, even though none of them "left the battle without his tunic soaked in blood". See Boyle, "Capture": p. 334.

121 Quiring-Zoche, *Isfahan*: p. 24. It is not clear what these Čahārdāngah and Dūdangāh are. Quiring-Zoche supposes it could refer to villages around Isfahan, or to the countryside in general, while Manz (*Power*: p. 160) speaks rather of "city section". The examples provided by Dihḫudā back the former hypothesis.

122 Quiring-Zoche, *Isfahan*: p. 40.

123 E.g. Quiring-Zoche, *Isfahan*: p. 37. See Doerfer, Gerhard, *Türkische und Mongolische Elemente im Neupersischen*, 4 vols (Wiesbaden: Franz Steiner Verlag, 1963-75): III, p. 65. See Aubin, *Deux sayyids*: p. 42.

This being said, while the local leaders did have the military means to be considered as full actors (not just as victims) of the so-called Turko-Mongol domination, their power should not be exaggerated. The first reason is that they were rarely united. There were opposing factions, each of them allied with players on the dynastic level. For example, in 583/1187, the Ḫujandī-s encouraged the (last) Saljuq sultan Ṭoġrïl b. Arslan to rebel against the Atabeg of Azarbayjan, the de facto ruler of Western Iran. In Isfahan, Ṭoġrïl governor retrenched in the Shafii part of the city, while the emir sent by the Atabeg enjoyed the support of the Hanafis. "The city was divided into two camps (*šahr do havā'ī šodand*)" says a contemporary chronicler.[124] A similar situation happened after the death of Jahān-Šāh Qara-qoyunlu in 872/1467, when local society was again divided. Inside Isfahan, some, like Ḫwāja Jalāl al-Dīn Ṣā'idī, were in favour of the Timurid ruler Abū Sa'īd; others, such as the amir Sayf al-Dīn, were partisans of Jahān-Šāh's son – with both Timurid princes vying for control of the 'Irāq-i 'ajam. Eventually Jalāl al-Dīn Ṣā'idī captured Sayf al-Dīn and took Ḫwāja Jamāl al-Dīn Turka hostage. A local chronicler reports a very tense atmosphere in the city, with each faction wary of what the other might do.[125]

In any case, the involvement of the elites on the political field often led to their demise. The *ra'īs* Ṣadr al-Dīn Muḥammad Ḫujandī had invited an Abbasid army into Isfahan after the fall of the Saljuq dynasty and the Khwarazmian conquest, but was beheaded within months by the Abbasid governor. Apparently, neither the *ra'īs* nor the governor were prepared to share what each considered to be their rightful control over Isfahan.[126] This is very similar to the relationship between the Ḫwāja Maḥmūd Ḥaydar, the leading Isfahani at the time of the Qara-qoyunlu takeover of the city, and the governor, who was the tutor (*lala*) of the ruler's son: Maḥmūd Ḥaydar was eventually killed in 857/1452.[127] Likewise, after having played a prominent role in Isfahan in the troubled years which followed Timur's death, often in the absence of the Timurid prince of Isfahan, Qadi Aḥmad Ṣā'idī was killed by this prince (Rustam) in 814/1412.[128] And when the Shafii notables got in touch with the Mongols and offered them

124 Durand-Guédy, *Iranian Elites*: p. 277.

125 Quiring-Zoche, *Isfahan*: p. 49. After the forced conversion to Shi'ism of all the Isfahanis under the Safavids, a new form of factional strife (*fitna*) raged between the Ḥaydarī and the Ni'matī Sufi orders, here again with the encouragement of the prince. See Perry, John, "Toward a Theory of Iranian Urban Moieties: the Ḥadyariyyah and Ni'matiyyah Revisited", *Iranian Studies* XXXII/1 (1999): pp. 51-70.

126 Durand-Guédy, *Iranian Elites*: pp. 283-4.

127 Quiring-Zoche, *Isfahan*: p. 44.

128 Quiring-Zoche, *Isfahan*: p. 25; Manz, *Power*: p. 161.

to open the gates of Isfahan in exchange for political supremacy in the city, things did not end well for them either.[129]

4 Elements of Change

4.1 *The Make-up of Local Society*

The previous section illustrated patterns of similar local community reaction in the Turko-Mongol period when facing similar situations (for example, when the imperial player was weak, excessive fiscal pressure often led to rebellion). This does not mean that the communities themselves remained the same. Indeed the makeup of the population underwent considerable change over the period. I will address this issue by commenting on Ibn Baṭṭūṭa's account, which contains many valuable insights on the social fabric of Isfahan.

During his two weeks stay, the Tangerine traveller was hosted in a Sufi lodge (*zāwiyya*, Ps.: *ḫānaqāh*) belonging to the Suhrawardiyya order. In Isfahan, Sufism had been present since ancient times: at the very beginning of our period, the local Sufi šayḫ Abū Nuʿaym Iṣfahānī (d. 430/1038) wrote the first biographical dictionary of mystics (his *Ḥilyat al-awliyā* was completed under the Ghaznavid occupation of Isfahan, in 422/1031). The vizier Niẓām al-Mulk, who was so fond of Sufis that his killer approached him in Sufi garb, founded a *ḫānaqāh* when Isfahan was the capital of the empire. However, the institutionalisation of Sufi orders dates to the sixth/twelfth century. Annemarie Schimmel remarks that "it is probably more than a sheer coincidence that the formation of Sufi orders sets in at the very moment that the power of the Ismailis began to fade away".[130] The troubled political context (as outlined in the introduction) surely played a major role. The life of Kamāl al-Dīn Ismāʿīl is emblematic of the situation during the chaotic decades prior to the Mongol conquest: this court poet turned ascetic was allegedly killed in his own *ḫānaqāh* by gold-thirsty Mongols.[131] The Suhrawardiyya attended by Ibn Baṭṭūṭa was a Sunni order established in Baghdad by the famous Abū Ḥafṣ ʿUmar Suhrawardī (d. 632/1234). Ibn Baṭṭūṭa met the third generation of *šayḫ*-s and from his account we under-

129 Durand-Guédy, *Iranian Elites*: pp. 295-7.

130 Schimmel, Annemarie, "The Ornement of the Saints: the Religious Situation in Iran in pre-Safavid times", *IrSt*, VII/1 (1974): pp. 88-111, here p. 103.

131 Durand-Guédy, David, "Kamāl al-Din Esmāʿil", in *EIr*, XV (2010): pp. 415-17. Kamāl al-Dīn's father, the equally famous ʿAbd al-Razzāq Iṣfahānī, speaks on several occasions about his desire to live a more ascetic life, but there is no evidence that he actually took the plunge, see Nafīsī, Saʿīd, "Jamāl al-Dīn ʿAbd al-Razzāq", *Armaġān*, VI (1921): pp. 109-18 and 153-63, here p. 117.

stand that the succession at the head of the Sufi order (*ṭarīqa*) was hereditary (similarly to the local positions of *ra'īs*, qadi, *ḫāṭib*, etc.).[132] One century later, the qadi of Isfahan, Ṣā'in al-Dīn 'Alī Turka, had to write an apology (the *Nafṯat al-maṣdūr*) to exonerate himself from Sufi-'deviancy' (*ṣūfīgarī*). Apart from the established Sufi orders (*ṭarīqat*), the eighth/thirteenth and ninth/fourteenth century saw the proliferation of *bābā*-s, the typical term for dervishes of the time (significantly the first *bābā*, Bābā Ṭāhir of Hamadan, was a contemporary of Ṭoġrïl Beg, the first Saljuq ruler). These *bābā*-s have left their mark in the landscape: Mahdavī has identified 36 places associated with them in Isfahan and its surroundings, the most famous of which was the Bābā Qāsim mausoleum (d. 740/1339-40), built for the mystic who taught at the nearby eponymous madrasa (now madrasa Imāmiyya).[133]

Ibn Baṭṭūṭa's account offers another interesting piece of information about the religious divide: "the strife between the Sunni and the Shi'is (*rawāfiḍ*) continues until today and they never stop fighting."[134] For someone who has studied pre-Mongol Isfahan, this sentence comes as a surprise, not about the existence of factional strife, a phenomenon which can be observed from the Buyid to the Safavids, but rather the identity of the actors: Shi'is versus Sunnis. Indeed, since Islam had become the mainstream faith, Isfahan was the quintessential Sunni town. Muqaddasī's account about revering 'Alī's nemesis Mu'awiya is often quoted to highlight the ultra-Sunnism of the Isfahanis in the pre-Saljuq period. Although Rasūl Ja'fariyān has gathered much evidence showing that the situation was more balanced, it is clear that in Isfahan, Shi'ism never achieved the grassroots support it did in Baghdad or in cities of northern 'Irāq-i 'ajam such as Āva, Qum or Kashan.[135] In the sixth/twelfth century, the Shi'i author Qazwīnī-Rāzī (fl. around 565/1170) still explicitly says that Isfahan was *not* a Shi'i city (Isfahan is mentioned as *dār al-sunna wa l-jamā'a*, and the Isfahanis are referred in several places as anthropomorphists, *mušabbīhān*).[136] At that time, the local political field was divided between Shafiis (led by the

132 Sobieroj, Florian, "Suhrawardiyya", *EI*², IX (1998): pp. 784-6. The building of the Suhrawardiyya *ḫānaqāh* of Isfahan is no longer visible, but it might have been similar to that of Naṭanz, where Ibn Baṭṭūṭa also stayed. See Blair, Sheila, "Sufi Saints and Shrine Architecture in the Early Fourteenth Century", *Muqarnas* VII (1990): pp. 35-49.

133 Mahdawī, Musliḥ al-Dīn, *Dānišmandān wa buzurgān-i Iṣfahān*, 2nd ed. with supplements by Raḥīm Qāsimī and Muḥammad Riḍā Nīlūfarišān, 2 vols (Isfahan: Šahrdārī-yi Iṣfahān, 1384š.[/2005-6]): I, pp. 297-301. On the *bābā*-s of Tabriz see Zakrzewski in this volume.

134 Ibn Baṭṭūṭa, trans. Gibb: p. 295.

135 Ja'fariyān, Rasūl, "Pīšīna-yi tašayyu' dar Iṣfahān", *Mīrāṯ-i jāvīdān*, XIX-XX (1376š.): pp. 17-31, here p. 19-24.

136 Ja'fariyān, "Pīšīna": pp. 24-5; Durand-Guédy, *Iranian Elites*: p. 202, note 78.

Ḫujandī-s) and Hanafis (led by the Ṣāʿid-s) and this was still the case at the eve of the Mongol conquest.

Many sayyids had settled in Isfahan when the city developed (the most famous was Ibn Ṭabāṭabā, the author of the *Muntaqila al-Ṭālibiyya*) and in the sixth/twelfth century, some of them are mentioned by Shiʿi authors such as Qazwīnī-Rāzī (*al-Naqż*) and Muntaḫab al-Dīn (*al-Fihrist*). But it is only after the Mongol period that the sources show sayyids playing a significant role in political events.[137] The author of the *Tarjuma-yi Maḥāsin Iṣfahān* demonstrates great respect for the Family of the Prophet. In the eighth section of the book, which lists the great men of Isfahan, Husayn Āvī starts with the sayyids, dedicating a long and florid eulogy to the most famous of them, one Muḥammad al-Ḥusaynī al-ʿAlawī.[138] In the next two centuries, the sayyids are everywhere in the sources. We have mentioned the role played by Šāh ʿAlāʾ al-Dīn in the events of 850/1446 under Šāhruḫ. Similarly, the Mīr-i Mīrān family, who played a major role during the Injuid and later the late Timurid period, were sayyids.[139] Being a sayyid does not imply a Shiʿi following, but the two are often correlated, at least in post-Abbasid Iran.[140]

Given what we have just said about the Saljuq period, the Shiʿi community must have become stronger during the first 70 years of Mongol rule. This is also the period for which the written sources are scarcer. Generally speaking, we know that the new context played in favour of the Shiah. Because (Twelver) Shiʿis were already living without rightful authority (since the great occultation of their last Imam in 260/874), they were better prepared to live in the new political and religious framework brought about by the destruction of the Abbasid caliphate (hated by the Shiʿis) and the rule of non-Muslim kings. In the following decades Shiʿism, with its insistence on lineage, would be

137 An exception is the *raʾīs* of Isfahan in the mid-fifth/eleventh century, Abū ʿAbd-Allāh al-Qāsim al-Ṯaqafī, who was a sayyid and also a Shiʿi but it was a relic from the Buyid period left untouched by the Saljuqs, until he his death, after which the function was given to a Shafii. See Durand-Guédy, *Iranian Elites*: pp. 116-17.

138 Ḥusayn Āvī, *Tarjuma*: p. 154. Ibid.: p. 25 (reference to *āluhu al-ṭāhirūn*), p. 32 (quotation from ʿAlī): pp. 46-7 (on sayyid Maḥmūd b. Aḥmad al-Ḥusaynī al-ʿAlawī al-Madīnī).

139 Quiring-Zoche, *Isfahan*: pp. 228-9. Aubin, “Etudes safavides”: p. 40, note 1.

140 On this, see Pfeiffer, Judith, “Confessional Ambiguity vs. Confessional Polarization: Politics and the Negotiation of Religious Boundaries in the Ilkhanate”, in J. Pfeiffer (ed.), *Politics, Patronage and the Transmission of Knowledge in 13th-15th Century Tabriz* (Leiden: Brill, 2014): pp. 129-167, pp. 144-5, and p. 159: “The Shiʿa exploited the ‘symbolic capital’ of sayyidism to their advantage.” The spread of an “‘Alid-loyalism” after the Mongol conquest was first noted by Marshall Hodgson. Melvin-Koushki, *Quest*, p. 73, prefers the more generic terms of “Shii-Sunnism” or “imamophilism”.

presented as good for the Chinggisids.[141] In any case, Shiʿis were now more free to practice their faith openly, and to engage in proselytism. The Shiʿi opuscule *Maṭlaʿ al-ṣabāḥtayn*, written in early Mongol Isfahan, is a proof of this vitality (Abu'l-Saʿādāt Šufurva aims to emphasise the fraternity, *uḫuwwat*, between ʿAlī b. Abī Ṭālib and the Prophet, through a comparison of two compilations of their sayings).[142]

In the case of Isfahan, other factors may have been in play. As a working hypothesis, I mention the repopulation of the city after the massacre that followed the Mongol conquest. Indeed, repopulation on such a scale implied an arrival of people from the outside (this is explicitly mentioned for Baghdad after its destruction by the Mongols),[143] and we have several observations and sources of evidence that the countryside was not as orthodox as the city. Niẓām al-Mulk tells us that in the third/ninth century, the (non-Muslim) Ḫurramdīniyya movement was strong in the province of Isfahan, which fits well with Patricia Crone's thesis about a geographic analysis of heterodoxy in Iran.[144] In the fourth/tenth century, Ṣāḥib b. ʿAbbād, the Shiʿi vizier of the Buyids, came from Ṭāliqān, in the hinterland of Isfahan.[145] During the crisis which broke out at the end of the fifth/eleventh century, there was a clear contrast between the city, from where the pro-Saljuq networks launched the counteroffensive, and the countryside, under the sway of the Ismailis. (This divide is best seen during the violent assault of the rurals (*sawādiyya*) against Isfahan in 495/1102).[146] We have less information about the century before the Mongol conquest, but the neo-urbans of the seventh/thirteenth century may have been more prone to becoming Shiʿis.[147]

The strengthening of the Shiʿi community over the seventh/thirteenth century can be explained with more certainty by the same factors as those at work in the strengthening of the Hanafi community in the Saljuq period: those of political backing, which involves 1) the foundation of richly endowed

141 Ibid.: pp. 145, 148.

142 Abu'l-Saʿādāt Šufurva compares the *Šihāb al-aḫbār* written by the Shafii Egyptian qadi al-Quḍāʿī (d. 454/1062) with the *Nahj al-Balāġa* of the Qumi sayyid al-Sharif al-Raḍī.

143 The author of the *Marāṣid* writes that "when the Tatars came, most of it was ruined. They killed its people and few were left. Then people from outside came", quoted by Duri, Abd al-Aziz, "Baghdād", in *EI*², I (1960): pp. 894-908, here p. 903.

144 Crone, Patricia, *The Nativist Prophets of Early Islamic Iran: Rural Revolt and Zoroastrism* (Cambridge: Cambridge Univ. Press, 2012).

145 Durand-Guédy, *Iranian Elites*: p. 32, note 57.

146 Durand-Guédy, "Ismailism": pp. 253-4.

147 Cf. Yāqūt's first-hand account (II, p. 893, lines 20-1), explicitly stating that in Rayy, "the Shiʿi part of town was the largest because half of the inhabitants of Rayy were Shiʿi, while in the countryside they were all Shiʿi, with a few Hanafi".

institutions; 2) direct material benefits (such as land grant and pensions); 3) protection and, in case of strife, military support for the leaders of the community. After they took Isfahan, the Saljuqs had awarded the position of qadi to a Hanafi, even though Hanafism was by then almost inexistent. In the sixth/twelfth century, with some delay compared to their Iranian servants, they engaged in religious patronage and sided more and more openly with the Hanafis. (ʿImād al-Dīn al-Iṣfahānī deplores their *ʿaṣabiyya* against the Shafiis.) In the early eighth/fourteenth century, the Hanafi madrasa (founded two centuries earlier by the Saljuq Muḥammad b. Malik-Šāh) was apparently still active,[148] but the religious community on the rise was the Shiah.

The governor Bahāʾ al-Dīn Juwaynī, introduced above, played a key role in the strengthening of the Shiah in Isfahan. For Rasūl Jaʿfariyān, the Shiʿi leaning of Bahāʾ al-Dīn is not only proven by the fact that he was the dedicatee of *al-Muʿtabar fī šarḥ muḫtaṣar al-nāfiʿ* of the great Shiʿi jurist Muḥaqqiq Ḥillī (d. 676/1277); still stronger proof is that Bahāʾ al-Dīn backed the Shiʿi scholar ʿImād al-Dīn Ḥasan Ṭabarī, who wrote the *Kāmil-i Bahāʾī* for him.[149] Ṭabarī also wrote the *Manāqib al-Ṭāhirīn* in Isfahan in 673/1274, and in the introduction he recalls that he came from Qum to Isfahan at the invitation of the governor, to engage in religious polemics with the local Sunnis. He explicitly says that the era of persecution of the Shiis (and therefore of their dissimulation, *taqiyya*) is over:

> For 600 years, the Umayyad and Abbasid caliphs had concealed the virtues of the *ahl al-bayt*, killed those who transmitted them, and burned their books and collections of hadiths. Shiʿi viziers and emirs who were defenders of the divine law (*nāṣir-i šariʿat*) in ancient times are mentioned in the book *Maṯālib al-navāṣib* (*"The Vices of the Sunnis"*).
>
> But in our times, thanks to the glory of the administration (*farr-i dawlat*) of Nāṣir al-Dīn-allāh Bahāʾ al-Ḥaqq wa'l-Dīn Muḥammad, the servant (*mawlā*) of the greatest vizier (*al-ṣāḥib al-aʿẓam*) Muḥammad Ṣāḥib Dīvān, the dissimulation (*taqiyya*) which was necessary for this notorious community has become forbidden, and the glory of the *ahl al-bayt* have been extended all over the world. And in the protection of this administration, the sects of the Yazīdiyya and of the Marwāniyya [Umayyads], of the Jabriyya [deniers of the human free will], of the Mujassama

148 Ḥamd-Allāh Mustawfī, *Geographical Part*: p. 49, trans. p. 51.

149 Jaʿfariyān, "Pīšīna": pp. 25-7.

> [anthropomorphists] and of the Nāṣiba [Sunnis] have become miserable and humiliated.[150]

It was also in Isfahan that the same ʿImād al-Dīn Ṭabarī wrote a treatise on the *imāmat*, in which he critiqued a group of Sunni scholars who openly manifested joy during the mourning of Ašurāʾ, the most important Shii ceremony. In Isfahan, such criticism would have been unthinkable in the previous century. In another of his books, the *Tuḥfat al-abrār* (written in Persian), he recounts how he successfully engaged in a religious debate with a group of Shafiis (the dominant madhhab in Isfahan according to Ḥamd-Allāh Mustawfī) and he even mentions conversion to Shiism.[151] Here again this proselytism shows how much things had changed for the Shiah in Isfahan.

After their conversion to Islam, the Ilkhans also supported the Shiah. Ġāzān (d. 703/1304) established *dār al-siyāda* with major endowments (*waqf*) in nine of the major towns of the Ilkhanate, among these Isfahan.[152] These institutions were founded, says Judith Pfeiffer, "in explicit comparison and contrast to the already existing institutions, such as *madrasas* for the scholars, *khānqāhs* for the Sufis, and *zāviyas* for the dervishes."[153] The political support for the Shiah was even stronger under Öljeitü (d. 716/1316) who officially adopted Twelver Shiʿism. An analysis of his dreams, as recounted in the *R. Favāʾid-i Ūljāytū*, shows that "he was not only supportive of the Shiah, as they had been under Ġāzān, they also become distinctly anti-Sunni." The message, concludes Judith Pfeiffer, was that "royal support for the offspring of ʿAlī is legitimate, and so is fighting the Sunnis, even with violent means."[154] The strife between Sunnis and Shiʿis recorded by Ibn Baṭṭūṭa was probably a consequence of Öljeitü's decision, in 710/1310, to have the prone at the Friday prayer (*ḫutba*) read in the Twelver Shiʿi way.

The Friday mosque itself bears traces of this evolution. The Shiʿi vizier of Öljeitü, Saʿd al-Dīn Muḥammad of Sāwa (d. 711/1311), not only supervised the building of the famous stucco *miḥrāb* with a Shiʿi profession of faith ("ʿAlī walī Allāh"), he also financed the erection of several new sections in the mosque,

150 Ibid. The Yazīdiyya refers to the Sufyanids, the first line of the Umayyad dynasty, succeeded by the Marwanid line in 64/683. "*Nāṣibī*" is a term of deprecation, stigmatising those who "founded" (*naṣaba*) Abū Bakr's caliphate, depriving ʿAlī of his right.

151 Ibid.: p. 27.

152 Rashīd al-Dīn: 40, 190-1. The *waqf-nāma* of the *dār al-siyāda* of Isfahan was probably similar to the one of nearby Kashan, which is extant and was signed by a larger number of notables close to the Ilkhan, see Pfeiffer, "Confessional Ambiguity": pp. 148-9.

153 Ibid.: p. 146.

154 Ibid.: pp. 151-55.

FIGURE 10
The *miḥrāb* of Öljeitü built by Sa'd al-Dīn Sāwajī, Friday Mosque of Isfahan
PHOTO D. DURAND-GUÉDY, 2015

including the room in which the *miḥrāb* is located (Figs. 9 and 10). The new chambers may have been reserved for the Shi'i worshippers, in the same way that two centuries earlier, the *īwān*-s opened around the central courtyard served to accommodate the various madhhabs in a religiously strained city. The distinctly Shi'i decorative program executed in the shrine of Pīr Bakrān, in 712/1312, shows that the hinterland was not isolated from what was going on in the metropolis.[155] The successors of the Ilkhans were not Shi'i (with exceptions, such as the Timurid prince Iskandar),[156] but out of affection for the Family of the Prophet they patronised the building of shrines for the descendants of the Imams (*imām-zāda*). In Isfahan, the Darb-i Imām, commissioned by Jahān-Šāh Qara-qoyunlu in 857/1453 to contain the tombs of Ibrāhim Baṭḥā and Zayn al-'Ābidīn, is a masterpiece of tile decoration (see Fig. 11).

In others words, similar processes were at work during both the Saljuq and Mongol periods, although the institutions changed. During the Saljuq period, the preeminent role was played by the (Shafii) *ra'īs*, and increasingly by the (Hanafi) qadi. From the Mongol period onward, the *ra'īs* and qadi still played important roles as political actors, but also had to reckon with the *naqīb*. The

155 Majeed, Tehnyat, "The Role of the Qur'anic and Religious Inscriptions in the Buq'a Pīr-i Bakrān, Isfahan: the Shī'ī Reign of Öljeytü Khudābande in Īlkhānid Iran", *Journal of Qur'anic Studies*, X/2 (2008): pp. 111-123.

156 Richard, "Un témoignage inexploité": p. 47.

FIGURE 11 The entrance of Darb-i Imām, commissioned by Jahān-Šāh Qara-qoyunlu in 857/1453
PHOTO A. SHOJAEE ESFAHANI

disappearance of the office of *ra'īs* (*sardar*) after the Safavid conquest heralded the end of an era: the emerging Safavid power, which transformed Isfahan into royal domain, succeeded in creating new offices, e.g. *kalāntar* and *kadḫudā*, more dependent of the central administration, to replace the old ones.[157]

Even without a source as detailed as the *Ta'rīḫ Nīšābūr*, analysed by Richard Bulliet in his seminal study on the *Patricians of Nishapur*, we can draw on the vast resources of the biographical dictionaries, poetical divans and chronicles, to piece together the structure of some of the Isfahan greatest families over several generations. For the Saljuq period, in addition to the Ḫujandī-s and the Ṣā'id-s, a handful of other families can be identified with some degree of precision (the Alūh, Ḫatībī, Kūšīdī and Mašāḏāh). For the period prior to Shah 'Abbās, Quiring-Zoche has distinguished eight leading families.[158] Because of their longevity, the emblematic family of Isfahan during the Turko-Mongol period was the Ṣā'id-s. Having arrived (from Khurasan) in the first decades of the Saljuq rule, they controlled the position of qadi from the reign of Muḥammad b. Malik-Šāh's onwards. During the sixth/twelfth century, during which Saljuq

157 Quiring-Zoche, *Isfahan*: pp. 158-69.

158 Apart from the Ṣā'id, we have already seen the Šahšahānī (the descendants of the martyred *naqīb* 'Alā' al-Dīn Muḥammad), and there were also the Jābirī Anṣārī (whose descendant wrote the *Tārīḫ-i Iṣfahān*). Studies on prosopography posterior to Quiring-Zoche's book include Haneda, Masashi, "La famille Khūzānī d'Isfahan (15e-17e siècle)", *StIr*, XVIII/1 (1989): pp. 77-92.

authority was on the wane, they were often in conflict with the Ḫujandī-s, leader of the Shafiis. The demise of the Saljuq dynasty and the Khwarazmian rule consecrated their position, and they were made the official representatives of the new imperial power in replacement of the Ḫujandī-s. We can surmise that by this time, the Hanafi community was considerably stronger than one century earlier.[159]

The Ṣāʿid-s survived the Mongol conquest and continued to played a role up to the Timurid period. Their genealogical tree is broken for most of the seventh/thirteenth century, but the Ṣāʿid-s of the post-Mongol period belonged to the same family.[160] In the eighth/fourteenth century, when Ḥusayn Āvī speaks of Niẓām al-Dīn Qiwām al-Islām Ṣāʿid (who defended Isfahan against the rulers of Fars), he extols a family in charge "from a very old time", and "hereditarily" (*ab*^{an} *ʿan jadd*^{an}). Even more significant, the vizier Saʿd al-Dīn Muḥammad al-Sāwajī deemed it unnecessary to have qasidas written for the Ṣāʿid-s of his time given that Isfahan's most brilliant poet, Kamāl al-Dīn Ismāʿīl, had already composed unsurpassable qasidas for their forefathers (Kamāl al-Dīn Ismāʿīl died at the very beginning of the Mongol period).[161] Contrary to the Ṣāʿid-s, the Ḫujandī-s appear to have been wiped out by the bloody Mongol capture of the city. However the most famous Isfahani of the ninth/fifteenth century, the qadi and scholar Ṣāʾin al-Dīn Turka Iṣfahānī, belonged to a Shafiʿi family originating from Khujand which had settled in Isfahan during the pre-Mongol period. There is little doubt that they were patronised by the powerful Ḫujandī family, and the Āl-i Turka could therefore be seen as a kind of survival of their network.[162]

4.2 *New Forms of Association?*

What was the social basis for leadership? The mechanisms of patron-client relationship analysed by Roy Mottahedeh for the Buyid period are valid for the following centuries. Antonio Jurado Aceituno, and more recently Jürgen Paul, have provided very detailed studies of the function of *ḫidma* (service/benefit)

159 But Ḥamd-Allāh Mustawfī tells us that in the early eighth/fourteenth century, the Sunnis of Isfahan were still Shafiis.

160 Quiring-Zoche, *Isfahan*: pp. 230-1.

161 Ḥusayn Āvī, *Tarjuma*: pp. 54-5. Durand-Guédy, "Kamāl al-Din Esmāʿil". I thank Daniel Zakrzewski for drawing my attention to an Ilkhanid *taqrīḍ*, for the theological treaties of the vizier Rašīd al-Dīn, signed by one "Jalāl al-Islām Muḥammad b. Muḥammad b. Ismāʿīl Ibn Ṣāʿid, qadi of Isfahan". This piece of information fills a gap between the Saljuq and the Timurid period.

162 It could be coincidental that the earliest known ancestor of Ṣāʾin al-Dīn bore the honorific "Ṣadr al-Dīn", which was the privilege of the senior members of the Ḫujandī family.

at the imperial level,[163] but it also operated at the local level. The documents of the *al-Muḫtārāt min al-rasāʾil* demonstrate this well for sixth/twelfth century Isfahan: the requests for direct aids, tax exemptions, and other, are worded in the paradigm of *ḫidma*. These documents offer an insight into the development of a local network, showing how an urban community (Shafiis), from weakest members (widows) to leaders, were bound together through a vast array of social bonds: friendship, kinship, and patron-client relationships.[164] It is this network, which ran deep in the different layers of the population, which made the Ḫujandī family indispensable during the first two thirds of the sixth/twelfth century.

For the Saljuq period, such social bonds were usually framed along doctrinal lines (the Shafii faction headed by the Ḫujandī-s – we do not have so much information about the Hanafis). This fits quite well Makdisi's interpretation of the madhhab as an "elementary form of association in a Muslim context".[165] But what about other forms of association? Once again, Ibn Baṭṭūṭa provides a useful description regarding Isfahan:

> The members of each craft (*ahl kulli l-ṣanāʿat*) appoint one of their own number (*kabīr minhum*) as headman over them, whom they call *kulū*, and so do the leading citizens from outside the ranks of the craftsmen (*kibār al-madīna min ġayr ahl al-ṣanāʿāt*). One company (*jamāʿat*) [for example] will be composed of bachelors (*al-šubbān al-aʿzāb*). These companies try to outdo one another and invite one another to banquets, displaying all the resources at their disposal and making a great show in regard to the dishes and everything else. I was told that one company of them (*ṭāʾifa*) invited another, and cooked their viands with lighted candles, then the second company returned the invitations and cooked their viands with silk.[166]

The beginning of the text is a clear characterisation of a corporation, meaning "a group of town people engaged in the same occupation, who elect their own

163 Paul, Jürgen, "*Khidma* in the social history of pre-Mongol Iran", *JESHO* LVII (2014): pp. 392-422.
164 Durand-Guédy, *Iranian Elites*: pp. 230-55 (chap. 9, The power of the notables).
165 Makdisi, George, "La corporation à l'époque classique de l'Islam", in *The Islamic world from classical to modern times. Essays in honor of Bernard Lewis*, ed. by C.E. Bosworth and al. (Princeton: Darwin Press, 1989): pp. 193-209.
166 Ibn Baṭṭūṭa, trans. Gibb: pp. 295-6. Cf. Quiring-Zoche, *Isfahan*: p. 208.

chief and officers, who pay guild taxes."[167] The election of a headmen shows that it was an autonomous association organizing the life of its members, and not a top-down organization, emanating from the state.[168] (The earliest mention of guild taxes in Isfahan is in 971/1563-4, when Shah Ṭahmāsp remitted them).

The term *kulū*, which is held by the headmen of the guilds, is frequent in post-Ilkhanid sources on western Iran. It probably derives from *kalān*, "big", and was synonymous with the polysemic term *ra'īs*. (In Injuid Shiraz, a few decades after Ibn Baṭṭūṭa travelled through the region, the head of the city districts who played a decisive role during the siege of the city by Mubāriz al-Dīn Muḥammad in 754/1353, are also called *kulū*.)[169]

Significantly *kulū*, says Dihḫudā, may refer to a guild leader, but also to a position among the *fityān* and the *aḫīs*, i.e. members of the *futuwwa*. And this is indeed an association of the *futuwwa* type which is described by Ibn Baṭṭūṭa in the rest of the text. The explicit mention of "*jamā'at* of unmarried young men" is reminiscent of the *šuṭṭār* and the *fityān* (Ps. *javānmard*), meaning young men originating from diverse horizons, living in community and joining forces to live the best they can in an atmosphere of solidarity and mutual devotion.[170] In Ibn Baṭṭūṭa's account, the different groups use hospitality to outdo one another. This is in keeping with the pages dedicated to the subject in each treaty of *futuwwa*, starting with the most famous of all, the *Futūwat-nāma-yi sulṭānī* of Ḥusayn Kāšifī (d. 910/1504-5). As wrote Jean-Claude Vadet, the rules of communal life in this *futuwwa* culminate in philoxenia, meaning to practice hospitality as generously as did the Prophet Abraham.[171] The hadith "Honour the

---

167 Floor, Willem, "Guilds and futuvvat in Iran", *ZDMG* CXXIV/1 (1984): pp. 106-14, here p. 100; same author "Aṣnāf", in *Enc. Iranica*, II (1987): pp. 772-78, here p. 773. Floor, ibid., says that reference to guild taxe is mentioned by Ibn Baṭṭūṭa elsewhere.

168 This key distinction is made by Cahen in "Mouvements populaires": p. 28. Cf. Cahen, Claude, "Y a-t-il eu des corporations professionnelles dans le monde musulman classique? Quelques notes et réflexions", in A. Hourani and S. Stern (eds), *op. cit.*: pp. 51-64, here p. 52.

169 On the *kulū*-s of Shiraz, see Limbert, John, *Shiraz in the Age of Hafez: the Glory of a Medieval Persian City* (Seattle: University of Washington Press, 2004), esp. pp. 89-91 (Limbert vocalises *kalu*). On the meaning of *kulū*, see Afšārī, Mihrān, *Čahārdah risāla dar bāb-i futūwat* (Tehran: Čašma, 1381š.): introduction, pp. 34-37. See also Quiring-Zoche, *Isfahan*: p. 206 (*ra'īs* as guild headmen, "Oberhaupt").

170 This is Cahen's definition in "Mouvements populaires": pp. 34-5.

171 Vadet, Jean Claude, "La Futuwwa, morale professionnelle ou morale mystique", *REI*, LXXVI (1978): pp. 57-90, here p. 65. Vadet insists on the importance of the notion of *kasb* (economic gain) at the core of the *futuwwa*, as described by Kāšifī, which distinguishes it from the Sufi *ṭarīqa*. Mysticism is a common point but members of *futuwwa* have never taken the world for negligible and instead endeavour to live within it, and to engage in

guest even if he is an unbeliever" is often quoted in that respect.[172] Of course, the food cooked with lighted candles mentioned by Ibn Baṭṭūṭa seems counter to the caution issued to the *futuwwa* members against "eating the food prepared by someone for fame and renown".[173] By using silk as a combustible, the other group managed to surpass its rival by showing abstention and restraint (silk clothing was one of the prohibited materials). In any case, this competitive atmosphere is perfectly in line with what the *futuwwa* treaties say about the diverse competitions (single combat, hand-to-hand, horse races, etc.) in which the *pahlavān* can prove his valour.[174]

Thus, Ibn Baṭṭūṭa's text refers at the same time to guilds, *futuwwa*, and competition. These bellicose virtues bring to mind the military organizations described by Nasawī for the early seventh/thirteenth century, as well as the fame of Isfahan for its resistance in the face of *fitna*. The *futuwwa* predates Ibn Baṭṭūṭa's travel. Before the Saljuq period, Ṣāḥib Ibn 'Abbād writes in one of his letters that he has defeated the *fityān* of Isfahan. But what about the guilds? Two interpretations are possible here. The first is that it took the acute eye of a non-Iranian to notice the phenomenon of the guilds, which had existed for some time. If it is so, Ibn Baṭṭūṭa would merely have 'revealed' their existence.[175] Another interpretation is to consider that the corporations described by Ibn Baṭṭūṭa are new, and for this reason required new words, such as *kulū*.[176] In this case, the new context created by the Mongol domination, 'the time of un-

---

economic activities (without seeking profit since what is earned must be given back to the "brothers"). This fits well Ibn Baṭṭūṭa's text linking *futuwwa* and the artisan guilds. The famous *futūwat-nāma* of Ibn Zarkūb, written in seventh/thirteenth Tabriz, distinguishes three types of *futuwwa* (*sayfī, qawlī, šurbī*), which apply, according to Mihrān Afšārī, *Čahārdah risāla*: p. 12, to the fighters, artisans and merchants, and finally the Sufis. There is a vast debate about the links between *futuwwa* and Sufism, see Ridgeon, Lloyd, *Morals and Mysticism in Persian Sufism: A History of Sufi-futuwwat in Iran* (London: Routledge, 2014).

172 Afšārī, *Čahārda risāla*: p. 109.

173 Kāšifī Sabzawārī, *Futūwat-nāma-yi sulṭānī*, ed. Maḥjūb, Muḥammad Ja'far (Tehran: Intišārāt-i bunyād-i farhang-i Īrān, 1971), trans. Jay R. Crook, *The Royal Book of Spiritual Chivalry* (*Futūwat nāma-yi sulṭānī*) (no place of publication: Great Books of the Islamic World, 2000): p. 225.

174 Ibid.: pp. 269-333 (part 6 on the *ma'ruka*, field of trial). One of the *pahlavān*-s of the tenth/eleventh century is buried in the mausoleum of Bābā Qāsim, see Hunarfar, *Ganjīna*: p. 314.

175 I use this expression in reference to Jüergen Paul's article "L'invasion mongole comme « révélateur » de la société iranienne", in Denise Aigle (ed.), *L'Iran face à la domination mongole* (Tehran: IFRI, 1997): pp. 37-53. I had pondered on this possible 'revelation effect' about Nasawī's account on the militia of Isfahan, see Durand-Guédy, "Iranians at war": p. 596.

176 *Kulū* appears only in one verse at the pre-Mongol period (see Dihḫudā, q.v.), and to my knowledge it never appears in all our vast documentation on pre-Mongol Isfahan. But I

certainty', with the disinterest of rulers for urban life, and perhaps also the strengthening of Shi'ism inside the city, stimulated new forms of work associations in the bazaars. We know that after Tabriz was taken by the Mongols, the Mongol general Čormaġun, appointed one Mengü-Bolad, also a Mongol, as *basqaq* over the artisans of Tabriz (*bar sar-i muḥtarifa ba ism-i basqāqī*).[177] This is a classical top-down attempt to supervise and tax the artisans. Given the troubled times after the fall of the Ilkhanate, it is not hard to imagine that the artisans would have taken things into their hands (their autonomy mirroring the increased power of the local leaders vis-à-vis the imperial actors, as shown in the third part of this article).[178] The multiplication of trade-oriented *futūwat-nāma* in the Late Middle Ages strengthen this theory. In Europe also, the formation of guilds has been interpreted as "a kind of historical response to the disorganisation of the powers".[179]

Further information on this issue would be help us to assess key 'political' events in the pre-Safavid history of Isfahan, in particular the revolutionary movements that challenged the existing order: the Ismailis and, to a lesser degree the Hurufis. Isfahan played a special place in the history of both movements. At the end of the fifth/eleventh century, it was the only major city in which the Ismailis (under the leadership of Aḥmad Ibn 'Aṭṭāš) had a large number of followers, and in the ninth/fourteenth century it had also been the centre of the earliest Hurufi community.[180] The founder, Faḍl-Allāh Āstārābādī (d. 896/1394), who claimed to be the promised messiah, came to Isfahan three times (significantly Faḍl-Allāh settled in the Ṭuqčī district, at the exact opposite of the citadel). It was in this city that he found his more devoted partisans, such as in 'Alī al-A'lā, and where he began to divulge his doctrine. After Faḍl-Allāh's death, the Timurid prince Iskandar, who had made Isfahan his capital, surrounded himself with scholars of Shi'i and even Hurufi leaning, such as

must admit that the *muḥtasib*, who symbolised the government's role in the economic sphere, is also invisible in our sources, even though he was there.

177 Juwaynī, *Tārīḫ-i Jahān-gušā*, ed. Muḥammad Qazwīnī, 3 vols (Leyden: Brill, 1912-37): II, p. 247, line 19; trans. John A. Boyle, *The History of the World Conqueror*, 2 vols (Manchester: Manchester University Press, 1958): II, p. 511. Later, Juwaynī (II, p. 249, line 14) says that Mengü Bolad's orders were not obeyed in Tabriz, which confirms that he was not seen as a legitimate representative of the people.

178 The antiquity of the guilds is still debated. Cahen, "Mouvements populaires": p. 27, opines that they did not exist in the early centuries of Islam. More recently scholars have tried to find guilds in Sasanid Iran, see Floor, "Aṣnāf": p. 773.

179 Boucheron and Menjot, "La ville médiévale" : p. 445.

180 On the Ismailis of Isfahan, see Durand-Guédy, "Ismailism". On the Hurifiyya in Isfahan, see Ažand, Ya'qūb, *Ḥurūfiyya dar tārīḫ* (Tehran: Našr-i nay, 1369š.): pp. 19-21, 84-90.

Ṣāʾin al-Dīn ʿAlī Turka.[181] Both movements challenged the Turko-Mongol order. The Ismaili leader Aḥmad Ibn ʿAṭṭāš (independently of Ḥasan-i Ṣabbāḥ) challenged the Saljuq order during the dynastic crisis following the death of Malik-Šāh. After a first period of success, when he held part of the city under his sway, he was defeated by Muḥammad b. Malik-Šāh with the help of the Sunni local elites, in 500/1107. As for the Hurufis, they rose in Isfahan in 835/1431, five years after the failed assassination attempt of Šāhruḫ by Aḥmad Lur and the subsequent repression (Ṣāʾin al-Dīn ʿAlī b. Turka was deported to Herat).[182] Their leader, Ḥājjī Surḫ, was one of the disciples of Āstārābādī. In both cases, the rebels attacked the symbols of the state (the greatest emirs were killed: Bursuq for the Saljuqs, ʿAbd al-Ṣamad for the Timurids), they failed to take the citadel of Ṭabarak, and eventually their leader was skinned alive.[183]

There are obvious similarities between both movements, and according to Henry Corbin doctrinal lineage can be traced. But what Corbin does not discuss is the social basis of these movements. Massignon maintained that the Ismailis had a strong interest in artisanal trades and *futuwwa* groups, and tried to infiltrate them (as one source regarding fifth/eleventh century Baghdad may show). To this, Cahen answered that such interest was in fact limited, and in no way exceptional.[184] We can remark that in both cases, the rebels could present themselves as fighting foreign rulers, and more specifically Turko-Khurasanians.[185]

181 Richard, "Un témoignage inexploité": pp. 47, 59-60. On Ṣāʾin al-Dīn ʿAlī Turka's thought, see Corbin, Henry, *En Islam iranien. Aspects spirituels et philosophiques, t. III, Les fidèles d'amour. Shiisme et soufisme* (Paris: Gallimard, 1972): pp. 233-74. Melvin-Koushki, *Quest*: p. 15, has reassessed Ibn Turka's religious attitude and showed that "he presented his system to the Timurid elite precisely as corrective to both the stultifying conservativism of the scholarly establishment and the antinomian decadence of groups such as the Ḥurūfiyya". Melvin-Koushki's brilliant research succeeds in highlights the occultist tenor of Turka's project, but is not concerned with the social context in which this project took place.

182 Melvin-Koushki, *Quest*, concludes that his association with the Hurufi movement compromised his efforts "to promote lettrism as the universal science for a new age".

183 Durand-Guédy, *Iranian Elites*: pp. 153-81; Ažand, *Ḥurūfiyya*: pp. 87-90.

184 Cahen, "Corporations professionnelles": p. 56.

185 On this see Durand-Guédy, *Iranian Elites*: p. 148. The Turko-Khurasanian dimension of the Timurid state is even more evident as the head of the state remained based in the East.

5 Conclusion

Isfahan declined significantly over the Turko-Mongol period, but unlike many of its rivals during the Saljuq period (Rayy, Nishapur) or the urban centres in the region (Ardistān, Qumša) that were reduced to mere villages, it maintained the characteristics of a city, in terms of demography, urbanism, economic and cultural functions, and the presence of specific social groups. This decline was not irreversible, as demonstrated by the rapid growth of Isfahan after it became the capital of Shah ʿAbbās, while many other cities would only recover in the twentieth century.

Certain phenomena appear remarkably constant over these five centuries. The important role played by the local elite on the political field is striking, but also the contestation of the social order through heterodox movements (Ismaili under the Saljuqs, Hurufi under the Timurids). The presence of warring factions also contributed to the city's reputation over the period, and since there was no denying this fact, local authors such as Kamāl Iṣfahānī transformed it into a title of bravura. The fact is that the factional strife which episodically ruined the city, and the relative autonomy enjoyed by the local elite of Isfahan at many moments of its history, are the two faces of the same phenomenon.

At the same time, the social fabric kept evolving: the rise of the Hanafis under the Saljuqs, and then of the Shiʿis under the Mongols were particularly remarkable, although in the final analysis both effects seem to have arisen from the same cause, of political backing. There is ample material available to write one or several monographs on the premises we have laid out. Each piece of information can and should be compared to what we know for other periods and other places – the only way possible to proceed beyond sterile generalizations. It is our hope that through its dynamic approach, this contribution will serve as a constructive complement to the existing articles in the various encyclopaedias. Even more, that it might demonstrate that the history of medieval Isfahan still offers rich grounds for research, with many sources, and can give rise to projects addressing new questions, taking new perspectives.

Bibliography

Abu'l-Saʿādāt Šufurwa al-Iṣfahānī, *Maṭlaʿ al-ṣabāḥtayn wa majmaʿ al-faṣāḥtayn*, ed. Sayyid Ṣādiq Aškivarī (Qum: Pažūhišgāh-i ʿulūm-i insānī, 1385š.).

Abu'l-Qāsim, ed. by Adam Mez, *Abulkasim, ein bagdâder Sittenbild* (Heidelberg: Carl Winter's Universitätsbuchhandlung, 1902); translated by René Khawam, *24 heures de la vie d'une canaille* (Paris: Phébus, 1998).

Afšārī, Mihrān, *Čahārda risāla dar bāb-i futūwat* (Tehran: Čašmah, 1381š.).

Album, Stephen, "The Coinage of Nūr-Āward, Atabeg of Lur Buzurg, 751-57 H./AD 1350-56", *Museum Notes* 22 (1977): pp. 213-39.

Āl-i Dāvūd, 'Alī, "Bahā' al-Dīn Juvaynī", *Dāyirat al-ma'ārif-i buzurg-i islāmī*, (Tehran), XIII (1383š.): pp. 64-5.

Asnād-i mawqūfāt-i Iṣfahān, 11 vols, ed. Sayyid Ṣādiq Ḥusaynī Aškivarī (Qum: Majma'-i ḏaḫā'ir-i islāmī, 1388š./2009).

Aubin, Jean, *Deux sayyids de Bam au XVe siècle. Contribution à l'histoire de l'Iran timouride* (Wiesbaden: Franz Steiner Verlag, 1956).

Aubin, Jean, "Etudes safavides. I. Šāh Ismā'īl et les notables de l'Iraq persan", *JEHSO*, II (1959): pp. 37-81.

Aubin, Jean, "Comment Tamerlan prenait les villes?", *StIsl*, XIX (1963): pp. 83-122.

Aubin, Jean, "La survie de Shīlāu et la route du Khunj-ō-Fāl", *Iran* VII (1969): pp. 21-37.

Aubin, Jean, "Eléments pour l'étude des agglomérations urbaines dans l'Iran médieval", in Hourani, Albert and Stern, Samuel (eds.), *The Islamic City, Papers on Islamic History I* (Carbondale: Southern Illinois University Press, 1970): pp. 65-75.

Aubin, Jean, "Chiffres de population urbaine en Iran occidental autour de 1500", *Moyen Orient & Océan Indien*, IV (1976-7): pp. 37-54.

Ažand, Ya'qūb, *Ḥurūfiyya dar tārīḫ* (Tehran: Našr-i nay, 1369š.).

Blair, Sheila, "Sufi Saints and Shrine Architecture in the Early Fourteenth Century", *Muqarnas* VII (1990): pp. 35-49.

Boucheron, Patrick and Menjot, Denis, "La ville médiévale", in Jean-Luc Pinol (ed.), *Histoire de l'Europe urbaine, I. de l'Antiquité au XVIIIe siècle. Genèse des villes européennes* (Paris: Seuil, 2003): pp. 287-592.

Boyle, John, "The Capture of Isfahan by the Mongols", reprinted in *The Mongol World Empire* (London: Variorum reprints, 1977).

Browne, Edgar, "Account of a rare manuscript History of Iṣfahán", *Journal of the Royal Asiatic Society of Great Britain and Ireland*, no volume mentioned/1-2 (1901): pp. 411-6 and 661-704.

Bulliet, Richard, *The Patricians of Nishapur. A Study in Medieval Islamic Social History* (Cambridge, Mass., Harvard University Press, 1972).

Cahen, Claude, "Mouvements populaires et autonomisme urbain dans l'Asie musulmane du Moyen-Age", *Arabica* V (1958) and VI (1959): pp. 1-91 (the article has its own pagination).

Cahen, Claude, "Y a-t-il eu des corporations professionnelles dans le monde musulman classique? Quelques notes et réflexions", in Hourani, Albert and Stern, Samuel (eds), *The Islamic City, Papers on Islamic History I*, (Carbondale: Southern Illinois University Press, 1970): pp. 51-64.

Corbin, Henry, *En Islam iranien. Aspects spirituels et philosophiques, t. III, Les fidèles d'amour. Shiisme et soufisme* (Paris: Gallimard, 1972).

Crane, Mary, "A Fourteenth-Century Mihrab from Isfahan", *Ars Islamica*, VII/1 (1940): pp. 96-100.

Crone, Patricia, *The Nativist Prophets of Early Islamic Iran. Rural Revolt and Zoroastrism* (Cambridge: Cambridge Univ. Press, 2012).

de Planhol, Xavier, *Les fondements géographiques de l'histoire de l'Islam* (Paris: Flammarion, 1968).

Doerfer, Gerhard, *Türkische und Mongolische Elemente im Neupersischen*, 4 vols (Wiesbaden: Franz Steiner Verlag, 1963-75).

Drechsler, Andreas, *Die Geschichte der Stadt Qom im Mittelalter (650-1350): politische und wirtschaftliche Aspekte* (Berlin: Klaus Schwarz Verlag, 1999).

Duri, Abd al-Aziz, "Baghdād", in *EI*², I (1960): pp. 894-908.

Durand-Guédy, David, "Iranians at War under Turkish Domination: the Example of Pre-Mongol Isfahan", *IrSt* XXXVIII/4 (2005): pp. 587-606.

Durand-Guédy, David, "The Political Agenda of an Iranian *adīb* at the Time of the Great Saljuqs: Māfarrūkhī's*K. Maḥāsin Iṣfahān* Put into Context", *Nouvelle Revue des Etudes Iraniennes*, I (2008): pp. 67-106.

Durand-Guédy, David, "Kamāl al-Din Esmāʿil", in *EIr* XV (2010): pp. 415-17.

Durand-Guédy, David, *Iranian Elites and Turkish Rulers: A History of Iṣfahān in the Saljūq Period* (London & New York: Routledge, 2010).

Durand-Guédy, David, "An Emblematic Family of Seljuq Iran: The Khujandīs of Iṣfahān", in C. Lange and S. Mecit (eds), *The Seljuqs. Politics, Society and Culture* (Edinburgh: Edinburgh Univ. Press, 2011): pp. 182-202.

Durand-Guédy, David, "Location of Rule in a Context of Turko-Mongol Domination", in D. Durand-Guédy (ed.) *Turko-Mongol Rulers, Cities and City Life* (Leiden-Boston: Brill, 2013): pp. 1-20.

Durand-Guédy, David, "1147. The Battle of Qara-Tegin and the Rise of Azarbayjan", *Der Islam*, XCII/1 (2015): pp. 161-96.

Durand-Guédy, David, "Ismailism in Saljuq Isfahan. A Reassement", in Eva Orthmann & Petra G. Schmidl (eds.), *Science in Context: the* Dustūr al-munajjimīn *and its World* (Berlin: EB-Verlag Dr. Brandt, 2017): pp. 223-61.

Fishel, Walter Joseph, *Jews in the Economic and Political Life of Mediaeval Islam* (London: Franck Cass., 1937).

Fishel, Walter Joseph, "Isfahan: The Story of a Jewish Community in Persia", in *Joshua Starr Memorial Volume* (New York: Conference on Jewish Relations, 1953): pp. 111-28.

Floor, Willem, "Guilds and Futuvvat in Iran", *ZDMG* CXXIV/1 (1984): pp. 106-14.

Floor, Willem, "Aṣnāf", in *EIr* II (1987): pp. 772-78.

Galdieri, Eugenio, *Iṣfahān: Masǧid-i Ǧumaʿ. Photographs and Preliminary Report* (Rome: Istituto italiano per il medio ed estremo oriente, 1972).

Godard, André, "Historique du Masdjid-é Djumʿa d'Iṣfahān", *Athār-é Īrān* (Paris: Geuthner) I/2 (1936): pp. 213-82.

Grabar, Oleg, *The great Mosque of Isfahan* (New York: New York University Press, 1990).

Ibn Baṭṭūṭa, translated by Gibb, H.A.R., *The travels of Ibn Baṭṭūṭa* (Cambridge: Hakluyt Society, 5 vols, 1958-2000).

Jaʿfariyān, Rasūl, "Pīšīna-yi tašayyuʿ dar Iṣfahān", *Mīrāṯ-i jāvīdān*, XIX-XX (1376š.): pp. 17-31, reprint in R. Jaʿfariyān (ed.) *Maqālāt-i tārīḫī* (Qum: Daftar Daftar-i Našr al-Hādī), II (1376š.): pp. 305-38.

Jaʿfariyān, Rasūl, "Abu'l-Saʿādāt Asʿad b. Qāhir-i Šufurwa, ʿālim-i shīʿa iṣfahānī dar qarn-i haftum-i hijrī wa yād-dāštanī dar barā-yi ḥamla-yi Muġūl bi Iṣfahān dar sāl-i 634", in R. Jaʿfariyān, *Maqālāt-i tārīḫī* (Qum: Intišārāt-i Dalīl-i mā), XVI (1386š): pp. 171-87.

Javeri, Mohsen, "The Excavations of Atiq square, Isfahan", in Genito, Bruno and Anaraki, Fariba Saiedi (eds.) ADAMJI *Project. From the Excavation (1972-1978) to the Archive (2003-2010) in the Masjed-e Jom'e, Isfahan* (Tehran: Embassy of Italy, 2011): pp. 117-122.

Juwaynī, *Tārīḫ-i Jahān-Gušā*, ed. Muḥammad Qazwīnī, 3 vols (Leyden: Brill, 1912-37); trans. John A. Boyle, *The History of the World Conqueror*, 2 vols (Manchester: Manchester University Press, 1958).

Ḥamd-Allāh Mustawfī, edited and translated by Guy Le Strange, *The Geographical Part of Nuzhat-al-qulūb* (Leyden and London: Brill & Luzac, 1915 for the text, 1919 for the translation).

Ḫāqānī, *Dīvān*, edited by Ḍiyā Sajjādī (Tehran: Zavvār, 1357š.).

Haneda, Masashi, "La famille Khūzānī d'Isfahan (15e-17e siècle)", *StIr*, XVIII/1 (1989): pp. 77-92.

Hillenbrand, Robert, "Architecture and Politics: the North and South Dome Chambers of the Isfahan *jāmiʿ*" in Edmund Herzig and Sarah Stewart (eds), *The Idea of Iran, VI: The Age of the Seljuqs* (London and New York: IB Tauris, 2014): pp. 149-74.

Holod, Renata (ed.) *Studies on Isfahan: Proceedings of the Isfahan Colloquium sponsored by the Fogg Museum of Art, held at Harvard University, January 21-24, 1974*, special issue of *Iranian Studies*, VII/1-2 (1974).

Humā'ī, Jalāl al-Dīn, *Silsila-yi sādāt wa mušajarāt wa musaṭṭaḥāt. Insāb wa nasab-i imām-zāda-hā-yi-Iṣfahān* (Tehran, Pažūhišgāh-i ʿulūm-i insānī wa muṭāliʿāt-i farhangī, 1390š./2011).

Hunarfar, Luṭf-Allāh, *Ganjīna-yi āṯār-i tārīḫī-yi Iṣfahān* (Isfahan: Imāmī, 1344š.).

Ḥusayn Āvī, *Tarjuma-yi maḥāsin-i Iṣfahān*, edited by ʿAbbās Iqbāl (2nd edition, Isfahan: Sāzmān farhangī va tafrīḥī-yi šahrdārī-yi Iṣfahān, 1385š.),

Kamāl Iṣfahānī, ʿAbd al-Salām b. Abī l-Mujāhid, *Munāẓara-yi Baġdād wa Iṣfahān*, ed. Aḥmad-Riḍā Yalamih-hā (Dihāqān: Dānišgāh-i āzād-i islāmī vāḥid-i Dihāqān, 1391š.).

Kāšifī Sabzawārī, *Futuwwat-nāma-yi sulṭānī*, ed. Maḥjūb, Muḥammad Jaʿfar (Tehran: Intišārāt-i bunyād-i farhang-i Īrān, 1971), trans. Jay R. Crook, *The Royal Book of Spiritual Chivalry = (Futūwat nāma-yi sulṭānī)* (no place: Great Books of the Islamic World, 2000).

Kamalizad, Jaleh, "The Landscape Archaeology in Lanjān Plain of Iṣfahān, Iran, during the Islamic Period (650-1722 AD) (Preliminary report)", *Archaeologies*, XII/1 (2016): pp. 91-109.

Kamaly, Hosseyn, *Four Moments in the Early Islamic History of Isfahan* (unpublished Ph.D. dissertation: Columbia University, 2004).

Kamaly, Hosseyn, "Iṣfahān – vi. Medieval Period", in *Enc. Iranica*, XIII (2006): pp. 641-50.

Khanikof, N. de, "Mémoire sur Khâcâni, poëte persan du XIIe siècle. Seconde partie", *Journal Asiatique*, Sixième série, V (1865): pp. 295-367.

Lane, George E. *Early Mongol Rule in Thirteenth-Century Iran. A Persian Renaissance* (London: Routledge, 2003).

Lambton, Ann, "Iṣfahān. History", in EI^2 IV (1978): pp. 97-105.

Lambton, Ann, "Yazd. 2 History", EI^2 XI (2011): pp. 302-9.

Lambton, Ann, "Economy. v. From the Arab Conquest to the End fo the Il-Khanids", *Enc. Iranica* VIII (1998): pp. 107-32,

Lambton, Ann, "Eqṭāʿ", in *Enc. Iranica* VIII (1998): pp. 521-32.

Limbert, John, *Shiraz in the Age of Hafez. The Glory of a Medieval Persian City* (Seattle: University of Washington Press, 2004).

al-Māfarrūḫī, *K. Maḥāsin Iṣfahān*, ed. J. Ṭihrānī (Tehran: Maṭbaʿa majlis, 1933).

Makdisi, George, "La corporation à l'époque classique de l'Islam", in *The Islamic World from Classical to Modern Times. Essays in Honor of Bernard Lewis*, ed. by C.E. Bosworth and al. (Princeton: Darwin Press, 1989): pp. 193-209.

Mahdawī, Musliḥ al-Dīn, *Dānišmandān va buzurgān-i Iṣfahān*, 2nd ed. with supplements by Raḥīm Qāsimī and Muḥammad Riḍā Nīlūfarišān, 2 vols (Isfahan: Šahrdārī-yi Iṣfahān, 1384š.[/2005-6]).

Manz, Beatrice Forbes, *Power, Politics and Religion in Timurid Iran* (Cambridge: Cambridge Univ. Press., 2007).

Melvin-Koushki, Matthew S., *The Quest for a Universal Science:the Occult Philosophy of Ṣāʾin al-Dīn Turka Iṣfahānī (1369-1432) and Intellectual Millenarianism in Early Timurid Iran* (PhD, Yale Univ., 2012).

Morgan, David, *Medieval Persia 1040-1797* (London & New York: Longman, 1988).

al-Muḫtārāt min al-rasāʾil, ed. fac-simile Īraj Afšār (Tehran: Intišārāt-i anjūman-i āṯār-i millī, 2535š. šahānšāhī.), ed. Īraj Afšār & Ġulām-Riḍā Ṭāhir (Tehran: Bunyād-i mawqūfāt-i duktur Maḥmūd Afšār Yazdī, 1378š.).

Nafīsī, Saʿīd, "Jamāl al-Dīn ʿAbd al-Razzāq", *Armaġān* VI (1921): pp. 109-18 and 153-63.

Ohlander, Erik S. "Inner-worldly Religiosity, Social Structuring and Fraternal Incorporation in a Time of Uncertainty: The *Futuwwat-nāma* of Najm al-Dīn Zarkūb of Tabriz", *British Journal of Middle Eastern Studies* XL/1 (2013): pp. 14-35.

Paone, Rosario, "The Mongol Colonization of the Isfahan region", *Quaderni del seminario di iranistica uralo-altaistica e caucasologia dell' universita' degli studi di Venezia* X (1981): pp. 1-33.

Paul, Jürgen, *Herrscher, Gemeinwesen, Vermittler: Ostiran und Transoxanien in vormongolischer Zeit* (Stuttgart: Franz Steiner Verlag, 1996).

Paul, Jürgen, "L'invasion mongole comme « révélateur » de la société iranienne", in Denise Aigle (ed.), *L'Iran face à la domination mongole* (Tehran: IFRI, 1997): pp. 37-53.

Paul, Jürgen, "Histories of Isfahan", *IrSt* XXXIII/1-2 (2000): pp. 117-32.

Paul, Jürgen, "*Khidma* in the Social History of pre-Mongol Iran", *JESHO* LVII (2014): pp. 392-422.

Petrushevsky, Ilya Pavlovic, "The Socio-Economic Condition of Iran under the Īl-K͟hāns", *Cambridge History of Iran*, V, edited by John Boyle (Cambridge: Cambridge Univ. Press, 1968): pp. 483-537.

Perry, John, "Toward a Theory of Iranian Urban Moieties: the Ḥadyariyya and Ni'matiyya revisited", *IrSt* XXXII/1 (1999): pp. 51-70.

Pfeiffer, Judith, "Confessional Ambiguity vs. ConfessionalPolarization: Politics and the Negotiation of Religious Boundaries in the Ilkhanate", in J. Pfeiffer (ed.), *Politics, Patronage and the Transmission of Knowledge in 13th-15th century Tabriz* (Leiden: Brill, 2014): pp. 129-167.

Pūrjavādī, Naṣr-Allāh, *Zabān-i ḥāl dar 'irfān va adabiyāt-i pārsī* (Tehran: Hirmis, 1385š.).

Qazwīnī Rāzī, *K. al-Naqḍ*, edited by J. Muḥaddiṯ (Tehran: Silsila-yi intišārāt-i anjuman-i āṯār-i millī, 1378š.).

Quiring-Zoche, Rosemarie, *Isfahan im 15 und 16 Jahrhundert. Ein Beitrag zur persischen Stadtgeschichte* (Freiburg: Schwarz, 1980).

Rashīd al-Dīn, ed. Karl Jahn, *Tārīḫ-i-Mubārak-i-Ġāzānī des Rašīd al-Dīn Faḍl Allāh Abī-l-Ḫair: Geschichte der Ilḫāne Abāġā bis Gaiḫātū (1265-1295)* (Prague: Verlag der Deutschen Gesellschaft der Wissenschaften und Künste, 1941).

Richard, Francis, "Un témoignage inexploité concernant le mécénat d'Eskandar Solṭān à Eṣfahān", *Oriente Moderno* XV (1996): pp. 45-72.

Ridgeon, Lloyd, *Morals and Mysticism in Persian Sufism: a History of Sufi-futuwwat in Iran* (London: Routledge, 2014).

Sa'īdī, 'Abbās, "Iṣfahān. I. Juġrāfiyā wa juġrāfiyā-yi tārīḫī", *Dāyirat al-ma'ārif-i buzurg-i islāmī*, IX, ed. by Muḥammad-Kāẓim Bujnūrdī (Tehran: Markaz-i Dāyirat al-ma'ārif-i buzurg-i islāmī, 1379š.): pp. 156-81.

Schimmel, Annemarie, "The Ornement of the Saints: the Religious Situation in Iran in Pre-Safavid Times", *IrSt* VII/1 (1974): pp. 88-111.

Selove, Emily, *Ḥikāyat Abī al-Qāsim. A Literary Banquet* (Edinburgh: Edinburgh Uni. Press, 2016).

Siroux, Maxime, *Anciennes voies et monuments routiers de la région d'Ispahan* (Cairo: IFAO, 1971).

Shojaee, Ali, "Excavations at the Jahānnamā Complex: Urban Archaeology at Isfahan, Iran", *Antiquity* XIC/355 (2017), online publication accessible since 20/01/2017.

Sobieroj, Florian, "Suhrawardiyya", *EI*[2] IX (1998): pp. 784-6.

Ṭihrānī, Abū Bakr, edited by Lugal, Necati and Sümer, Faruk, *Kitab-i Diyarbakriyya. Ak-Koyunlular tarihi* (Ankara: Türk Tarih Kurumu Basımevi, 1964).

Smith, Myron Bement, "Material for a Corpus of Early Iranian Islamic Architecture, II. Manār and Masdjid, Barsīān (Iṣfahān)", *Ars Islamica* IV (1937): pp. 7-41.

Tehnyat, Majeed, "The Role of the Qur'anic and Religious Inscriptions in the Buq'a Pīr-i Bakrān, Isfahan: the Shī'ī Reign of Öljeytü Khudābande in Īlkhānid Iran", *Journal of Qur'anic Studies* X/2 (2008): pp. 111-123.

Wilber, Donald, *The Architecture of Islamic Iran. The Il-Khanid Period* (Princeton: Princeton Univ. Press, 1955).

Yāqūt, *Mu'jam al-buldān*, ed. F. Wüstenfeld, *Jaqut's Geographisches Wörterbuch*, 6 vols (Leipzig: Brockhaus, 1866-73).

Ẓahīr al-Dīn Nīšāpūrī, edited by A. Morton, *The Saljūqnāma of Ẓāhir al-Dīn Nīshāpūrī* (Antony Rowe UK, 2004).

Biographical Note

David Durand-Guedy, PhD 2004 (Aix-en-Provence University), is an historian specialized on eleventh-twelfth-century Iran. He is the author of the award winning monograph *Iranian Elites and Turkish Rulers: A History of Iṣfahān in the Saljūq Period* (Routledge, 2010) as well as many scholarly articles on pre-Mongol Iran. He has also edited several collective volumes, including *Turko-Mongol Rulers, Cities and City-Life* (Brill, 2013). He is currently based in Tehran.

Balkh, from the Seljuqs to the Mongol Invasion

Jürgen Paul
Martin-Luther-Universität Halle-Wittenberg
juergen.paul@orientphil.uni-halle.de

Abstract

The current article deals with the forms of local rule at Balkh in the Seljuq and post-Seljuq period up to the Mongol invasion. At all times we observe relatively high degrees of regional autonomy, in which local rulers were far more than "governors". Balkh was a regional state, or "minimal beylik", at times included within larger imperial structures and at others continuing more on its own. Military manpower was frequently provided by Turkish nomads (Ghuzz), who are seen throughout the period as a powerful regional force. Urban notables (*aʿyān*) played a decisive role in local rule, in particular the qadis, who judged according to sharia rules, and the *raʾīs*, in charge of fiscal and administrative affairs. Besides these office holders the sayyids must also have been important, alongside other Muslim scholars, mostly Hanafi. Balkh therefore is another example for the *amīr-aʿyān* system, as has been described in detail for other Iranian cities of the Seljuq period. In the post-Seljuq era the situation continues but becomes more unstable. Hereditary lines of emirs emerge again and again, but the sources do not offer a clear picture of the chronological and prosopographic details.

The paper draws on general historiography, the extant city history and other narrative and non-narrative sources, as well as numismatic evidence.

Keywords

amīr-aʿyān system – local rule – hagiography – urban notables – Ghuzz – minimal beylik

1 Introduction

One of the most important questions in local history is about power. Who is really in power at the city level: is it the notables or the representative of

 | DOI:10.1163/9789004434332_013

the sultanic government, most commonly the amir? 'Notables' often means Muslim scholars, and Muslim scholars profit from a literary genre of their own – the biographical dictionary, in both its general and regional forms.[1] In some cases, this abundant biographical material permits complex prosopographic research. One of the earliest studies of the kind was Bulliet's book on the notables of Nishapur, or as he calls them, 'patricians'.[2] Roy Mottahedeh in turn studied the notables of Qazvīn and their relationships with Būyid rulers, while David Durand-Guédy analyzed the Isfahan case, where notables were at times the driving factor.[3] Marshall Hodgson has spoken of the *aʿyān-amīr* system as terminological shorthand for a situation where the sultan's representative holds ultimate power, due to his superior military resources, but where it is effectively the notables who do most of the ruling, and indeed the amir cannot hope to stay in control without the consensus of the *aʿyān*.[4] Put the other way round, it is impossible to understand imperial rule without understanding what is going on locally, and continuing from this point of view, an empire can be seen as a conglomeration of local entities which would at times be sufficiently autonomous to be considered local or regional states.[5]

The present contribution is a discussion of urban politics, in particular the relationship between the urban notables and the representatives of sultanic power. The study focuses on the city of Balkh, over the period where this *aʿyān-amīr* system was the rule in much of Iran, meaning from the Seljuq conquest of around 431/1040 to the Mongol one of around 616-7/1220. The examination includes attention to the question of how the presence of nomads, in this case Ghuzz, altered this particular context. The city is viewed along with its hinterland, including the agricultural oasis and pastures. The thesis is that for much of the period under study, Balk was indeed the urban centre of a regional state, which in turn would be part of a larger imperial structure.

---

1 For a recent review of the state of the art, see al-Qāḍī, Waḍād, "In the Footsteps of Arabic Biographical Literature: A Journey, Unfinished", *JNES*, LXVIII/4 (2009): pp. 241-52.

2 Bulliet, Richard, *The Patricians of Nishapur. A Study in Medieval Islamic Social History* (Cambridge: Harvard University Press, 1972).

3 Mottahedeh, Roy, "Administration in Būyid Qazwīn", in Richards, D.S. (ed.), *Islamic Civilisation, 950-1150* (Oxford: Cassirer, 1973): pp. 33-45; Durand-Guédy, David, *Iranian Elites and Turkish Rulers. A history of Iṣfahān in the Saljūq period* (London: Routledge, 2010).

4 Hodgson, Marshall, *The Venture of Islam*, 3 vols. (Chicago: University of Chicago Press, 1974): II, pp. 67-69 and *passim*.

5 Paul, Jürgen, *Lokale und imperiale Herrschaft im Iran des 12. Jahrhunderts. Herrschaftspraxis und Konzepte*, (Wiesbaden: Reichert, 2016) with an extensive discussion of imperial, regional, and local power structures.

Unlike some other cities of Khurasan, Seljuq and post-Seljuq Balkh have as yet received very little specialized study.[6] Recent research on the city has concentrated on the transition from the pre-Islamic to early Islamic period, and has been facilitated by the discovery of sets of documents, in both Arabic and Bactrian, which shed important light on this span of regional history.[7] The central concerns in this research have been religious history in general and the processes of Islamisation in particular. The matter of local rule has as yet been very little discussed, including for the case of Balkh. The current article therefore takes on the task of assessing the autonomy of the local rulers, whether members of the ruling dynasty (in this case the Seljuq family) or other men, who would normally be considered as 'governors'. What I hope to show is that many such personages were far more than what this administrative term might suggest. As stated, the continuous presence of warlike nomads in the region adds a further distinct element to the picture.

A brief remark on the sources is in order. Examining the narrative materials for Balkh, we find them not particularly revealing concerning the matters of interest. In particular, the extant city history, the *Fażā'il-i Balḫ*,[8] does not offer the kind of material that has made prosopographic studies possible for other cities. The relations also concentrate on the earlier periods, and show evident religious-political biases in favor of a militant Hanafi stance.[9] While this work offers less than what we might hope, it still has much to say about the city, and evidently offers advantages compared to the cases for which no city history is known, such as for Marw. However, the researcher is inevitably forced to turn to general historiography, supported at times by glimpses of political

6 Studies on Khurasanian cities include: Pritsak, Omeljan, "Āl-i Burhān", *Der Islam*, XXX (1952): pp. 81-96; Aubin, Jean, "L'aristocratie urbaine dans l'Iran seljoukide: l'exemple de Sabzavâr", in Gallais, P., Riou, Y.J. (eds.), *Mélanges offerts à René Crozet*, I (Poitiers : Société des études médiévales, 1966) : pp. 323-32 ; Bulliet, *Patricians*; Paul, Jürgen, "The Histories of Herat", *IrSt*, XXXIII/1 (2000): pp. 93-115. An earlier study – albeit for the Mongol period – is Petruševskii, Ilya P., "Gorodskaia znat' v gosudarstve Khulaguidov", *Sovetskoe Vostokovedenie*, V (1948): pp. 85-110.

7 Azad, Arezou, *Sacred Landscape in Medieval Afghanistan. Revisiting the Faḍā'il-i Balkh* (Oxford: Oxford University Press, 2013), and de la Vaissière in this volume. For the newly found documents, in Bactrian, see the publications by Nicholas Sims-Williams in the *Corpus Inscriptionum Iranicarum*; for Arabic ones, see Khan, Geoffrey, "Documents arabes du début de l'Islam, récemment découverts dans le Khorassan", in Regourd, Anne (ed.), *Documents et histoire. Islam, VIIe-XVIe siècle* (Geneva: Droz, 2013): pp. 175-93.

8 Balḫī, 'Abdallāh b. 'Umar, Persian rendering of 'Abdallāh Ḥusaynī Balḫī, *Fażā'il-i Balḫ*, ed. 'Abd al-Ḥayy Ḥabībī (Teheran: Intišārāt-i bunyād-i farhang-i Īrān, 1350š.).

9 The source has been presented by Radtke, Bernd, "Theologen und Mystiker in Ḫurāsān und Transoxanien", ZDMG, CXXXVI (1986): pp. 536-69 and recently and extensively by Azad, *Sacred Landscape*.

relationships derived from numismatic evidence. Even then the picture remains patchy, with many unresolved uncertainties.

The presentation of the current research takes a chronological approach, and for this reason the paper opens with a discussion of major events and the presentation of a timeline.

Balkh was one of the four urban centres of medieval Khurasan, the others being Herat, Nishapur and Marw. Balkh was the name of both the city and its oasis, formed by the '18 river (or canal) system' *Haždah Nahr*.[10] The city and oasis are situated between the foothills of the Hindukush mountain range, rising quite steeply from the southern side of the plain to elevations of more than 2000 meters, and the Amu Darya watercourse to the north. Between the northern fringes of the oasis and the river banks is a zone of desert and steppe, in places just 20 kilometers wide. On the other side of the river lies the city of Tirmiḏ: a centre closely related to Balkh, mentioned frequently in the subsequent analyses.

The current paper considers the relationship between the city of Balkh and its hinterland, positing that for much of its history, and particularly during the Seljuq and post-Seljuq periods, it assumed the role of urban centre for a regional polity. Under the Seljuqs, the city first was a dynastic appanage, but later developed into a kind of amirate within the Great Seljuq empire. After the fall of the empire, and some years later the amirate, Balkh then continued as a small state, now ruled by a coalition of Türkmen Ghuzz and their Qarakhanid leaders, under Qarakhitay suzerainty. With the takeover of the Ghurids towards the end of the twelfth century, a kind of local amirate again developed, this time within the Ghurid empire. The Khwarazmians ended this amirate at the beginning of the thirteenth century. The city later fell to the Mongols, ostensibly through conquest, however as was the case in many places, some of the local leaders had gone over to the Mongols.

The region included good grazing grounds and was therefore attractive to nomads. Although there is little direct information on the annual migration routes of the eleventh and twelfth centuries, it can be surmised that the nomads followed the patterns still observed in the nineteenth and twentieth

10 See maps in McChesney, Robert, *Waqf in Central Asia: Four Hundred Years in the History of a Muslim Shrine*, (Princeton: Princeton University Press, 1991). McChesney also describes the geographical situation of the city and the oasis. General information on the geography in Planhol, Xavier de, s.v. "Balkh. i. Geography", in *EIr*, III (1989): pp. 587b-588a. See also Vaissière, Étienne de la, "Early Medieval Central Asian Population Estimates", *JESHO*, LX/6 (2017): pp. 788-817, and Wordsworth, Paul, "The hydrological network of the Balkh Oasis: a landscape archaeological perspective", *Afghanistan*, I/1 (2018): pp. 182-208.

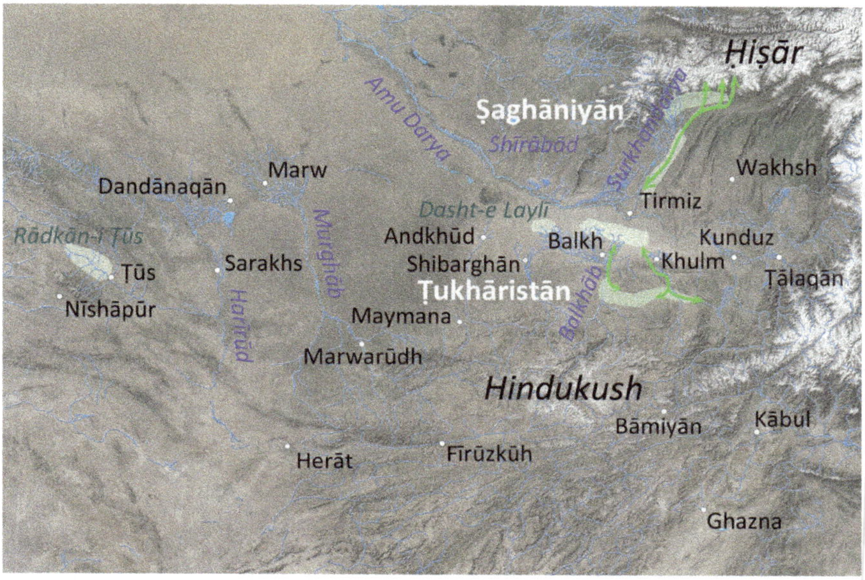

FIGURE 1 Khurasan situation map (pastoral migrations are indicated with green arrows)
DRAWING: PETER VERKINDEREN (BASED ON GOOGLE EARTH)

centuries, involving summer pasturage in the hills and a return to the plains for winter grazing.[11] On the northern side of the Amu Darya, where Tirmiḏ served

11 In particular the Dašt-i Laylī between Andḫūd and Šibarġān, west of Balkh, offers good winter pasture, Jentsch, Christoph, *Das Nomadentum in Afghanistan. Eine geographische Untersuchung zu Lebens- und Wirtschaftsform im asiatischen Trockengebiet* (Meisenheim/Glan: Hain, 1973): p. 40; Balland, Daniel, *Afghanistan: Nomadismus und Halbnomadismus* (Wiesbaden: Reichert, 1989 [Tübinger Atlas des Vorderen Orients A 10.12.7]). However, at least in the recent past, this area was quite distinct from Balkh and its immediate neighborhood. Consideration of areas closer to the city requires accounting for the large areas of arable lands left in fallow. Under crop rotation practices of the mid-twentieth century, one half to two thirds of the lands were kept in fallow, and this ratio could have been even higher in earlier periods. These fallow lands were used for winter grazing. Grötzbach, Erwin, *Afghanistan. Eine geographische Landeskunde* (Darmstadt: Wissenschaftliche Buchgesellschaft, 1990): p. 299; de la Vaissière, "Population Estimates" sees the proportion of fallow as high as three quarters in some oases, including Balkh. By the 1970s only a few nomads wintered near Balkh, a state of affairs that is said to be due to the Afghan conquest of the region after 1849, McChesney, *Waqf*: pp. 257-64; Jentsch, *Nomadentum*: pp. 73-6. Seasonal migrations took the nomads as far as the Balḫāb and other rivers, around 100 km as the crow flies; Jentsch, *Nomadentum*, map *hors texte*, "Weidegebiete und Wanderwege der Nomaden"; see also Balland, *Afghanistan*.

as the urban and military centre, similar nomadic patterns held, with winter pasturage in the riverine lowlands, and summer camps in the higher country.[12]

The current paper proposes Balkh as a case of the regional state which I have elsewhere defined as a 'minimal beylik'.[13] Such a regional state needed an urban centre, an agricultural hinterland, and a sufficient amount of pasture. Rulers of minimal beyliks could recognise other men as overlords, but even in that case would have to rely on their own means to maintain their role. The resources at the disposal of such regional rulers were generally not plentiful, and military manpower in particular could be in short supply. This meant that the ruler frequently had to come to terms with groups that might otherwise represent a threat of uncontrolled military manpower – the Turkic nomads, in the case of Balkh. Regional states centered on Balkh could therefore pursue alternate strategies of fighting the nomads, or of pursuing profitable alliances.[14]

Such regional states could be integrated into larger polities. The regional state could then be a dynastic appanage, in which a member of the central ruling family would act as regional ruler or governor. Other figures, such as leaders with a background in military slavery, also strove to establish hereditary rule, and a principle of heredity was accepted and indeed preferred from the twelfth century down to the Mongol invasion.

12 Adamec, Ludwig, *Historical and Political Gazetteer of Afghanistan. IV: Mazar-i Sharif and North-Central Afghanistan* (Graz: Akademische Druck- und Verlagsanstalt, 1979): pp. 588-9, quotes a British report of 1886, saying that in this region north of the Amu, Uzbek Qungrat nomads raise sheep, camels and horses and that the horses are famous for military value. These Qungrat are described as full nomads and there is only little agriculture in the area. Bregel, Yuri, *An Historical Atlas of Central Asia* (Leiden: Brill, 2003): map 35 "The Özbek tribes in the 19th century" shows Qungrat in this area. While winter pastures were around Tirmiḏ, summer pastures were up the Surkhandarya and Shirabad rivers.

13 Paul, Jürgen, "Mongol Aristocrats and Beyliks in Anatolia. A Study of Astarābādī's *Bazm va Razm*", in Büssow, J., Durand-Guédy, D., and J. Paul (eds.), *Nomads in the Political Field*, special issue of *ES*, IX (2011): pp. 103-56.

14 Fedorov, Michael, "Qarakhanid Coins of Tirmidh and Balkh as a Historical Source. New Numismatic Data on the History of the Qarakhanid Dominions of Tirmidh and Balkh", *The Numismatic Chronicle*, CLXIII (2003): pp. 261-85, sees a much too simple dichotomous relationship between state and nomads: for Fedorov, states are intrinsically bound to fight the nomads. The author never seems to consider alliances, even when explicitly mentioned in the sources, such as the choice of a Qarakhanid prince (Muḥammad b. Maḥmūd) as leader of the Ghuzz.

TABLE 1 Rulers of Balkh from the Seljuqs to the Mongols

Shortly after 431/1040	Seljuq conquest
438/1046-490/1097	Seljuq princely appanage; Various princes, some of them 'rebellious', with nomad support
490/1097-ca. 524/1130	Sanjar, direct rule?
ca 524/1130-558/1163	Qumaj amirate, within the empire under Sanjar, after 552/1157 as a separate entity; Ghuzz alliances with Qumaj and others; Brief occupation by the Qarakhitay after the battle of Qaṭwān (536/1141)
From 545-555/1150s	Ghurid aspirations, with a short Ghurid conquest in 546/1151-2 (perhaps with Ghuzz help)
558/1163-586/1190	Qarakhanid regional state, probably based on Ghuzz; Qarakhitay overlordship; Second Ghurid conquest probably in the early 570s/later 1170s.
586/1190-602/1206	Ghurid period; Perhaps establishment of another line of hereditary amirs
602/1206 or 605/1208-618/1221	Khwarazmian period; forms of regional rule not visible
618/1221	Mongol conquest; Perhaps short-lived return of Ghurid line of hereditary amirs

2 A Seljuq Dynastic Appanage (ca. 431/1040-ca. 524/1130)

Under the Ghaznavid empire Balkh had been an important city, in which the young sultan Masʿūd, as a prince, maintained a palace. The city was known for its fortifications, and the Ghaznavids used it as a starting point for expeditions beyond the Amu Darya.[15] Following the Arab conquest, Balkh no longer served as the urban centre of a larger empire.

15 For the history of Balkh in the early Islamic period, see Azad, *Sacred Landscape*, and de la Vaissière in this volume.

Balkh was not easily conquered by the Seljuqs. After the defeat of Masʿūd at Dandānaqān, in Ramaḍān 431/May 1040, the city resisted under a Ghaznavid commander, until it became clear that no support was forthcoming from his successor Mawdūd. Militiamen from the countryside are known to have joined in the resistance,[16] however there is no specific information on the role of the city residents at Balkh or of the reaction of the urban notables to the Seljuq advance. At Herat the resistance was led by notables;[17] in Nishapur the patriciate was divided, with one faction opposing the Seljuqs and another one siding with the newcomers and eventually prevailing; for Marw the picture is unclear.[18] Since Balkh had a noted citadel, but it is said that the Ghaznavid commander led the defense using militiamen drawn from the countryside, then perhaps the Balkhi notables were unable to reach a decision. In general, as we will see below, Balkhi notables (at least those holding religious offices and other Muslim scholars) had a quietist outlook.

Balkh very quickly became a Seljuq dynastic appanage, as can be seen from the coinage. Coins minted at Balkh indicate the local ruler, alongside the head of the Seljuq family as overlord, and of course the caliph.

The first Seljuqid coins surviving from Balkh were minted with the name of Čaġrï Beg Dā'ūd, starting in 438/1046-7, leaving some years at the onset of the Seljuq period for which we have no material. After this are the coins of

16 Paul, Jürgen, "The Seljuq Conquest(s) of Nishapur: A Reappraisal", *IrSt*, XXXVIII/4 (2005): p. 575. – The Ghaznavid commander who led the defense at Balkh was Altuntaq; Ibn al-Aṯīr, *al-Kāmil*: IX, p. 484. He was invited to join the Seljuqs (but refused); Ḥusaynī, Abu'l-Ḥasan ʿAlī b. Abi'l-Fawāris Nāṣir b. ʿAlī (1933), *Aḫbār al-dawla al-saljūqīya*, ed. Muḥammad Iqbāl (Lahore: University of the Panjab, 1933): p. 12. Amīrak-i Bayhaqī is said to have been the postal and intelligence officer at Balkh (the *ṣāḥib-i barīd*), Bayhaqī, Abu'l-Faḍl Muḥammad b. al-Ḥusayn, *Tārīḫ-i Bayhaqī*, ed. Ġanī/Fayyāż (Tehran: Bānk-i Millī-yi Īrān, 1324): p. 643; *The History of Beyhaqi (The* History *of Sultan Masʿud of Ghazna, 1030-1041) by Abu l-Fażl Beyhaqi*, trans. Clifford Edmund Bosworth and Mohsen Ashtiany, 3 vols. (Boston: Ilex, 2011): II, pp. 342-3. Tirmiḏ is credited with (an unbelievable) fifteen years of resistance; Bosworth has a more reasonable – but still impressive – five years: Bosworth, Clifford Edmund, s.v. "Amīrak-e Bayhaqī", in *EIr*, I (1985): p. 972; the same Amīrak-e Bayhaqī, is reported as commanding at Tirmiḏ, in Ibn Funduq Bayhaqī, Ẓahīr al-Dīn Abu'l-Ḥasan ʿAlī, *Tārīḫ-i Bayhaq*, ed. A. Bahmanyār (Tehran: Furūġī, 1317: p. 120). It is not quite clear how Amīrak managed to take care of two places at the same time. However, in all reports – Ḥusaynī, Bū l-Fażl and Ibn Funduq – the defenders at Balkh and Tirmiḏ surrender after losing hope of any support from the remaining Ghaznavid power.

17 Paul, Jürgen, "The Histories of Herat", *IrSt*, XXXIII/1 (2000): pp. 93-115.

18 Paul, Jürgen, "The Seljuq Conquest(s)"; for Marw: Peacock, Andrew, *Early Seljuq History: A New Interpretation* (London: Routledge, 2010): pp. 90-94.

Alp Arslan, the second sultan, from 440-450/1048-9 to 1058-9, and then Ilyās b. Čaġrï, 454 to 465/1062-3 to 1072-3.[19]

The Seljuq prince Tekeš b. Alp Arslan, at Balkh from 465-477/1072 to 1084-5 (with coins extant from 465 and 466/1072-4),[20] is particularly interesting. This prince arrived in 465/1072-3, as a replacement for Ayāz b. Alp Arslan (for whom no coins are on record). He served as subordinate ruler without problems until 473/1080-1. It is not quite clear what brought him to revolt at that time, but a case can be made for the allegations of Sibṭ b. al-Jawzī, who writes that the Türkmens incited the uprising.[21] This author also vividly describes the events and reasons for the failure of the revolt:

> Tekeš's followers had hidden their belongings and their flocks in the inaccessible mountains, but nevertheless, the sultan's army found them, routed them and drove them away [as booty]. Then Tekeš's followers came to him and told him: "Our belongings have been taken away. If you now asked for a truce they would be returned to us. If not, we'll leave you for the sultan and serve him".[22]

Given this text, it seems that Tekeš relied on Türkmen warriors as military manpower for his revolt, but that these ultimately proved just as interested in their flocks as they were in fighting for the prince (or probably any other local ruler). Tekeš eventually asked for pardon from the sultan in question, who was his brother Malik-Šāh, and indeed received the pardon. The coins show something

19 Schwarz, Florian, *Sylloge Numorum Arabicorum Tübingen. Balḫ und die Landschaften am oberen Oxus. XIVc Ḫurāsān III* (Tübingen-Berlin: Wasmuth, 2002): p. 92. Album, Stephen, *A Checklist of Islamic Coins* (Santa Rosa: Album, 1998) is not helpful because of omitting data on the mints. Ayāz b. Alp Arslan, mentioned as appanage prince at Balkh in the narrative sources, is not known from the coins. It is Sulaymān b. Čaġrï Beg Dā'ūd (one of Alp Arslan's brothers) who received the Balkh appanage in 458/1066, whereas Ilyās b. Čaġrï Beg was given Ṣaġāniyān and Ṭuḫāristān, Ibn al-Aṯīr, *al-Kāmil fi'l-ta'rīḫ*, IX-XII, ed. Tornberg (reprint, Beyrouth: Dār Ṣādir, 1982): X, p. 50. It was uncommon to separate Balkh from Ṭuḫāristān. The appanages were awarded at Rādkān-i Ṭūs, a place noted for its summer pastures, in the beginning of Alp Arslan's reign (1065-72). These grounds became imperial pastures in the Khwarazmian and Ilkhanid periods (late twelfth-early fourteenth centuries).

20 Schwarz, *Sylloge*: p. 94, no. 759 dated 465/1072-3; coinage attested until 476/1083-4.

21 Paul, Jürgen, "Arslān Arghūn – nomadic revival?", in Lange, Christian and Songül, Mecit (eds.), *The Seljuqs. Politics, Society and Culture* (Edinburgh: Edinburgh University Press, 2011): p. 103; Sibṭ b. al-Jawzī, *Mir'āt al-zamān fī ta'rīḫ al-a'yān*, ed. Ali Sevim (Ankara: Türk Tarih Kurumu Basımevi, 1968): p. 209.

22 Sibṭ, *Mir'āt*: p. 209.

of this movement: after an interval for which we have no coins, Tekeš reappears in 476/1083-4, as Šihāb al-Dawla Tekeš Arslan.

In spite of his initial defeat, Tekeš led a second revolt in 477/1084-5, described as a typical Ghuzz raid with unrestricted looting of every town the rebels could possibly seize, including Marw al-rūḏ and Marw. The final scenes of the drama once more took place around Balkh, leading to the supposition that the military basis of the uprising was again the Türkmens, with their pastures around Balkh.[23]

On this occasion Tekeš was sent far from Balkh, and did not escape with a pardon. He was blinded, and imprisoned for the remainder of his life at Tikrit, in central Iraq, where his nephew Barkyaruq had him killed in 487/1094.

The reason for this execution was another 'rebellion' in that year (487/1094), when yet another of Alp Arslan's sons succeeded in taking large parts of Khurasan, probably including Balkh. The rebellious prince was Arslan Arġu or Arslan Arġun b. Alp Arslan, who founded an ephemeral regional state in Khurasan, centered on Marw, which lasted from 486/1093 to 490/1097. He had coins minted in Balkh in 487/1094, indicating only the caliph al-Muqtadir without mention of any Seljuq overlord,[24] and also in Nishapur, although the written sources are explicit that he never took that city and did not lay claim to it.[25]

Arslan Arġu seems to have in some way continued his brother Tekeš's initiatives. In fact the reason that Barkyaruq had the latter killed was due to suspicions that Tekeš might go to Balkh "because the people there wanted him".[26] 'People' is *ahl* in the Arabic original, and it is hard to determine which group of people the source has in mind: it could have been the urban notables, but also the military commanders with a basis in the region, including those relying on the Türkmens.

The revolt led by Arslan Arġun was very quickly quelled after this second prince was murdered in what appears to have been a private crime. His son was far too young to be of any consequence, but Barkyaruq then had to fight another Seljuqid prince. Barkyaruq was personally engaged at Balkh for seven months, in suppressing Muḥammad b. Sulaymān b. Čaġrï Beg, called Amīr-i

23 No coins are attested for Tekeš b. Alp Arslan later than 476/1083-4; another Seljuqid is mentioned on coins from 483/1090-1: Maḥmūd b. Alp Arslan; Schwarz, *Sylloge*: p. 94, no. 763-5.

24 Arslan Arġū (name as indicated on coins) is attested on currency from 487/1094, Schwarz, *Sylloge*: p. 94, no. 766-7. Album does not give dates for the coins he quotes from "eastern mints", Album, *Checklist*: p. 85, no. 1681.

25 Album, *Checklist*; Paul, "Arslan Arghūn".

26 Ibn al-Aṯīr, *al-Kāmil*: x, p. 239.

Amīrān, whose father had at one time been governor of Balkh and who received military help from the Ghaznavids.[27]

3 The Qumaj Amirate (ca. 524/1130-558/1163)

The next Seljuqid to rule at Balkh and over all of Khurasan was Sanjar b. Malik-Šāh, first as a subordinate appanage prince 490-511/1097-1118, and then as supreme sultan from 511/1118 until his death in 552/1157.[28] At some point during Sanjar's reign he appointed one of his leading generals as governor of Balkh. This man is known in the sources by his *laqab* 'Imād al-Dīn and the name Qumaj; his father had been appointed as Sanjar's *atabek* by Barkyaruq when Sanjar was made regional lord in Khurasan in 490/1097.[29] It is hard to pinpoint exactly when 'Imād al-Dīn Qumaj took control over Balkh city and province. Ibn Funduq Bayhaqī mentions a shaykh whom this Qumaj had appointed, but who perished in the battle of Qaṭwān, where Sanjar's army suffered a disastrous defeat at the hands of the Qarakhitay, in 536/1141.[30] Qumaj must have therefore been in office some time prior to the battle with the death of the appointed shaykh.[31] Moreover, Ġarnāṭī reports a date of 530/1135-6 for the discovery of the tomb of 'Alī b. Abī Ṭālib, at the village where the shrine of Mazār-i Šarīf would later be erected. If this chronology is correct, then Qumaj must have already been in power by that year, since in the Ġarnāṭī story he is the man in charge at Balkh.[32] It would therefore probably not be far off the mark to presume that Qumaj had been installed at Balkh by around 524/1130.

Qumaj was one of Sanjar's most important amirs, responsible for key commands in major battles. At Qaṭwān, he was taken prisoner together with his

27 Bosworth, Clifford Edmund, s.v. "Balkh. ii. History from the Arab Conquest to the Mongols", in *EIr*, III (1989): pp. 588a-591a.

28 Sanjar is first mentioned on the coinage minted at Balkh in 491/1098, together with Barkyaruq and the caliph al-Mustaẓhir billāh.

29 Ibn al-Aṯīr, *al-Kāmil*: X, p. 235; Nīšāpūrī, Ẓahīr al-Dīn, *Saljūq nāma*, ed. A.H. Morton (Warminster: Gibb Memorial Trust, 2004): p. 38.

30 Biran, Michal, *The Empire of the Qarakhitai in Eurasian History. Between China and the Islamic World* (Cambridge: Cambridge University Press, 2005).

31 Ibn Funduq Bayhaqī, Ẓahīr al-Dīn Abū 'l Ḥasan 'Alī, *Tatimma ṣiwān al-ḥikma*, ed. Muḥammad Šafī' (Lahore: Ishwar Das, 1351š.): p. 125. After Qaṭwān, the Qarakhitay occupied Balkh briefly, see Köymen, Mehmed Altay, *Büyük Selçuklu İmparatorluğu Tarihi. Cilt II: İkinci İmparatorluk Devri* (Ankara: Türk Tarih Kurumu Basımevi, 1954): p. 333, not in Biran, *Qarakhitai*.

32 McChesney, *Waqf*: pp. 27-8; Ġarnāṭī, Abū Ḥāmid al-Andalūsī, *Tuḥfat al-albāb*, ed. Gabriel Ferrand, *JA*, CCVII (1925): pp. 145-8.

son, 'Alā' al-Dīn Abū Bakr, and ransomed at high cost. But 'Imād al-Dīn Qumaj is best known from the sources and in scholarly literature for his role in the conflict with the Ghuzz, in the 1150s.

As is well known, the conflict with the Ghuzz led to the capture of Sanjar, in 548/1153, with the sultan remaining a prisoner until the spring of 551/1156. During this time the Ghuzz ravaged many cities and towns in Khurasan. Among these was Balkh, although there is no detailed report of this particular event.[33] There are various versions of the beginning and background of the conflict. All of them involve the Qumaj amirs of Balkh, and all describe these amirs as acting very independently against the Ghuzz, at times not even informing the sultan of what was going on. Ibn al-Aṯīr is the only author to report that an otherwise unknown local lord, Zangī b. Ḫalīfa, invited the Ghuzz with the intention of using them in his fights against Qumaj. But that misfired, because the Ghuzz changed sides and went over to Qumaj. Qumaj in turn tried to integrate them in his army when he went to war against the Ghurids, however this again went wrong, as the Ghuzz once more joined the erstwhile enemy. The resulting victory by the Ghuzz-Ghurid coalition allowed them to briefly occupy Balkh. But Sanjar came in person to defeat the Ghurid sultan 'Alā' al-Dīn Jahānsūz in 547/1152, and the city was then reintegrated in the empire.[34] The sultan thus had to come to rescue his amir from a war he had started on his own, without permission or order, and possibly without even consulting the sultan.

The final catastrophe started in a similar manner. Once more the Qumaj amirs began the fight against the Ghuzz, again without clear record as to whether Sanjar had been consulted. It was only after the amirs had been defeated and killed that the sultan himself went to war in person, resulting in his eventual capture.

After defeating the Qumaj amirs, but before Sanjar took to the field against them, the Ghuzz established themselves at Balkh. It is not known what exactly they did there, nor do we have any information about the position of the urban notables in this regard. However Sanjar again appointed a young Qumaj as amir. This man was called Abu'l-Fatḥ,[35] and the reasons for his appointment

33 Köymen, *Büyük Selçuklu*: pp. 430-45 has a very detailed report on Ghuzz movements, but does not mention Balkh in particular.

34 Zangī b. Ḫalīfa: Ibn al-Aṯīr, *al-Kāmil*: XI, p. 179. The Ghurids had taken Ghazna in 1150, and their advance to the north should be dated after that. Another story: Ibn al-Aṯīr, *al-Kāmil*: XI, p. 164. Sanjar's governor at Herat, 'Alī Čatrī, had joined the Ghurid at this point, and whereas the Ghurid was treated honorably, 'Alī Čatrī was executed. See Tor, Deborah, "Mamlūk Loyalty. Evidence from the late Seljūq period", *Asiatische Studien*, LXV/3 (2011): pp. 767-96.

35 The appointment deed is preserved in Juwaynī's collection of official documents, Juwaynī, Muntajab al-Dīn Badī' Atabik, *'Atabat al-kataba*, ed. Muḥammad Qazwīnī (Teheran: Širkat-i Sihāmī-yi Čāp, 1329š.): pp. 73ff. See also Horst, Heribert, *Die Staatsverwaltung der*

are clearly stated in the deed of appointment: he had a hereditary right to the position. Soon after the appointment came the final battle of 548/1153, in which the Ghuzz took the sultan prisoner. Over the following next years, the Ghuzz continued to raid larger Khurasan, taking the sultan with them. They retained Balkh as one of their centres, and in a letter allegedly written by the Khwarazmshah Atsïz to the leader Ṭūṭī Beg, he discusses the possibility of the Ghuzz gathering and settling there. Ṭūṭī Beg must have been a central figure around Balkh because Anwarī, Sanjar's panegyrist who was then based at Balkh, dedicated poems to this Ghuzz leader.[36]

The city of Balkh itself suffered from Ghuzz depredations. One of the private letters in Waṭwāṭ's collection offers a literary treatment of the events. The sender apparently lived in Balkh and had been asked how things were going. He answered that the situation was deplorable. Fiendish warriors (*junūd-i šayṭān*, i.e. the Ghuzz) had established themselves, and the consequences for the city and region were described in vivid terms. The author paints a series of scenes of the destruction wrought by the nomads, all from spheres of sedentary life:

> You do not see dwellers of houses there any longer, you do not meet anybody who kindles a fire [such as for smithy or bakery], you do not see anybody kneading dough for bread, or putting a kettle on the stove. The beautiful gardens have become wilderness, places of mourning, the schools and places where nobles gathered have become caves where wolves live, and lairs of lions. The situation which has befallen the quarters and villages (*maḥallāt*) of Balkh is known to the settled people and the nomads.[37]

Other authors describe the disasters of the Ghuzz in the same vein, although it is the destruction of mosques and madrasas and the killing of Islamic scholars that is most often deplored.[38] Samʿānī always gave the names of scholars

---

Großselǧūqen und Ḫōrazmšāhs (1038-1231) (Wiesbaden: Steiner, 1964); Köymen, *Büyük Selçuklu*: p. 455 note 1 gives this man's name as 'Imād al-Dīn Abu'l-Fatḥ Aḥmad b. 'Alā'al-Dīn Qumaj.

36 *Aḥkām-i salāṭīn-i māżī*. MS St Petersburg, Institute for Oriental Manuscripts, C-816: fol. 56b; Anwarī, *Dīwān*, ed. Saʿīd Nafisī (Tehran: Intišārāt-i Sikka, 1364š./1985): p. 99.

37 Waṭwāṭ, Rašīd al-Dīn, *Nāme-hā-yi Rašīd al-Dīn-i Waṭwāṭ*, ed. Qāsim Tūysirkānī (Tehran: Čāpḫāne-yi Dānišgāh-i Tihrān, 1338š.): p. 139, no. 20. This is a *muraṣṣaʿ* letter, a text where the author uses rhetoric language throughout. Therefore, Anwarī's famous poem called "The Tears of Khurasan" in Western scholarship was possibly written from personal experience. Anwarī, *Dīwān*, p. 105.

38 Samarqandī, *Aʿrāż al-siyāsa*, quoted in Bartol'd, V.V., *Turkestan v ėpokhu mongol'skogo našestviia*, I. Teksty (St Petersburg: Tipografiia Imperatorskoi Akademii Nauk, 1900): p. 72; and *Aḥkām* 33b-34a.

killed, enabling Köymen to use his data to reconstruct the swath of destruction reaped by the Ghuzz across Khurasan.[39] The learned authors thus construct the Khurasani history of this period in terms of a sedentary-nomad dichotomy, a perspective that continues to inform modern scholarly literature.

Sanjar escaped from Ghuzz custody in Ṣafar 551/March 1156, when he was either in the city of Balkh or close enough to the Amu Darya to make a river crossing feasible.[40] From here he went to Tirmiḏ, which was not far away and was one of the most important fortresses of Greater Khurasan at the time. Moreover a Qumaj amir was in command there, probably Abu'l-Fatḥ. We can thus conclude that the Qumaj amirate on the upper Oxus still existed, but that with the loss of Balkh to the Ghuzz, the amirs were now restricted to Tirmiḏ and its region. Sanjar left Tirmiḏ for Marw in fall of the same year, and died there in Rabīʿ I 552/May 1157.

At some point during the next years, the Qumaj ʿAlāʾ al-Dīn (who may or may not be identical with Abu'l-Fatḥ) recovered Balkh from the Ghuzz, certainly by 558/1163, because it was from here that he marched on the Ghurid capital Fīrūzkūh, in an attempt to intervene in a succession struggle among the Ghurids, occurring in that year.[41] ʿAlāʾ al-Dīn was killed in battle, and we do not hear of any successor, marking this as the end of the Qumaj amirate at Balkh.

Summarizing the situation for this period, we can conclude that the Qumaj amirs ruled Balkh in the same way a Seljuq prince would have done, with the exception that they were never mentioned on coins. It is indeed remarkable that we have no coins minted at Balkh for the entire arc of time from 532/1137-8 to the early 560s/later 1160s.[42] We have seen that Qumaj rule at Balkh is attested to exactly this period from which we have no coins – from the early 530s/mid-1130s (the year in which a Qumaj amir first appears in Ġarnāṭī's story)

39 Best shown in Köymen's map of the Ghuzz raids across the province; Köymen, *Büyük Selçuklu*: 433. Ghuzz destructions at Balkh are not mentioned in Samʿānī, Abū Saʿd ʿAbd al-Karīm, *al-Ansāb*, ed. Yamānī, 13 vols. (Hyderabad/D.: Dāʾirat al-maʿārif al-ʿuṯmāniyya, 1962-82): s.v. "al-Balḫī".

40 Since this was in March, the Ghuzz were probably still in their winter quarters on the banks of the Amu Darya, which would explain the possibility for Sanjar to escape to the Transoxanian side.

41 Jūzjānī refers to the Qumaj amir ʿAlāʾ al-Dīn-i Sanjarī, the *ṣāḥib-i Balḫ*; Jūzjānī, Minhāj al-Dīn ʿUṯmān b. Sirāj al-Dīn Muḥammad, *Ṭabaqāt-i nāṣirī*, ed. ʿAbd al-Ḥayy Ḥabībī Qandahārī (Quetta: no publisher, 1328): pp. 418-22; 454. Ġarnāṭī, *Tuḥfat al-albāb*, also cites the Qumaj amir by the same name.

42 Schwarz, *Sylloge*: p. 98 no. 793-6. Schwarz dates the earliest coin after Sanjar to "later than 562" (1166-7) because it follows a type known from Samarqandi coins of that date.

until 558/1163 when the Qumaj amirate ended.[43] Balkh had been a very active mint through all the preceding centuries, so this long lack of Balkhi coins cannot be explained by chance. For whatever reason, it seems the Qumaj amirs did not mint coins, not even ones that would be limited to mentioning Sanjar, alongside the caliph.

Apart from the lack of coinage under Qumaj rule, Balkh continued in virtually the same way as a dynastic appanage. Sanjar had no sons, and so had to rely on military slaves for positions that otherwise would have gone to princes. In official correspondence, depending on the generations, the Qumaj amirs are called 'brothers' or 'sons'. We can gain further insight into the status of the amirs from several clues. One of these is their right to use a *sarāparda*, a textile used to mark off a royal enclosure, considered one of the insignia of kingship. This particular right is mentioned in the deed of appointment for Abu'l-Fatḥ, and the context makes it clear that his father and grandfather had been honored in the same way.[44] It is also telling that in writing of his victories over the Ghurids and the Ghuzz in the Kabul region, the Ghaznavid ruler Bahrāmšāh addressed his letter not to Sanjar, but to the Qumaj amir.[45]

The Qumaj amirs also were free to wage their own wars, as all the Ghuzz narratives show, however the source reports can differ in other respects. The sultan was apparently not even consulted, not when the Qumaj amirs went to war against the Ghurids together with the Ghuzz, and not when the amirs went to fight the Ghuzz a bit later. Therefore, for all practical purposes besides the minting of coins, the Qumaj were rulers of a regional state, an amirate (or, a 'minimal beylik') with its centres at Balkh and Tirmiḏ, always struggling to come to terms with the Ghuzz who without doubt were the single most important reservoir of military manpower in the region at that point.

Given that the Qumaj amirate continued until 558/1163, what became of the Ghuzz themselves after 552/1157? They had established themselves in Balkh (the province probably rather than the city itself), and under the year 553/1158-9, Ibn al-Aṯīr relates that they decided to invite the Qarakhanid Maḥmūd b. Muḥammad as their leader. Although Maḥmūd refused, his son Muḥammad accepted and their leadership then continued, until the moment when both

43 Fedorov is not troubled by this absence of coins from Balkh, in his numismatic review beginning with 567/1171-2. Schwarz has the last Sanjari coin at Balkh in the year 532/1137-8 (Schwarz, *Sylloge*: p. 98 no. 791 and 792), but also with little representation from the 520s (ca. 1126-1136).

44 Juvaynī, *'Atabat* : pp. 76-7. *Sarāparda*: Durand-Guédy, David, "The Tents of the Saljuqs", in Id. (ed.), *Turko-Mongol Rulers, Cities and City-Life* (Leiden: Brill, 2013): pp. 159-63.

45 *Aḥkām* 99b-102b. This letter is dated to mid-Muḥarram 544 (around May 20, 1149), thus some years before the Ghuzz defeated the Qumaj amirs and then Sanjar himself.

father and son were blinded and killed by the new lord of Nishapur, al-Mu'ayyid Ay Aba.[46] Ibn al-Aṯīr reports that in the year 558/1163, Marw, Balkh, Herat and Saraḫs were all in the hands of the Ghuzz, and says that the Friday sermon in these cities was held in the name of Sanjar, and secondly in the name of the ruling amir.[47] However, we have seen that 558/1163 is the closing year of the continuation of the Qumaj amirate in Balkh. Given the mention of the ruling amir, one of the Qumaj must have been chosen by the Ghuzz at Balkh as their leader, or imposed himself in that role – whatever this may have meant in a Ghuzz context.[48] The Qumaj amirate in these later years therefore must have been a joint venture between the Qumaj amir and a certain group of Ghuzz.

We have seen that the Ghuzz were highly flexible regarding partners and lords in the years leading up to 1153, changing sides ever so often and accepting Sanjar as well as the Ghurid ruler as lords. After Sanjar's death, their preferred choice was the Qarakhanid Maḥmūd, and failing him they were willing to follow his son Muḥammad. Maḥmūd had been Sanjar's official heir, so this succession must have made sense, particularly to inveterate legitimists. However, what we can infer is that the Ghuzz, in entirety or as separate groups, had very little ongoing commitment to any of the ruling houses in the region. They apparently chose their lords in view of the momentary situation. However, they must have felt that they did need a lord: very rarely do we see one of their own take command.

4 The *aʿyān-amīr* System in the Reign of Sanjar

Before continuing chronologically, a look at the city itself is on order. How was it administered, and what about the urban notables who were so prominent in other Iranian cities in pre-Mongol times?

46 Ibn al-Aṯīr, *al-Kāmil*: XI, pp. 230; 232. Ay Aba was not a Ghuzz captain, as Bartol'd thought and Fedorov repeats: he originated from the corps of military slaves. See Bartol'd, V.V., "Turkestan v ėpokhu mongol'skogo našestviia", in *Sočineniia*, I (Moscow: Nauka, 1963): p. 398 and Fedorov, "Qarakhanid Coins": p. 263. Bartol'd's error in Album, *Checklist*, as well, and also in Bosworth, Clifford Edmund, s.v. "Il Arslān", in *EIr*, XII (2004): pp. 643-4. For al-Mu'ayyid Ay Aba, see Schwarz, Florian, *Der Sulṭān von Ḫurāsān* (unpublished M.A. thesis, Tübingen 1992).

47 Ibn al-Aṯīr, *al-Kāmil*: XI, p. 292.

48 Around the same time (meaning earlier than 555/1160), the Ghuzz at Marw and Saraḫs came under the leadership of Malik Dīnār, one of their own captains, and stopped ravaging the countryside. Malik Dīnār was in fact a vassal of Ay Aba.

The *Fażā'il-i Balḫ*, the extant city history, is rather disappointing on these matters,[49] but offers snippets of information that can be combined with bits from other sources to develop a useful, if imprecise, picture.

In Seljuq times there was a division of power between three personages of the city of Balkh: the amir, the *ra'īs* and the qadi. We can gather this implication from the appointment deed for Abu'l-Fatḥ, in which the young Qumaj is told to listen to complaints carefully, to investigate all matters brought before him, to consult well-informed and reliable men (among the notables), and to send *šar'ī* matters to the qadi's court. Meanwhile, "official matters as well as financial and fiscal ones" were to go to the office of the *ra'īs*.[50]

The office of the qadi and the supreme qadi (*qāżī l-qużāt*) is explained well in the *Fażā'il-i Balḫ*, but this source focuses on the ideal (Hanafi) Muslim, not on the qadi as judge or urban notable. Azad states that during the first centuries of Islam, Balkhi qadis tried to maintain some distance from the Abbasid administration. A certain quietism seems to linger through time, even in situations where the qadi could no longer avoid being seen as a public figure.[51]

The figure of the *ra'īs* is recorded from Ghaznavid times, with one source mentioning that the man filling this role would be an expert in financial and fiscal matters.[52] At Balkh, it is probable that the *ra'īs* was similarly involved in processes of tax assessment and extraction, and the deed of appointment for Abu'l-Fatḥ indicates this. The *Fażā'il-i Balḫ* names only one person in this role:[53] Abu'l-Ḥasan Muḥammad b. al-Ḥusayn al-Ḥusaynī was from a wealthy family of sayyids (possibly with deep local roots), praised for the observance of an ascetic lifestyle in spite of his wealth. The family held the position of *naqīb* (head of the sayyids) on a hereditary basis but in Balkh, Abu'l-Ḥasan was also *ra'īs*. The source's only notation on his behavior in office is that "nobody ever saw him make a mistake or introduce an innovation".[54] An 'innovation' in fiscal affairs would mean the introduction of new taxes or dues. This shows that Abu'l-Ḥasan's activity as a *ra'īs* was mainly about fiscal administration. He died

49 Azad, *Sacred Landscape*, concentrates on earlier periods, and, as her title indicates, the "sacred landscape", partly because the *Fażā'il-i Balḫ* does not really offer much for later times (and nothing at all after ca. 576/1180).

50 Juvaynī, *'Atabat*: p. 79, *šar'īyāt ba-majlis-i qażā firistad wa-rasmīyāt wa-mu'āmalātī wa-dīvānī ba-dīvān-i riyāsat*.

51 Azad, on the quietism of Balkhi ulema: Azad, *Sacred Landscape*: pp. 150-2. The scholar provides a list of qadi, to which can be added: Ḥamīd Abū Bakr b. Muḥammad b. 'Umar b. 'Alī al-Maḥmūdī, appointed in Jumādā I, 547 (August 1152). *Aḥkām* 115b-117a.

52 Paul, Jürgen, *Herrscher, Gemeinwesen, Vermittler: Ostiran und Transoxanien in vormongolischer Zeit* (Beirut-Stuttgart: Steiner, 1996): p. 80; see also Bosworth, "Balkh".

53 Radtke, "Theologen": p. 549.

54 Balḫī, *Fażā'il*: p. 357.

in 537/1142 at a very advanced age, leaving two sons who are both are cited as *naqīb*, and as *ra'īs* of Khurasan.

As elsewhere, the sayyids were an important factor in the urban life. When Malik-Šāh (465-85/1072-92) came to Balkh at the beginning of his reign, the notables came out to meet him. Among these were the sayyids, possibly including Abu'l-Ḥasan as a child or youth. The notables were intent on complaining about Qarakhanid raids: "the Qarakhanid warriors, they said, are evil people, they do not keep contracts and are constantly violating the agreements",[55] which is exactly what we would expect to hear about nomads with whom they had repeatedly tried to come to terms. The nomads are essentially unpredictable, and do not play according to the rules.[56] It was at this time that Malik-Šāh's brother, Tekeš, became amir at Balkh. It is not clear what his brother and predecessor Ayāz had done about Qarakhanid incursions; for their part the Qarakhanids had sent an embassy to Malik-Šāh asking him to respect the Amu Darya as a frontier between the Seljuq and their own domain, so that Tirmiḏ would come under their rule.[57] As we have seen, Tekeš succeeded in winning popular support, but fearing that people in Balkh would favor his claim to the sultanate, he was then killed by his nephew Barkyaruq. It is important to note a surviving record of Tekeš saying funeral prayers for a deceased scholar, in 471/1078-9.[58] An amir who participates in the funeral prayers for scholars, or would even go so far as to say them, does this to show respect. This had been a tradition under the Samanids, and also the Balkhi governors of the tenth century.[59] Perhaps Tekeš was informed of the practice and therefore hastened to the city do the same. At the very least the story is evidence that Tekeš was at times so close to the city as to observe the event of a funeral procession.

Thus, even when ostensibly structured as a province within the well-ordered Great Seljuq empire, Balkh and its hinterland constituted a separate political sphere. Urban affairs, including fiscal and legal matters, were left to the city

---

55 Ḥusaynī, *Aḫbār*: p. 59. This raid is the one mentioned by Bosworth, "Balkh", and dated to 464/1072.

56 See above about the Ghuzz. It is not evident that these raids were conducted by regular Qarakhanid troops; these could have been Qarluq warriors acting with or without Qarakhanid permission. From a Balkhi perspective, all raiders arriving from the northern shore of the Amu Darya would have looked like 'Qarakhanids'. For context, it is important to recall that just a few weeks earlier, the preceding sultan, Alp Arslan, had been killed by a Transoxanian noble during a campaign against the Qarakhanids.

57 Ḥusaynī, *Aḫbār*: pp. 58-9.

58 Radtke, "Theologen": p. 548; Azad, *Sacred Landscape*: p. 163; the index refers to Tekeš as "sultan of Balkh and Ṭukhāristān" (s.v. "Saljūqs"), Azad, *Sacred Landscape*: p. 211. In correct Seljuq nomenclature, he would have been *malik*.

59 Paul, Jürgen, "The Histories of Samarqand", *StIr*, XXII (1993): pp. 88-9.

government, the *raʾīs* and the qadi. When the need arose, the urban notables, including the *naqīb* of the sayyids, led negotiations with the state and military representatives.

We have no accounts of how Balkh behaved under siege following those of the 430s/1040s. The reports of assumption of Seljuq domination after Dandānaqān do not indicate a prominent role for the city notables. Defenses were organized jointly by the Ghaznavid commander and militiamen from the countryside. Of course we cannot automatically assume that the townspeople declined to defend the city walls against siege, or that there were no urban militias.

Still, only the famous report from Nishapur by Abu'l-Faḍl Bayhaqī mentions the engagement of Balkhis in fighting. When the Seljuqs were about to take Nishapur during their first advance in the late 420s/1030s, the old Hanafi qadi of Nishapur spoke in favor of a kind of neutrality, explaining himself on the basis of a story from Balkh. In 396/1006, when the Qarakhanids had briefly taken Nishapur and Balkh, Balkh had resisted, and as a result a bazaar owned personally by the sultan had been burnt down. The sultan was about to ask for compensation, but in magnanimity chose to refrain, simply reminding the Balkhis of the political principle that subjects do not fight (although he would have known that this ideal was not always possible).[60] The qadi closes his reasoning by citing the sultan's own conclusion: that when enemies appeared before the city walls, he was himself in charge of defending it, and if he were incapable of doing so then it would be better to surrender. We cannot verify that this account, attributed to the qadi in Nishapur, corresponds to actual events of the early eleventh century. However Azad asserts that Balkhi Ḥanafīs had a tradition of quietism,[61] which would help explain why the author of the *Fażāʾil-i Balḫ* does not mention military activity against all the oppressors who ruled at Balkh in his time.

---

60 Bosworth, Clifford Edmund, *The Ghaznavids. Their Empire in Afghanistan and Eastern Iran (994-1040)* (Edinburgh: Edinburgh University Press, 1963): pp. 27ff; Bosworth, "Balkh"; Paul, Jürgen, "The Seljuq Conquest(s)".

61 When individuals venture into fighting it was for private jihad, as reported for early heroes of Balkh such as Ibrāhīm b. Adham and Šaqīq al-Balḫī; Azad, *Sacred Landscape*: pp. 128-9 and Paul, Jürgen, "Islamizing Sufis in pre-Mongol Central Asia", in de la Vaissière, Etienne (ed.), *Islamisation de l'Asie centrale. Processus locaux d'acculturation du VIIe au XIe siècle* (Paris: Association pour l'avancement des études iraniennes, 2008): pp. 297-317. The *Fażāʾil-i Balḫ* evidences a certain militant tone in spite of the dominant quietism. It seems no accident that as one of the role models, the text includes ʿAbdallāh b. Mubārak, a central proto-Sufi and prominent advocate of jihad.

5 Qarakhanid-Qarakhiṭay-Ghuzz Alliances (558-586/1163-1190)

As we have seen, the Qumaj amirate ended in or around 558/1163. Coins minted at Balkh reappear after this, now following different models. The rulers mentioned for the decades up to 586/1190 all seem linked to the Qarakhanid house.[62] The Qarakhanid princes also minted coins at other places in the region along the upper Oxus, such as Tirmiḏ and Waḫš, and the names of the princes differ, meaning that there was no single unified Qarakhanid principality or amirate, rather several.

The Qarakhanids evidently came to Balkh as vassals of non-Muslim Qarakhitay overlords, under a working agreement implying that like other subject dynasties, they would be left to administer their realms as they saw fit. At that point the Qarakhitay would simply send embassies to collect their tribute.[63]

The Qarakhitay-Qarakhanid takeover at Tirmiḏ and Balkh seems to be the result of two military campaigns, but under closer scrutiny it results that this may have been only one. Bartol'd quotes an eyewitness report about a Qarakhitay raid on Balkh, said to have taken place in 560/1165,[64] but without any further details. Given that the defeat and death of the last Qumaj lord of Balkh had occurred only shortly earlier (558/1163), there was probably still no military and political leadership among the Ghuzz, and the situation would have been one of a power vacuum. Did the Ghuzz at Balkh then simply go over to the Qarakhanids and the Qarakhitay, who at any rate were by then the most powerful players in the region?

A second report is provided in al-Kātib al-Samarqandī's collection of biographical anecdotes, in this case concerning a winter campaign led by the Qarakhanid Masʿūd (Qïlïč Ṭamġač Khan) across the frozen Amu Darya. Balkh is not mentioned as a specific target, but the report appears in a context where the Qarakhanid ruler is lauded for his efforts in fighting two groups of nomads:

62 Fedorov, "Qarakhanid Coins", Kochnev (Kotchnev), Boris D., "La chronologie et la généalogie des Karakhanides du point de vue de la numismatique", *CaCe*, IX (2001 [Études karakhanides]): pp. 49-75; Kochnev, Boris D., *Numizmatičeskaia istoriia karakhanidskogo kaganata (991-1209). Čast' 1: Istočnikovedčeskoe issledovanie* (Moscow: Sofiia, 2006); Schwarz, *Sylloge*.

63 Biran, *Qarakhitai*.

64 Bartol'd, "Turkestan": p. 399. The source presented by Bartol'd is the *Tārīḫ-i ḫayrāt* by Mūsāvī, which is described as quoting eyewitness testimony from Yūsuf b. ʿAbdallāh al-Andḫūdī. (MS London, Rieu Suppl. Pers. 423, 162a). Azad also mentions this event, but does not give a source for her report and provides a different date. Azad, *Sacred Landscape*: p. 25. – See also Biran, *Qarakhitai*: p. 55.

the Ghuzz who had risen in rebellion against the sultan of Islam (Sanjar),[65] and the Qarluq who had killed the ruler's father Ibrāhīm, near Bukhara, in 551/1156. An inference from the text is that this campaign took place near the beginning of Masʿūd's rule, meaning soon after the year of the father's death.[66] Given that the second report provides no date, it could then be that this campaign is the very one of 560/1165, as indicated in the eyewitness report.[67]

Whatever the circumstances of the takeover, a new Qarakhitay-Qarakhanid government was established in Balkh and Tirmiḏ, taking up where the Qumaj amirs had left off. From the numismatic evidence for Balkh, this period lasted until ca. 586/1190. In fact the numismatic evidence suggests details of who ruled where and when, and the specialists of the material have recently been exploring how the Qarakhanids at Balkh, most of them obscure, might be linked to better known figures.[68] However our interest in this paper is to continue our inquiry into urban politics, in this case to ask what could have been the basis of Qarakhitay-Qarakhanid rule at Balkh, after the Qarakhitay army and their Qarakhanid vassals had withdrawn across the Amu Darya, not too long after 560/1165. In particular, what of the Ghuzz who had supported the Qumaj amirate? Did they all leave for other provinces and easier pastures?

This is improbable, and in fact a report in Jūzjānī shows that many of them must have stayed on. When the Ghurids took Herat in 571/1175-6, followed two years later by Pūšang/Fūšang, not far downstream from Herat along the Harī-rūd, this caused a sea change in the allegiances of local lords. The regional rulers of Sīstān chose to align with the new Ghurid dispensation. The Ghuzz captains (*mulūk-i Ġuzz*) of Gurgān came to the Ghurid court and did obeisance. Likewise, all the local lords of the Khurasan districts dependent on Herat and Balkh went over to the Ghurids.[69] Even at Waḫš, to the north of the Amu Darya,

65 Bartol'd, *Turkestan-Teksty*: pp. 71-2.

66 The victories he gained are likened to adornments on a bride's gown.

67 Fedorov takes this for granted, Fedorov, "Qarakhanid Coins": p. 264.

68 Details in Fedorov, "Qarakhanid Coins", partly in Kochnev, "Chronologie" and *Numizmatičeskaia istoriia*. See Schwarz, *Sylloge*, for the coins themselves. Kochnev, *Numizmatičeskaia istoriia* and "Chronologie" for lists of Qarakhanid rulers. The Balkhi Qarakhanids are mentioned already in Pritsak, Omeljan, "Die Karachaniden", *Der Islam* XXXI (1953): pp. 17-68 (on the basis of written sources). – Ṭoġrïl Tegin, a ruler to whom Anwarī dedicated a number of poems, probably was a Qarakhanid, but no coins seem to be extant mentioning him. Anwarī, *Dīwān*: pp. 79, 82, 257. Ṭoġrïl Tegin is styled *malik* (local ruler) and is credited with having the fanfare played for him three times (p. 257 line 6365). He is perhaps identical with the Ḫāqān Tegin, mentioned as ruler at Balkh, p. 298. Could it be that the form "Ṭoġrïl Tegin" in fact is mistaken for "Tuġan Tegin"? See below at note 87.

69 Jūzjānī, *Ṭabaqāt*: p. 422. The source names the following districts: Ṭālaqān (today Ṭāluqān), Andḫūd, Maymana, Fāryāb, Panjdih, Marw al-rūḏ, Dīžaq, Ḫulm. This covers virtually all

coins were minted in 573/1177-8 in the name of the Ghurid of Bāmiyān, Šams al-Dīn Muḥammad b. Sām (558-588/1163-1192).[70] This does not mean that Waḫš was conquered or occupied by the Ghurids, but that whatever Ghuzz was the local ruler, he preferred a Ghurid as overlord. Indeed the relationship between the Ghurids of Bāmiyān and the Waḫš local rulers must have remained stable through time, because we learn that Bahā' al-Dīn Sām (588-602/1192-1206) was married to a daughter of the ruler at Waḫš, Malik-Šāh.[71]

The Ghuzz around Marw soon found another lord, Sulṭānšāh b. Il Arslan the Khwarazmshah,[72] but one cannot be so sure about the Ghuzz in the region of Balkh.[73] The continued minting of coins with Qarakhanid rulers shows that they probably came back to a Qarakhitay-Qarakhanid allegiance quite soon after the Ghurids had again withdrawn. The last Qarakhanid coins minted at Balkh were for Ibrāhīm b. Ḥusayn, a well-known ruler who is also on record for Samarqand and Bukhara (574-600/1178-1204; in Ferghana 559-574/1164-1178).[74]

Since it is reasonable to assume that Ghuzz nomads stayed in the province in relatively large numbers, the whole of this coinage evidence demonstrates that, at the close of the Qumaj amirate, they allied or adapted themselves to Qarakhanid-Qarakhitay rule. What we have in this period is then not simply a Qarakhanid principality, but a Qarakhanid principality based on the one hand in Qarakhitay overlordship and from the other on Ghuzz support.

---

districts dependent on Balkh in the entire region between the mountains and the Amu Darya.

70 Fedorov, "Qarakhanid Coins": p. 283; Schwarz, *Sylloge*: p. 144 no. 1245-7, coins minted for Abū Bakr Qaraṭuz in 597/1200-01. Khwarazmian coinage at Vaḫš begins only in 612/1215-6 or thereabouts, Schwarz, *Sylloge*: p. 145 no. 1249-1250.

71 Jūzjānī, *Ṭabaqāt*: p. 460.

72 For Sulṭānšāh, see Paul, Jürgen, "Forces and Resources. The failing state of Sulṭānshāh b. Il Arslān Khwārazmshāh", in Bemmann, Jan and Schmauder, Michael (eds), *The Complexity of Interaction along the Eurasian Steppe Zone in the first Millennium AD. Empires, Cities, Nomads and Farmers* (Bonn: Vor- und frühgeschichtliche Archäologie, 2015): pp. 597-622. Sulṭānšāh appears as heir to the Ghuzz leader Malik Dīnār in his regional state centered on Saraḫs and Marw.

73 Some coins from Balkh do not name any ruler other than the caliph, al-Mustaḍī' (r. 566-75/1170-80), Schwarz, *Sylloge*: no. 805-7. It is tempting to link these coins to an interval of interruption in Qarakhanid-Qarakhitay rule. The coins would have been minted without a local ruler because of the unclear situation: the city notables would have hesitated between the Qarakhanid and the new Ghurid allegiance, probably not trusting that the rule of the latter would last.

74 Schwarz, *Sylloge*: p. 100.

6 The *aʿyān-amīr* System in the Second Half of the Sixth/Twelfth Century

What befell the city in the period of Qarakhanid-Qarakhiṭay-Ghuzz alliances? We have only bits of evidence, in this case all from the *Fażāʾil-i Balḫ*.

One of the list of noted shaykhs who form the subject matter of the book is al-Ṣadr al-Imām al-ajall al-kabīr [...] Abū Bakr Muḥammad b. Aḥmad (d. 584/1188-9), who also happened to be a personal teacher of the author. Abū Bakr always treated the rulers and their representatives with respect and honor, even though they were unbelievers, given that in those days, our source reports, Balkh was ruled by unbelievers and wrongdoers (*kafara wa-fajara*). One day, Abū Bakr was brought to Tirmiḏ along with other scholars, notables, amirs and fiscal officials from Balkh.[75] When they came close to Tirmiḏ, their military escort made them take off their headgear and proceed bareheaded, with their hands in chains. Being forced to uncover one's head is of course an attack on one's honor, just as much as being chained at the wrist, so these scholars and notables entered Tirmiḏ enormously humiliated. However, the warriors escorting the cortège had not dared to touch Abū Bakr, and so only he came before the commander of the unbelievers with his hands free, and still in his exquisite garment. And the commander said:

> If I suspected that anybody under the blue sky was greater than you, I'd give order to put you in chains as well. But I know that nobody on earth is greater than you. Therefore, they all have refrained from doing you violence.[76]

Sometime later, back in Balkh, the shaykh returned to these events in one of his Friday sermons. By this time he had apparently been asked many times why the Khitay had treated him differently and with respect, so that his honor was not endangered. He explained:

---

75 Balḫī, *Fażāʾil*: p. 372.

76 Balḫī, *Fażāʾil*: p. 372; Azad, *Sacred Landscape*: p. 152. The cosmological metaphor "under the blue sky" for "entire world" conveys a distinct flavor of the steppe. This would probably mean that the commander was intended as a Khitay, and that scholars at Balkh were familiar with steppe ways of thinking, sufficiently so that the author could use this figure as a cultural marker. The episode is also in Biran, *Qarakhitai*: p. 184; the scholar wonders whether the story might refer to the beginning of Qarakhitay rule in the region, thus to 1165 or shortly thereafter. We cannot be certain, however the fact that the source does place datable stories about this shaykh in the late 1170s could point towards a later date for the events.

> Everybody knows that the oppressors (*ẓālimān*) are scorpions and serpents. And you go out and seek the company of such animals, go to war against them, but you do not put on any weapons when doing so. No wonder that you carry wounds of snakes' bites and scorpions' stings. But I have an impenetrable armour, namely God's word: "Whoever relies on God is under His custody".[77] So I relied on His bounty and turned my face towards the sky which is the *qibla* of all ardent prayers and said: "Praise be to God the Destroyer, He who suffices.[78] Lord God, in Your bounty, turn their evil away from me".[79]

The shaykh makes a strong case for confidence in God (*tawakkul*) in an eminently quietist way. It is not up to humans to fight the evildoers, God himself will do everything that is necessary. On the contrary, who goes out to fight them should do so with the appropriate weapons. Material weapons evidently have no value, and it is spiritual purity alone that brings protection from God. Whoever goes to fight evildoers and unbelievers with material weapons therefore mingles with them in a way that is improper for good Muslims. In conclusion, it is fervent and sincere prayer, pronounced in complete sincerity (*iḫlāṣ*) and a state of total confidence in God, which brings effects.[80]

Abū Bakr was called to Tirmid̲ on a further occasion. There, the shaykh had to justify himself for certain sayings (we are not informed of which ones)

---

77 *wa-man yatawakkalu ʿalā llāhi fa-huwa ḥasbuhu,* Qurʾān 65 (*al-ṭalāq*) 3; "Whoso places his trust in God, God shall suffice him", tr. Khalidi.

78 *subḥān allāh al-qāhir al-kāfī.*

79 Balḫī, *Fażāʾil*: pp. 372-3.

80 The prayer in this case is of the *duʿā* type in which God's names can be used depending on the circumstance. In this case, the shaykh uses *al-qāhir al-kāfī*. That could be typical for a *duʿā-yi bad*, praying for somebody's end or downfall. Paul, *Herrscher*: p. 230.

This story is mentioned briefly in Azad, *Sacred Landscape*: p. 152, however the scholar fails to recognize the significance of the spiritual morale given by the shaykh. There is no basis for her statement that "It did not matter whether the rulers were Muslims or not, as long the *ʿulamāʾ* were treated with respect, and their authority remained unchallenged" (ibid.) – this is not mentioned in the story. And the text does not say that this shaykh "was close to kings and sultans even if they were idolaters (*kāfir*)" (ibid.), it has: *wa-dar ġāyat-i taʿẓīm wa-tabjīl būd nazdīk-i mulūk va-salāṭīn va-vulāt agarči kāfir būdandī* (Balḫī, *Fażāʾil*: p. 371), perhaps more properly translated as "Rulers, lords and governors held him in high esteem and fully respected him even if they were idolaters". This shaykh certainly would have avoided the company of rulers as far as possible, all the more so since they were idolaters in his time.

Radtke has a very short notice: he says that this man was deported to Tirmid̲ in the period of the Ghuzz raids, Radtke, "Theologen": p. 549. The period of Ghuzz raids was long over at that point.

which had been reported to the ruler by a slanderer. However, the man who had slandered the shaykh was found dead the following morning. Then, the shaykh was taken to the ruler in the *sarā-yi imārat* (place where the amir was sitting in office). The ruler offered him his own mount (for the return trip) and apologized.[81] Again, it was not an intervention of the shaykh that killed the slanderer. It was God Himself who punished the wrongdoer, even if the messenger who brings the news says: "Arise from your prostration, shaykh, you have killed who needed to be killed".[82] This way of meting out punishment is well known from hagiographic texts.[83]

Another report tells of a confrontation between this same shaykh and a vizier at Tirmiḏ, called Muḥammad-i Ṭabbāḫ (the cook or baker).[84] The two men were apparently old enemies, and one day the vizier swore that he would see to it that the shaykh would never be able to set foot on the *minbar*. "That's easy", the shaykh is reported to have replied. "If you keep me from the *minbar*, I'll get me a basket and some loaves of bread and roam the streets and sell them". This was meant and taken as an insult, and so the matter was brought before the governor (*vālī*) at Tirmiḏ. The governor called on Ṭuġan Khan, a ruler who did nothing without consulting the Ṭabbāḫ. However the dispute was again decided by the shaykh's fervent prayer, this time at the shrine of al-Ḥakīm al-Tirmiḏī.[85] The result of the prayer was that the Turkish ruler was stricken with an eye ailment, which became so bad that the Ṭabbāḫ finally asked to be allowed to return to his original profession (presumably baking and selling bread): it seems he understood that it was he that was the target of the shaykh's activities, and for this his lord would not recover as long as he served as vizier, so he decided to sacrifice his position. The ruler treated the shaykh with respect and let him return to Balkh in all honors. Later, in one of his sermons, the shaykh said that the Ṭabbāḫ was not his enemy, and could not have been so because he was not his equal. It was the Ṭabbāḫ who wanted the fight, so he himself was responsible for the unhappy outcome.[86]

Two rulers known from the coinage could qualify for identification with the Ṭuġan Khan mentioned in the story. One of them appears as Ṭuġan Tegin,

81 Balḫī, *Fażā'il*: pp. 374-5. Not used by Azad.

82 Ibid.: p. 374.

83 E.g. in the post-Mongol stories about Bahā' al-Dīn Naqšband, see Paul, Jürgen, "Hagiographische Texte als historische Quelle", *Saeculum*, XLI (1990): p. 40.

84 By the late twelfth century, 'vizier' no longer means 'minister', instead denoting a kind of general representative of the sultan in a given city or region.

85 See Radtke, "Theologen": p. 549. Radtke comments that this passage demonstrates the existence of such shrine at Tirmiḏ in the pre-Mongol period.

86 Balḫī, *Fażā'il*: pp. 375-7.

together with the caliph al-Mustanjid (r. 555-566/1160-70). No other ruler is cited on the relevant coins, and some can be precisely dated to 561/1166 and later. Fedorov reports the observation of the title "Tirmiẕšāh" on one specimen. Fedorov also points out that this man must be the ruler who ordered the decoration of the shrine of al-Ḥakīm al-Tirmiḏī's.[87] The other ruler known from coinage and open to identification with Ṭuġan Khan is Nāṣir al-Mulk Ṭuġan Khan, who is cited along with the caliph al-Mustaḍī' (566-75/1170-80), and whose name could also be read as Yaġan Khan. Because of the dates in the stories that follow in the source, I think that the later king is more probable. The title of this man also corresponds more exactly to the one given in the story.[88]

The stories in the *Fażā'il-i Balḫ* make clear that responsibility for Balkh lay with the governor at Tirmiḏ. Radtke is therefore correct in stating that Balkh was ruled from Tirmiḏ in the later part of the century.[89] It is interesting to see how, time and time again, notables of Balkh were summoned to Tirmiḏ where evidently the Qarakhanid or other representative of Qarakhitay rule was sitting.

Another story is set precisely in 573 or 574 (1177-1179), the period when the Ghurid expansion into the upper Oxus region was so successful. It is about the king of Kāsān, a town and region in the Ferghana Valley[90] who carried out a raid against Balkh and apparently succeeded in establishing his rule there.[91] Elsewhere, this king is called Naṣr Khan "known as Kāsān".[92] "In the time of the sultan [the sultan of Kāsān] all affairs came under the purview of the Ṣadr-i ajall Šaraf al-Zamān"[93] who was an enemy of the hero of these stories, shaykh Abū Bakr Muḥammad b. Aḥmad, and the author sides with his hero: the author's father was afraid of the Ṣadr-i ajall (who therefore must have been a wicked man). The enmity between Šaraf al-Zamān and Abū Bakr was due

87 Fedorov, "Qarakhanid Coins": pp. 265-6.

88 Fedorov, "Qarakhanid Coins": p. 266 discusses the question of whether the *laqab* Nāṣir al-Dīn belonged to Ṭuġan Khan (Fedorov uses 'Yaġan Khan', following Kochnev's reading of that name on the coins) or Sanjar b. al-Ḥasan. He decides that it belonged to the latter. The evidence in Balḫī would instead speak in favor of Ṭuġan Khan, and one is tempted to take him as the ruler in all the stories related in this part of the *Fażā'il-i Balḫ*.

89 *wa-dar Tirmiẕ vālī būd ki Balḫ-rā ū dāštī wa-nām-i ū Ṭuġān Ḫān būd*, Balḫī, *Fażā'il*: p. 375 *in fine*. The editor proposes to identify this Ṭuġan with a ruler from Marġinān, who is quoted in 'Awfī, Muḥammad, *Lubāb al-albāb*, ed. Sa'īd Nafīsī (Teheran : Ibn Sīnā, 1335): p. 54. He also refers to a Ṭuġan Ḫaqan praised in a verse by Sūzanī, Balḫī, *Fażā'il*: p. 375 note 7.

90 The Qarakhanids had several distinct hereditary lines in that region.

91 None of the rulers of Balkh mentioned on the coins fit this description. Could the raid have been directed against the Ghurids or the local lords who had gone over to them? The Ghurids had considerably expanded their zone of influence in 573/1177-8, see above. Balḫī, *Faẓā'il*: pp. 377-8.

92 Balḫī, *Fażā'il*: p. 37, see Azad, *Sacred Landscape*: p. 152.

93 He also appears as "Šaraf al-Dīn" and as "Sayyid-i ajall", Balḫī, *Fażā'il*: pp. 378 and 37.

to the fact that Šaraf al-Zamān was not of the same religious school – he is blamed for *ḫubṯ-i ʿaqīdat*, "evil belief".[94] Abū Bakr's methods for fighting his enemy are the same as in the stories previously seen. Our hero retires to a sacred spot in the mountains, where he prays fervently through the night. After returning to the city and offering the morning prayer in community, he goes to see the sultan and preach for him; since no trip across the river is mentioned, we can suppose the setting is close to or in Balkh. The sermon so impresses the sultan that he prefers it to that of Šaraf al-Zamān; he explicitly states that the hero reminds him of the brilliant Balkhi scholars of the past, and that Abū Bakr excels even further. The story does not end with the punishment or fall from favor of Šaraf al-Zamān. Instead it concludes with a maxim, said to have been spoken by the hero's son, Tāj al-Dīn:

> People do not know how to approach the courts of sultans and kings and are unable to present petitions to them. Because of that, they return from there unsuccessful and frustrated. – The right way to approach creatures (sultans) is to first approach one's Creator, and it is He who at the end will arrange things according to His bounty.[95]

We understand that the Ṣadr-i ajall really is in charge from an aside about his activities: "He [the *ṣadr*] started repairing the citadel and began convoking dictating sessions in his madrasa".[96] Repairing the Balkh citadel – most probably fortifying it against the Ghurids – is what a story about the last of the three iron gates at the Nawbahār gate is about;[97] the *Fażāʾil-i Balḫ* has it that Šaraf al-Zamān sent it to "Hindūstān" in 575/1179-80. But this is clearly a scribal error for 'Hinduvān' – the name for the citadel, as well as for the gate to the city proper.[98]

What can we make of these stories? They are told in the style of hagiographic writings, and qualify as *tawakkul* stories, in which the shaykh wins a contest or gets out of a difficult situation unharmed because he relies on God alone. All the stories of the *Fażāʾil-i Balḫ* are structured in similar manner, around essentially the same personages. We have the hero, the *šayḫ al-islām*, the author's teacher, then his enemies, who are the sultan's henchmen, and

94 Ibid.: p. 379. *ʿAqīdat* refers to schools in speculative theology rather than schools of law.

95 Ibid.: p. 379.

96 Ibid.: p. 378.

97 This gate was named after the famous Nawbahār, the ancient Buddhist monastery at Balkh, see de la Vaissière's contribution in this volume.

98 Juwaynī, ʿAṭā Malik, *Tārīḫ-i Jahān-gušā*, II, ed. Muḥammad Qazwīnī (Leiden: Brill, 1916): p. 63. For the location of the Hinduwān gate, see Azad, *Sacred Landscape*: plate 12. Azad states (without giving a source) that this gate was taken to Qunduz, see ibid.: p. 153.

finally the person of the sultan himself. The sultan is not an enemy. He is a man capable of listening to the message, who values Islamic learning and scholars, and who as a rule gives good judgment – in the end siding with the hero. The methods which the shaykh uses in fighting his enemies are always the same: he wins because he is not occupied with worldly affairs, and confidence in God is his impenetrable armor. He does not act himself, nor curse his enemies, he simply prays God to let good prevail over evil. God is the one who punishes the wrongdoers, invariably according to the hero's wishes. The stories therefore exemplify the hero's quality of being *mustajāb al-duʿāʾ*, one whose prayers are answered. This is the point in the stories and the reason why the author transmits them.

The background of these narratives – which is what matters when drawing on hagiographic stories as historical sources – is the political situation at Balkh and Tirmiḏ in the 1170s, perhaps specifically the second half of the decade. We must recall that the region was under Qarakhitay (therefore non-Muslim) overlordship, with the immediate rulers being Qarakhanids (Muslims), both of these ruling from north of the Oxus river. It appears that Balkh itself was ruled from Tirmiḏ, at least at times. That should not surprise, given that Balkh would have been on the frontiers of Qarakhitay-Qarakhanid dominions. The Ghurid regions began not much farther south, and we have seen that the latter even managed a foothold north of the Amu Darya (at Vaḫš and Čaġāniyān), and expanded briefly into the upper Oxus region around 573/1177-8. In the west, towards Saraḫs and Marw, we can assume that the Ghuzz captains ruled, probably by and large under Malik Dīnār's overlordship until roughly 575-6/1180, later under Sulṭānšāh.

The Qarakhitay-Qarakhanid rulers were using representatives to govern the city. The Seljuqs had done the same, so this was not new – but the structure was now a bit different. In the Seljuq city, *šarʿī* legal affairs went to the qadi and fiscal and administrative ones to the *raʾīs*, however we no longer see these figures collaborating with the government. Individuals with remarkably disparate backgrounds are cited, and where the source associates them with positions these are court offices such as vizier, or with the title of *ṣadr*.[99] We no longer see a *raʾīs*. As occurred in other cities, such as Bukhara, and also Isfahan, it could be that *ṣadr* replaces *raʾīs* as the title for the central figure in the city's civil life.

99 See Floor, Willem, s.v. "Ṣadr": www.iranicaonline.org/articles/sadr. The title was still new in the twelfth century. It was used for men of high position in the civilian administration, not only for religious dignitaries.

In the story of the Ṭabbāḫ, the baker, we see a man disqualified because of his plebeian background (there were also other unclear accusations). In another story it is the Ṣadr-i ajall Šaraf al-Zamān, a leading scholar and true member of the city notables, who is set aside. We also see cases of city notables summoned as a group to the sultan's court, in Tirmiḏ, including military and civilian-bureaucratic officials, and holders of religious and judicial offices. It seems that the Balkh citadel needed repair, and it was the Ṣadr-i ajall who took the necessary measures. We do not learn whether he wanted to use the citadel for his own purposes, or whether he was simply doing his lord's bidding. Could it be that there was no amir for the citadel, and that the Ṣadr-i ajall was also in charge of military matters? At the very least, the source reports that he undertook repair work for the citadel, and at the same time organized dictation sessions at his madrasa, meaning he managed affairs that earlier would have fallen to the amir, and at the same time was active as a scholar.

The *ṣadr-i ajall* is an intriguing character – but we never read his name, and he cannot be confidently identified in other sources. Azad remarks that Qarakhitay rule over Māwarānnahr was "like a protectorate",[100] and cites Bukhara as an example of the religious establishment acting in partnership with the Qarakhitay. In Bukhara, the family in question were the Āl-i Burhān, who carried the title of *ṣadr*.[101] Azad theorizes that "the same may have been the case in Balkh",[102] and hints that here it might have been the hero of the stories, Muḥammad al-Balḫī, who held the role equivalent to that of the Bukharan *ṣadr*. Muḥammad al-Balḫī is cited with the title *ṣadr* in one place, but normally is called the *šayḫ al-islām*. However Azad's proposal does not match the information given in the source. The *šayḫ al-islām*, the hero of many stories in the *Fażā'il-i Balḫ*, is not particularly close to the rulers, and never acts as a military or political leader. If anything we could infer that it was his enemy, the *ṣadr-i ajall* Šaraf al-Zamān, who served as representative of the Qarakhitay-Qarakhanid lords in Balkh: he is generally cited with the title of *ṣadr*, and had either usurped or been delegated responsibilities in the military sphere. The other possibility is that the city was ruled from a distance, without a Qarakhitay or even Qarakhanid garrison at Balkh (while still having a presence at Tirmiḏ). Such forms of local government were observed in many parts of the Persianate world after the coming of the Seljuqs.[103]

---

100 Qarakhitay administration is best described in Biran, *Qarakhitai*: e.g. pp. 114-128.

101 Azad, *Sacred Landscape*: p. 152 note 177. For the Āl-i Burhān, see Pritsak, "Āl-i Burhān".

102 Azad, *Sacred Landscape*: p. 152 note 177.

103 Durand-Guédy, David, "Ruling from the outside: A new perspective on early Turkish kingship in Iran", in Mitchell, Lynette and Melville, Charles (eds.), *Every inch a king:*

After the raid of 560/1165, which appears to have installed the Qarakhitay, the city experienced a period without further sieges or plundering. The Ghurids made advances in the early 570s/later 1170s but no fights for Balkh are recorded, and any territorial gains would have been mostly due to local lords switching sides, with Ghurid troops perhaps not even showing up in the region. Yet it is doubtful that this period was truly peaceful, and likely that it was not one of prosperity.

There were probably frequent demands for tribute and other contributions. The biography of shaykh Abū Bakr Muḥammad b. al-Muʿtaṣim (d. 572/1176-7; not to be confused with our previous Abū Bakr) reports that in his lifetime, Balkh was subject to all kinds of adversities, including unjust and irregular taxation. This shaykh asked the people of his neighborhood to include him among the numbers who would bear the tax burdens; after first being refused,[104] he eventually succeeded in changing their minds. However these details are not our concern: it is again the background information that counts. What appears is that some taxes were demanded on a neighborhood basis, and that the households could decide who would pay how much. However we cannot discern the purpose of these demands: one possibility would be for repairs to the fortress, although these are indicated as beginning some years after the death of this Abū Bakr.

Although the *Fażāʾil-i Balḫ* was written in the early thirteenth century, and translated into Persian in the Mongol period, its anecdotes and relations extend only to around 576/1180. For our study of the subsequent periods the only sources available are then general historiography and coins, making our interpretation of the urban politics at Balkh still more challenging.

7 Ghurid and Khwarazmian Rule: Rise and Fall of a New Amirate? (586-618/1190-1221)

The threshold year for the next period in the history of city is 586/1190, when the first Ghurid coins were minted at Balkh.[105] They bear the name of the ruler

Comparative studies in kings and kingship in the ancient and medieval worlds (Leiden: Brill, 2013): pp. 325-42.

104 Balḫī, *Fażāʾil*: p. 359.

105 Azad, *Sacred Landscape*, sees the Ghurids as close to the Ghuzz (she hyphenates "Oghuz-Ghūrids" on page 6 and in the index s.v. "Ghūrids" and "Oghuz-Ghūrids", p. 206). Coins: Schwarz, *Sylloge*: p. 100 note 26, correcting the statement made at coin no. 824 that the earliest Ghurid coins from Balkh were minted in 592/1195-6. The last Qarakhanid on Balkhi coinage is Nuṣrat al-Dīn Ibrāhīm b. al-Ḥusayn, a well-known ruler (at Samarqand

of Bāmiyān, Šams al-Dīn Muḥammad b. Mas'ūd (r. 558-588/1163-1192), however we do not know how this person came to power. Jūzjānī credits him with the conquest of the city[106] – and therefore, Ghurid rule at Balkh must have begun before 588/1192 which matches the numismatic evidence. The scholarly literature indicates a later start to the Ghurid period, in 594/1198, evidently because this is the year in which Ibn al-Aṯīr reports them as taking Balkh.[107] Based on the evidence from the coins, confirmed by Jūzjānī, we can conclude that Ibn al-Aṯīr (or his source) recorded an incorrect date.

The Ghurid's enemy at Balkh was a Turk – more precisely a Ghuzz, who had given tribute (*ḫarāj*) to the Qarakhitay. It is interesting to see a Ghuzz as master of the city and the citadel; this confirms the thesis that Qarakhanid rule at Balkh, with Qarakhitay backing, relied ultimately on Ghuzz manpower and support.

We do not learn what became of this Ghuzz emir following the arrival of the Ghurid, marked by the coinage date of 586/1190. However the former Qarakhitay overlords were not prepared to give the Ghurids free rein over Balkh. When the first Ghurid ruler Muḥammad b. Mas'ūd died, the Qarakhitay asked his son and successor Bahā' al-Dīn Sām b. Muḥammad (ruler of Bāmiyān, 588-602/1192-1206) to either vacate Balkh or pay tribute in amount at least equal to the tribute of earlier rulers.[108] The son refused, but we do not learn of any Qarakhitay attempts to pursue the claim.

At some point the Ghurid made one 'Imād al-Dīn 'Umar governor at Balkh. In Ḏu'l-Qa'da of 601/June-July 1205, this 'Imād al-Dīn captured Tirmiḏ from the Qarakhitay, and in this city installed his son as governor.[109] Soon after, in 602/1205-6, the Khwarazmshah Muḥammad b. Tekeš tried to take Tirmiḏ. Now Khwarazmshah Muḥammad was still formally a vassal of the Qarakhitay, competing with the Ghurids over hegemony in eastern Iran, and 'Imād al-Dīn refused to surrender Tirmiḏ, saying that he would turn over the fortress only to its lord, the Ghurid ruler of Bāmiyān. The Khwarazmians told 'Imād al-Dīn that the two Ghurid rulers of Bāmiyān (installed after the death of Bahā' al-Dīn Sām in 602/1205-6) had been taken prisoner, and so were no longer active. 'Imād al-Dīn then agreed to accept Khwarazmian overlordship, under which he was first was allowed to keep his position. Later he was accused of conspiracy and taken to Khwarazm, joining the crowd of regional lords held

1178-1204), on specimens dating 582-5/1186-7 to 1189-90: Ghurid coinage thus follows immediately on Qarakhanid production, without hiatus.

106 Jūzjānī, *Ṭabaqāt*: p. 455.

107 Ibn al-Aṯīr, *al-Kāmil*: XII, p. 134.

108 Ibid.: p. 136.

109 Ibid.: p. 206.

captive there. Later still during the Mongol invasion, Terken Khatun, mother of the Khwarazmshah Muḥammad's, had ʿImād al-Dīn killed.[110] His son, who had maintained a position at Tirmiḏ under similar conditions, met the same fate.

Nasawī's chronicle of Jalāl al-Dīn Mingburnī, the last Khwarazmshah, gives the names of father and son in different forms, in one instance saying that Tāj al-Dīn was the *ṣāḥib Balḫ* and naming a son al-Malik al-aʿẓam as *ṣāḥib Tirmiḏ*,[111] in another indicating ʿImād al-Dīn as the *ṣāḥib Balḫ* and naming Bahrāmšāh as the *ṣāḥib Tirmiḏ*.[112] Did the same *ṣāḥib Balḫ* have more than one son, and could one of these have survived the executions of Terken Khatun?

Nasawī's account of the Mongol invasion also mentions another man, Aʿẓam Malik, as *ṣāḥib Balḫ*. This man joined Jalāl al-Dīn Mingburnī, as he began to build up resistance against the Mongols around Ghazna. Leading a body of Ghurid warriors, he was one of the most important commanders in the celebrated victory over a Mongol army at Parwān in 1221. Immediately afterwards Jalāl al-Dīn's commanders quarreled, ostensibly about booty, but also about precedence. The Qipchaq Turks who were Jalāl al-Dīn's kin made it clear that they were not prepared to treat other warriors as equals. For this, Aʿẓam Malik abandoned Jalāl al-Dīn and departed to his mountain fortress, between Ghazna and India, where a little later the Mongols besieged and probably killed him.[113]

The form of the name "Aʿẓam Malik" leads Grachev to identify this man with a temporary ruler at Balkh, Muḥammad al-Ḥusaynī, Abu'l-Maḥāmid (probably r. 617-8/1220-1), named on the coins as "al-Malik al-aʿẓam". The basis for the identification is the title on these coins, and the fact that Nasawī has Aʿẓam Malik as *ṣāḥib Balḫ*.[114] Even if we accept that this Aʿẓam Malik is the son of the ruler of Balkh Tāj al-Dīn, this does not give much basis to argue for a hereditary line of amirs under Ghurid rule of Balkh. At most we could see this as a possibility. Among other considerations, it is unclear how and when Aʿẓam Malik could have claimed the heritage of Balkh, given the account of the

110 Ibid.: p. 229; Jūzjānī, *Ṭabaqāt*: p. 363; Juwaynī, *Jahāngušā* 2:61ff. For the killing: Nasawī, Šihāb al-Dīn Muḥammad, *Sīrat al-sulṭān Jalāl al-Dīn Mingburnī*, ed. Z.M. Buniyatov (Moscow: Nauka, 1996): p. 78 (chapter 18).

111 Nasawī, *Sīra*: p. 61 (chapter 11).

112 Ibid.: p. 78 (chapter 18).

113 Ibid.: p. 120 (chapter 36), 123 (chapter 37). Juwaynī, *Jahāngušā*, 2:195ff.

114 Grachev, Anton, "The Mongol Conquest of Balkh According to Numismatic Sources", *Journal of the Oriental Numismatic Society*, CCVII (2010): p. 16. Nasawī does not have *malik Balḫ*, but *ṣāḥib*; in Khwarazmian terminology, a regional ruler is *ṣāḥib* (whereas *malik* tended to be a military rank in the Khwarazmian army, equivalent roughly to Mongol *noyan*). Nasawī, *Sīra*: p. 119 (chapter 36). Buniyatov does not identify the son of the Ghurid *ṣāḥib Balḫ* (called al-Malik al-Aʿẓam) and this *ṣāḥib Balḫ* of the Mongol period (called Aʿẓam Malik).

imprisonment and death of his father. However he would not have been the only one to succeed in such claims: another family of local lords ruling at Nasā are known to have carried out such operations.[115]

With the Khwarazmian conquest of around 602/1206, Balkh was integrated into a new imperial framework. The Khwarazmians established other governors at Balkh and Tirmiḏ after imprisoning ʿImād al-Dīn and his son. Tirmiḏ was clearly restored to the Qarakhanids – indeed Qarakhanid coinage of the year 604/1207-8 is extant from here.[116] At Balkh, the Ghurids appear on coinage until the end of Bahāʾ al-Dīn Sām's rule, in 602/1205-6. Khwarazmian coinage starts only in 606/1209-10,[117] however it is possible that Khwarazmians began their rule at Balkh earlier. The last coins minted at Balkh for the Khwarazmians are from 616/1219-20:[118] taking the maximum dates, we would have a period of barely ten to fifteen years for Khwarazmian rule at Balkh. We do not have names for Khwarazmian governors at Balkh, however we know that Muḥammad the Khwarazmshah tried to eliminate local and regional ruling families and replace them with his own men. Some out of his appointees in fact were more beholden to Terken Khatun than to the sultan.[119]

The Mongols reached Balkh in the spring of 618/1221, after taking Samarqand. The military situation had evolved in such a way that many Khwarazmian leaders and regional and local rulers had defected to the Mongols, and this is also what happened at Balkh. Nasawī records an amir called Māhrūy, introduced as "one of the leading men at Balkh", as going over to the victorious Mongols. However, being an amir, this man certainly did not belong to the city notables.[120] The regional ruler of Qunduz also switched to Mongol allegiance at this moment.

---

115 The hereditary rulers of Nasā had likewise been deported and kept captive at Khwarazm; Muḥammad b. Tekeš replaced them with a man whom he moved to Nasā from Utrār. During the Mongol invasion, members of the earlier ruling family of Nasā managed to get back to Nasā and reinstate themselves. The same thing might thus have happened at Balkh.

116 Fedorov, "Qarakhanid Coins": p. 272.

117 Schwarz, *Sylloge*: p. 102. Grachev has 608/1211-2 as the year in which Balkh was "incorporated into the Khwārizmshāh state", Grachev, "The Mongol Conquest": p. 16.

118 Schwarz, *Sylloge*: p. 104; Grachev, "The Mongol Conquest": p. 16.

119 Hanstein, Sebastian, Terken-Ḫatun von Ḫvārazm. Zur Person der Mutter Sultan ʿAlāʾ ad-Dīn Muḥammads und ihrer Darstellung in der arabischen sowie persischen historio-und biographischen Literatur des 13. Jahrhunderts (Unpublished MA thesis, Leipzig 2013).

120 Nasawī, *Sīra*: p. 83, Arabic text 57. *min qudamāʾ Balḫ*.

According to reports in Juwaynī and Ibn al-Aṯīr, Balkh surrendered to the Mongols in the usual fashion.[121] However, soon afterwards there must have been a rebellion, since the city was taken a second time and the inhabitants slaughtered.[122] The city then recovered within some decades.[123]

8 Conclusion

Balkh was the urban center of a region comprising not only the greater oasis irrigated by the Balḫāb and the *Haždah Nahr* system, but also the summer and winter pastures in the mountainous regions and along the banks of the Amu Darya. The whole of the region was one of the major provinces in Khurasan. In the Seljuq period, Balkh and its hinterland was at first a princely appanage, and then under Sanjar developed as a hereditary amirate of the Qumaj family. From the evidence, we can discern that the city itself was governed largely by its notables: the qadi or *qāżī l-qużāt* and the *ra'īs* are indicated explicitly. It is not clear who resided in the citadel, if anyone. The sources show that the Seljuq princes and Qumaj amirs were close enough to the city to come into contact with the townspeople and notables from the oasis villages, however this is not proof that they resided in the citadel. Under Seljuq rule, therefore, Balkh was an example of the *aʿyān-amīr* system, common through much of Iran.[124] This system implied that the city – like so many others – was ruled from afar.

The Qumaj amirate did not collapse during the 548/1153-552/1157 implosion of the eastern Seljuq empire, nor does it seem that the Ghuzz left the region. Ghuzz presence, and intermittently Ghuzz amirs, are attested at Balkh until the end of the twelfth century. Balkh was still ruled from afar – the major centre of the regional state which formed in 1165 was located across the Amu Darya, at Tirmiḏ. The numismatic evidence points to Qarakhanid rule from 1165 to 1190, but information from narrative sources illustrates the qualifications to

121 Juwaynī, *Jahāngušā*: p. 113; Ibn al-Aṯīr, *al-Kāmil*, XII, p. 390. Juwaynī has the *mašāhīr*, the city notables, come out of the city for surrender, with subsequent installation of a Mongol *šiḥna* (governor). In Ibn al-Aṯīr it is the *ahl Balḫ* who accept Mongol rule. *Ahl* can refer to notables, but also to military commanders. Paul, Jürgen, "L'invasion mongole comme 'révélateur' de la société iranienne", in Aigle, Denise (ed.), *L'Iran face à la domination mongole* (Tehran: IFRI, 1997): p. 43 note 24.

122 Grachev, "The Mongol Conquest" has a long discussion about the chronology.

123 Grachev, "The Mongol Conquest": p. 16; Akhmedov, Bori, *Istoriia Balkha* (*XVI-pervaia polovina XVIII v.*) (Tashkent: Fan, 1982): pp. 21-2 with reports from later sources about how the Mongols took Balkh. Both also quote Marco Polo's travel account; he reportedly came to Balkh in 1262 and described it as a great city with many beautiful buildings.

124 Hodgson, *Venture*: II, pp. 62-6.

this rule: the Qarakhanids were backed by their Qarakhitay overlords and had the support of Ghuzz warriors.

Affairs within the city are difficult to assess during this period. The office of *ra'īs* is no longer mentioned, and the list of qadis cannot be extended to the time of Qarakhanid-Qarakhitay rule. New positions emerge. On the one hand, the *šayḫ al-islām* is the hero of the hagiographic stories of our main source. However his interventions in public affairs are achieved only by supplication in private prayer, and thus, at Balkh, the quietist tradition of the Hanafi ulema continues. On the other hand in the 1170s we have the office of *ṣadr*, and descriptions of the person holding it as cooperating very closely with the Qarakhanid-Qarakhitay lords. He is also shown repairing the citadel, meaning he was active in the military sphere. This office thus continues the earlier *ra'īs*.

The role of the *a'yān* as a group would thus seem reduced. Individuals with *a'yān* background could still establish brilliant religious and administrative careers, but this meant that that they severed their links with the *a'yān* milieu. The urban notables return in view during the Mongol invasion, but not very prominently. In sum, Balkhi notables, and Hanafi scholars in particular, seem to have followed a different course from their counterparts in other cities in Khurasan and larger Iran. Nishapur, for example, was beset with factional strife to the point that the continued existence of the city was in danger,[125] but Balkh apparently maintained a still solid Hanafi majority. The main conflicts in Balkh involved those who were prepared to cooperate with any government, and for this were prepared to disregard what others considered to be essential Muslim standards. Balkh was also different from Herat, where city notables took a much more activist stance concerning the political affiliation of their city, in the twelfth and early thirteenth centuries.[126] Nothing of the sort seems to have occurred at Balkh. Instead the city notables preferred a much more quietist attitude. Nonetheless, like other cities in Khurasan, Balkh was the centre for a regional state, sometimes included in a larger imperial structure, sometimes existing more or less on its own.

Bibliographical References

Adamec, Ludwig, *Historical and Political Gazetteer of Afghanistan. IV: Mazar-i Sharif and North-Central Afghanistan* (Graz: Akademische Druck- und Verlagsanstalt, 1979).

Aḥkām-i salāṭīn-i māżī. MS St Petersburg, Institute for Oriental Manuscripts, C-816.

125 Bulliet, *Patricians*.

126 Paul, "Histories of Herat".

Akhmedov, Bori, *Istoriia Balkha (XVI-pervaia polovina XVIII v.)* (Tashkent: Fan, 1982).

Album, Stephen, *A Checklist of Islamic Coins* (Santa Rosa: Album, 1998).

Anwarī, *Dīwān*, ed. Saʿīd Nafīsī (Tehran: Intišārāt-i Sikka 1364š./1985).

Aubin, Jean, "L'aristocratie urbaine dans l'Iran seljoukide: l'exemple de Sabzavâr", in Gallais, P., Riou, Y.J. (eds.), *Mélanges offerts à René Crozet*, I (Poitiers: Société des études médiévales, 1966): pp. 323-32.

ʿAwfī, Muḥammad, *Lubāb al-albāb*, ed. Saʿīd Nafīsī (Teheran: Ibn Sīnā, 1335š.).

Azad, Arezou, *Sacred Landscape in Medieval Afghanistan. Revisiting the Faḍāʾil-i Balkh* (Oxford: Oxford University Press, 2013).

Balḫī, ʿAbdallāh b. ʿUmar, Persian rendering of ʿAbdallāh Ḥusaynī Balḫī (1350), *Fażāʾil-i Balḫ*, ed. ʿAbd al-Ḥayy Ḥabībī (Teheran: Intišārāt-i bunyād-i farhang-i Īrān, 1350š.).

Balland, Daniel, *Afghanistan: Nomadismus und Halbnomadismus* (Wiesbaden: Reichert, 1989 [Tübinger Atlas des Vorderen Orients A 10.12.7]).

Bartol'd, V.V., *Turkestan v ėpokhu mongol'skogo našestviia*. I. *Teksty* (St Petersburg: Tipografiia Imperatorskoi Akademii Nauk, 1900).

Bartol'd, V.V., "Turkestan v ėpokhu mongol'skogo našestviia", in *Sočineniia*, I (Moscow: Nauka, 1963).

Bayhaqī, Abu'l-Faḍl Muḥammad b. al-Ḥusayn, *Tārīḫ-i Bayhaqī*, ed. Ġanī/Fayyāż (Tehran: Bānk-i Millī-yi Īrān, 1324). Trans. by Clifford Edmund Bosworth and Mohsen Ashtiany, *The History of Beyhaqi (The* History *of Sultan Masʿud of Ghazna, 1030-1041) by Abu l-Fażl Beyhaqi*, 3 vols. (Boston: Ilex, 2011).

Ibn Funduq Bayhaqī, Ẓahīr al-Dīn Abu'l-Ḥasan ʿAlī, *Tārīḫ-i Bayhaq*, ed. A. Bahmanyār (Tehran: Furūġī, 1317).

Ibn Funduq Bayhaqī, Ẓahīr al-Dīn Abu'l-Ḥasan ʿAlī, *Tatimma ṣiwān al-ḥikma*, ed. Muḥammad Šafīʿ (Lahore: Ishwar Das, 1351š.).

Biran, Michal, *The Empire of the Qarakhitai in Eurasian History. Between China and the Islamic World* (Cambridge: Cambridge University Press, 2005).

Bosworth, Clifford Edmund, *The Ghaznavids. Their Empire in Afghanistan and Eastern Iran (994:1040)* (Edinburgh: Edinburgh University Press, 1963; 2nd ed. Beirut: Librairie du Liban, 1973).

Bosworth, Clifford Edmund, s.v. "Amīrak-e Bayhaqī", in *EIr*, I (1985): p. 972.

Bosworth, Clifford Edmund, s.v. "Balkh. ii. History from the Arab Conquest to the Mongols", in *EIr*, III (1989): pp. 588a-591a.

Bosworth, Clifford Edmund, s.v. "Il Arslān", in *EIr*, XII (2004): 643-4.

Bregel, Yuri, *An Historical Atlas of Central Asia* (Leiden: Brill, 2003).

Bulliet, Richard, *The Patricians of Nishapur. A Study in medieval Islamic Social History* (Cambridge: Harvard University Press, 1972).

Durand-Guédy, David, *Iranian Elites and Turkish Rulers. A history of Iṣfahān in the Saljūq period* (London: Routledge, 2010).

Durand-Guédy, David, "Ruling from the outside: A new perspective on early Turkish kingship in Iran", in Mitchell, Lynette, Melville, Charles (eds.), *Every inch a king:*

Comparative studies in kings and kingship in the ancient and medieval worlds (Leiden: Brill, 2013): pp. 325-42.

Durand-Guédy, David, "The Tents of the Saljuqs", in Durand-Guédy, David (ed.), *Turko-Mongol Rulers, Cities and City-Life* (Leiden: Brill, 2013): pp. 149-89.

Fedorov, Michael, "Qarakhanid Coins of Tirmidh and Balkh as a Historical Source. New Numismatic Data on the History of the Qarakhanid Dominions of Tirmidh and Balkh", *The Numismatic Chronicle*, CLXIII (2003): pp. 261-85.

Floor, Willem, s.v. "Ṣadr": www.iranicaonline.org/articles/sadr (accessed March 5, 2015).

Ġarnāṭī, Abū Ḥāmid al-Andalūsī, *Tuḥfat al-albāb*, ed. Gabriel Ferrand, *JA*, CCVII (1925): pp. 1-148.

Grachev, Anton, "The Mongol Conquest of Balkh According to Numismatic Sources", *Journal of the Oriental Numismatic Society*, CCVII (2010): pp. 15-19.

Grötzbach, Erwin, *Afghanistan. Eine geographische Landeskunde* (Darmstadt: Wissenschaftliche Buchgesellschaft, 1990).

Hanstein, Sebastian, Terken-Ḫatun von Ḫvārazm. Zur Person der Mutter Sultan 'Alā' ad-Dīn Muḥammads und ihrer Darstellung in der arabischen sowie persischen historio-und biographischen Literatur des 13. Jahrhunderts (Unpublished MA thesis, Leipzig 2013).

Hodgson, Marshall, *The Venture of Islam*, 3 vols. (Chicago: University of Chicago Press, 1974).

Horst, Heribert, *Die Staatsverwaltung der Großselǧūqen und Ḫōrazmšāhs (1038-1231)* (Wiesbaden: Steiner, 1964).

Ḥusaynī, Abu'l-Ḥasan 'Alī b. Abi'l-Fawāris Nāṣir b. 'Alī, *Aḫbār al-dawla al-saljūqīya*, ed. Muḥammad Iqbāl (Lahore: University of the Panjab, 1933). Trans. by C.E. Bosworth, *The History of the Seljuq State*. A Translation with commentary of the *Akhbār al-dawla al-saljūqiyya* (Abingdon-New York: Routledge, 2011).

Ibn al-Aṯīr, *al-Kāmil fi'l-ta'rīḫ*, IX-XII, ed. Tornberg (reprint, Beyrouth: Dār Ṣādir, 1982).

Jentsch, Christoph, *Das Nomadentum in Afghanistan. Eine geographische Untersuchung zu Lebens- und Wirtschaftsform im asiatischen Trockengebiet* (Meisenheim-Glan: Hain, 1973).

Juwaynī, 'Aṭā Malik, *Tārīḫ-i Jahān-gušā*, II, ed. Muḥammad Qazvīnī (Leiden: Brill, 1916).

Juwaynī, Muntajab al-Dīn Badī' Atabik, *'Atabat al-kataba*, ed. Muḥammad Qazvīnī (Teheran: Širkat-i Sihāmī-yi Čāp, 1329š.).

Jūzjānī, Minhāj al-Dīn 'Uṯmān b. Sirāj al-Dīn Muḥammad, *Ṭabaqāt-i nāṣirī*, ed. 'Abd al-Ḥayy Ḥabībī Qandahārī (Quetta: no publisher, 1328).

Khan, Geoffrey, "Documents arabes du début de l'Islam, récemment découverts dans le Khorassan", in Regourd, Anne (ed.), *Documents et histoire. Islam, VIIe-XVIe siècle* (Geneva: Droz, 2013): pp. 175-93.

Kochnev (Kotchnev), Boris D., "La chronologie et la généalogie des Karakhanides du point de vue de la numismatique", *CaCe*, IX (2001; Études karakhanides): pp. 49-75.

Kochnev, Boris D., *Numizmatičeskaia istoriia karakhanidskogo kaganata (991-1209). Čast' 1: Istočnikovedčeskoe issledovanie* (Moscow: Sofiia, 2006).

Köymen, Mehmed Altay, *Büyük Selçuklu İmparatorluğu Tarihi. II: İkinci İmparatorluk Devri*, (Ankara: Türk Tarih Kurumu Basımevi, 1954).

McChesney, Robert, *Waqf in Central Asia: Four Hundred Years in the History of a Muslim Shrine*, (Princeton: Princeton University Press, 1991).

Mottahedeh, Roy, "Administration in Būyid Qazwīn", in Richards D.S. (ed.), *Islamic Civilisation, 950-1150* (Oxford: Cassirer, 1973): pp. 33-45.

Nasawī, Šihāb al-Dīn Muḥammad, *Sīrat al-sulṭān Jalāl al-Dīn Mingburnī*, ed. Z.M. Buniyatov (Moscow: Nauka, 1996).

Nīšāpūrī, Ẓahīr al-Dīn, *Saljūq nāma*, ed. A.H. Morton (Warminster: Gibb Memorial Trust, 2004).

Paul, Jürgen, "Hagiographische Texte als historische Quelle", *Saeculum*, XLI (1990): pp. 17-43.

Paul, Jürgen, "The Histories of Samarqand", *StIr*, XXII (1993): pp. 69-92.

Paul, Jürgen, *Herrscher, Gemeinwesen, Vermittler: Ostiran und Transoxanien in vormongolischer Zeit* (Beirut-Stuttgart: Steiner, 1996).

Paul, Jürgen, "L'invasion mongole comme 'révélateur' de la société iranienne", in Aigle, Denise (ed.), *L'Iran face à la domination mongole* (Tehran: IFRI, 1997): pp. 37-53.

Paul, Jürgen, "The Histories of Herat", *IrSt*, XXXIII/1 (2000): pp. 93-115.

Paul, Jürgen, "The Seljuq Conquest(s) of Nishapur: A Reappraisal", *IrSt*, XXXVIII/4 (2005): pp. 575-85.

Paul, Jürgen, "Islamizing Sufis in pre-Mongol Central Asia", in de la Vaissière, Etienne (ed.), *Islamisation de l'Asie centrale. Processus locaux d'acculturation du VIIe au XIe siècle* (Paris: Association pour l'avancement des études iraniennes, 2008): pp. 297-317.

Paul, Jürgen, "Mongol Aristocrats and Beyliks in Anatolia. A Study of Astarābādī's Bazm va Razm", in Büssow, J., Durand-Guédy D., and J. Paul (eds.), *Nomads in the Political Field*, special issue of ES, IX/1-2 (2011): pp. 103-56.

Paul, Jürgen, "Arslān Arghūn-nomadic revival?", in Lange, Christian and Mecit, Songül (eds.), *The Seljuqs. Politics, Society and Culture* (Edinburgh: Edinburgh University Press, 2011): pp. 99-116.

Paul, Jürgen, "Forces and Resources. The failing state of Sulṭānshāh b. Il Arslān Khwārazmshāh", in Bemmann, Jan and Schmauder, Michael (eds.), *The Complexity of Interaction along the Eurasian Steppe Zone in the first Millennium AD. Empires, Cities, Nomads and Farmers* (Bonn: Vor- und frühgeschichtliche Archäologie, 2015): pp. 597-622.

Paul, Jürgen, *Lokale und imperiale Herrschaft im Iran des 12. Jahrhunderts. Herrschaftspraxis und Konzepte* (Wiesbaden: Reichert, 2016).

Peacock, Andrew, *Early Seljuq History: A New Interpretation* (London: Routledge, 2010).

Petruševskii, Ilya P., "Gorodskaia znat' v gosudarstve Khulaguidov", *Sovetskoe Vostokovedenie*, v (1948): pp. 85-110.

Planhol, Xavier de, s.v. "Balkh. i. Geography", in *EIr*, III (1989): pp. 587b-588a.

Pritsak, Omeljan, "Āl-i Burhān", *Der Islam*, XXX (1952): pp. 81-96.

Pritsak, Omeljan, "Die Karachaniden", *Der Islam*, XXXI (1953): pp. 17-68.

al-Qāḍī, Wadād, "In the Footsteps of Arabic Biographical Literature: A Journey, Unfinished", *JNES*, LXVIII/4 (2009): pp. 241-52.

Radtke, Bernd, "Theologen und Mystiker in Ḫurāsān und Transoxanien", *ZDMG*, CXXXVI (1986): pp. 536-69.

Samʿānī, Abū Saʿd ʿAbd al-Karīm, *al-Ansāb*, ed. Yamānī, 13 vols. (Hyderabad/D.: Dā'irat al-maʿārif al-ʿuṯmāniyya, 1962-82).

Schwarz, Florian, Der Sulṭān von Ḫurāsān (unpublished M.A. thesis, Tübingen 1992).

Schwarz, Florian, *Sylloge Numorum Arabicorum Tübingen. Balḫ und die Landschaften am oberen Oxus. XIVc Ḫurāsān III* (Tübingen, Berlin: Wasmuth, 2002).

Sibṭ b. al-Jawzī, *Mir'āt al-zamān fī ta'rīḫ al-aʿyān*, ed. Ali Sevim (Ankara: Türk Tarih Kurumu Basımevi, 1968).

Tor, Deborah, "Mamlūk Loyalty. Evidence from the late Seljūq period", *Asiatische Studien*, LXV/3 (2011): pp. 767-96.

Vaissière, Étienne de la, "Early Medieval Central Asian Population Estimates", *JESHO*, LX/6 (2017): pp. 788-817.

Waṭwāṭ, Rašīd al-Dīn, *Nāma-hā-yi Rašīd al-Dīn-i Waṭwāṭ*, ed. Qāsim Tūysirkānī (Tehran: Čāpḫāna-yi Dānišgāh-i Tihrān, 1338š.).

Wordsworth, Paul, "The hydrological network of the Balkh Oasis: a landscape archaeological perspective", *Afghanistan*, I/1 (2018): pp. 182-208.

Biographical Note

Jürgen Paul is Emeritus Professor of Islamic Studies at Martin-Luther-Universität Halle-Wittenberg. He specialises in the history of medieval Iran and Central Asia, focusing on forms of local rule and the relations between local and imperial rule. Recent publications include: *Lokale und imperiale Herrschaft im Iran des 12. Jahrhunderts: Herrschaftspraxis und Konzepte* (Wiesbaden, Reichert, 2016); "Nomads and Bukhara; A Study in Nomad Migrations, Pasture, and Climate Change (11th Century CE)", in *Der Islam* XCIII/2 (2016); and "Khidma in the Social History of Pre-Mongol Iran", in *JESHO* LVII/3 (2014).

Local Elites and Dynastic Succession: Tabriz prior to, under and following Mongol Rule (Sixth/Twelfth to Ninth/Fifteenth Centuries)

Daniel Zakrzewski
Center for Near and Middle Eastern Studies, Philipps-Universität Marburg
zakda@web.de

Abstract

This article analyses the history of Tabriz from the late sixth/twelfth to the end of the ninth/fifteenth century. It develops the thesis that the local elites played an active and important role in determining a specific sequence of dynasties passing through the Mongols to the Safavids. Through a focus on two leading families, the analysis elucidates how Mongol rule transformed local society. Urban elites generally retained their status throughout the period, while rural elites gained new influence beginning with the time of Mongol rule.

Keywords

Tabriz – Leading Families – Mongols – Social Change – Sufism

1 Introduction

Mongol impact on the cities of medieval Iran has long been intensely debated. Specialists in Iranian history emphasise that many centres were destroyed during the first invasion of 617-8/1220-1 and that urban life continued to decline during the decades of non-Muslim rule following soon after.[1] Specialists in

1 Lambton, Ann K.S., *Continuity and Change in Medieval Persia. Aspects of Administrative, Economic and Social History, 11th-14th Century* (London: I.B. Tauris, 1988): pp. 15-20, 25-6; Avery, Peter, *The Spirit of Iran. A History of Achievement from Adversity* (Costa Mesa: Mazda Publishers, 2007): pp. 412-3; Katouzian, Homa, *The Persians. Ancient, Medieval and Modern Iran* (New Haven and London: Yale University Press, 2009): pp. 100-5; Pistor-Hatam, Anja, "History and its Meaning in the Islamic Republic of Iran: The Case of the Mongol invasion(s)

 | DOI:10.1163/9789004434332_014

Mongol history recognise the destruction of cities and exploitation of the settled population, but now tend to place greater stress on the positive aspects of economic, cultural and scientific exchanges triggered by Iran's incorporation into the empire of Chinggis Khan (d. 624/1227) and his descendants.[2]

A point of accord is that Tabriz first developed as a major city under Mongol rule in Iran. Tabriz is well known as the principal urban centre of the Ilkhanid dynasty (654/1256-736/1335). Chinggis Khan's grandson Hülegü (d. 663/1265) founded this ruling house at the time of the middle-eastern campaign that led to the Mongol conquest of Baghdad (656/1258) and the elimination of the Abbasid Caliphate. The Ilkhan most closely associated with Tabriz was the conqueror's great-grandson Ġāzān (d. 703/1304), who embraced Islam shortly before ascending to the throne in 694/1295, and ultimately secured the conversion of the dynasty. Ġāzān had a massive pious endowment (*waqf*) complex erected outside the city, including the mausoleum of the ruler and a number of other structures. After the collapse of the house of Hülegü in the mid-eighth/fourteenth century, several successor dynasties of the Ilkhans likewise accorded Tabriz the distinction of being the principal urban centre of their realms. Finally the Safavids became one of these successor dynasties, by taking possession of the city in 907/1501.

This article concentrates on the local elites or 'notables' of Tabriz from the late sixth/twelfth century to the end of the ninth/fifteenth, and in particular on two leading families. The intention is to shed new light on the history of the city prior to, under and following Mongol rule, and thereby also on more general issues, such as the history of Sufism. The article advances the specific thesis that the choices made by the local elites to achieve their political interests were crucial in determining the sequence of dynastic succession passing through the Mongols to the Safavids.

At this point, it is necessary to introduce the two leading families of Tabriz who will be in the focus of the discussion. One, which came to bear the name of Malikān in the course of the seventh/thirteenth century, produced some of

---

and rule", in Ansari, A. M. (ed.), *Perceptions of Iran. History, Myths and Nationalism from Medieval Persia to the Islamic Republic* (London: I.B. Tauris, 2014): pp. 147-62, here pp. 150-4.

2 Allsen, Thomas T., *Culture and Conquest in Mongol Eurasia* (Cambridge: Cambridge University Press, 2001): pp. 54-6, 193-5; Biran, Michal, *Chinggis Khan* (Oxford: Oneworld Publications, 2007): pp. 85-93; Lane, George, "Persian Notables and the Families Who Underpinned the Ilkhanate", in Amitai, R. and Biran, M. (eds.), *Nomads as Agents of Cultural Change: The Mongols and Their Eurasian Predecessors* (Honolulu: The University of Hawaii Press, 2015): pp. 183-213, here p. 208.

the city's most eminent notables, prior to and under Mongol rule. The other, which became known as the Kujujīs, rose to prominence under the Mongols and remained highly influential through the subsequent centuries. The discussion traces the overall chronology of events, paying particular attention to the involvement of members of the Malikān and the Kujujī families in politics. The next section sets out the scholarly framework for the study.

2 Guiding Questions, Scholarship, Sources

Historical studies of Iran, including those regarding the periods before and after Mongol rule, have shown that the notables of major urban centres purposefully interceded in politics at both the local and imperial levels.[3] They would render loyal service to established rulers and dynasties, but also selectively support or oppose rival claimants in times of conflict and disputed succession. The Mongol conquests occurred in just such a period, when intra and inter-dynastic struggles were taking place throughout the Iranian plateau and adjacent lands. Jürgen Paul has demonstrated that during the first invasion, in the time of Chinggis Khan, reactions to the non-Muslim armies varied according to local conditions.[4] The case of Tabriz could be very instructive in this respect. Yet with few exceptions, Tabriz, and its connections with the Mongols in Iran, have received serious scholarly attention only as concerns the period after Hülegü. Among recent works is the collective volume reputedly devoted to Tabriz from the seventh/thirteenth to the ninth/fifteenth century, edited by Judith Pfeiffer, but this again chooses the end of the Abbasid Caliphate as the

3 There are several monographs devoted to specific cities. See Bulliet, Richard W., *The Patricians of Nishapur. A Study in Medieval Islamic Social History* (Cambridge MA.: Harvard University Press, 1972): pp. 61-75; Quiring-Zoche, Rosemarie, *Isfahan im 15. und 16. Jahrhundert. Ein Beitrag zur persischen Stadtgeschichte* (Freiburg: Klaus Schwarz Verlag, 1980): pp. 157-61, 208-9; Szuppe, Maria, *Entre Timourides, Uzbeks et Safavides. Questions d'histoire politique et sociale de Hérat dans la première moitié du XVIe siècle* (Paris: Association pour l'avancement des études Iraniennes, 1992): pp. 143-60; Durand-Guédy, David, *Iranian Elites and Turkish Rulers. A History of Iṣfahān in the Saljūq Period* (Abingdon: Routledge, 2010): pp. 58-70, 275-97. Beatrice Manz also discussed the political significance of urban notables in her study of ninth/fifteenth-century Iran, *Power, Politics and Religion in Tīmūrid Iran* (Cambridge: Cambridge University Press, 2007): pp. 156-7, 271-81.

4 Paul, Jürgen, "L'invasion mongole comme "révélateur" de la société iranienne", in Aigle, D. (ed.), *L'Iran face à la domination mongole* (Tehran: Institut français de recherche en Iran, 1997): pp. 37-53.

point of departure.[5] In addition, the volume offers scarce consideration of the post-Ilkhanid period and leaves the local elites almost completely out of the picture.

Non-Muslim domination and the demise of the Abbasids mark a watershed in Islamic and Iranian history. A key question of the present study is therefore: what immediate effects and long-term consequences did Mongol rule have among the notables of Tabriz? In particular, what were the resulting transformations in local society regarding its religious makeup, and otherwise?

An important factor for the history of the city during the long period under study is the availability of excellent grazing grounds in the hinterland of Tabriz, as well as in the broader region of Azarbaijan. The pastures of Azarbaijan and the adjoining lands in southern Caucasia and eastern Anatolia attracted the nomadic Chinggisids, as well as many nomadic Turkic and Turco-Mongol rulers before and after them. Some recent scholarship has stressed the increasing geopolitical centrality of this region from the Seljuq period onward.[6] David Durand-Guédy has further stressed that the nomadic Seljuqs tended to reside outside of cities and to rule from suburban and rural spaces.[7] The Mongols have long been viewed as even more nomadic than the Seljuqs. Putting aside the religious aspect of the question above one may then ask: did Mongol rule increase the effects of such a nomadic mode of domination and, if so, how did this affect the local balance of power between urban and rural populations? Given that the rulers of the period under study tended not to use cities as places from which they exercised power, another question that arises is whether Tabriz itself, especially under the Ilkhans and their successors, could be called a 'capital' in any meaningful sense.

5 Pfeiffer, Judith, "Introduction. From Baghdad to Marāgha, Tabriz and Beyond: Tabriz and the Multi-Cephalous Cultural, Religious and Intellectual Landscape in the 13th to 15th Century Nile-to-Oxus Region" in Ead. (ed.), *Politics, Patronage and the Transmission of Knowledge in 13th-15th century Tabriz* (Leiden: Brill, 2014): pp. 1-11.

6 Peacock, A. C. S., "Nomadic Society and the Seljūq Campaigns in Caucasia", *Iran and the Caucasus*, IX/2 (2005): pp. 205-30; Yildiz, Sara Nur, "Post-Mongol Pastoral Polities in Eastern Anatolia during the Late Middle Ages", in Beyazit, D. (ed.), *At the Crossroads of Empires: 14th-15th Century Eastern Anatolia* (Paris: Institut français d'études Anatoliennes, 2012): pp. 27-48; Durand-Guédy, David, "1147: The Battle of Qara-Tegin and the Rise of Azarbayjan", *Der Islam*, XCII/1 (2015): pp. 161-96.

7 Durand-Guédy, David, "Where did the Saljuqs live? A case study based on the reign of Sultan Mas'ūd b. Muḥammad (1134-1152)", *StIr*, XL (2011): pp. 211-58; Id. "Ruling from the outside: a new perspective on early Turkish kingship in Iran", in Mitchell, L. and Melville, C. (eds.), *Every Inch a King. Comparative Studies on Kings and Kingship in the Ancient and Medieval Worlds* (Leiden: Brill, 2012): pp. 325-42.

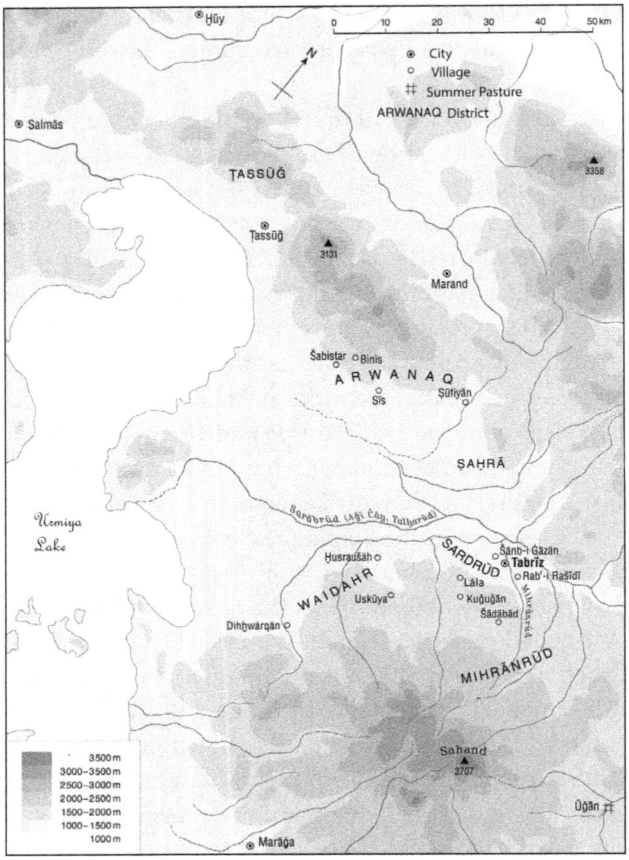

FIGURE 1 Hinterland of Tabriz
DRAWING: BORLEIS/WEISS, LEIPZIG.

In considering the Mongol period, most scholars tend to dwell on the general religious change intensified by temporary non-Muslim rule and the end of the caliphate. They stress that from the conquests onwards, traditional Sunni Islam declined, while Sufism and Shiʿism (often coupled with messianic expectations) rose to dominance in society and politics.[8] The Safavids, who emerged

8 Mazzaoui, Michel, *The Origins of the Ṣafawids. Šīʿism, Ṣūfism and the Ġulāt* (Wiesbaden: Franz Steiner Verlag, 1972); Arjomand, Said A., "Conceptions of Authority and the Transition of Shiʿism from Sectarian Religion to National Religion in Iran", in Daftary, F. and Meri, J.W. (eds.), *Culture and Memory in Medieval Islam. Essays in Honour of Wilferd Madelung* (London and New York: I.B. Tauris, 2003): pp. 388-409; Bashir, Shahzad, *Messianic Hopes and Mystical Visions. The Nūrbakhshīya between Medieval and Modern Islam* (Columbia SC: University of South Carolina Press, 2003); Pfeiffer, Judith, "Reflections on a 'Double Rapprochement':

as Sunni Sufis not far from Tabriz in the early eighth/fourteenth century, and adopted some form of messianic Shiʿism in the ninth/fifteenth, are often cited as an example. Interesting research questions then arise concerning the similarities and differences between the historical trajectory of the Safavids and that of the Kujujī family. Indeed, several scholars have noted roles of the Kujujīs as both Sufis and court officials.[9] Christoph Werner has recently directed the publication of the *Kujujī-waqfiyya*, accompanied by a German translation of the document and a thorough study of the family.[10] The *waqfiyya* is the deed endowing a pious foundation, established in Tabriz by an outstanding family member in 782/1380, and is one of a few relatively new sources for the history of Tabriz and its leading families over the centuries.

Other sources have been known for some time but have not been systematically analysed in regards to the specific research questions raised here. Among these are many works of dynastic and universal historiography written within and beyond Iran, offering scattered references to notables of Tabriz in politics and providing important details about members of the Kujujī and Malikān families. Two other quite well-known, fundamental sources pertain to different genres. One is the Sufi pilgrimage guide of Ibn Karbalāʾī (d. 997/1589) and the other is the biographical dictionary of Ibn al-Fuwaṭī (d. 723/1323). Both offer extensive information on family ties among the local elites of Tabriz and allow at least partial reconstruction of the web of social relations encompassing the actions of the relevant individuals.

Another recently discovered source has brought the Malikān family to scholarly attention. A family member compiled the precious manuscript collection known as *Safīna-yi Tabrīz* in the early eighth/fourteenth century.[11]

Conversion to Islam among the Mongol elite during the early Ilkhanate", in Komaroff, L. (ed.), *Beyond the Legacy of Genghis Khan* (Leiden: Brill, 2006): pp. 369-89; Tucker, William F., "The Kūfan *Ghulāt* and Millenarian (Mahdist) Movements in Mongol-Türkmen Iran", in Mir-Kazimov, Orkhan (ed.), *Unity in Diversity. Mysticism, Messianism and the Construction of Religious Authority in Islam* (Leiden and Boston: Brill, 2014): pp. 177-95.

9 Aubin, Jean, "Etudes safavides. I. Šāh Ismāʿīl et les notables de l'Iraq persan", *JESHO*, II (1959): pp. 37-81, here pp. 60-3; Gronke, Monika, *Derwische im Vorhof der Macht. Sozial- und Wirtschaftsgeschichte Nordwestirans im 13. und 14. Jahrhundert* (Stuttgart: Franz Steiner Verlag, 1993): pp. 276-82; Lewisohn, Leonard, "Palāsī's Memoir of Shayk Kujujī, a Persian Sufi Saint of the Thirteenth Century", *JRAS*, VI/3 (1996): pp. 345-66; Woods, John E., *The Aqquyunlu. Clan, Confederation, Empire* (2nd revised and expanded edition, Salt Lake City: The University of Utah Press, 1999): pp. 151-61. Lane, *Early Mongol Rule*: pp. 250-2.

10 Werner, Christoph, Zakrzewski, Daniel, and Tillschneider, Hans-Thomas, *Die Kuǧuǧī-Stiftungen in Tabriz: Ein Beitrag zur Geschichte der Ǧalāyiriden (Edition, Übersetzung, Kommentar)* (Wiesbaden: Reichert Verlag, 2013).

11 Seyed-Gohrab, Ali-Asghar, "Casing the Treasury", in Seyed-Gohrab, Ali-Asghar and McGlinn, S. (eds.), *The Treasury of Tabriz. The Great Il-Khanid Compendium* (Amsterdam

The *Safīna* contains some copies of official documents, but mainly works in science, religion and literature, in particular Persian poetry. Some of these works were authored by either the compiler or other family relatives. Finally, there is a versified universal history written in emulation of the *Šāh-nāma*, by a poet of mid-seventh/thirteenth century Tabriz known as Zajjājī. This work is also related to the Malikān family and entitled *Humāyūn-nāma*. Although it is a universal history, the author adopts a fundamentally local and regional perspective. His work thus offers unique insights into the historical consciousness of notables of Tabriz at the moment when the first two Ilkhans were consolidating their power in Iran. The poet Zajjājī's original patron was Malik Ṣadr al-Dīn Muḥammad Tabrīzī (d. 668/1269-70), and it was from this individual that the later Malikān probably took their family name. However the editor of the *Humāyūn-nāma* did not identify Malik Ṣadr al-Dīn as Zajjājī's original patron, and the scholars studying the *Safīna* remained unaware that the former belonged to the Malikān family.[12] It is the combination of genealogical information preserved in these two sources with other details, given by Ibn Karbalāʾī and Ibn al-Fuwaṭī, that makes it possible to identify Malik Ṣadr al-Dīn and other local political leaders in pre and early Mongol Tabriz as members of the family.

---

and West Lafayette: Rozenberg Publishers and Purdue University Press, 2007): pp. 15-42.

12 Zajjājī, *Humāyūn-nāma*, ed. ʿAlī Pīr-niyā, 2 vols. (Tehran: Farhangistān-i zabān va adab-i fārsī/Mīrāṯ-i maktūb, 1383-90š./2004-11): I, p. 19 (Intr.), p. 854, II, p. 15, 18 (Intr.), 306, 461-3, 857-8, 1274, 1323-4. The author consistently calls his original patron Malik Ṣadr al-Dīn but in publishing the second part, the editor ʿAlī Pīr-niyā wrote an introduction in which he mistakenly identifies the patron as the later Ilkhanid vizier Ṣadr al-Dīn Zanjānī (executed 696/1297). In the introduction to the first part, which was actually published later, Pīr-niyā noted that it was copied on the order of Šams al-Dīn Juwaynī (executed 684/1283) but did not correct the initial mistake. Zajjājī deplores the death of his original patron in both parts, each of which exist as a single manuscript. Malik Ṣadr al-Dīn's date of death and the detail relating him to the family are given by Ibn al-Fuwaṭī, *Majmaʿ al-ādāb fī muʿjam al-alqāb*, ed. Muḥammad al-Kāẓim, 6 vols. (Tehran: Sāzmān-i čāp va intišārāt-i Wizārat-i Farhang va Iršād-i Islāmī, 1995): II, p. 163. For relevant genealogical information, also see Afšār, Īraj, "Safīna-yi Tabrīz: Anjāma-hāy Safīna-yi Tabrīz", *Nāma-yi Bahāristān*, XIII/14 (1387š./2008): pp. 285-330, here p. 294; Ibn Karbalāʾī, *Rawżāt al-jinān va jannāt al-janān*, ed. Jaʿfar Sulṭān al-Qurrāʾī, 2 vols. (Tehran: Bungāh-i tarjuma va našr-i kitāb, 1344-9š./1965-70, reprint Tabriz: Sutūda, 1383š./2004): I, p. 469. For a biographical study of Malik Ṣadr al-Dīn, see Zakrzewski, Daniel, "Malik Ṣadr al-Dīn Tabrīzī and the Establishment of Mongol Rule in Iran", *Asiatische Studien / Études Asiatiques*, LXXI/4 (2017): pp. 1059-73.

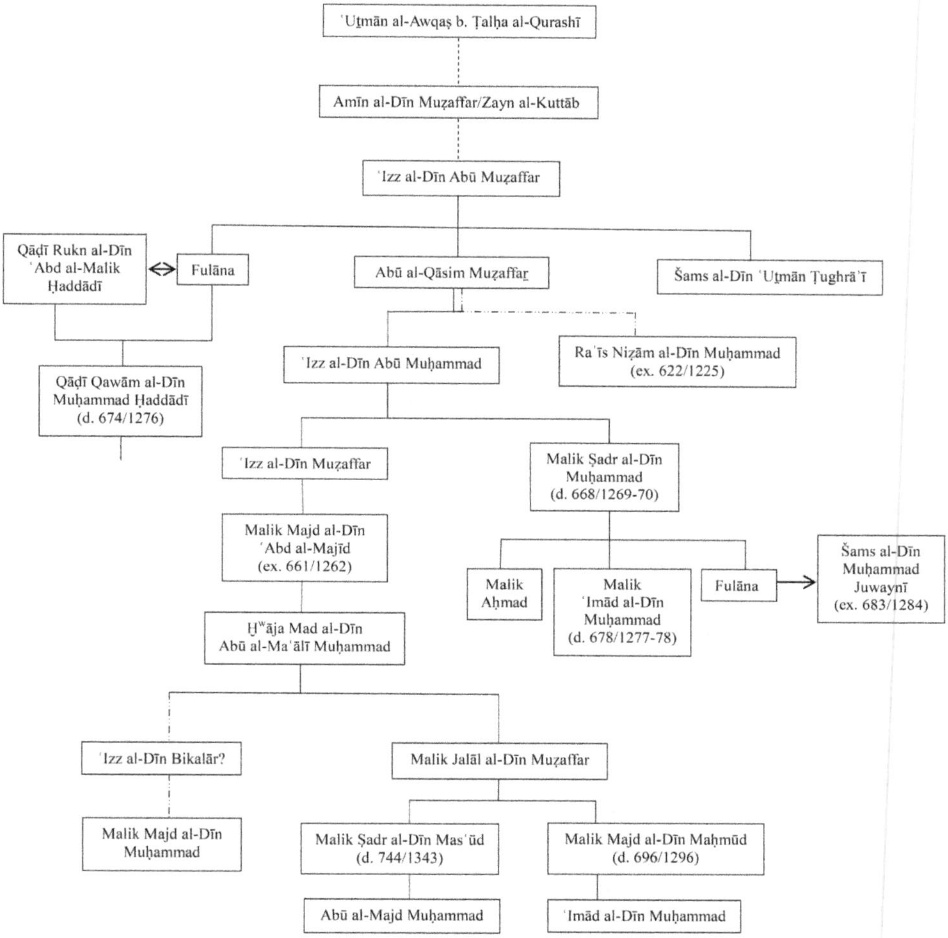

FIGURE 2 The Malikān Family

3 The Malikān and Dynastic Succession prior to Mongol Rule

As the authority of the Seljuq sultans waned in the sixth/twelfth century, two powerful atabeg dynasties vied for supremacy in northwestern Iran.[13] The Aḥmadīlīs were based in Marāġa, on the southern side of Mount Sahand, but also controlled Tabriz on the northern side. The Eldigüzids were based still

13 Luther, Kenneth A., s.v. "Atābakān-e Marāġa", in *EIr*, II (1987): pp. 898-900; Id., s.v. "Atābakān-e Āḏarbāyjān", in *EIr*, II (1987): pp. 890-894.

further north, in Naḫjawān, and took control of the Seljuq Sultanate and its *dār al-mulk* Hamadan in the middle of the century. In 572/1176, shortly after the death of the dynastic eponym, Jahān-Pahlawān b. Eldigüz then wrested Tabriz from the Aḥmadīlīs. His brother Qïzïl Arslan set up a subordinate court at the city, where he ruled as deputy of Jahān-Pahlawān and tutor of the latter's eldest son Abū Bakr.[14]

In this period, Malik Ṣadr al-Dīn's great-grandfather, ʿIzz al-Dīn Abu'l-Muẓaffar, appears to have been *ra'īs* of Tabriz. This was a personage analogous to a mayor,[15] representing the local community and mediating between the population and the ruler. It is not clear either how long the Malikān had been based in Tabriz, or how they had risen to prominence in the city. However, around 579/1183, ʿIzz al-Dīn led a delegation of notables (*buzurgān-i Tabrīz*), including his son Šams al-Dīn Ṭuġrā'ī, to Naḫjawān. Jahān-Pahlawān also travelled from Hamadan to Nahjawan, making only a brief stop at Tabriz.[16]

The decades following Jahān-Pahlawān's death, in 582/1186, were marked by endemic fighting, with effects on Tabriz. The poet-historian Zajjājī mentions Šams al-Dīn Ṭuġrā'ī, together with the last Seljuq Sultan Ṭoġrïl, in the entourage of Qïzïl Arslan, who imposed himself as leader of the Eldigüzid dynasty.[17] Conflict soon erupted between the sultan and this new atabeg, and after battles involving various internal and external forces, Qïzïl Arslan imprisoned Sultan Ṭoġrïl in a castle in Azarbaijan, in 585/1189. Two years later Qïzïl Arslan himself mounted the Seljuq throne at Hamadan, but was almost immediately found dead. Released from prison, Ṭoġrïl first went to Tabriz but was denied entry to the city.[18] Obviously, the notables did not recognise the Seljuq sultan and opponent of the Eldigüzid atabeg as their lord.

Yet Šams al-Dīn Ṭuġrā'ī accompanied Ṭoġrïl in 590/1194 when the sultan was killed in battle against Khwarazmshah Tekeš (d. 596/1200) near Rayy.[19] At this point Seljuq rule in Iran ended, and Ṭuġrā'ī must then have returned to Tabriz to join the court of Abū Bakr b. Jahān-Pahlawān. Immediately upon Qizil Arslān's death, Abū Bakr had hurried to seize the Eldigüzid core territories in

---

14 Ibn al-Aṯīr, *al-Kāmil fi'l-tā'rīḫ*, ed. Carl Johan Tornberg, 13 vols. (Leiden; reprint with a different pagination Beirut: Dār Ṣādir, 1965-7): XI, p. 423; Nīšāpūrī, *Saljūqnāma*, ed. A.H. Morton (Warminster: Gibb Memorial Trust, 2004): pp. 120-1; Ḥusaynī, *Aḫbār al-dawlat al-saljūqiyya*, ed. Muḥammad Iqbāl (Lahore: The University of the Panjab, 1933, reprint Beirut: Dār al-āfāq al-jadīda, 1984): p. 173; Zajjājī, *Humāyūn-nāma*: II, p. 1319.

15 Havemann, Axel and Bosworth, C.E., s.v. "Ra'īs", in *EI*², VIII (1995): pp. 402-3.

16 Zajjājī, *Humāyūn-nāma*: II, pp. 1323-4.

17 Ibid.: p. 1334.

18 Ḥusaynī, *Aḫbār*: p. 182.

19 Zajjājī, *Humāyūn-nāma*: II, pp. 1268-77.

Azarbaijan and the northern region of Arrān, facing opposition from two of his half-brothers. Sources disagree as to whether these half-brothers of Abū Bakr's entered Tabriz.[20] They had weak ties to the city at best, and in fact Abū Bakr defeated both rivals, who died soon after. Zajjājī summarises the result:

Finally, he was victorious against the bold ones, through manliness he gained kingship, the kingdom of Iran. In Tabriz was Abū Bakr Šāh, he established the royal court in that place.[21]

The author does not conceal that Eldigüzid power diminished on all fronts under Abū Bakr but the affirmative nature of this cited statement illustrates the local focus of his work. Indeed, covering post-Seljuq history in his *Humāyūn-nāma*, Zajjājī adheres to a dynastic structure only for the Abbasid caliphs and otherwise switches to writing chapters recounting the successive rulers of Tabriz. He indicates that Šams al-Dīn Ṭuġrā'ī served as vizier of Abū Bakr around the turn of the century but was dismissed in the context of intensifying court rivalries, after having spent some time in Georgian captivity.[22] Abū Bakr was able to repel an Aḥmadīlī attack in 603/1207 and to add Marāġa to the dynastic realm, thanks to the help of Ayṭoġmiš, who was one of the former Eldigüzid slave commanders based in central Iran together with Abū Bakr's brother Özbeg. Georgian incursions reached a peak in the following years,[23] and so Aytoqmiš brought Özbeg to Tabriz as Abū Bakr's successor. However the inhabitants fought them, in opposition to this development.[24] By 607/1210, Abū Bakr was dead and his vizier, a former *ra'īs* of the city, then convinced unnamed notables to accept the succession of Özbeg. The precise mechanisms used by Ṭuġrā'ī and his relatives to adapt to this change remain obscure, however it is clear that the Malikān were able to continue their prominence among the local elites.

---

20 Ibid.: II, pp. 1192-7; Ḥusaynī, *Aḫbār*: p. 185; Brosset, Marie-Félicité, *Histoire de la Géorgie depuis l'antiquité jusqu'au XIXe siècle*, 4 vols. (St. Petersburg: Imprimerie de l'Académie Impériale des Sciences, 1849-57): I, pp. 435-45.

21 Zajjājī, *Humāyūn-nāma*: II, p. 1189 (*sar-anjām dast az dilīrān bi-burd, bi-mardī mulk, mulk-i Īrān bi-burd; bih Tabrīz būdī Abū Bakr Šāh, zadī andar ān būm-ū-bar bārgāh*).

22 Ibid.: II, pp. 1196-1204.

23 Ibn al-Aṯīr, *al-Kāmil*: XII, pp. 236-7, 275. Zajjājī, *Humāyūn-nāma*: II, p. 1206; Brosset, *Histoire*: I, pp. 466-73; Ḥusaynī, *Aḫbār*: p. 189. On Abū Bakr's former slave and the situation in central Iran, see Luther, Kenneth A., "Rāvandī's Report on the Administrative Changes of Muḥammad Jahān Pahlavān", in Bosworth, C.E. (ed.), *Iran and Islam, in memory of the late Vladimir Minorsky* (Edinburgh: Edinburgh University Press, 1971): pp. 393-406, here pp. 397-401.

24 Zajjājī, *Humāyūn-nāma*: II, pp. 1205-9.

At Tabriz, Özbeg installed a harsher regime than that of his brother. This caused unrest in the city, to which he reacted with arrests and confiscations.[25] Beyond Azarbaijan, the Eldigüzid forged alliances against Khwarazmshah Muḥammad, but the latter marched westwards in 614/1217, forcing Özbeg to acknowledge him as overlord.[26] The Khwarazmshah returned to Khurasan to face the Mongols shortly afterwards, and in 617/1220 he was hunted to death by two generals dispatched by Chinggis Khan. The armies of these generals reached Azarbaijan towards the end of this year, then remained in the area until the following spring. During this time they repeatedly came to the gates of Tabriz. Although Özbeg may have been absent, all the encounters were peaceful and on at least one occasion, Šams al-Dīn Ṭuġrā'ī negotiated the withdrawal of the Mongols. In this way he avoided a military confrontation with the powerful Mongols which would have resulted in large scale destruction, although according to Ibn al-Aṯīr and several later authors, the inhabitants of Tabriz were ready to fight the infidels under his leadership.[27]

Šams al-Dīn Ṭuġrā'ī again played a key role when Khwarazmshah Jalāl al-Dīn conquered Tabriz in the summer of 622/1225. Özbeg (who would die shortly after the Khwarazmian conquest) had abandoned the city. Unlike the Mongols, the Khwarazmshah was forced to launch a siege lasting about a week. This time Ṭuġrā'ī led the resistance, assisted by the *ra'īs* and the qadi, both of whom were his nephews. Meanwhile, a renowned jurist named ʿIzz al-Dīn Qazwīnī led the pro-Khwarazmian faction inside the city.[28] ʿIzz al-Dīn's father had settled in Tabriz but his family was also prominent in Marāġa, where the Eldigüzids, to whom the Malikān were so closely linked, enjoyed little support.[29] An in-depth discussion of the complexities of the situation would be too

25 Ibid.: II, pp. 1213-4.

26 Ibid.: II, pp. 934-6, 1212-5; Ibn al-Aṯīr, *al-Kāmil*: XII, pp. 306-8, 316-8; Nasawī, *Sīrat al-Sulṭān Jalāl al-Dīn Minkbirnī*, ed. Z. M. Buniatov (Moscow: Izdatelskaya firma «Vostočnaya literatura» RAN, 1996): pp. 17-22; Juwaynī, ʿAṭā-Malik, *Tārīḫ-i Jahān-gušāy*, ed. Mīrzā Muḥammad Qazwīnī (Leiden: Gibb Memorial Trust, 1912-37): II, pp. 97, 120-1; III, pp. 245-6.

27 Ibn al-Aṯīr, *al-Kāmil*: XII, pp. 374-83; Ibn Abi'l-Ḥadīd, *Šarḥ Nahj al-balāġa. Les invasions mongoles en Orient vécues par un savant médiéval arabe*, part. ed. and transl. Moktar Djebli (Paris: L'Harmattan, 1995): pp. 42-6; Zajjājī, *Humāyūn-nāma*: II, pp. 1217-20; Rašīd al-Dīn, *Jāmiʿ al-tavārīḫ*, ed. Muḥammad Rawšan and Muṣṭafā Mūsawī (Tehran: Našr-i Alburz, 1373š./1994): I, pp. 532-3; Ibn Karbalā'ī, *Rawżāt al-jinān*: I, pp. 452-3. Minorsky, Vladimir [C. E. Bosworth], s.v. "Tabriz", in *EI* ², X (2010): pp. 41-9, here p. 43.

28 Ibn al-Aṯīr, *Kāmil*: XII, pp. 432-7. Nasawī, *Sīrat*: pp. 133-4, 137-9, 141-4; Juwaynī, *Tārīḫ-i Jahān-gušāy*: II, pp. 156-7; Zajjājī, *Humāyūn-nāma*: II, pp. 1222-5; Rašīd al-Dīn *Jāmiʿ al-tavārīḫ*: I, p. 552.

29 al-Rāfiʿī al-Qazwīnī, *al-Tadwīn fī aḫbār Qazwīn*, ed. ʿAzīz Allāh al-ʿAṭāridī (Hydarabad, 1984, reprint Beirut: Dār al-kutub al-ʿilmiyya, 1987): III, pp. 138-9; Ibn al-Fuwaṭī, *Majmaʿ al-ādāb*: II, p. 91; IV, p. 234.

lengthy for current purposes, but both ʿIzz al-Dīn's hope for a mighty Muslim ruler and his own personal ambition appear to have determined his stance toward the Khwarazmshah. In any case, the leaders of both factions attempted to protect the city populace.

After Jalāl al-Dīn conquered Tabriz, the Malikān at first retained their positions. Šams al-Dīn Ṭuġrāʾī seems to have been appointed as a local or regional vizier, without authority to administer matters of royal finance, as may have been possible for him as vizier under the atabegs.[30] However, ʿIzz al-Dīn, who taught as a professor in a madrasa founded by his rival, conspired against the Malikān to replace Ṭuġrāʾī's maternal nephew as qadi of Tabriz. The Qazwīnī family most likely adhered to the Šāfiʿī school of law (*maḏhab*), and there is evidence that this was also the case for the Malikān.[31] Their qadi relatives may have been Ḥanafīs, but the Šāfiʿīs were dominant in Tabriz and Azarbaijan in this period. and remained so later.[32] If *maḏhab* antagonism played any role in the rivalry, this was probably less important than matters of local attachment and loyalty to the house of Eldigüz.

In any case, ʿIzz al-Dīn Qazwīnī achieved his goal in autumn 622/1225, and even managed to have Šams al-Dīn Ṭuġrāʾī's paternal nephew, the *raʾīs*, executed, while driving Ṭuġrāʾī himself into exile. But within a year, this formerly fervent advocate of dynastic change was in turn dismissed for insults against the Khwarazmians. After Jalāl al-Dīn had conquered Tabriz the new ruler continued to spend most of his time campaigning and concerned himself very little with his principal city.

Ṭuġrāʾī was back in Tabriz by summer 626/1229, and may even have joined Jalāl al-Dīn's court. Meanwhile the local economy had begun to decline, and the opposition had grown stronger.[33] It is uncertain whether Ṭuġrāʾī was still alive two years later when the notables peacefully surrendered to the Mongols once again, having little reason to lament the imminent demise of

---

30 Nasawī, *Sīrat*: pp. 134, 253 does not mention any specific position in his description of the immediate post-conquest arrangements. In a later passage, the author refers indirectly to Ṭuġrāʾī as being vizier at this time. On the appointment of local or regional viziers under the Khwarazmshahs in western Iran, also see Durand-Guédy, *Iranian Elites*: p. 290.

31 Ibn al-Fuwaṭī, *Majmaʿ al-ādāb*: 1, pp. 278-9 mentions that ʿIzz al-Dīn studied with the celebrated Imām al-Rāfiʿī in Qazvin, but generally lacks explicit references to the *maḏhab* affiliation of family members. The *Safīna* contains several works of standard Šāfiʿī literature in *fiqh* and related disciplines; see Seyed-Gohrab, "Casing the Treasury": pp. 35-7.

32 Mustawfī Qazwīnī (Ḥamd-Allāh), *Baḥš-i naḥust az maqāla-yi sivvom-i Nuzhat al-qulūb*, ed. Muḥammad Dabīr Siyāqī (Tehran: Ṭahūrī, 1336š./1958): pp. 88, 91-6, 99-100, indicates Ḥanafī dominance only for Marāġa.

33 Nasawī, *Sīrat*: pp. 178, 197, 253.

the Khwarazmshah.[34] The armies sent west by Chinggis Khan's son and successor Ögödei established their main bases in Azarbaijan and Arrān, and the Malikān were only one among the local leading families who entered willingly into Mongol service.

4 The Malikān and the Continuity of Urban Elites under Mongol Rule

ʿIzz al-Dīn Qazwīnī was again qadi of Tabriz in 630/1233, when he dealt with a case that Ṭuġrāʾī's brother-in-law had already adjudicated at the time of Eldigüzid rule. ʿIzz al-Dīn probably remained in office until his death in 648/1250.[35] During the initial stages of Mongol rule in Iran he must have collaborated closely with Malik Ṣadr al-Dīn, a grandnephew of his erstwhile rival, Šams al-Dīn Ṭuġrāʾī. One of the established meanings of the term *malik* was as a designation of indigenous local or regional governors under the Mongols.[36] Their formal competencies and duties are difficult to assess; however, it is obvious that Malik Ṣadr al-Dīn fulfilled administrative and military functions, both before and after the arrival of Hülegü.

Malik Ṣadr al-Dīn's career coincided with a time of heightening tensions within the Chinggisid dynasty, from which the house of Hülegü emerged. Peter Jackson has analyzed these tensions, with attention to regional contexts, and has generally highlighted the role of Batu (d. 653/1255),[37] a son of Chinggis Khan's eldest son Jochi. Batu established his base north of the Caucasus, around the time of Ögödei's death in 639/1241, and thus led the only branch of the imperial dynasty based in relatively close proximity to Tabriz. Batu very soon became the senior and most powerful Chinggisid prince, while his relations with the house of Ögödei deteriorated. The Jochid therefore offered decisive support to the claim of Möngke, a son of Chinggis Khan's youngest son Tolui, who ascended the imperial throne in 650/1252. Möngke dispatched his

34 Ibn al-Aṯīr, *al-Kāmil*: XII, pp. 502-3; Rašīd al-Dīn, *Jāmiʿ al-tavārīḫ*: I, p. 655.

35 Gronke, Monika, *Arabische und persische Privaturkunden des 12. und 13. Jahrhundert aus Ardabil (Aserbeidschan)* (Berlin: Klaus Schwarz Verlag, 1982): pp. 414-45; Ibn al-Fuwaṭī, *Majmaʿ al-ādāb*: I, pp. 278-9; II, p. 175; III, p. 360. Relatives of ʿIzz al-Dīn had been and remained qadis of Maraġa.

36 Aigle, Denise, *Le Fārs sous la domination mongole. Politique et fiscalité* (XIIIe-XIVe s.) (Paris: Association pour l'avancement des études Iraniennes, 2005): p. 90. A discussion of Arabic-Persian and Turko-Mongolian terminology concerning governmental and administrative personnel could be helpful here and more generally but is unfortunately beyond the scope of this article.

37 Jackson, Peter, "The Dissolution of the Mongol Empire", *CAJ*, XXXII (1978): pp. 186-243.

brother Hülegü to the west, but by selecting Azarbaijan and Arrān as the base for his own house, the first Ilkhan threatened Jochid interests in the region.

The case of Malik Ṣadr al-Dīn's nephew, Malik Majd al-Dīn, offers an example of the ties between the notables of Tabriz and the Jochids at the time of early Mongol rule. In describing a meeting with Malik Majd al-Dīn in Tabriz in 659/1261, Ibn al-Fuwaṭī mentions him as a clerk (*kātib*) of Berke, Batu's eventual successor.[38] The uncle, Malik Ṣadr al-Dīn, certainly had connections to Mongol amirs and Persian administrators acting on behalf of Batu, and active in Azarbaijan and Arrān prior to the arrival of Hülegü. Malik Ṣadr al-Dīn thus took part in the Mongol campaign against the Rum Seljuqs of Anatolia, in 641/1243.[39] And Malik Ṣadr al-Dīn's client, the above-mentioned poet-historian Zajjājī, commemorated the death of an important tax official who had set up his headquarters at Tabriz as a representative of Batu in 642/1244, and who died in Khurasan in the following year.[40]

During the brief reign of Ögödei's son Güyüg (644/1246-646/1248), a Mongol official who had served as supervisor of the artisans of Tabriz, and an Eldigüzid scion who had been hiding in Anatolia, challenged Malik Ṣadr al-Dīn's authority.[41] In spite of enjoying court backing, both soon disappeared from the political scene, while Malik Ṣadr al-Dīn continued to increase his influence. Indeed, in 649/1251 he joined a host of dignitaries travelling to Mongolia, where the newly enthroned Great Qa'an Möngke confirmed him as governor of Azarbaijan and Arrān. He was also charged with conducting a census and introducing a poll tax in the region.[42] Soon after returning from court in 651/1253, he completed this under supervision of two high-ranking Mongols, and assisted by his nephew Malik Majd al-Dīn.

---

38 Ibn al-Fuwaṭī, *Majmaʿ al-ādāb*: IV, p. 457 (*wa ḏakkarū ʿanhu annahu kātib Berke*). The author confused the name of Malik Majd al-Dīn's son with that of his father. The identifications otherwise match exactly with the genealogies of later family members, notably the compiler of the *Safīna*. Afšār, "Safīna-yi Tabrīz": pp. 287, 294-5, 309, 315; Ibn Karbalāʾī, *Rawżāt al-jinān*: I, p. 469.

39 Zajjājī, *Humāyūn-nāma*: II, p. 1088 (*Malik Ṣadr-i Dīn andar ān ḫayl būd, firistād mīraš bih Sīvās zūd*).

40 Juwaynī, *Tārīḫ-i Jahān-gušāy*: II, pp. 244-5, 274-81.

41 Ibid.: II, pp. 247-53; Aubin, Jean, *Émirs mongols et vizirs persans dans les remous de l'acculturation* (Paris: Association pour l'avancement des études iraniennes, 1995): p. 16. Also see Zakrzewski, "Malik Ṣadr al-Dīn": pp. 1065-6, for a slightly more detailed discussion.

42 Juwaynī, *Tārīḫ-i Jahān-gušāy*: II, pp. 254-8; *Humāyūn-nāma*: II, pp. 1098-1101; Allsen, Thomas T., *Mongol Imperialism. The Policies of the Grand Qan Möngke in China, Russia and the Islamic Lands, 1251-1259* (Berkeley: University of California Press, 1987): pp. 131-2, 163, explained the poll tax as a centralizing policy of Möngke, and stressed the role of the Mongols who supervised the Malikān, but did not mention the local officials.

Subsequently, the Malikān performed various tasks in Hülegü's service. For storage of booty, including from Baghdad, Malik Majd al-Dīn erected a fortified building on a mountain rising over Lake Urmia.[43] Malik Ṣadr al-Dīn commanded auxiliary troops in the 659/1261 siege of Mosul, where the ruler had defected to the rising Mamluk sultans of Egypt.[44] In the following year, in the run-up to war between Hülegü and Berke, both Malikān and a number of other notables were subjected to a Mongol legal trial (*yarġū*). Malik Ṣadr al-Dīn escaped with a lashing but Malik Majd al-Dīn was among those sentenced to death, presumably having been found too close to the Jochids.[45] As Jackson correctly asserts, the war between Berke and Hülegü "may be said to signify the dissolution of the Mongol Empire".[46] Following the death of Möngke in 657/1259, a struggle for succession raged through Central and East Asia, in which the Ilkhan and Jochid supported rival candidates. Berke (d. 665/1267), as the first Muslim Mongol ruler, also disapproved of Hülegü's order to kill the last Abbasid caliph, and allied with the Mamluks against his Chinggisid cousin. Hülegü suffered a humiliating defeat, but it was only after the collapse of the Ilkhanid house that the Jochids of the Golden Horde could reassert their claim over the core area of Mongol Iran with any measure of success.

Hülegü reconfirmed Malik Ṣadr al-Dīn as governor of Tabriz, and the city was again assigned to him in 663/1265 when the conqueror's son Abaqa (d. 680/1282) rose to the Ilkhanid throne.[47] Although many scholars stress the nomadic lifestyle of the Mongol rulers, it is also generally believed that following the enthronement of Abaqa, Tabriz became the 'capital' of the Ilkhans. Birgitt Hoffmann addressed the issue with caution, and yet her discussion illustrates the confusion in speaking of a particular city as 'capital' of nomadic rulers. Hoffmann rightfully remarks that the mobile royal encampment (*ordu*) was the effective capital (*wirkliche Kapitale*) of the Ilkhan Empire, while also stating that under Abaqa, Tabriz obtained the rank of a capital (*Hauptstadt*), and adding the epithet *dār al-mulk* to clarify this German term.[48]

David Durand-Guédy has stressed that the understanding of epithets such as *dār al-mulk* requires careful analysis on several levels, and that simple English

---

43 Rašīd al-Dīn, *Jāmiʿ al-tavārīḫ*: II, p. 1022, Allsen, *Mongol Imperialism*: p. 182.

44 Rašīd al-Dīn, *Jāmiʿ al-tavārīḫ*: II, p. 1041.

45 Ibid.: II, p. 1045; Ibn al-Fuwaṭī, *Majmaʿ al-ādāb*: IV, p. 457.

46 Jackson, "Dissolution": p. 238.

47 Rašīd al-Dīn, *Jāmiʿ al-tavārīḫ*: II, pp. 1049, 1061.

48 Hoffmann, Birgitt, *Waqf im mongolischen Iran. Rašīduddīns Sorge um Nachruhm und Seelenheil* (Stuttgart: Franz Steiner Verlag, 2000): p. 107.

translations such as 'capital' should possibly be abandoned.[49] This assertion is supported by a brief statement of the famous Ilkhanid vizier and historian Rašīd al-Dīn (d. 718/1318), concerning Tabriz under Abaqa. Contrary to general understanding, his Persian text literally says of the second Ilkhan that "he made the *dār al-mulk* Tabriz the seat of the royal throne".[50] From this wording, Tabriz was clearly already a *dār al-mulk* at the time of Abaqa's enthronement. Moreover, this historian uses the same epithet in passages on the pre-Mongol history of the city, in particular under the Eldigüzids.[51] One could therefore express doubts that the status of Tabriz underwent any 'official' change at all with the coming of the Ilkhans, doubts that would be strengthened by Malik Ṣadr al-Dīn's continuing administrative record.

As noted, Malik Ṣadr al-Dīn had been charged under Möngke with conduct of a census and introduction of a poll tax in northwestern Iran, and so it seems that governorship of Tabriz implied some responsibilities concerning imperial finances. It is noteworthy that it was only after Malik Ṣadr al-Dīn's death in 668/1269-70 that there was a substantial change in the coinage of the city. The local monetary standards were eventually imposed as imperial standards throughout the Ilkhanid realm, in the reign of Ġāzān.[52] However, the political landscape had already undergone a decisive change with the arrival of Hülegü in Azarbaijan. It was at this time that Tabriz became associated with a major court, based in the region. The village of Šam, west of Tabriz, later the site of Ġāzān's mausoleum, had been a preferred royal campsite in the immediate city environs going back to at least the second half of the seventh/thirteenth century. There were other major royal campsites in the hinterland, such as at Ujān, clearly used since the time of pre-Mongol rulers,[53] but under the early Ilkhans, an unprecedented courtly entourage was attracted to Tabriz. With

49 Durand-Guédy, David, "Location of Rule in a Context of Turko-Mongol Domination", in Durand-Guédy, David (ed.), *Turko-Mongol Rulers, Cities and City Life* (Leiden and Boston: Brill, 2013): pp. 1-20: here p. 13.

50 Rašīd al-Dīn, *Jāmiʿ al-tavārīḫ*: II, p. 1061 (*va dār al-mulk Tabrīz-rā maqarr-i sarīr-i pādišāhī sāḫt*).

51 Ibid.: I, pp. 553, 557.

52 Album, Stephen, *Sylloge of Islamic Coins in the Ashmolean, IX, Iran after the Mongol Invasion* (Oxford: Ashmolean Museum, 2001): pp. x-xi; Kolbas, Judith, *The Mongols in Iran. Chingiz Khan to Uljaytu, 1220-1309* (Abingdon: Routledge, 2006): pp. 151-4, credits Malik Ṣadr al-Dīn with having devised the *imperial hexagon coinage* under Möngke. But she remained unaware of his date of death and local origin, in addition to other shortcomings in her analysis. Rašīd al-Dīn, *Jāmiʿ al-tavārīḫ*: II, p. 1463.

53 Eldigüzid Atabeg Özbeg and Khwarazmshah Jalāl al-Dīn both laid camp at Ujān. See Zajjājī, *Humāyūn-nāma*: II, pp. 1206-7; Nasawī, *Sīrat*: pp. 133, 152, 193.

the death of Sadr al-Dīn, the Malikān soon lost political pre-eminence. This is made clear by a note from Ibn al-Fuwaṭī on the son, Malik ʿImād al-Dīn:

> He took up the position of his father Malik Ṣadr al-Dīn in the year 668 [1269-70]. But in reality, the governor in Tabriz and elsewhere was the master Šams al-Dīn Muḥammad b. Muḥammad al-Juwaynī, ʿImād al-Dīn's sister being with the master. He died young in the year 676 [1277-8] and was buried next to his father in the madrasa Ṣadriyya.[54]

This Šams al-Dīn Juwaynī, from a notable family of Khurasan, was the most influential immigrant to early Ilkhanid Tabriz. His predecessor as Hülegü's vizier had been executed along with Malik Majd al-Dīn.[55] After Malik Ṣadr al-Dīn's death, Juwaynī patronised the poet-historian Zajjājī, but only for the first part of the *Humāyūn-nāma*, consisting of the biography of the prophet Muḥammad, and not for the second, the actual universal history.[56] This could be explained by the historical vision of Zajjājī, set forth in this second part. Here, we have seen that he celebrates the Atabeg Abū Bakr as "king of Iran" and the Eldigüzids as undisputed heroes. Juwaynī was a newcomer from Khurasan, with no connection to the last Muslim regional dynasty of Azarbaijan, and Zajjājī's vision would have lacked appeal. On the other hand Malik Ṣadr al-Dīn had certainly endorsed this vision, and may indeed have commissioned the work in entirety.

Juwaynī's influence in Tabriz was dependent on his position at the Ilkhanid court. When he fell from grace in 684/1283, his family was virtually extinguished.[57] Apart from praise of the Juwaynīs, Zajjājī makes no reference to the events following Hülegü's sack of Baghdad, and it would appear that the

54 Ibn al-Fuwaṭī, *Majmaʿ al-ādāb*: II, p. 163, (*rataba makān ʾabīhi al-Malik Ṣadr al-Dīn sana 668. wa-kāna al-ḥākim fī'l-ḥaqīqa bi-Tabrīz wa ġayrihā al-ṣāḥib Šams al-Dīn Muḥammad b. Muḥammad al-Juwaynī wa-kānat ʾuḫt ʿImād al-Dīn ʿinda al-ṣāḥib. wa-tuwuffiya sana 676 wa-dufina ʿinda wālidihi fī'l-madrasa al-ṣadriyya*). Aubin, *Emirs mongols*: p. 22, asserted that Juwaynī married the daughter of Malik Ṣadr al-Dīn but this does not appear evident from Ibn al-Fuwaṭī's wording.

55 Rašīd al-Dīn, *Jāmiʿ al-tavārīḫ*: II, p. 1045.

56 Zajjājī, *Humāyūn-nāma*: II, pp. 19-21 (Intr.), 306-9, 534-41. The editor ʿAlī Pīr-niyā remarked, in his 2004 introduction to the second part, that Zajjājī sought Juwaynī's patronage without success and showed in his 2011 introduction to the first part that the author obtained it for this; see also note 12. On Juwaynī, see also Biran, Michal, s.v. "Jovayni, Ṣāḥeb Divān", in *EIr*, XV (2009): pp. 71-4. See Zakrzewski, "Malik Ṣadr al-Dīn": pp. 1070-1, for a preliminary comparison between the Juwaynīs and the Malikān.

57 Rašīd al-Dīn, *Jāmiʿ al-tavārīḫ*: II, pp. 1157-60.

author died before the demise of his second patron's family.[58] The fate of the house of Hülegü was all but certain at the time when Zajjājī was composing the *Humāyūn-nāma*. Given the local focus of his work, the author must have considered it more important that the principal city of the Ilkhans had been under Mongol rule well before this dynasty even began to take shape. So Zajjājī posited a special relation between Tabriz and the conquerors, beginning in the time of Chinggis Khan himself. He discusses the generals of Chinggis, who supposedly negotiated with Šams al-Dīn Ṭuġrā'ī:

> They said that this pleasant city has peacefully surrendered to us, supporting our army and cavalry; this golden city here forms private property of the khan, for no [city] is more amiable than it in the world.[59]

With Tabriz under Mongol rule, the urban elites continued traditions of Sunni Islamic learning in the city. The *Safīna*, most of which was compiled between 721/1321 and 723/1323 by a descendant of Malik Majd al-Dīn, includes selections from the *Maṣābīḥ al-sunna*, a highly influential hadith collection compiled by the Khurasani transmitter Rukn al-Dīn Baġawī (d. 516/1122).[60] A renowned student of Baġawī, known as Imām Ḥafda, left his native Khurasan and settled in Tabriz, where he died in 571/1175.[61] He probably brought the *Maṣābīḥ* with him. Local transmissions of another major hadith collection compiled by Baġawī himself, the *Šarḥ al-sunna*, definitely derived from a copy by Imām Ḥafda. The most famous local Mongol-era transmitter of the *Šarḥ al-sunna* was Muḥyī al-Dīn Qazwīnī, a son of the aforementioned 'Izz al-Dīn, who had succeeded his father as qadi of Tabriz prior to the arrival of Hülegü. Muḥyī al-Dīn taught the *Šarḥ al-sunna* to the celebrated polymath Quṭb al-Dīn Šīrāzī (d. 711/1311), and to Ṣadr al-Dīn Ḥamuwayi (d. 722/1322), who presided over Ilkhan Ġāzān's conversion to Islam shortly before the latter rose to the throne in 694/1295.[62]

---

58 Zajjājī, *Humāyūn-nāma*: II, pp. 20 (Intr.), 289-301, 307, 537. The editor 'Alī Pīr-niyā points out that there is no reference to Juwaynī's execution.

59 Ibid.: II, p. 1219 (*bi-guftand ka-īn šahr-i ḫūsh īl-i mā-st, ka yārī-dah-i laškar va ḫayl-i mā-st; čīnīn šahr-i zarīn būd ḫāṣṣ-i ḫān, kah ḫūshtar nabāšad az īn dar jahān*).

60 Seyed-Gohrab, "Casing the Treasury": p. 35; Robson, J., s.v. "al-Baghawī", in *EI* 2, I (1960): p. 893.

61 Ibn Ḫallikān, *Wafayāt al-a'yān wa anbā' al-zamān*, eds Yūsuf 'Alī Ṭawīl and Maryam Qāsim, 6 vols. (Beirut: Dār al-kutub al-'ilmiyya, 1998): IV, pp. 238-9; Ibn al-Fuwaṭī, *Majma' al-ādāb*: II, p. 199; Ibn Karbalā'ī, *Rawżāt al-jinān*: I, pp. 285-90.

62 Ibn al-Fuwaṭī, *Majma' al-ādāb*: III, pp. 95, 530; V, pp. 54, 78-9; al-Ḏahabī, *Tārīḫ al-islām wa wafayāt al-mašāhīr wa'l-a'lām*, ed. 'Umar 'Abd al-Salām Tadmurī, 61 vols. (Beirut: Dār al-kitāb al-'arabī, 1987-2004): LXI, pp. 100-1, 202-3, 322. On Šīrāzī, see Anwār, Sayyed 'Abd-Allāh, s.v. "Qoṭb-al-Din Širāzi", in *EIr*, online edition (2005). On Ḥamuwayi and his family, see Elias, Jamal J., "The Sufi Lords of Bahrabad: Sa'd al-Din and Sadr al-Din Hamuwayi",

For some time prior to 674/1275 Muḥyī al-Dīn shared the post of qadi with the same relative of the Malikān who, at the time of the Khwarazmian conquest, had been temporarily replaced by Muḥyī al-Dīn's own father. He then held the position alone, until his death in 697/1298.[63] Muḥyī al-Dīn is usually referred to as chief qadi (*qāḍī al-quḍāt*) of Tabriz, but in fact we know very little about the organization of the city's judiciary. His son was one of four individuals who, as qadi of Tabriz, signed a laudatory statement (*taqrīẓ*) concerning the theological works of Rašīd al-Dīn.[64] Such a *taqrīẓ* is generally issued on request, but not one of the four signed as chief qadi. They may thus have been equal in rank, probably representing not more than two schools of law, and may even have all been Šāfiʿīs. It is quite certain that all four belonged to families based in Tabriz for at least one generation.

Other members of the Qazwīnī family served in various functions under the Ilkhan Ġāzān, and under Öljeitü, his brother and successor (d. 716/1316).[65] Although Öljeitü briefly adopted Shiʿi Islam, a distant cousin of Muḥyī al-Dīn's, who was leader of the Sunni-Šāfiʿī party in his court, remained chief qadi of the Ilkhanid dominions (*qāḍī quḍāt al-mamālik*).[66] Following this last period there is no longer any detailed information on the Qazwīnīs, but it is clear the Malikān were still flourishing in 735/1336, at the time of the death of Öljeitü's son Abū Saʿīd, and the subsequent collapse of the Mongol dynasty. Indeed the *Safīna* contains a letter dealing with the problem of succession to Abū Saʿīd,

---

IrSt, XXVII/1 (1994): pp. 53-75. Among other matters, Elias discusses the vocalization Ḥamuwayi, rather than Ḥamawayh or Ḥamūya.

63 Ibn al-Fuwaṭī, *Majmaʿ al-ādāb*: III, p. 530; V, p. 54.

64 van Ess, Josef, *Der Wesir und seine Gelehrten. Zu Inhalt und Entstehungsgeschichte der theologischen Schriften des Rašīduddīn Fażlullah (gest. 718/1318)* (Wiesbaden: Franz Steiner Verlag, 1981): pp. 24, 28, 30. Muḥyī al-Dīn's son is only attested through his *taqrīẓ*. On the other three, including a relative of the Malikān, as well as on their families, also see Hoffmann, *Waqf*: pp. 48-50; Ibn al-Fuwaṭī, *Majmaʿ al-ādāb*: III, p. 97; Mustawfī Qazwīnī (Ḥamd-Allāh), *Tārīḫ-i guzīda*, ed. ʿAbd al-Ḥusayn Nawāʾī (Tehran: Amīr Kabīr, 1339š./1960): p. 803.

65 Rašīd al-Dīn, *Jāmiʿ al-tavārīḫ*: II, p. 1309; Qāšānī, *Tārīḫ-i Uljāytū*, ed. Mahīn Hambalī (Tehran: Bungāh-i tarjuma va našr-i kitāb, 1349š./1970): p. 42; Ibn al-Fuwaṭī, *Majmaʿ al-ādāb*: I, pp. 175-6; III, pp. 424-5.

66 Qāšānī, *Tārīḫ-i Uljāytū*: pp. 76, 96-108, 137, 221. Mustawfī Qazwīnī, *Tārīḫ-i guzīda*: pp. 810-1, names the family as Kiyāʾān, and lists only ʿIzz al-Dīn, Muḥyī al-Dīn, and Öljeitü's chief qadi Niẓām al-Dīn Marāġī Qazwīnī (d. 716/1316), as members. Niẓām al-Dīn and a nephew of Muḥyī al-Dīn also wrote *taqārīẓ* for Rašīd al-Dīn's theological works; van Ess, *Der Wesir*: pp. 23, 28.

and a teacher of the compiler, named Jalāl al-Dīn ʿAtīqī (d. 744/1343), likewise outlived the house of Hülegü.[67]

Jalāl al-Dīn ʿAtīqī pertained to another old leading family of Tabriz. Like his father he was best known as a poet at the courts of Öljeitü and Abū Saʿīd, by the middle of the eighth/fourteenth century.[68] But in fact the ʿAtīqīs were first of all Islamic scholars, preachers and Sufis, who had close ties to the Abbasids in the first half of the seventh/thirteenth century.[69] It is possible that these ʿAtīqīs had contracted a matrimonial relationship with the famous Suhrawardī family, whose members lent crucial support to the efforts for promotion of scholarly Sufism under the last caliphs in pre-Mongol Baghdad.[70] This does not necessarily imply that the ʿAtīqīs belonged to a distinct *Suhrawardiyya* 'brotherhood' or 'order'. However they continued to author commentaries on the Quran and transmit hadith, even in the times when the rulers of Tabriz were not Muslim. This was one of several long-standing urban families still giving rise to notable individuals in the post-Ilkhanid period. However, under Mongol rule, families from the hinterland villages of Tabriz also began rising to prominence.

5 The Kujujīs and the Rise of Rural Elites under Mongol Rule

The Kujujīs, originally from the village Kujujān (about ten kilometres south of Tabriz), developed as the most prominent of these rural families. The spiritual ancestor, Ḫwāja Muḥammad Kujujānī, died in 677/1279.[71] This individual

67 Seyed-Gohrab "Casing the Treasury": pp. 20-1; Al-e Davud, Sayyed Ali, "A Review of the Treatises and Historical Documents in *Safīna-yi Tabrīz*", in Seyed-Gohrab, A.-A. and McGlinn, S. (eds), *The Treasury of Tabriz. The Great Il-Khanid Compendium* (Amsterdam and West Lafayette: Rozenberg Publishers and Purdue University Press, 2007): pp. 79-89, here pp. 85-8.

68 ʿAtīqī, Jalāl al-Dīn, *Dīvān-i ʿAtīqī*, fac simile edition by Naṣr Allāh Pūrjawādī and Saʿīd Karīmī (Tehran: Farhangistān-i zabān va adab-i fārsī, 1388š./2009): pp. 11-9 (Intr.). Mustawfī Qazwīnī, *Tārīḫ-i guzīda*: pp. 725-6, 745.

69 Ibn al-Fuwaṭī, *Majmaʿ al-ādāb*: III, p. 396; IV, p. 192; Ibn Karbalāʾī, *Rawżāt al-jinān*: I, pp. 354-8; van Ess, *Der Wesir*: pp. 23, 31.

70 Ibn Karbalāʾī, *Rawżāt al-jinān*: I, p. 565, wrote down a genealogy of the ʿAtīqīs (*nasab-nāma-yi ʿAtīqiyān*), positing a matrimonial relationship with the Suhrawardīs. The editor Jaʿfar Sulṭān al-Qurrāʾī included this genealogy, which has some flaws, in the appendix to his edition of the main work. The scope of the current article does not permit discussion of other hints to the matrimonial relationship between the two families. See Ohlander, Erik S., *Sufism in an Age of Transition. ʿUmar al-Suhrawardī and the Rise of the Islamic Mystical Brotherhoods* (Leiden: Brill, 2006): pp. 57-106, on Šihāb al-Dīn ʿUmar Suhrawardī, his family, and their links to the Abbasid court.

71 Ibn Karbalāʾī, *Rawżāt al-jinān*: II, pp. 9-10, 38.

and his tomb in Kujujān are recorded in the two major early eighth/fourteenth century works by Ḥamd-Allāh Mustawfī.[72] But the basic source is a hagiographical account by a disciple, in Arabic and later translated into Persian. (I will return to this translation below.) Ḫʷāja Muḥammad's disciple had studied an esteemed commentary of the Quran with its author in Mosul. He then came to Tabriz and Kujujān to visit the master on a number of occasions, including during Ramaḍān 675/February-March 1277.[73] The fact that such a learned man would seek the mystical guidance of Ḫʷāja Muḥammad is another example disproving the existence of any rift between traditional scholars and Sufis.

Ḫʷāja Muḥammad belonged to a loose group of local Sufi šayḫs, which Ibn Karbalā'ī introduces as the "70 Bābās of Tabriz".[74] Most remain obscure, however some tentative remarks are possible. First, it is unclear when and how the individuals listed by Ibn Karbalā'ī came to be considered as a group. Second, these local Sufis did not constitute a community of any kind. Yet several bābās were connected in a complex web of master-disciple relationships, in subsequent generations often intersecting with matrimonial ties. Particularly important with regard to the generally accepted history of Sufism is that some of the bābās, as well as later members of their families and other local Sufis, were direct or indirect disciples of representatives of widespread traditions (usually called 'brotherhoods' or 'orders') or were at least acquainted with these representatives. Among these 'brotherhoods' or 'orders' there were the aforementioned *Suhrawardiyya*, as well as the *Kubrāwiyya*, however there is no evidence that such distinct groups existed in Tabriz between the sixth/twelfth and ninth/fifteenth centuries. Instead, local cults and corresponding interrelated Sufi lineages are well attested, prior to, under and following Mongol rule.

The main distinction among the 70 Bābās of Tabriz would be between those based in the city or its immediate suburbs and those of the villages. The urban and the rural Sufi milieus were not neatly separated. Nevertheless, the relations within the urban milieu itself, and within the rural milieu on the other hand, appear to have been much closer than those across the city-village divide. The chain of initiation (*silsila*) of Ḫʷāja Muḥammad Kujujānī, for instance,

---

72 Mustawfī Qazwīnī, *Tārīḫ-i guzīda*: p. 672; Mustawfī Qazwīnī, *Nuzhat al-qulūb*: p. 89.

73 Ibn Karbalā'ī, *Rawżāt al-jinān*: II, pp. 15, 36. On the teacher of Ḫʷāja Muḥammad's disciple, see Ibn al-Fuwaṭī, *Majmaʿ al-ādāb*: V, pp. 593-4.

74 Ibn Karbalā'ī, *Rawżāt al-jinān*: I, pp. 49-50; Lewisohn, Leonard, *Beyond Faith and Infidelity. The Sufi Poetry and Teachings of Maḥmūd Shabistarī* (Richmond: Curzon Press, 1995): pp. 126-35 discussed some of the bābās based mainly on their presentation in an early eighth/fourteenth work by the famous Sufi poet Maḥmūd Shabistarī.

was traced through the šayḫs of the neighbouring village of Šādābād.[75] Ḫwāja Muḥammad is said to have gone to Tabriz once as a young boy, when he received an inspiring glance (*naẓar*) from Bābā Ḥasan Surḫābī, the original leader of the 70 Bābās, who had a lodge (*zāwiya*) in the suburb of Surḫāb.[76] On the other hand, a preferred venue for the most noted city bābās of the mid-seventh/thirteenth century was the *zāwiya* of Ḫwāja Ṣāyin al-Dīn Yaḥyā Tabrīzī (d. 683/1284), inherited from his father,[77] and there are no explicit mentions of a village bābā attending a session there.

A tale about Ḫwāja Yūsuf Ḥayrān Dihḫwārqānī (d. 670/1271) and his disciple Ḫwāja ʿAlī Bādāmyārī (d. 699/1299) further strengthens the argument that there was a relatively pronounced urban-rural divide. Ḫwāja Yūsuf once prompted a gathering of eminent village bābās, included Ḫwāja Muḥammad Kujujānī, at his house in Dihḫwārqān, to secure their blessings for his unborn nephew.[78] However no city bābā is named as present. One of the rural Sufis in attendance, Bābā Faqīh Aḥmad Asbustī, had migrated from Konya to Tabriz and settled in Asbust. Almost all others were natives of the villages they lived in, and were eventually buried in these same villages in what developed into family shrines.[79] The majority probably had rather modest formal education. Aḥmad's additional designation as *faqīh* identifies him as a scholar, and indeed among these 70 Bābās he passed as the expert in Islamic law.

Some among the second generation of prominent village bābā families frequented urban scholarly milieus. Ḫwāja Yūsuf's nephew, known as Pīr Čūpān Walī (d. 724/1324), studied the most authoritative hadith collection, the *Ṣaḥīḥ al-Buḫārī*, as transmitted from the copy brought to Tabriz by Imām Ḥafda. He was eventually buried in Bādāmyār, next to his mystical master Ḫwāja ʿAlī.[80] Šihāb al-Dīn ʿAtīqī (d. ca. 663/1265) represented an important link between the urban and the rural milieus even farther into Azarbaijan, as early as the first generation. Šihāb al-Dīn's mausoleum is located in Ahar, between Tabriz and Ardabil, where the ʿAtīqīs had maintained a second base since pre-Mongol

75 Ibn Karbalāʾī, *Rawżāt al-jinān*: II, pp. 4-10.

76 Ibid.: I, pp. 50-2.

77 Ibid.: I, pp. 121-2, 296-9.

78 Ibid.: II, pp. 71-9.

79 Ibid.: II, pp. 44-8, 61-5. Asbust is located between the district of Sardrūd and the village of Uskū or Uskūya.

80 Ibid.: II, pp. 78-9. The author reproduces an *ijāza* issued by Pīr Čūpān whose teacher was an opponent of Rašīd al-Dīn, see al-Birzālī, *al-Muqtafā ʿalā kitāb al-rawḍatayn al-maʿrūf bi-tārīḫ al-Birzālī*, ed. ʿUmar ʿAbd al-Salām Tadmurī, 4 vols. (Beirut: al-Maktaba al-ʿaṣriyya, 2002-06): IV, pp. 318, 331, 364. Bādāmyār is located in the immediate environs of Dihḫwārqān.

times. He occupies a key position in the Safavid *silsila*, in that of Aḥmad Asbustī and of two other bābās from villages of Tabriz, most notably Bāla Ḥasan Binīsī.[81]

Ṣafī al-Dīn Ardabīlī (d. 735/1334), the Safavid eponym, had forged ties to Ilkhanid court circles by the time of the reign of Abū Saʿīd, however experienced difficulties in gaining acceptance among the scholars and Sufis of Tabriz.[82] These difficulties are illustrated by the fact that only one member of a village bābā family is specifically indicated as a follower of Ṣafī al-Dīn.[83] Indeed in the first decades after his death, his community deemed it necessary to connect Ṣafī al-Dīn Ardabīlī, and also his master Šayḫ Zāhid Gīlānī, with Bāla Ḥasan Binīsī and Ḫʷāja Muḥammad Kujujānī. This Bāla Ḥasan was credited with a vision predicting the future appearance of Šayḫ Zāhid, as well as another vision jointly with Ḫʷāja Muḥammad, spiritual ancestor of the Kujujīs, predicting the future appearance of Šayḫ Ṣafī.[84]

Ḫʷāja Muḥammad, and another outstanding village bābā of Tabriz, Bābā or Pīr Ḥāmid Sardrūdī (d. 690/1291), had clearly impressed even the learned elites of the city. Pīr Ḥāmid, as well as Muḥammad Kujujī, had authored poems in ancient regional Iranian tongues (*fahlawiyyāt*) which the compiler of the *Safīna* included in his portable private library.[85] Later members of the Kujujī family were descended from the brother of Ḫʷāja Muḥammad, named Ibrāhīm. By the ninth/fifteenth century, these had intermarried with the Ḥāmidī šayḫs of Sardrūd, and with the family of another village bābā.[86] All remained firmly rooted in the Sufi milieus of Tabriz and its villages. In the eighth/fourteenth century, the Kujujīs appear to have led a Sufi movement comparable to that of the Safavids. This certainly involved rivalries between the two families, for

81 Ibn Bazzāz, *Ṣafvat al-ṣafā dar tarjuma-yi aḥvāl va aqvāl va karāmāt-i Šayḫ Ṣafī al-Dīn Isḥāq Ardabīlī*, ed. Ġulām-Riḍā Ṭabātabāʾī Majd (Ardabil, 1372š./1993): p. 180; Ibn Karbalāʾī, *Rawżāt al-jinān*: I, pp. 249-50, 357; II, pp. 48, 81-2; Ibn al-Fuwaṭī, *Majmaʿ al-ādāb*: II, pp. 195-6, 343; V, p. 276. Evidence on family members in Ahar, and on Šihāb al-Dīn in particular, is not conclusive. For a general study on the latter personage, see Īmānī, Bihrūz, "Bargī az Šarḥ-i ḥāl-i Šayḫ Šihāb al-Dīn Maḥmūd Aharī", *Payām-i Bahāristān* XXX (1383š./2004): pp. 79-83.

82 Gronke, *Derwische im Vorhof der Macht*: pp. 132-6; Ibn Bazzāz, *Ṣafvat al-ṣafā*: pp. 646-9, 808-16; Ibn Karbalāʾī, *Rawżāt al-jinān*: I, pp. 336-8.

83 Ibn Bazzāz, *Ṣafvat al-ṣafā*: pp. 272-3, 740-3; Ibn Karbalāʾī, *Rawżāt al-jinān*: II, pp. 64-5.

84 Ibn Bazzāz, *Ṣafvat al-ṣafā*: pp. 67-8, 190-2; Ibn Karbalāʾī, *Rawżāt al-jinān*: II, pp. 83-7.

85 Seyed-Gohrab, "Casing the Treasury": p. 18. On Bābā Ḥāmid, also see DeWeese, Devin, "'Stuck in the Throat of Chingīz Khān:' Envisioning the Mongol Conquests in Some Sufi Accounts from the 14th to 17th Centuries", in Pfeiffer, J. and Quinn, S. A. (eds.) *History and Historiography of Post-Mongol Central Asia and the Middle East. Studies in Honor of John E. Woods* (Wiesbaden: Harrassowitz Verlag, 2006): pp. 23-60, here pp. 37-9.

86 Ibn Karbalāʾī, *Rawżāt al-jinān*: I, pp. 486-7; II, pp. 39-42, 667. The second village bābā family with which the Kujujīs intermarried was based in Ḫusrawšāh.

instance for followers, but also common concerns, such as shielding their respective movements from persecution. Further research on the early Safavids and the Kujujīs would be needed to enhance our understanding of similarities and differences between these movements.

The recent publication of the *Kujuji-waqfiyya*,[87] provides new evidence, showing that the descendants of Ḫʷāja Ibrāhīm and spiritual heirs of Ḫʷāja Muḥammad were the equal of the Safavids in terms of social status, and that they disposed of similar material resources. The large part of the portfolio consisted of agricultural estates in the rural hinterland, supplemented by suburban fields and gardens, and urban structures such as shops in city markets. The large Kujujī *waqf* complex in Tabriz was established in 782/1380 by a great-grandson of Ḫʷāja Ibrāhīm, named Ġiyāṯ al-Dīn Muḥammad, also known as Ḫʷāja Šayḫ Kujujī.[88] However some possessions had been given in endowment a generation earlier. Indeed Ḫʷāja Šayḫ's father had not only possessed a *zawiya* in Tabriz, but very likely also already founded one in Damascus, where family members remained present as Sufi masters for at least several decades.[89] However, among the Kujujī family, it was Ḫʷāja Šayḫ who would first rise as a major political figure in Tabriz and as a highly influential courtier beyond the local level.

6 The Kujujīs and Dynastic Succession following Mongol Rule

After the death of the Ilkhan Abū Saʿīd in 736/1335, the two principal contenders to succession pertained to the houses of the Chupanids and the Jalayirids, both Mongol military aristocratic families related through various ways to the Ilkhanid dynasty and to each other. For some time, their leaders raised true or pretended descendants of Hülegü to the throne, left it vacant, or recognised other Chinggisids as nominal sovereigns.[90] In 738/1338, the Chupanid Šayḫ Ḥasan expelled his Jalāyirid namesake from Azarbaijan, who then ruled in Baghdad until his death in 757/1356.

The Chupanid were the first serious successors of the Ilkhans at Tabriz, but troubles increased at the time when the conqueror's wife had him murdered, in 744/1343. Šayḫ Ḥasan's brother, Malik Ašraf, then met with resistance when

87 Werner et al., *Die Kuğuğī-Stiftungen*: pp. 37-9.

88 The complex was located in the vicinity of the ʿAlišāh-Mosque; the latter is indicated in the map in fig. 4.

89 Werner et al., *Die Kuğuğī*-Stiftungen: pp. 24-6, 88-90.

90 Melville, Charles and Zaryāb, ʿAbbās, s.v. "Chobanids", in *EIr*, V (1991): pp. 496-202; Jackson, Peter, s.v. "Jalayerids", in *EIr*, XIV (2008): pp. 415-9.

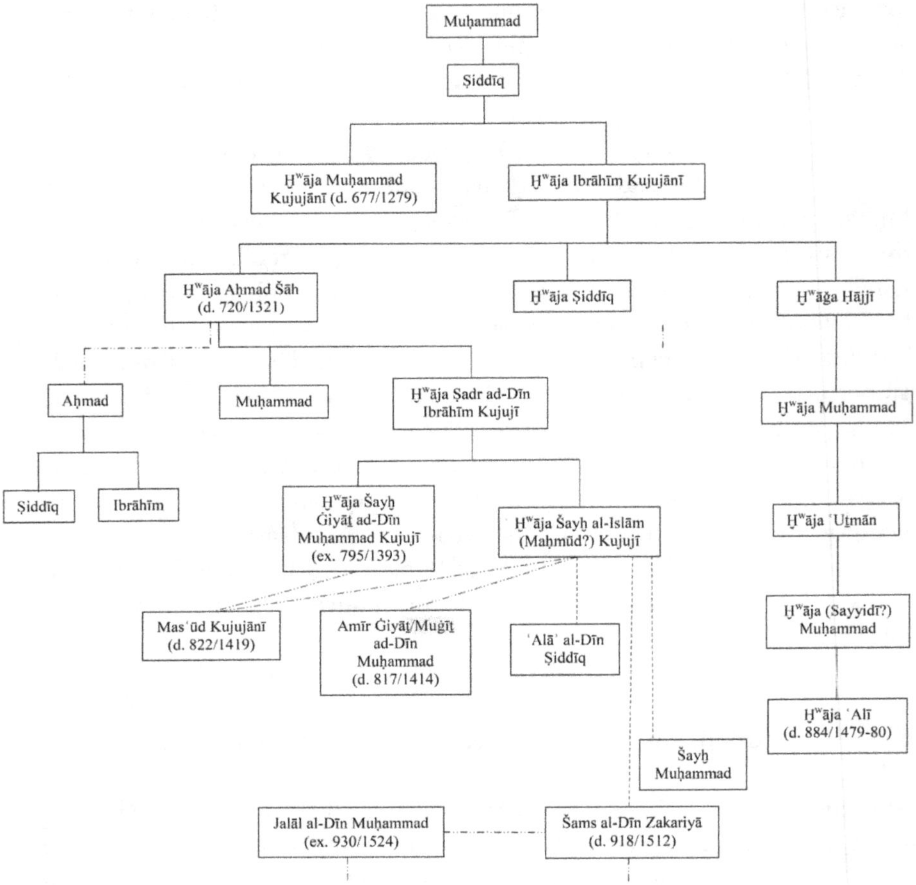

FIGURE 3 The Kujujī family

he attempted to take the city, so two prominent Sufis ultimately led a delegation for negotiation of terms of surrender.[91] Both of these had settled in Tabriz under the Mongols. Malik Ašraf was devoted to one of them, who was named Niẓām al-Dīn Ġūrī (d. 752/1351) and had become a promoter of the local cult of Bābā Mazīd Surḫābī in Tabriz. Bābā Mazīd had emerged as leader of the 70 Bābās, in succession to Bābā Ḥasan, even though he was not a disciple of the

91 Aharī, *Tavārīḫ-i Šayḫ Uways*, ed. Īraj Afšār (Tabriz: Sutūda, 1389š./2010): pp. 229-30; Mustawfī Qazwīnī, *Ẕayl-i Tārīḫ-i guzīda* ed. Īraj Afšār (Tehran: Mawqūfāt-i Duktur Maḥmūd Afšār Yazdī, 1372š./1993): p. 31.

original leader.[92] Malik Ašraf's vizier was, like Niẓām al-Dīn Ġūrī, a notable of Tabriz, but all in all the new Chupanid ruler enjoyed little support among the inhabitants, and by 751/1350 even his Sufi master Niẓām al-Dīn no longer trusted him.[93] The historian Zayn al-Dīn b. Ḥamd-Allāh Mustawfī reports, under the year 757/1356:

> In the realm of Malik Ašraf his tyranny reached extreme proportions, making the people leave their homeland. Ḫwāja Šayḫ Kujujī went to Shiraz and on to Damascus where he erected lofty buildings consisting of Sufi lodges. Ḫwāja Ṣadr al-Dīn [b. Ṣafī al-Dīn] Ardabīlī, mercy upon him, went to Gilan.[94]

Ḫwāja Šayḫ Kujujī must also have intensified his contacts with the Jalāyirids in Baghdad at the time, while another exile encouraged Jānī Beg Khan of the Golden Horde to invade Azarbaijan. The Jochid ruler was welcomed in Tabriz, and had Malik Ašraf executed in 758/1357. He then returned to his dominions, and soon died.[95] The Muẓaffarids of Fars then briefly occupied the city. In 761/1360, Ḫwāja Šayḫ Kujujī secured the succession of the Jalāyirid Sultan Uways as ruler of Tabriz.[96] Ḫwāja Šayḫ remained close to Uways, and, in 776/1374, he was even admitted to the ruler's deathbed to partake in deciding on his successor.[97] Uways was buried in Šādābād at the shrine complex of the local family of village šayḫs.[98] Their connection to the Kujujīs has already been noted.

Uways' successor, Sultan Ḥusayn Jalāyir, was subject to both internal and external attacks. Ḫwāja Šayḫ Kujujī, for his part, did not hesitate to welcome another Muẓaffarid pretender to Tabriz in 778/1376, and may even have incited him to take over the city.[99] The resulting short-term occupation would not

92 Ibn Karbalā'ī, *Rawżāt al-jinān*: I, pp. 59, 96-7, 108-13, 339-45.

93 Mustawfī Qazwīnī, *Ẕayl-i Tārīḫ-i guzīda*: pp. 51-2; Aharī, *Tavārīḫ-i Šayḫ Uways*: p. 231, is the only one to give the vizier's full name 'Abd al-Ḥayy Ḥammāmī Tabrīzī.

94 Mustawfī Qazwīnī, *Ẕayl-i Tārīḫ-i guzīda*: p. 57 (*dar mamlakat-i Malik Ašraf čūn ẓulm-i ū bih ġāyat rasīd mardum julāy vaṭan kardand. Ḫwāja Shaykh Kujujī bih Šīrāz raft va az ānjā bih Šām raft va 'imārāt-i 'āliya az zāviya va ḫānaqāh dar Šām bi-sāḫt; va Ḫwāja Ṣadr al-Dīn Ardabīlī, 'alayhi al-raḥma, bih Gīlān raft*).

95 Aharī, *Tavārīḫ-i Šayḫ Uways*: pp. 236-9; Mustawfī Qazwīnī, *Ẕayl-i Tārīḫ-i guzīda*: pp. 58-65.

96 Mustawfī Qazwīnī, *Ẕayl-i Tārīḫ-i guzīda*: pp. 66-70.

97 Naṭanzī, *Muntaḫab-i tavārīḫ-i mu'īnī* (*ta'līf-i 816 va 817 h.q.*), ed. Jean Aubin (Tehran: Kitābfurūšī-yi Ḫayyām, 1336š./1957): p. 167; Ḥāfiẓ-i Abrū, *Zubdat al-tavārīḫ*, ed. Kamāl Ḥājj Sayyid Jawādī, 4 vols. (Tehran: Sāzmān-i čāp va intišārāt-i Wizārat-i farhang va iršād-i islāmī, 1380š./2001): I, p. 489.

98 Ibn Karbalā'ī, *Rawżāt al-jinān*: II, pp. 6-8.

99 Mustawfī Qazwīnī, *Ẕayl-i Tārīḫ-i guzīda*: p. 95. Ḥāfiẓ-i Abrū, *Zubdat al-tavārīḫ*: 1:506-7.

endure. Nonetheless Ḫʷāja Šayḫ's influence at the Jalāyirid court was curtailed while Sultan Ḥusayn's position remained precarious.[100] This situation appears to have motivated the Kujujī leader to transfer a sizable part of his possessions to a pious foundation, in 782/1380. Two years later, Ḫʷāja Šayḫ procured the assassination of Sultan Ḥusayn and brought Ḥusayn's brother, Sultan Aḥmad, to the throne.[101] The new ruler duly reconfirmed the Kujujī-*waqf* in 784/1382, and Ḫʷāja Šayḫ supported him against rival Jalāyirid claimants.[102] Still, even greater troubles than these loomed for Sultan Aḥmad and the principal city of Mongol Iran.

In 787/1385, Toqtemiš Khan of the Golden Horde (d. 809/1406) attacked Tabriz during his struggle with Timur, who had raised his claim in the name of the Chaghadaid branch of the Chinggisid dynasty.[103] Mongol dynastic claims were certainly of limited relevance to the local elites in such situations. Negotiations with the armies of Toqtemiš failed, and the notables decided to defend their city by military means. One of them, Qadi Quṭb al-Dīn ʿUbaydī, died while fighting the Jochid assailants.[104] Since as early as the time of the Ilkhans, the ʿUbaydīs had been known as astronomers, but primarily as Šāfiʿī jurists.[105] Qadi Quṭb al-Dīn's father had penned the *Kujujī-waqfiyya*, and Qadi Quṭb al-Dīn himself, his brother Faḍl-Allāh and a member of the aforementioned ʿAtīqī family, had all testified to the legality of the endowment.[106] Another notable, Sayyid Riḍā Naqīb Ḥasanī, was humiliated by the troops of Toqtemiš, and then in 788/1386, joined the delegation welcoming Timur to Tabriz.[107] Sayyid Riḍā's ancestors were reportedly the first Alids to settle in the city and most probably held the post of *naqīb* on a hereditary basis. Family

100 Werner et al., *Die Kuǧuǧī-Stiftungen*: p. 33.

101 Ibn Ḥajar, *Inbāʾ al-ġumr bi-abnāʾ al-ʿumr*, ed. Ḥasan Ḥabašī, 4 vols. (Cairo, 1969): I, p. 246, 263.

102 Werner et al., *Die Kuǧuǧī-Stiftungen*: p. 33-4.

103 Manz, Beatrice Forbes, "Mongol history rewritten and relived", in Aigle, D. (ed.), *Figures mythiques de l'Orient musulman*, REMM, LXXXIX/XC (2000): pp. 129-49, here pp. 137-9.

104 Mustawfī Qazwīnī, *Ẕayl-i Tārīḫ-i guzīda*: pp. 166-7, 177. Ibn Karbalāʾī, *Rawżāt al-jinān*: II, pp. 648-9, 656.

105 Ragep, F. Jamil, "New Light on Shams: The Islamic Side of Σάμψ Πουχάρης", in Pfeiffer, J. (ed.), *Politics, Patronage and the Transmission of Knowledge in 13th-15th century Tabriz* (Leiden: Brill, 2014): pp. 231-47, here p. 244; Seyed-Gohrab, "Casing the Treasury": p. 17; van Ess, *Der Wesir*: pp. 22, 28-9; Hoffmann, *Waqf*: pp. 50-1.

106 Werner et al., *Die Kuǧuǧī-Stiftungen*: pp. 51-2, 104-6.

107 Mustawfī Qazwīnī, *Ẕayl-i Tārīḫ-i guzīda*: p. 177; Ibn Karbalāʾī, *Rawżāt al-jinān*: II, p. 656; Yazdī, *Ẓafar-nāma*, ed. Sayyid Saʿīd Mīr Muḥammad Ṣādiq [ʿAbd al-Ḥusain Nawāʾī], 2 vols. (Tehran: Kitābḫāna-yi mūza va markaz-i asnād-i Majlis-i šūrā-yi islāmī, 1387š./2008): I, p. 562.

members are well attested from the time of the Muslim Ilkhans to the early ninth/fifteenth century.[108]

Prior to the Jochid assault on Tabriz and the arrival of Timur at the city, Ḫwāja Šayḫ Kujujī had apparently departed for Baghdad with the Jalāyirid Sultan Aḥmad. It is possible that Ḫwāja Šayḫ instigated a reported plot against Timur's life, and indeed it is quite likely that he attempted to return Tabriz to the hands of his Jalāyirid protégé.[109] Shortly afterwards, Timur returned to Samarqand. Numerous contenders, including the Qara Qoyunlu, then fought for control of Tabriz over the coming years.[110] The conqueror returned in 795/1393, and installed his son Mīrān-Šāh as governor. Timur expelled Sultan Aḥmad from Baghdad, but the latter had Ḫwāja Šayḫ Kujujī executed, prior to his escape.[111] The Kujujīs in Tabriz certainly sought accommodation with the Timurids. Their associate, who translated the above-mentioned hagiographical account of Ḫwāja Muḥammad Kujujānī, definitely served the governor.[112]

Timur removed Mīrān-Šāh from the governorship of Tabriz during his last great western campaign, which began in 802/1399, when the latter attempted to assert his independence. But Timur then reassigned the region to the family of Mīrān-Šāh.[113] The notables would certainly have been unhappy at having their city subordinate to distant Samarqand, and had probably endorsed Mīrān-Šāh's attempt at independence. Yet after Timur's death in 807/1405, Mīrān-Šāh and his sons were unable to provide sufficient security and stability, and the Kujujīs instead initiated the succession of the Qara Qoyunlu. A year later, Ḫwāja Sayyidī Muḥammad Kujujī went to the camp of Qarā Yūsuf, near Naḫjawān, and urged him to protect Tabriz.[114] Qarā Yūsuf soon defeated the quarrelling Timurids. The Jalāyirid Sultan Aḥmad reoccupied the city in 813/1411, and was then likewise defeated. He was executed, and his corpse was

108 Ibn Karbalā'ī, *Rawżāt al-jinān*: I, pp. 155-7; Qāšānī, *Tārīḫ-i Uljāytū*: p. 42; van Ess, *Der Wesir und seine Gelehrten*: p. 30; Hoffmann, *Waqf*: p. 355.

109 Ḥāfiẓ-i Abrū, *Zubdat al-tavārīḫ*: II, p. 654; Metsobets'i, trans. Bedrosian, *History of Tamerlane and his Successors* (New York: Sources of the Armenian Tradition, 1987): p. 4.

110 Mustawfī Qazwīnī, *Ẕayl-i Tārīḫ-i guzīda*: pp. 124-39.

111 Manz, Beatrice Forbes, *The Rise and Rule of Tamerlane* (Cambridge: Cambridge University Press, 1989): p. 72; Ibn Ḥajar, *Inbā'*: I, p. 463; Naṭanzī, *Muntaḫab-i tavārīḫ-i mu'īnī*: p. 168.

112 Lewisohn, "Palāsī's Memoir": p. 346; Ḫwāndamīr, *Tārīḫ-i Ḥabīb al-siyar fī aḫbār-i afrād-i bašar*, ed. Jalāl al-Dīn Humā'ī and Muḥammad Dabīr Siyāqī, 4 vols. (Tehran: Kitābfurūšī-yi Ḫayyām, 1362š./1982): III, p. 550.

113 Manz, *Tamerlane*: pp. 72-3.

114 Mīrḫwānd, *Rawżat al-ṣafā*, ed. Naṣr Allāh Sabūḥī, 7 vols. (Tehran: Intišārāt-i markazī Ḫayyām-Pīrūz, 1338š./1959): VI, pp. 555-6; Ḫwāndamīr, *Ḥabīb al-siyar*: III, p. 569.

displayed in the madrasa of Ḫwāja Šayḫ Kujujī, demonstrating to the inhabitants that he was indeed dead.[115]

Qarā Yūsuf died in 823/1420, during the approach of Timur's successor Šāhruḫ. Without resisting, Tabriz surrendered to the huge Timurid army. Ḫwāja Sayyidī Kujujī provided for Qarā Yūsuf's corpse, and must have been among those notables who favoured the Qara Qoyunlu, and who ensured that they regained the city as soon as the Timurid had left for his base in Khurasan.[116] Šāhruḫ undertook two more major campaigns to the Ilkhanid heartland, but as Beatrice Manz convincingly argues, "real control over Azerbāijān was beyond his grasp, as it had been beyond Temür's".[117] Iskandar Qara Qoyunlu imposed himself as lord of Tabriz, against other sons of Qarā Yūsuf, and survived both these invasions. The Kujujīs preserved their pre-eminence among the local elites.

Šāhruḫ's first major campaign against Iskandar had in part been provoked by an assassination attempt, in 830/1426-27, allegedly by an adherent of the messianic Ḥurūfī movement. A son of the founder of the movement lived in Tabriz. He was suspected of having sent the assassin, and told Timurid officials that Ḫwāja Sayyidī Kujujī would exculpate him.[118] Although the Ḥurūfīs had gained a sizable following in the city, Ḫwāja Sayyidī's acquaintance with the individual in question should not be seen as evidence of messianic orientations among the Kujujīs. In the aftermath of Šāhruḫ's last invasion, in 835-36/1432-33, Jahān-Šāh b. Qarā Yūsuf prevailed over his brother Iskandar, and the latter was then killed by one of his sons.[119] Jahān-Šāh crushed the Ḥurūfīs of Tabriz in the wake of an 845/1441 uprising.[120] Qadi Najm al-Dīn Uskū'ī (d. 879/1474-75), from the village of Uskū or Uskūya, sanctioned this measure. Being a native of the rural hinterland he had at least indirect ties to some notable village bābā lineages and families.[121] With their deep roots in local Sufi milieus, especially those of the villages, the Kujujīs would have had little objection to crushing the Ḥurūfī movement, which originated from eastern Iran and had only gradually established a foothold in Azarbaijan.

---

115 Ḥāfiẓ-i Abrū, *Zubdat al-tavārīḫ*: III, pp. 169-72, 222-9, 401-4; Mīrḫwānd, *Rawżat al-ṣafā*: VI, pp. 579-83.

116 Ḥāfiẓ-i Abrū, *Zubdat al-tavārīḫ*: IV, pp. 735, 752; Ṭihrānī, *Kitāb-i Diyarbakriyya*, ed. Necati Logal and Faruk Sümer (Ankara: Türk Tarih Kurumu Basımevi, 1964): p. 89.

117 Manz, "Mongol History": p. 146.

118 Binbaş, İlker Evrim, "The Anatomy of a Regicide Attempt: Shāhrukh, the Ḥurūfīs, and the Timurid Intellectuals in 830/1426-27", *JRAS*, XXIII/3 (2013): pp. 391-428, here pp. 406-11.

119 Ṭihrānī, *Kitāb-i Diyarbakriyya*: pp. 138-9, 144-5.

120 Ibn Karbalā'ī, *Rawżāt al-jinān*: I, pp. 478-81.

121 Ibid.: I, pp. 410-1; II, pp. 49-54, 58-61, 95-9.

However in the post-Mongol era, most members of the Kujujī family were better known for involvement in court administration and government than as prominent Sufis. Another example of a Kujujī man of politics is ʿAlāʾ al-Dīn Ṣiddīq, a vizier of Jahān-Šāh in the years following Šāhruḫ's death (850/1447), when the Qara Qoyunlu ruler expanded his dominions eastwards at the expense of the Timurids.[122] ʿAlāʾ al-Dīn was dismissed and imprisoned around 860/1455-56. A certain Šams al-Dīn Ḥusayn Tabrīzī replaced both ʿAlāʾ al-Dīn and his co-vizier, but he too soon lost favour. In 863/1458, when Jahān-Šāh occupied the Timurid principal city of Herat, his son tried to seize power in Tabriz, supported by the same Šams al-Dīn as vizier. The ruler immediately returned and ordered the execution of the vizier, but pardoned his son.[123] Jahān-Šāh died during a campaign against his great enemy Uzun Ḥasan Aq Qoyunlu, in 872/1467. ʿAlāʾ al-Dīn Ṣiddīq Kujujī played a leading role in the Qara Qoyunlu succession struggle that erupted at Tabriz.[124] However in 874/1469, Uzun Ḥasan took over the city after defeating the Timurid Abū Saʿīd, who was taken captive in Azarbaijan and handed over to a disgruntled Timurid prince for execution.[125]

A relative of ʿAlāʾ al-Dīn Ṣiddīq, named Ḫʷāja ʿAlī Kujujī (d. 884/1479-80), was the best-known family member still active as a Sufi. Ḫʷāja ʿAlī further exemplifies the proximity of the family to the Qara Qoyunlu, having allegedly inspired Jahān-Šāh's wife to found the royal *waqf*-complex around the famous *Blue Mosque* of Tabriz, in 870/1465.[126] Qadi Najm al-Dīn Uskūʾī was the judge responsible for legalizing the endowment, and Qadi Quṭb al-Dīn, of the ʿAtīqī family introduced above, took an administrative position in the complex, which was intended as an inheritance for his descendants. The deed of endowment stipulated that officials working in the complex were to adhere to Sunni Islam.[127] For his part, Ḫʷāja ʿAlī was intimately associated with an eminent Sunni Sufi of the time, suggesting that the Kujujīs also adhered to Sunni Islam.

The name of Ḫʷāja ʿAlī's associate, who in a way represented the broad *Kubrāwī* Sufi tradition, was Sayyid Badr al-Dīn Aḥmad (d. 912/1507). Many scholars of Sufism regard this individual as the founder of a local Sunni branch of an increasingly Shiʿa-leaning *Kubrāwiyya* 'order',[128] however this interpretation

122 Ṭihrānī, *Kitāb-i Diyarbakriyya*: p. 258.
123 Ibid.: pp. 359-60.
124 Ibid.: pp. 434-9.
125 Woods, *The Aqquyunlu*: pp. 96-100.
126 Ibn Karbalāʾī, *Rawżāt al-jinān*: II, p. 43.
127 Werner, Christoph, "Ein Vaqf für meine Töchter. Ḫātūn Jān Bēgum und die Qarā Quyūnlū Stiftungen zur ‚Blauen Moschee' in Tabriz", *Der Islam*, LXXX (2003): pp. 94-109, here pp. 99-101; Maškūr, Muḥammad Jawād, *Tārīḫ-i Tabrīz tā pāyān-i qarn-i nuhum-i hijrī* (Tehran: Anjuman-i Aṯār-i Millī, 1352š./1973): pp. 675-6.
128 Arjomand, Said A., *The Shadow of God and the Hidden Imam. Religion, Political Order and Societal Change in Shiʿite Iran from the Beginning to 1890* (Chicago-London: The University

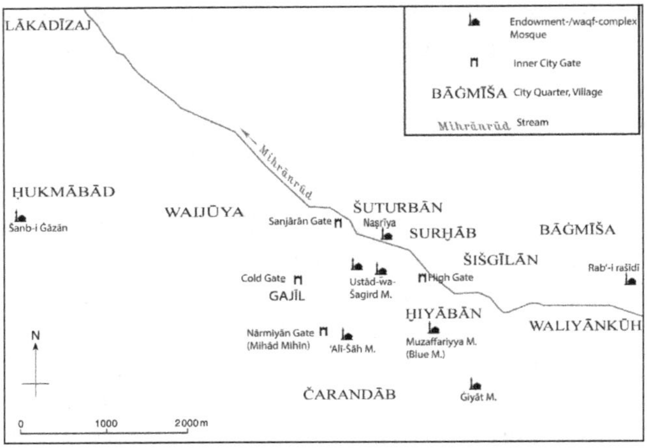

FIGURE 4 Tabriz and its suburbs
DRAWING: BORLEIS/WEISS, LEIPZIG

should be rejected. Ḫwāja ʿAlī helped Sayyid Badr al-Dīn settle in the village of Lāla, when the latter returned to Tabriz at the time of Jahān-Šāh's death. Lāla is situated about five kilometres south of the city. Here, Badr al-Dīn's Sufi lodge, named *Darwīšābād*, attracted mystics of various affiliations.[129] As with the Safavids and the Kujujīs, family members had also emigrated from Azarbaijan under the Chupanid Malik Ašraf. Badr al-Dīn's ancestors had Sufi ties to the Safavids and other communities. However, decades after the Safavids had seized power as self-declared Shiʿi rulers, Badr al-Dīn's descendants, known as the "Sayyids of Lāla", remained the main proponents of Sunnism in Tabriz.[130]

Given their proximity to the Qara Qoyunlu, the Kujujīs probably lost much prominence at court when Uzun Ḥasan Aq Qoyunlu took control over Tabriz.

---

of Chicago Press, 1984): p. 114; Algar, Hamid, s.v. "Kobrawiya ii. The Order", in *EIr*, online edition (2009); Lingwood, Chad G., *Politics, Poetry and Sufism in Medieval Iran. New Perspectives on Jāmī's* Salāmān va Absāl (Leiden and Boston: Brill, 2014): pp. 84-5. Algar criticised Arjomand for using the term *Ḏahabiyyah* to designate that supposed branch due to its absence in Ibn Karbalāʾī's work. Algar himself mistook a designation naming individual spiritual ancestors of Sayyid Badr al-Dīn which the editor Jaʿfar Sulṭān al-Qurrāʾī added in the table of contents as an explanation for a designation used by the author. Ibn Karbalāʾī, *Rawżāt al-jinān*: II, p. 207, uses the formula *silsilat al-ḏahab wa rābiṭat al-adab* to designate Sayyid Badr al-Dīn's chain of mystical initiation.

129 Ibn Karbalāʾī, *Rawżāt al-jinān*: II, pp. 150-2.

130 Ibid.: II, pp. 109-13; Šuštarī, *Majālis al-muʾminīn*, ed. Aḥmad ʿAbd al-Munāfī (Tehran, 1955): I, p. 82. For a study of the Kujujīs and the Sayyids of Lāla after the coming to power of the Safavids, see Zakrzewski, Daniel, "Local Aristocrats in a Time of Critical Dynastic Change. Two Notable Families of Tabriz Facing the New Safavid Order", DYNTRAN Working Papers, no. 26, online edition, July 2017 (dyntran.hypotheses.org/1972).

Meanwhile, Sayyid Badr al-Dīn of Lāla supported Uzun Ḥasan (d. 882/1478), gaining some favour with the conqueror, and presumably with his son and eventual successor Yaʿqūb.[131] Recent scholarship specifically devoted to Sunni Sufis at the courts of these two rulers still adheres to the interpretative framework of distinct organised 'orders', and tends to overlook Sayyid Badr al-Dīn.[132] The latter was directly connected to the broad *Kubrāwī* Sufi tradition through his *silsila*, but also had an indirect and possibly more important connection through the thriving local cult of Bābā Faraj Tabrīzī (d. 568/1172-73).[133] The family of Ibn Karbalāʾī were the custodians of the tomb of Bābā Faraj beginning at least in the time of Uzun Ḥasan, and Ibn Karbalāʾī was himself a disciple of a son of Sayyid Badr al-Dīn.[134] In his writing, the author of the local pilgrimage guide took great care to stress Sayyid Badr al-Dīn's integration in the Sufi milieus of Tabriz. In particular, Ibn Karbalāʾī emphasised Sayyid Badr al-Dīn's close association with Ḫʷāja ʿAlī Kujujī and with other eminent local Sufis, only one of whom was a representative of the Safavids. Hence the Sufis of Ardabil were far from dominating the religious scene in Tabriz.

The foresaid Bābā Faraj played a key role in the spiritual biography of Najm al-Dīn Kubrā (d. 618/1221). These two individuals met in Tabriz when the latter came to study the *Šarḥ al-sunna*, with Imām Ḥafda.[135] In the late ninth/fifteenth century, Uzun Ḥasan and prominent courtiers of Yaʿqūb Aq Qoyunlu are recorded as visiting the tomb of Bābā Faraj, demonstrating that the ruling elites of Tabriz still honoured his local cult.[136] In contrast, it seems that the

131 Ibn Karbalāʾī, *Rawżāt al-jinān*: II, pp. 154-5, 167.

132 Lingwood, *Politics*: pp. 81-110, summarises earlier scholarship highlighting the *Ḫalwatiyya* and focuses on the *Naqšbandiyya*. One of the supposed Naqšbandīs, Darwīš Sirāj al-Dīn Qāsim who was executed in 891/1486, had married one of his daughters to Sayyid Badr al-Dīn; that Darwīš Qāsim was a Naqšbandī is in fact all but clear as no source known to me states his *silsila* affiliation or anything comparable.

133 Ibn Karbalāʾī, *Rawżāt al-jinān*: I, pp. 376-7; II, pp. 167-8.

134 Ibid.: I, pp. 388-9, 520-2; II, pp. 146-54. English-language scholarship also claims that Ibn Karbalāʾī was a grandson or great-grandson of Sayyid Badr al-Dīn. Arjomand, *Shadow of God*: p. 114. Algar, "Kobrawiya"; Lewisohn, Leonard, s.v. "Ḥosayn Karbalāʾī", in *EIr*, XII (2004): pp. 512-3. There is no evidence for this and the introduction to Ibn Karbalāʾī's work by Jaʿfar Sulṭān al-Qurrāʾī is still the most thorough and accurate study of the author and the Sayyids of Lāla.

135 Ibn Karbalāʾī, *Rawżāt al-jinān*: I, pp. 377-9; Ibn al-Fuwaṭī, *Majmaʿ al-ādāb*: III, pp. 117, 351, 537; IV: pp. 76, 391, 477, 506. The author recorded several individuals who studied the *Šarḥ al-sunna* with Kubrā as transmitted from Imām Ḥafda adding in some cases parts of the details that this happened in Khwarazm in Ṣafar 615/May 1218. This should dissipate doubts among scholars of Sufism about the credibility of the one version of Kubrā's biography that features the visit to Tabriz and the encounter with Bābā Faraj.

136 Ibn Karbalāʾī, *Rawżāt al-jinān*: p. 381; Gülşenī, *Menāḳib-i Ibrāhīm Gülşenī*, ed. Tahsin Yazıcı (Ankara: Türk Tarih Kurumu Basımevi, 1981): pp. 121-2.

works of hadith brought to the city by Imām Ḥafda were no longer available in high quality transmissions based on his copies. The aforementioned Qadi Najm al-Dīn Uskū'ī, one of the most renowned local hadith teachers, transmitted the *Ṣaḥīḥ al-Buḫārī* from a copy that had arrived in Tabriz much more recently, and which this scholar had perhaps brought in himself. One of Sayyid Badr al-Dīn's sons studied this copy with a student of Qadi Najm al-Dīn.[137]

The Kujujīs reappear on the political scene after the death of Ya'qūb in 896/1490, when Aq Qoyunlu rule gradually collapsed. A certain Šayḫ Muḥammad Kujujī served as vizier of Ya'qūb's infant son, Baysunġur, but the latter was captured and executed by his rivals in 898/1493. A year earlier this same Šayḫ Muḥammad had already been instrumental in engineering the succession of the relatively effective Rustam, who likewise fell victim to factional fighting and was executed in 902/1497.[138] Šams al-Dīn Zakariyā Kujujī (d. 918/1512) served as vizier of Alwand (d. 910/1504), the last Aq Quyunlu ruler of Tabriz. He was the first high official to switch allegiance to the future Safavid ruler, Šāh Ismā'īl (d. 924/1530), a maternal grandson of Uzun Ḥasan, having changed sides prior to the decisive Safavid defeat of Alwand, meaning before Ismā'īl took the city in 907/1501.[139]

The sources vary in explaining the motivation for Šams al-Dīn Zakariyā's shift in allegiance, some of them citing religious convictions. This is probably the basis on which John Woods considered the Kujujīs as Shi'is.[140] Jean Aubin rightfully remarked long ago that there is no adequate definition of the precise contents of the categories of *Sunni* and *Shi'i* in the ninth/fifteenth century.[141] While acknowledging this, it seems that the Kujujīs were indeed not Shi'is. Also, unlike the Safavids, they had not contracted a matrimonial alliance with a royal family, nor formulated a claim to kingship on their own. However, like all notables of Tabriz and other cities, the Kujujīs faced an enormous political problem in the last years of the Aq Qonunlu, as was aptly summarised by the tenth/sixteenth century historian Budāq Munšī Qazwīnī: "In the realm of Persian Iraq (i.e. western Iran) and Azarbaijan chaos prevailed and every two days someone became king."[142]

---

137 Ibn Karbalā'ī, *Rawżāt al-jinān*: II, pp. 59-60, 190-2.

138 Budāq Munšī Qazwīnī, *Javāhir al-aḫbār, baḫš-i tārīḫ-i Īrān az Qarā Quyūnlū tā sāl-i 984 h.q.*, ed. Muḥsin Bahrāmnižād (Tehran: Mīrāṯ-i maktūb, 1378š./1999): pp. 92-3.

139 Ibid.: p. 117.

140 Aubin, "Etudes safavides I": p. 62; Woods, *The Aqquyunlu*: p. 152.

141 Aubin, "Etudes safavides I": p. 55.

142 Budāq Munšī, *Javāhir*: p. 112 (*dar mamlakat-i 'Irāq-i 'ajam va Azarbaijan harj-ū-marj būd va har dū rūz yakī pādišāh mīšud*).

Summary of dynastic history concerning Tabriz

Period	Ruler/Dynasty	Events, Remarks
Much of 6th/12th c.	Aḥmadīlī Atabegs of Marāġa	
572/1176-622/1225	Eldigüzid Atabegs (of Azarbaijan): Qïzïl Arslan, Abū Bakr, Özbeg	590/1194: end of Seljuq dynasty in Iran 617-18/1220-21: three times peaceful surrender to Mongols
622/1225-629/1231	Khwarazmshah Jalāl al-Dīn	Local resistance to conquest, then peaceful surrender to Mongols again
629/1231-654/1256	Mongol amirs and governors	Gradual development into center of imperial administration in Iran
654/1256-736/1335	Mongol Ilkhans	
736/1336-738/1338	Šayḫ Ḥasan (Buzurg) Jalāyir	Formally on behalf of alleged Hülegüid puppet sovereign
738/1338-758/1357	Chupanids: Šayḫ Ḥasan (Kučik), Malik Ašraf	Formally on behalf of alleged Hülegüid puppet sovereigns; local resistance to Malik Ashraf; no resistance to conquest by Jānī Beg Khan of the Golden Horde
758/1357-761/1360		After withdrawal of Golden Horde control disputed between Chupanid amīr, Muzaffarid ruler of Fars and Uways Jalāyir
761/1360-788/1386	Jalayirids: Uways, Ḥusayn, Aḥmad	778/1376: no resistance to conquest by Muẓaffarid ruler of Fars, perhaps invited 787/1385: after failed negotiations local resistance to conquest by Toqtemish Khan of the Golden Horde 788/1386: peaceful surrender to Timur
788/1386-795/1393		Control disputed, dynasties involved include Timurids, Jalayirids and Qara Qoyunlu

TABLE (*cont.*)

Period	Ruler/Dynasty	Events, Remarks
795/1393-810/1408	Mīrān-Šāh b. Tīmūr (his sons ʿUmar and Abā Bakr)	Governor installed by Timur, temporary claim of independent rule then reassignment of governorship to ʿUmar 807/1405: death of Timur, intensifying conflicts between Mīrān-Šāh and his sons, then local appeal to Qara Qoyunlu
810/1408-875/1469	Qara Qoyunlu: Qarā Yūsuf, Iskandar, Jahān-Šāh, Ḥasan-ʿAlī	813/1411: no local opposition to return of Aḥmad Jalāyir, but then defeat and execution by Qarā Yūsuf 822/1420: no resistance to conquest by troops of Šāhruḫ b. Timur but to Aq Qoyunlu prince he installed as governor 833/1430: no resistance to conquest by troops of Šāhruḫ b. Timur, ousting of Qara Qoyunlu prince he installed as governor 836/1433: no resistance to conquest by troops of Šāhruḫ b. Timur who continued to claim formal suzerainty while virtually independent Qara Qoyunlu rule stabilized under Jahān-Šāh 874/1468: defeat and execution of claimant Abū Saʿīd b. Muḥammad b. Mīrān-Šāh by Uzun Ḥasan Aq Qoyunlu
875/1469-906/1501	Aq Qoyunlu: Uzun Ḥasan, Sulṭān-Ḫalīl, Yaʿqūb, Baysunghur, Rustam, Aḥmad, Alwand	906/1500: major Aq Qoyunlu courtier and local notable switches allegiance to Safavids

7 Conclusion

This paper has examined the Malikān and Kujujī families of Tabriz, and in so doing has revealed that the local elites participated actively in determining the sequence of dynastic succession from the sixth/twelfth to ninth/fifteenth century. In general, pretenders, rulers, and dynasties would gain support from the local notables if they fulfilled the following conditions. They needed to demonstrate themselves capable of providing a minimum of stability and were to base their courts in Azarbaijan, granting Tabriz the distinction of being the principal city. Also, the notables of Tabriz repeatedly refrained from resisting exceptionally powerful armies, and instead sought peaceful accommodation with the invaders.

The Malikān were outstanding political leaders under the Eldigüzid Atabegs of Azarbaijan, and subsequently, crucial players in the establishment of Mongol rule, which would then become Ilkhanid. The case of this family demonstrates that in Tabriz, urban life remained largely intact through this period, and that traditional Sunni Islam continued to flourish despite temporary non-Muslim domination. Conditions seem to have taken a serious turn for the worse in the period between the collapse of the Ilkhanate and the consolidation of the Qara Qoyunlu. In Tabriz, it was not the Mongol conquests, but rather Khwarazmian rule and the repeated Timurid invasions that led to significant negative impacts.

The Kujujīs were prominent Sufis who became eminent political leaders in association with the most important post-Ilkhanid dynasties, specifically the Jalāyirids and the Qara Qoyunlu. Their example indicates that the influence of Sufis in society and politics did indeed increase in Mongol and post-Mongol times. However, this growth was firmly embedded in a foundation of local pre-Mongol pious traditions. The cult of Bābā Faraj, and the Sufi lineages emanating from the 70 Bābās, developed over the entire three-century period under consideration. The family component of these lineages was particularly strong in the villages of the hinterland of Tabriz. The examination concerning this city fails to confirm and indeed counters the widely held assumption that Sufism in Azarbaijan became increasingly Shiʿi, in the century prior to the coming to power of the Safavids.

The transformation of local society under Mongol rule did not have very great effects in the religious domain, nor were the old urban elites replaced by newly arrived courtiers and their associates. The local elites of post-Mongol Tabriz instead pertained to rural backgrounds, and beginning in the time of Mongol rule, even notable immigrants to the city took up residence in the hinterland villages. This change in the local balance of power may be attributed to the endurance of a nomadic mode of domination, or perhaps its intensification beginning with Mongol rule. Of course, Ġāzān and subsequent lords of Tabriz, like their predecessors, made use of residences in the immediate suburbs of Tabriz. Yet they exercised power mainly from their mobile camps. Therefore Tabriz should not be called a 'capital', even when referring to the time under the Ilkhans and their successors. Finally, the tendency of the nomadic imperial elites to stay outside of the city possibly implied an advantage of access to them for those notables who had rural backgrounds.

Bibliographical References

Afšār, Īraj, "Safīna-yi Tabrīz: Anjāma-hāy Safīna-yi Tabrīz", *Nāma-yi Bahāristān*, XIII/14 (1387š./2008): pp. 285-330.

Aharī, *Tavārīḫ-i Šayḫ Uways*, ed. Īraj Afšār (Tabriz: Sutūda, 1389š./2010).

Aigle, Denise, *Le Fārs sous la domination mongole. Politique et fiscalité (XIIIe-XIVe s.)* (Paris: Association pour l'avancement des études Iraniennes, 2005).

Al-e Davud, Sayyed Ali, "A Review of the Treatises and Historical Documents in *Safīna-yi Tabrīz*", in Seyed-Gohrab, A.-A. and McGlinn, S. (eds.), *The Treasury of Tabriz. The Great Il-Khanid Compendium* (Amsterdam and West Lafayette: Rozenberg Publishers and Purdue University Press, 2007): pp. 79-89.

Album, Stephen, *Sylloge of Islamic Coins in the Ashmolean, IX, Iran after the Mongol Invasion* (Oxford: Ashmolean Museum, 2001).

Algar, Hamid, s.v. "Kobrawiya ii. The Order", in *EIr*, online edition (2009).

Allsen, Thomas T., *Mongol Imperialism. The Policies of the Grand Qan Möngke in China, Russia and the Islamic Lands, 1251-1259* (Berkeley: University of California Press, 1987).

Allsen, Thomas T., *Culture and Conquest in Mongol Eurasia* (Cambridge: Cambridge University Press, 2001).

Anwār, Sayyed 'Abd-Allāh, "Qoṭb-al-Din Širāzi", *EIr*, online edition (2005).

Arjomand, Said A., *The Shadow of God and the Hidden Imam. Religion, Political Order and Societal Change in Shi'ite Iran from the Beginning to 1890* (Chicago-London: The University of Chicago Press, 1984).

Arjomand, Said A., "Conceptions of Authority and the Transition of Shi'ism from Sectarian Religion to National Religion in Iran", in Daftary, F. and Meri, J. W. (eds.), *Culture and Memory in Medieval Islam. Essays in Honour of Wilferd Madelung* (London-New York: I.B. Tauris, 2003): pp. 388-409.

'Atīqī, Jalāl al-Dīn, *Dīvān-i 'Atīqī*, fac simile edition by Naṣr Allāh Pūrjawādī and Sa'īd Karīmī (Tehran: Farhangistān-i zabān va adab-i fārsī, 1388š./2009).

Aubin, Jean, "Etudes safavides. I. Šāh Ismā'īl et les notables de l'Iraq persan", *JESHO*, II (1959): pp. 37-81.
Aubin, Jean, *Émirs mongols et vizirs persans dans les remous de l'acculturation* (Paris: Association pour l'avancement des études iraniennes, 1995).
Avery, Peter, *The Spirit of Iran. A History of Achievement from Adversity* (Costa Mesa: Mazda Publishers, 2007).
Bashir, Shahzad, *Messianic Hopes and Mystical Visions. The Nūrbakhshīya between Medieval and Modern Islam* (Columbia SC: University of South Carolina Press, 2003).
Binbaş, İlker Evrim, "The Anatomy of a Regicide Attempt: Shāhrukh, the Ḥurūfīs, and the Timurid Intellectuals in 830/1426-27", *JRAS*, XXIII/3 (2013): pp. 391-428.
Biran, Michal, *Chinggis Khan* (Oxford: Oneworld Publications, 2007).
Biran, Michal, s.v. "Jovayni, Ṣāḥeb Divān", in *EIr*, XV (2009): pp. 71-4.
al-Birzālī, *al-Muqtafā 'alā kitāb al-rawḍatayn al-ma'rūf bi-tārīḫ al-Birzālī*, ed. 'Umar 'Abd al-Salām Tadmurī, 4 vols. (Beirut: al-Maktaba al-'aṣriyya, 2002-06).
Brosset, Marie-Félicité, *Histoire de la Géorgie depuis l'antiquité jusqu'au XIXe siècle*, 4 vols. (St. Petersburg: Imprimerie de l'Académie Impériale des Sciences, 1849-57).
Budāq Munšī Qazwīnī, *Javāhir al-aḫbār, baḫš-i tārīḫ-i Īrān az Qarā Quyūnlū tā sāl-i 984 h.q.*, ed. Muḥsin Bahrāmnižād (Tehran: Mīrāṯ-i maktūb, 1378š./1999).
Bulliet, Richard W., *The Patricians of Nishapur. A Study in Medieval Islamic Social History* (Cambridge MA.: Harvard University Press, 1972).
DeWeese, Devin, "'Stuck in the Throat of Chingīz Khān:' Envisioning the Mongol Conquests in Some Sufi Accounts from the 14th to 17th Centuries", in Pfeiffer, J. and Quinn, S.A. (eds.), *History and Historiography of Post-Mongol Central Asia and the Middle East. Studies in Honor of John E. Woods* (Wiesbaden: Harrassowitz Verlag, 2006): pp. 23-60.
al-Ḏahabī, *Tārīḫ al-islām wa wafayāt al-mašāhīr wa'l-'a'lām*, ed. 'Umar 'Abd al-Salām Tadmurī, 61 vols. (Beirut: Dār al-kitāb al-'arabī, 1987-2004).
Durand-Guédy, David, *Iranian Elites and Turkish Rulers. A History of Iṣfahān in the Saljūq Period* (Abingdon: Routledge, 2010).
Durand-Guédy, David, "Where did the Saljuqs live? A case study based on the reign of Sultan Mas'ūd b. Muḥammad (1134-1152)", *StIr*, XL (2011): pp. 211-58.
Durand-Guédy, David, "Ruling from the outside: a new perspective on early Turkish kingship in Iran", in Mitchell, L. and Melville, C. (eds.), *Every Inch a King. Comparative Studies on Kings and Kingship in the Ancient and Medieval Worlds* (Leiden: Brill, 2012): pp. 325-42.
Durand-Guédy, David, "Location of Rule in a Context of Turko-Mongol Domination", in Id. (ed.), *Turko-Mongol Rulers, Cities and City Life*, (Leiden and Boston: Brill, 2013): pp. 1-20.

Durand-Guédy, David, "1147: The Battle of Qara-Tegin and the Rise of Azarbayjan", *Der Islam*, XCII/1 (2015): pp. 161-96.

Elias, Jamal J., "The Sufi Lords of Bahrabad: Sa'd al-Din and Sadr al-Din Hamuwayi", *IrSt*, XXVII/1 (1994): pp. 53-75.

Gronke, Monika, *Arabische und persische Privaturkunden des 12. und 13. Jahrhundert aus Ardabil (Aserbeidschan)* (Berlin: Klaus Schwarz Verlag, 1982).

Gronke, Monika, *Derwische im Vorhof der Macht. Sozial- und Wirtschaftsgeschichte Nordwestirans im 13. und 14. Jahrhundert* (Stuttgart: Franz Steiner Verlag, 1993).

Gülşenī, *Menāḳib-i Ibrāhīm Gülşenī*, ed. Tahsin Yazıcı (Ankara: Türk Tarih Kurumu Basımevi, 1981).

Ḥāfiẓ-i Abrū, *Zubdat al-tavārīḫ*, ed. Kamāl Ḥājj Sayyid Jawādī, 4 vols. (Tehran: Sāzmān-i čāp va intišārāt-i Wizārat-i farhang va iršād-i islāmī, 1380š./2001).

Havemann, Axel and Bosworth, C.E., s.v. "Ra'īs", in *EI*², VIII (1995): pp. 402-3.

Hoffmann, Birgitt, *Waqf im mongolischen Iran. Rašīduddīns Sorge um Nachruhm und Seelenheil* (Stuttgart: Franz Steiner Verlag, 2000).

Ḥusaynī, *Aḫbār al-dawlat al-saljūqiyya*, ed. Muḥammad Iqbāl (Lahore: The University of the Panjab, 1933, reprint Beirut: Dār al-āfāq al-jadīda, 1984).

Ibn Abi'l-Ḥadīd, *Šarḥ Nahj al-balāġa. Les invasions mongoles en Orient vécues par un savant médiéval arabe*, part. ed. and transl. Moktar Djebli (Paris: L'Harmattan, 1995).

Ibn al-Aṯīr, *al-Kāmil fi'l-tā'rīḫ*, ed. Carl Johan Tornberg, 13 vols. (Leiden, reprint with a different pagination Beirut: Dār Ṣādir, 1965-67).

Ibn Bazzāz, *Ṣafvat al-ṣafā dar tarjuma-yi aḥvāl va aqvāl va karāmāt-i Šayḫ Ṣafī al-Dīn Isḥāq Ardabīlī*, ed. Ġulām-Riḍā Ṭabāṭabā'ī Majd (Ardabil, 1372š./1993).

Ibn al-Fuwaṭī, *Majma' al-ādāb fī mu'jam al-alqāb*, ed. Muḥammad al-Kāẓim, 6 vols. (Tehran: Sāzmān-i čāp va intišārāt-i Wizārat-i Farhang va Iršād-i Islāmī, 1995).

Ibn Ḥajar, *Inbā' al-ġumr bi-abnā' al-'umr*, ed. Ḥasan Ḥabašī, 4 vols. (Cairo, 1969).

Ibn Karbalā'ī, *Rawżāt al-jinān va jannāt al-janān*, ed. Ja'far Sulṭān al-Qurrā'ī, 2 vols. (Tehran: Bungāh-i tarjuma va našr-i kitāb, 1344-9š./1965-70, reprint Tabriz: Sutūda, 1383š./2004).

Ibn Ḫallikān, *Wafayāt al-a'yān wa anbā' al-zamān*, ed. Yūsuf 'Alī Ṭawīl and Maryam Qāsim, 6 vols. (Beirut: Dār al-kutub al-'ilmiyya, 1998).

Īmānī, Bihrūz, "Bargī az Šarḥ-i ḥāl-i Šayḫ Šihāb al-Dīn Maḥmūd Aharī", *Payām-i Bahāristān*, XXX (1383š./2004): pp. 79-83.

Jackson, Peter, "The Dissolution of the Mongol Empire", *Central Asiatic Journal* XXXII (1978): pp. 186-243.

Jackson, Peter, s.v. "Jalayerids", in *EIr*, XIV (2008): pp. 415-9.

Juwaynī, 'Aṭā-Malik, *Tārīḫ-i Jahān-gušāy*, ed. Mīrzā Muḥammad Qazwīnī (Leiden: Gibb Memorial Trust, 1912-37).

Katouzian, Homa, *The Persians. Ancient, Medieval and Modern Iran* (New Haven-London: Yale University Press, 2009).

Kolbas, Judith, *The Mongols in Iran. Chingiz Khan to Uljaytu, 1220-1309* (Abingdon: Routledge, 2006).

Ḫwāndamīr, *Tārīḫ-i Ḥabīb al-siyar fī aḫbār-i afrād-i bašar*, ed. Jalāl al-Dīn Humā'ī and Muḥammad Dabīr Siyāqī, 4 vols. (Tehran: Kitābfurūšī-yi Ḫayyām, 1362š./1982).

Lambton, Ann K.S., *Continuity and Change in Medieval Persia. Aspects of Administrative, Economic and Social History, 11th-14th Century* (London: I.B. Tauris, 1988).

Lane, George, *Early Mongol Rule in Thirteenth Century Iran. A Persian Renaissance* (London and New York: Routledge, 2003).

Lane, George, "Persian Notables and the Families Who Underpinned the Ilkhanate", in Amitai, R. and Biran, M. (eds.), *Nomads as Agents of Cultural Change: The Mongols and Their Eurasian Predecessors* (Honolulu: The University of Hawaii Press, 2015): pp. 183-213.

Lewisohn, Leonard, *Beyond Faith and Infidelity. The Sufi Poetry and Teachings of Maḥmūd Shabistarī* (Richmond: Curzon Press, 1995).

Lewisohn, Leonard, "Palāsī's Memoir of Shayk Kujujī, a Persian Sufi Saint of the Thirteenth Century", JRAS, VI/3 (1996): pp. 345-66.

Lewisohn, Leonard, s.v. "Ḥosayn Karbalā'ī", in *EIr*, XII (2004): pp. 512-3.

Lingwood, Chad G., *Politics, Poetry and Sufism in Medieval Iran. New Perspectives on Jāmī's* Salāmān va Absāl (Leiden and Boston: Brill, 2014).

Luther, Kenneth A., "Rāvandī's Report on the Administrative Changes of Muḥammad Jahān Pahlavān", in C.E. Bosworth (ed.), *Iran and Islam, in memory of the late Vladimir Minorsky* (Edinburgh: Edinburgh University Press, 1971): pp. 393-406.

Luther, Kenneth A., s.v. "Atābakān-e Marāḡa", in *EIr*, II (1987): pp. 898-900.

Luther, Kenneth A., s.v. "Atābakān-e Āḏarbāyjān", in *EIr*, II (1987): pp. 890-4.

Manz, Beatrice Forbes, *The Rise and Rule of Tamerlane* (Cambridge: Cambridge University Press, 1989).

Manz, Beatrice Forbes, "Mongol History Rewritten and Relived", in Aigle D. (ed.), *Figures mythiques de l'Orient musulman*, REMM, LXXXIX/XC (2000): pp. 129-49.

Manz, Beatrice Forbes, *Power, Politics and Religion in Tīmūrid Iran* (Cambridge: Cambridge University Press, 2007).

Maškūr, Muḥammad Jawād, *Tārīḫ-i Tabrīz tā pāyān-i qarn-i nuhum-i hijrī* (Tehran: Anjuman-i Aṯār-i Millī, 1352š./1973).

Mazzaoui, Michel, *The Origins of the Ṣafawids. Šī'ism, Ṣūfism and the Ġulāt* (Wiesbaden: Franz Steiner Verlag, 1972).

Melville, Charles and Zaryāb, 'Abbās, s.v. "Chobanids", in *EIr*, V (1991): pp. 496-202.

Metsobets'i, trans. Bedrosian, *History of Tamerlane and his Successors* (New York: Sources of the Armenian Tradition, 1987).

Minorsky, Vladimir [C. E. Bosworth], s.v. "Tabriz", in *EI*², X (2010): pp. 41-9.

Mīrḫwānd, *Rawżat al-ṣafā*, ed. Naṣr Allāh Sabūḥī, 7 vols. (Tehran: Intišārāt-i markazī Ḫayyām-Pīrūz, 1338š./1959).

Mustawfī Qazwīnī (Ḥamd-Allāh), *Baḫš-i naḫust az maqāla-yi sivvom-i Nuzhat al-qulūb*, ed. Muḥammad Dabīr Siyāqī (Tehran: Ṭahūrī, 1336š./1958).

Mustawfī Qazwīnī (Ḥamd-Allāh), *Tārīḫ-i guzīda*, ed. ʿAbd al-Ḥusayn Nawāʾī (Tehran: Amīr Kabīr, 1339š./1960).

Mustawfī Qazwīnī (Zayn al-Dīn b. Ḥamd-Allāh), *Ẕayl-i Tārīḫ-i guzīda*, ed. Īraj Afšār (Tehran: Hayʾat-i bar-rasī va guzīnaš-i majmūʿa-yi mawqūfāt-i Duktur Maḥmūd Afšār Yazdī, 1372š./1993).

Nasawī, *Sīrat al-Sulṭān Jalāl al-Dīn Minkbirnī*, ed. Z. M. Buniatov (Moscow: Izdatelskaya firma «Vostočnaya literatura» RAN, 1996).

Naṭanzī, *Muntaḫab-i tavārīḫ-i muʿīnī* (*taʾlīf-i 816 va 817 h.q.*), ed. Jean Aubin (Tehran: Kitābfurūšī-yi Ḫayyām, 1336š./1957).

Nīšāpūrī, *Saljūqnāma*, ed. A.H. Morton (Warminster: Gibb Memorial Trust, 2004).

Ohlander, Erik S., *Sufism in an Age of Transition. ʿUmar al-Suhrawardī and the Rise of the Islamic Mystical Brotherhoods* (Leiden: Brill, 2006).

Paul, Jürgen, "L'invasion mongole comme "révélateur" de la société iranienne", in Aigle, D. (ed.), *L'Iran face à la domination mongole* (Tehran: Institut français de recherche en Iran, 1997): pp. 37-53.

Peacock, A. C. S., "Nomadic Society and the Seljūq Campaigns in Caucasia", *Iran and the Caucasus*, IX/2 (2005): pp. 205-30.

Pistor-Hatam, Anja, "History and its Meaning in the Islamic Republic of Iran: The Case of the Mongol invasion(s) and rule", in Ansari, A.M. (ed.), *Perceptions of Iran. History, Myths and Nationalism from Medieval Persia to the Islamic Republic* (London: I.B. Tauris, 2014): pp. 147-62.

Pfeiffer, Judith, "Reflections on a 'Double Rapprochement': Conversion to Islam among the Mongol elite during the early Ilkhanate", in Komaroff, L. (ed.), *Beyond the Legacy of Genghis Khan* (Leiden: Brill, 2006): pp. 369-89.

Pfeiffer, Judith, "Introduction. From Baghdad to Marāgha, Tabriz and Beyond: Tabriz and the Multi-Cephalous Cultural, Religious and Intellectual Landscape in the 13th to 15th Century Nile-to-Oxus Region" in Ead. (ed.), *Politics, Patronage and the Transmission of Knowledge in 13th-15th century Tabriz* (Leiden: Brill, 2014): pp. 1-11.

Qāšānī, *Tārīḫ-i Uljāytū*, ed. Mahīn Hambalī (Tehran: Bungāh-i tarjuma va našr-i kitāb, 1349š./1970).

Quiring-Zoche, Rosemarie, *Isfahan im 15. und 16. Jahrhundert. Ein Beitrag zur persischen Stadtgeschichte* (Freiburg: Klaus Schwarz Verlag, 1980).

al-Rāfiʿī al-Qazwīnī, *al-Tadwīn fī aḫbār Qazwīn*, ed. ʿAzīz Allāh al-ʿAṭāridī (Hydarabad, 1984, reprint Beirut: Dār al-kutub al-ʿilmiyya, 1987).

Ragep, F. Jamil, "New Light on Shams: The Islamic Side of Σάμψ Πουχάρης", in Pfeiffer, J. (ed.), *Politics, Patronage and the Transmission of Knowledge in 13th-15th century Tabriz* (Leiden: Brill, 2014): pp. 231-47.

Rašīd al-Dīn, *Jāmiʿ al-tavārīḫ*, ed. Muḥammad Rawšan and Muṣṭafā Mūsawī (Tehran: Našr-i Alburz, 1373š./1994).

Robson, J., s.v. "al-Baghawī", in *EI*², I (1960): p. 893.

Seyed-Gohrab, Ali-Asghar, "Casing the Treasury", in Id. and McGlinn, S. (eds), *The Treasury of Tabriz. The Great Il-Khanid Compendium* (Amsterdam and West Lafayette: Rozenberg Publishers and Purdue University Press, 2007): pp. 15-42.

Šuštarī, *Majālis al-muʾminīn*, ed. Aḥmad ʿAbd al-Munāfī (Tehran, 1955).

Szuppe, Maria, *Entre Timourides, Uzbeks et Safavides. Questions d'histoire politique et sociale de Hérat dans la première moitié du XVIe siècle* (Paris: Association pour l'avancement des études Iraniennes, 1992).

Ṭihrānī, *Kitāb-i Diyarbakriyya*, ed. Necati Logal and Faruk Sümer (Ankara: Türk Tarih Kurumu Basımevi, 1964).

Tucker, William F., "The Kūfan *Ghulāt* and Millenarian (Mahdist) Movements in Mongol-Türkmen Iran", in Mir-Kazimov, Orkhan (ed.), *Unity in Diversity. Mysticism, Messianism and the Construction of Religious Authority in Islam* (Leiden and Boston: Brill, 2014): pp. 177-95.

van Ess, Josef, *Der Wesir und seine Gelehrten. Zu Inhalt und Entstehungsgeschichte der theologischen Schriften des Rašīduddīn Fażlullah (gest. 718/1318)* (Wiesbaden: Franz Steiner Verlag, 1981).

Werner, Christoph, "Ein Vaqf für meine Töchter. Ḫātūn Jān Bēgum und die Qarā Quyūnlū Stiftungen zur ‚Blauen Moschee' in Tabriz", *Der Islam*, LXXX (2003): pp. 94-109.

Werner, Christoph, Zakrzewski, Daniel, and Tillschneider, Hans-Thomas, *Die Kuǧuǧī-Stiftungen in Tabriz: Ein Beitrag zur Geschichte der Ǧalāyiriden (Edition, Übersetzung, Kommentar)* (Wiesbaden: Reichert Verlag, 2013).

Woods, John E., *The Aqquyunlu. Clan, Confederation, Empire* (2nd revised and expanded edition, Salt Lake City: The University of Utah Press, 1999).

Yazdī, *Ẓafar-nāma*, ed. Sayyid Saʿīd Mīr Muḥammad Ṣādiq [ʿAbd al-Ḥusain Nawāʾī], 2 vols. (Tehran: Kitābḫāna-yi mūza va markaz-i asnād-i Majlis-i šūrā-yi islāmī, 1387š./2008).

Yildiz, Sara Nur, "Post-Mongol Pastoral Polities in Eastern Anatolia during the Late Middle Ages", in Beyazit, D. (ed.), *At the Crossroads of Empires: 14th-15th Century Eastern Anatolia* (Paris: Institut français d'études Anatoliennes, 2012): pp. 27-48.

Zajjājī, *Humāyūn-nāma*, ed. ʿAlī Pīr-niyā, 2 vols. (Tehran: Farhangistān-i zabān va adab-i fārsī/Mīrāṯ-i maktūb, 1383-90š./2004-11).

Zakrzewski, Daniel, "Malik Ṣadr al-Dīn Tabrīzī and the Establishment of Mongol Rule in Iran", *Asiatische Studien / Études Asiatiques*, LXXI/4 (2017): pp. 1059-73.

Zakrzewski, Daniel, "Local Aristocrats in a Time of Critical Dynastic Change. Two Notable Families of Tabriz Facing the New Safavid Order", DYNTRAN Working Papers, no. 26, online edition, July 2017 (dyntran.hypotheses.org/1972).

Biographical Note

Daniel Zakrzewski is currently preparing a PhD thesis on the history of Tabriz between the sixth/twelfth and tenth/sixteenth centuries, focusing on the local elites and on the significance acquired by the city for Iranian royal tradition, in connection with Mongol rule. His research interests include Persian historiography and hagiography, forms of social distinction and political agency in medieval and early modern Iran, and especially the role of families in the transmission of knowledge and authority. He has recently published "Malik Ṣadr al-Dīn Tabrīzī and the establishment of Mongol Rule in Iran" (Asiatische Studien / Études Asiatiques LXXI/4, 2017) and co-edited with Chr. Werner and H.-Th. Tillschneider *Die Kuǧuǧī-Stiftungen in Tabrīz: Ein Beitrag zur Geschichte der Ǧalāyiriden (Edition, Übersetzung, Kommentar)* (Wiesbaden: Reichert 2013).

Medieval Kashan: Crossroads of Commerce and Culture

Mehrdad Amanat
University of California, Santa Barbara
mehrdad.amanat@gmail.com

Roy P. Mottahedeh
Harvard University
mottahed@fas.harvard.edu

Abstract

This essay first attempts to explain why the Iranian city of Kashan is where it is and then how it came to grow.

Extensive use of irrigation allowed an adequate and perhaps abundant agriculture in the surrounding region. The arrival of substantial numbers of Arab immigrants in Kashan in the early Islamic period played an important role in the city's development and its continuation as a center of Shiʿism. A strong educational tradition produced many talented Kashani officials, who served in the Saljuq and later administrations and sent some of their wealth back to Kashan.

It was also in the Saljuq period that Kashan gained a reputation for its production of luxury ceramics. Artisanal traditions were passed from generation to generation and contributed to exports of brass, and especially of textiles, which continued for centuries. Wealthy Kashanis (probably including a fair number of *sayyids*) invested heavily in charitable endowments, which served the poor and furthered learning in general. In the Timurid period, investments in mathematical education produced several outstanding mathematicians and astronomers. Tax yields from the medieval period may indicate the increasing prosperity of Kashan.

Keywords

Kashan – Shiʿism – madrasa – *qanāt* irrigation – luster and *mīnāʾī* ware

 | DOI:10.1163/9789004434332_015

1 Introduction

No city history of Kashan survives from the Middle Ages, and it is possible that no such history was written. The *Mir'āt al-Qāsān* is an extraordinarily fine example of local history produced in the nineteenth century, incorporating a fair amount of previous history and further amplified by very rich appendices, assembled by the late Iraj Afshar.[1] Kashan served only rarely as a seat of government, and most dynastic histories offer only passing references. Yet Kashan was an important artisanal center, and for this we have significant material evidence, which we can only hint at in the present essay. The history of smaller yet important centers such as Kashan represents a new frontier in city histories for Iran. Although smaller cities did not always escape sack and pillage, such settlements survived as alternate centers towards which displaced people could move. They could also serve as sites for refuge of minority beliefs, as is the case for the Shiʿism in Kashan.

2 Why Kashan Is Where It Is

Kashan arose at the center of the fan of an alluvial plain, approximately half way between a plentiful spring at Fīn and the distant town of Bīdgul, at the end of the plain (Fig. 1). Beyond Bīdgul lies the Kavīr,[2] one of Iran's extremely arid interior deserts. Kashan has its own seasonal river, the Šāhāb, known upstream as the Quhrūd. An elaborate system of underground water channels (sing. *qanāt* or *kārīz*), fed by the higher water table in the Karkas Mountains and streams in the foothills as well as waters from the spring at Fīn, distributes this asset to farms and towns on the alluvial plain. The primary components of this system are the underground channels. However, there are also surface

---

* Both authors regard Kashan as their ancestral home and are greatly interested in the history and culture of the city. We wish to thank Hossein Modarressi, Ali Mousavi, and Heidi Walcher.

1 The main appendix consists of the notes of Allāh-yār Ṣāliḥ.

2 Zanjani, Habibollah, s.v. "Kashan i. Geography", in *EIr*, XVI/1 (2012): pp. 1-2. Planhol, Xavier de, s.v. "Kashan ii. Historical Geography", in *EIr* XVI/1 (2012): pp. 2-5. De Planhol offers several strong arguments for the existence of Kashan at its present location. On surface channels see de Planhol, Xavier de, s.v. "Kārīz i. Terminology", in *EIr*, XV (2012): pp. 564-65. We were unable to obtain a copy of the geological thesis by Mehdi Nadji-Esfahani, *Geologie und Hydrogeologie des Gebietes von Kashan*, (Aachen: Technische Hochschule, Fakultät für Bergbau und Hüttenwesen, 1971), however his 1972 article is available, published under the name Mehdi Nadji: Nadji, Mehdi, "Luftbild: Kanate in der Ebene von Kashan, Iran", *Die Erde*, CIII (1972): pp. 209-15.

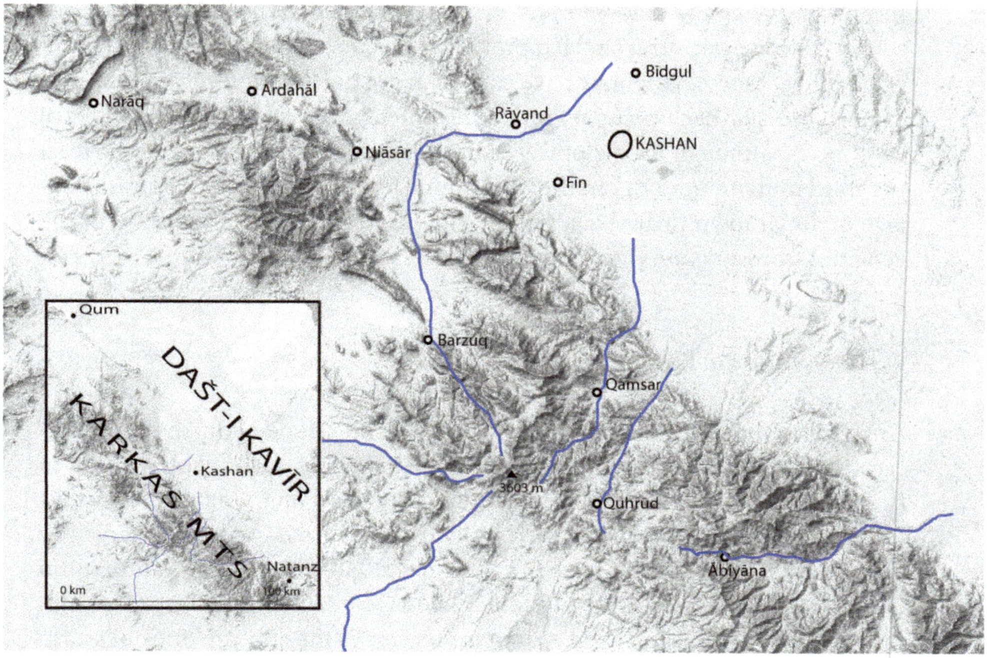

FIGURE 1 Map of Kashan and nearby villages
DRAWING D. DURAND-GUÉDY

channels, often the extension of the underground channels as they reach the settlements. The *qanāt*s are oriented in the same directions as the groundwater flows and at intervals are provided with air holes. The underground channels minimize losses to evaporation and aid in balancing the salinity of the waters. We can immediately understand that any upstream irrigation drawn from the *qanāt*s would decrease the downstream supplies of water, and in fact the *History of Qum*, written in 378/988-89, reports the elaborate attempts to calculate the fractions of water that landholders in Qum would be entitled to take from the channels passing by their various properties.[3] These agreements, as well as the extensive social cooperation necessary for construction and maintenance of the channels, indicate the levels of organization required to irrigate such arid lands, where the alluvial plain abuts the desert.[4] In more

3 Qumī, Ḥasan ibn Muḥammad, *Tārīḫ-i Qum*, ed. Muḥammad Riḍā Anṣārī Qumī, (Qum: Markaz-i Qumšināsī, 2006): pp. 26-43. See also the outstanding study by Mudarrisī Ṭabāṭabā'ī, Ḥusayn, *Turbat-i pākān: āsār wa banāhā-yi qadīmī*, 2 vols, (Qom: Čāpḫāna-yi Mihr, 1976) includes considerable reference to Kashan.

4 See Jomehpour, Mahmoud, "Qanat Irrigation Systems as Important and Ingenious Agricultural Heritage: Case Study of the *qanats* of Kashan, Iran", *International Journal of Environmental*

recent centuries the underground system is known to have included large rooms called *saypak* or *ū'ī*, which probably served as refuges from both invaders and heat-waves,[5] and may have existed in pre-Safavid times.

Kashan itself became a city only after the arrival of Islam in the seventh century CE. although the general area has been a site of cities since at the least the third millennium BCE. In the pre-Islamic period Kashan was at most a village or small town near the spring of Fīn; it did not merit a mention in the geographical works surviving from the Sasanian period.[6]

3 When did Kashan Emerge as a City?

Al-Ya'qūbī, who wrote his geography in 278/891, mentioned only the surrounding villages and completely omitted Kashan.[7] Ibn Rusta, probably writing in the period 290-300/903-13, also fails to mention Kashan.[8] One of the earliest references to Kashan is in the geography of Ibn Ḫurradāḏbih, who served as a postmaster for much of western Iran in the late third/ninth century. Ibn Ḫurradāḏbih refers to two adjacent regions, *rustāq Ṣard-Qāšān* (the sub-province of cold Kashan) and *rustāq Jarm Qāšān* (the sub-province of warm Kashan), but does not mention Kashan as a city.[9] (The district of Kashan continues to be divided into 'cold climate' and 'warm climate' zones up to the present day.)

Other third/ninth century references are set in the context of the Muslim conquest of Kashan, almost invariably attached to references to the prior conquest of Qum. We believe these references were added to the conquest narratives of Qum after Kashan emerged in the third/ninth century, at one hundred kilometers distance from Qum. Many of these colorful narratives have fanciful

---

Studies, LXVI/3 (June 2009): pp. 297-315, who has a good discussion of the social organization necessary to maintain *qanāts*; Nadji, "Luftbild": pp. 209-15. On the classification of *qanāt*s, see Mahdavī, M. and Anderson, E.W., "The Water-Supply System in the Margin of Dasht-e-Kavir", *Bulletin: British Society for Middle Eastern Studies*, X/2 (1983): pp. 131-47.

5 Bīdgulī, Maḥmūd Sādāt and Sārūḫānī, Zahrā, "Bar-rasī-yi 'ilal-i tārīḫ-iqlīmī-yi ījād-i ū'ī-hā dar manṭaqa-yi Kāšān", *Ganjīna-yi asnād*, LXXI (Fall 1387š/1967): pp. 25-36.

6 See *Šahrestān-i Ērānshahr* or *A Catalogue of the Provincial Capitals of Eranshahr*, ed. and trans. J.A. Markwart (Rome: Pontificio Istituto Biblico, 1931); see also Bīdgulī and Sārūḫānī, "Bar-rasī": p. 28.

7 Ya'qūbī, *Kitāb al-buldān*, ed. M.J. de Goeje (Leiden: Brill, 1892): p. 275.

8 However, he does mention a "wondrous spring" (Fīn?), and an antimony mine. Ibn Rusta, *Kitāb al-a'lāq al-nafīsa*, ed. M.J de Goeje (Leiden: Brill, 1892): p. 156.

9 Ibn Ḫurradāḏbih, *Kitāb al-masālik wa'l-mamālik*, ed. M.J. de Goeje (Leiden: Brill, 1889): p. 21.

details as to how the attackers and defenders of Kashan rained scorpions on each other.[10]

A small town called Kāšān (or, in its earlier spellings, Qāšān or Qāsān) may well have existed near the present site of Kashan. A fundamental factor in its development would have been the flight of proto-Shiʿi Arabs from anti-Shiʿi governors of Iraq to central Iran, especially to Qum and Kashan. (Twelver Shiʿism was not fully formed until the fifth/eleventh century.) A history by al-Ṭabarī (d. 310/923) mentions that an anti-Umayyad leader from Iraq, Muṭrif b. al-Muġīra, entered into open opposition with al-Ḥajjāj, the ferocious governor of Iraq, in 77/696-97. Muṭrif was killed but, according to al-Ṭabarī, his followers fled to Qum and Qāšān.[11] If al-Ṭabarī is not adding Kashan because of the customary habit of mentioning it after Qum, this would be a genuine early reference to the developing town of Kashan.

Another early mention of Kashan in the geographical literature is in the *al-Masālik wa-ʾl-mamālik* of al-Iṣṭaḫrī (d. 346/957). The only comments from this author are that Kashan is three stages (*marḥala*) or twelve *farsaḫ*s beyond Qum, that it is a part of Qūhistān, the Persian name for the province al-Jibāl, and that it is "a small city in which the houses are built of clay".[12] A geographer of the same century writing in Persian, the unknown author of *Ḥudūd al-ʿĀlam* (begun in 372/982-3) says that Kashan is "a city with excellences, in which there are numerous Arabs. It has produced many scribes (*dabīrān*) and *literati* (*adībān*)".[13] This important passage anticipates the role that sophistication in

---

10 For example, Qudāma b. Jaʿfar (fl. 320/922), a learned and generally reliable source, tells the story that after conquering Qum, Abū Mūsā al-Ašʿarī sent al-Aḥnaf b. Qays to conquer Qāšān (Kāšān), which he succeeded in doing "by force"; Qudāma b. Jaʿfar, *al-Ḫarāj wa-ṣināʿat al-kitāba*, ed. Muḥammad Ḥusayn Zabīdī (Baghdad: Dār al-Rašīd li l-našr, 1981): p. 373. This last detail is probably meant to establish the status of the Kashan area as owing the heavier *ḫarāj* tax levied on conquered areas. Many similar accounts exist, as in the earlier work by al-Balāḏuri (d. 279/892): Balāḏurī, *Futūḥ al-buldān*, ed. M.J. de Goeje (Leiden: Brill, 1866): p. 312.

11 Ṭabarī, *Taʾrīḫ al-umam wa ʾl-mulūk*, ed. M.J. de Goeje, 15 vols. (Leiden: Brill, 1879ff): VI, p. 293; see also ibid.: VII, p. 405. Some early references to Qāšān could relate to the town of that name in central Asia. Ṭabarī, *Taʾrīḫ*: IX, p. 186 also mentions Kashan after Qum, in his description of the arrangements of 235/850 regarding which sons of Hārūn al-Rašīd would inherit governorship in the Abbasid empire. This is clearly not a reference to Kashan in Central Asia.

12 Iṣṭaḫrī, *Kitāb al-masālik wa ʾl-mamālik*, ed. M. J. de Goeje (Leiden: Brill, 1870): p. 201 (*madīna ṣaġīra*). See also the early Persian translation: Iṣṭaḫrī, *al-Masālik wa-al-māmālik*, ed. Īraj Afshār (Tehran: Bungāh-i tarjama va-našr-i kitāb, 1340š./1961): p. 166, where Kashan is called *šahrak*, "a small city".

13 *Ḥudūd al-ʿālam min al-mašriq ilā al-maġrib*, ed. Manūčihr Sutūda (Tehran: Dānišgāh-i Tihrān, 1962): p. 143. Could this sentence be a later interpolation?

Arabic and scribal talents will play in Kashani history. It also testifies to the importance of the Arab element in Kashan, where the Arabs were proto-Shiʿis largely of the Ašʿarī tribe, as were the Arabs in Qum.

4 The Growing Importance of Shiʿi Orientation

During the ninth and tenth centuries many proto-Shiʿis gradually became Twelver Shiʿis, which as many geographers testify remained the dominant religious community in Kashan. However, some towns in the Kashan district were and continued to be Sunni, throughout this period.[14] Like many other central Iranian cities, Kashan changed hands several times until the Shiʿi dynasty of the Buyids took control in the fourth/tenth century. The Buyids showed favor to Qum because of their Shiʿi orientation and presumably did so to nearby Kashan,[15] until the conquest of central Iran by the Ghaznavids in 420/1029. In this period the legend grew that Kashan was founded by Zubayda bint Jaʿfar, wife of the Abbasid caliph Hārūn al-Rašīd, and the story of her secret Shiʿism developed correspondingly.[16]

It was no secret that Shiʿis favored *sayyids*, so many descendants of ʿAlī emigrated to Shiʿi cities such as Kashan. The fourth/tenth century history of Qum says that the first of the descendants of Imam al-Riḍā (d. 203/819) to arrive in Qum from Kūfa was Abū Jaʿfar Mūsā b. Muḥammad b. ʿAlī al-Riḍā. But because of the disrespect shown him by some of his fellow Arabs in Qum he then went to Kashan, and the Shiʿis of that city honored his coming. Eventually he returned to Qum and was provided with a handsome stipend.[17]

5 The Saljuq Period

The conquest by the Saljuqs in 443/1051 ushered in a long period of prosperity for the people of Kashan. The city flourished, in part for economic and in part

14 A few Ẓāhirīs may have come from Kashan. See Ibn al-Nadīm, *The Fihrist of al-Nadim*, trans. Bayard Dodge (New York: Columbia University Press, 1970): I, p. 523, concerning a Muḥammad b. Isḥāq al-Qāšānī. There was also an Isma'ili presence; see Virani, Shafique, *The Ismailis in the Middle Ages* (Oxford: Oxford University Press, 2007): pp. 107, 112, 117.

15 The destruction caused by an earthquake in the Buyid period may have left evidence in the form of the rubble of the original mosque. See Karīmī, Fātima, "Kāshān", in *Shahr-hā-yi Irān*: III, ed. Muḥammad Yusūf Kīyānī (Tehran: Vizārat-i Farhang, 1989): p. 211.

16 Le Strange, Guy, *Lands of the Eastern Caliphate* (London: Cass, 1966): p. 209.

17 Qumī, *Tārīḫ*: pp. 575-80.

for political reasons. One of the political reasons was the success of the people of Kashan and the surrounding district in acquiring positions in the Saljuq bureaucracy. A Kashani clerk of the very late Saljuq period, Muḥammad b. ʿAlī al-Rāwandī, writes in his *Rāḥat al-ṣudūr*:

> Holders of senior positions such as vizier and chief-treasurer and the majority of clerks in the government of the *sulṭān* were Kāšīs. Our origin and place of birth has been the district of Kashan [...] In Iraq [western Iran] whenever they see good handwriting they say it is the writing of a Kāšānī.[18]

The eminence of Kashani (often shortened to "Kāšī") administrators arose in large part from their excellent education, including in calligraphy. Scholarly excellence can also be a useful defense against charges of heresy, which Shiʿis continually encountered in the pre-Safavid period.

The Saljuq period was the golden age for the founding of *madrasas* or seminaries, and in Kashan several such schools were established. Two generations of scholarship have associated the founding of such seminaries with the 'Sunni Revival'. However, the fact that these foundations took place simultaneously in Shiʿi Kashan suggests that they were also expressions of more general advancement in teaching and scholarship. Ṣafī al-Dīn Abū Ṭāhir Ismāʿīl Kāšī was a high clerk in the service of Malik-Šāh (reg. 465-85/1072-92), whose famous vizier, Niẓām al-Mulk, hated everything Shiʿi. Ṣafī al-Dīn founded the Ṣafiyya Madrasa along with a worthy library in his native Kashan. In spite of his known Shiʿi adherence, Ṣafī al-Dīn was granted great favor in the Saljuq court.[19] His nephew, Muḫtaṣṣ al-Mulūk Muʿīn al-Dīn Aḥmad b. ʿAbd Allāh al-Kāšī, again a high official, also founded a much frequented *madrasa* in Kashan (this *madrasa* was still functioning in the ninth/fifteenth century).[20]

A generation later, Majd al-Dīn Abu'l-Qāsim 'Ubayd Allah b. Faḍl b. Maḥmūd al-Qāšānī[21] (d. 535/1141) founded the Madrasa Majdiyya. He was very rich and famous for his benefactions of every kind, including building a city wall

18 Rāwandī, Muḥammad b. ʿAlī, *Rāḥat al-Ṣudūr*, ed. Muḥammad Iqbāl (Leiden: Brill, 1921): p. 51. It is significant that local patriotism outweighs the very Hanafī (and anti-Shiʿi) sentiments of this Rāwandī.

19 See Marāğī, Maḥmūd Ṭayyār, "Madāris-i ʿilmīya-yi Kāšān", *Mīrāṣ-i jāvīdān*, VI/2 (1377š./1999): p. 69.

20 Ibid.: pp. 69-70.

21 Ibid.: pp. 70-1. See Rāwandī, Ḍiyāʾ al-Dīn Abū Riḍā Faḍl Allāh, *Dīwān*, ed. Jalāl al-Dīn Muḥaddiṯ Urmawī (Tehran: Maṭbaʿat al-Majlis, 1334š./1955): pp. 60 and 87 on the benefactions of Majd al-Dīn.

and hospitals, and the provision of a fund for dowries of poor women. Urban notables played significant roles in the public works of many Islamic Iranian cities, and this may have been even more important in smaller centers such as Kashan, which did not become capital cities. Their works could include militarily and politically important structures, including city walls. The Shiʿi Kashani vizier Šaraf al-Dīn Anūšīrwān b. Ḫālid (d. 532-3/1137-9), whose significant career is described below, founded a *madrasa* and library, the Šarafiyya, in his native town. Hindūšāh Naḫjavānī, a prominent official and author of the Ilkhanid period, writes:

> In the year 674 [1275-6] when this humble man governed Kashan on behalf of his brother, that *madrasa* and library [founded by Anūšīrwān] was prospering (*maʿmūr*) but now in Muḥarram 724 [Jan. 1323] I have heard that the *madrasa* is ruined.[22]

Another *madrasa* of the Saljuq era was the ʿAzīziyya, founded in the second half of the sixth/twelfth century by yet another vizier, ʿAzīz al-Dīn b. Raḍī al-Dīn Asʿad Kāšāni (d. 588/1192), who served the Sultan Ṭoġrïl b. Arslan. The *Nasāʾim al-asḥār* says of ʿAzīz al-Dīn that: "He was one of the heads of department who were extremely skilled and one of the masters among scribes of financial administration for the two Iraqs [meaning Arab Iraq (modern central Iraq) and Persian Iraq (the Iranian lands between Fars and Azarbaijan)]". Not surprisingly, the *madrasa* curriculum was said to involve deep study of varieties of accounting (*istīfāʾ*) as well as "multiplication, division, algebra, and other forms (*aškāl*) [to be understood through mathematics]".[23] It is a measure of the excellence of Kashani education that the future Sunni chronicler of the Saljuqs, ʿImād al-Dīn al-Iṣfahānī, as a talented boy, was sent to study in Kashan, along with his brother.[24]

22 Marāġī, "Madāris": p. 70. There is a half century between these dates, which is not impossible but slightly suspicious.

23 Ibid.: p. 71 and Kirmānī, Nāṣir al-Dīn Kirmānī, *Nasāʾim al-asḥār min laṭāʾim al-aḫbār dar tārīḫ al-vuzarāʾ*, ed. Mīr Jalāl ad-Dīn Urmavī (Tehran: Čāpḫāna-yi Dānišgāh, 1364š./1985): p. 91. *Aškāl* in mathematical texts is usually translated as "figures" but this translation does not work here. Perhaps *aškāl* should be translated as "operations". Marāġī, "Madāris", in his very full article on education in Kashan lists over thirty *madrasas* in Kashan, up to and including those of the Safavid period; only two of these were Sunni in orientation.

24 Klausner, Carla L., *The Seljuk Vezirate: A Study of Civil Administration, 1055-1194* (Cambridge, MA: Harvard University Press, 1973): p. 66. See also the outstanding recent study of the Saljuqs by Peacock, Andrew C.S., *The Great Seljuk Empire* (Edinburgh: Edinburgh University Press, 2015): p. 214.

Niẓām al-Mulk was fully aware that these well-educated Shiʿi clerks with 'excellent handwriting' were penetrating the Saljuq bureaucracy. He writes:

> In the days of Maḥmūd, Masʿūd, Ṭoġrïl and Alp Arslan (may Allah have mercy on them) no Zoroastrian or Jew or Rafidi [i.e., Shiʿi] would have had the audacity to appear in a public place or to present himself before a great man. Those who administered the affairs of the Turks were all professional civil servants and secretaries from Khurasan, who belonged to the orthodox Hanafi or Shafiʿi sects. The heretics of ʿIraq [i.e., western Iran] were never admitted as secretaries and tax-collectors; in fact the Turks never used to employ them at all; they said, 'These men are of the same religion as the Dailamites [i.e., Buyids] and their supporters; if they get a firm footing they will injure the interests of the Turks and cause distress to the Muslims. It is better that enemies should not be in our midst'. Consequently, they lived free from disaster. Now things have reached such a state that the court and divan are full of them, and every Turk has ten or twenty of these individuals running after him, and their object is to prevent even a few Khurasanis from entering the service of this court and earning a living here. In former times if a man offered himself for service to a Turk as an administrator or in any other capacity, and said that he was of the Hanafi or Shafiʿi sect and from a Sunni city, he was accepted; but if he said that he was a Shiʿi from Qum, Kashan or Aba' [Āva] he was refused and told, 'Be gone; we kill snakes not nourish them.'[25]

Early Saljuq rulers and viziers such as Niẓām al-Mulk, and their largely Khurasanian ministers, allowed comparatively few Shiʿis into government. However, the later rulers of this dynasty were lax, and even the sons and grandsons of Niẓām al-Mulk hired Shiʿis.

The Twelver Shiʿi scholar ʿAbd al-Jalīl Qazwīnī, in writing a long and rambling defence of Twelvers, circa 560/1165, offers a picture that is partially complementary to the comments of Niẓām al-Mulk. Throughout his book, Qazvīnī quotes a Sunni opponent (who had recently written a refutation of Shiʿism) as saying that under the sultans Malik-Šāh and Muḥammad, Shiʿis were not allowed to open schools; Qazvīnī counters with examples, including that of

25 Niẓām al-Mulk, *Siyar al-mulūk*, ed. Hubert Darke (Tehran: Bungāh-i tarjama va-našr-i kitāb, 1962): pp. 203-4; trans. Hubert Darke, *Siyāsat-nāma* (*The Book of Government or Rules for Kings*) (London: Routledge & Kegan Paul, 1960): pp. 164-6. For a similar quote, see Durand-Guédy, David, *Iranian Elites and Turkish Rulers. A History of Iṣfahān in the Saljūq Period* (London: Routledge, 2010): p. 189. This book is the most important study of the rivalry between Khurasanian secretaries and those from central and western Iran.

the schools opened by ministers from Kashan.[26] The Sunni opponent says: "In the ministries (*dīvānhā*) the clerks are all [Shiʿis] and things now are exactly as under [the Sunni Abbasid caliph] al-Muqtadir [who had Shiʿi viziers]". Qazvīnī replies that since God has given Turks rule of the world they are free to foster whomever they like.[27] Of course Twelver Shiʿis had ceased their political claims in 260/874, and were free to support any ruler until the return of the Twelfth Imam. Qazvīnī also mentions other Shiʿi clerks who were good viziers to the Saljuqs, including ʿAlī b. ʿImrān al-Kāšī and Anūšīrwān b. Ḫālid (discussed below), both from Kashan.

Qazvīnī devotes a passage to Kashan, its excellent schools (naming all four of them) and charitable endowments, with teachers like Ḍiyāʾ al-Dīn (the author of the *Dīwān*) who "is without rival in all the lands of the world for knowledge and religious devotion". He also mentions that the judges in Kashan are Shiʿi, which the government may have allowed because only a prominent Shiʿi could command obedience. He refers to the benefactions of Majd al-Dīn to the magnificent shrine of ʿAlī b. Muḥammad al-Bāqir at Ardahāl.[28]

The success of Kashani bureaucrats in the late Saljuq administration is well illustrated by the career of Anūšīrwān b. Ḫālid, founder of the Šarafiyya *madrasa* mentioned above. Anūšīrwān, supposedly a descendant of the Sasanian monarch, was one of the clerks who worked for the staunchly Sunni vizier Niẓām al-Mulk despite the vizier's extreme disdain for Shiʿis.[29] He was especially close to Niẓām al-Mulk's son, Muʾayyid al-Mulk, a vizier to the Saljuq sultan Barkyāruq (reg. 485-498/1092-1105). Anūšīrwān was treasurer in Isfahan for a time; subsequently, another son of Niẓām placed him in charge of the budget for the army. Under the Saljuq ruler Sultan Maḥmūd he was appointed vizier (518/1124-5), but Anūšīrwān could accomplish little under this greatly enfeebled sultan. Very surprisingly, this Shiʿi clerk was then appointed vizier to al-Mustaršid, an Abbasid caliph and a symbol of Sunnism. As noted above, there had previously been crypto-Shiʿi viziers to the Abbasid caliphs, such as the Banū Furāt under the caliph Muqtadir (reg. 295-320/908-932), but Anūšīrwān's Shiʿism was no secret. Anūšīrwān died in Baghdad in 532 or 3/1137-39 and was

26 Qazvīnī Rāzī, *Kitāb-i naqż*, ed. Jalāl al-Dīn Muḥaddiṣ (Tehran: Anjuman-i Āṣār-i Millī, 1358š./1980): p. 34.

27 Qazvīnī Rāzī, *Kitāb-i naqż*: pp. 78-9. See also p. 114, quoting the Sunni opponent as saying that everyone in the service of the Turks is Shiʿi, down to the cooks and cleaners.

28 Qazvīnī Rāzī, *Kitāb-i naqż*: pp. 198-9.

29 Niẓām al-Mulk, trans. *Siyāsat-nāma*: pp. 214, 238 and 1962, 263, 289, mentions Kashan twice for its Shiʿism, along with Qum.

buried near the tomb of ʿAlī in Najaf.[30] Besides a *madrasa* and a library he also had other important structures built in Kashan, such as a hospital (*dār al-šifāʾ*) and a caravanserai.

Two families of Kashani scholars of the Saljuq period had the place-name Rāwandī. Rāwand is a small town a few miles from Kashan. One set of Rāwandīs consisted of the Ḥanafī family related to Muḥammad b. ʿĀlī al-Rāwandī, author of *Rāḥat al-Ṣudūr*.[31] Although Muḥammad b. ʿAlī offered high praise to the Kashanis' education and handwriting, he distanced himself strenuously from their Shiʿism. He rebukes them for "rejecting the *ijtihād* of the [Sunni] schools of law, for praying only three times a day, for refusing to pay the *zakāt* [alms tax] and for going on pilgrimage to Ṭūs [near where Imam al-Riḍā is buried]".[32] The second family, who were *sayyid*s and Shiʿis, included the very wealthy Majd al-Dīn Rāwandī, founder of the Madrasa Majdiyya noted above, and nephew of Anūšīrwān b. Ḫālid. Ḍiyāʾ al-Dīn Abū Riḍā Faḍl-Allāh, another member of this Shiʿi family, was the director of a *madrasa* and a prolific poet in the Arabic language. His *Dīwān* is filled with references to contemporaries, and to Kashan, where his relative Majd al-Dīn founded "schools (*madāris*) and caravanserais (*ḫānāt*) and a hospital (*māristān*)".[33] In another poem Ḍiyāʾ al-Dīn also praises his relative for building mosques, water channels and bridges/aqueducts (*qanāṭir*).[34]

Perhaps the most dramatic of Ḍiyāʾ al-Dīn's poems is his description of the siege of Kashan in 532/1137-8. The work offers the names of forty nearby

---

30 Iqbāl, ʿAbbās, *Vizārat dar ʿahd-i salāṭīn-i buzurg-i Saljūqī* (Tehran: Čāpḫāna-yi dānišgāh, 1959): pp. 254-62. Another prominent Shiʿi family in Saljuq employ was that of Muḫtaṣṣ al-Mulūk, a vizier to Sanjar, and his son Faḫr al-Mulk, vizier to Sulaymānšāh, as well as his grandson, vizier to Ṭoġrïl III. Peacock, *Great Seljuk Empire*: p. 263. Among several short sketches of Anūšīrwān's life, see Bosworth, C.E., s.v. "Anūšervān Kāšānī", in *EIr*, II/2 (1987): p. 139. For the benefaction of the Saljuq governor Abū Naṣr al-Kāšī to Kashan, see Narāqī, Ḥasan, *Tārīḫ-i ijtimāʿī-yi Kāšān* (Tehran: Institut d'Études et de Recherches sociales, 1345š./1967): pp. 57-8.

31 Muḥammad b. ʿAlī was first helped by his maternal uncle, a member of the court of a Saljuq *atabeg* who ruled Hamadan. He was then fostered by another uncle, Zayn al-Dīn Maḥmūd Kāšī. He acknowledges the patronage of a Šihāb al-Dīn Aḥmad al-Qāšī. See Meisami, Julie Scott, "Rāwandī's Rāḥat aṣ-Ṣudūr", *Edebiyat*, V (1994): pp. 183-215, especially 183-4.

32 All of these actions or misperceptions apply to Twelver Shīʿīs. Rāwandī, *Rāḥat al-ṣudūr*: pp. 394-5.

33 Ḍiyāʾ al-Dīn Rāwandī, *Dīwān*: p. 33. On the prominent Rāwandī families, see Ṣādiqī, Muṣṭafā, "Dīdgāh-i tārīḫī-yi Rāwandīyān", *Nāma-yi tārīḫ-pažūhišān*, X (1386š./2008): pp. 115-38. A deed of pious endowment by Majd al-Dīn dated Rabīʿ II 565 (January 1170) survives. Wāṯiqī, Husayn, "Barā-yi tārīḫ-i Kāšān", *Mīrāṯ-i Šahāb*, LXXXIV-LXXXV (1395š.): pp. 147-51.

34 Ḍiyāʾ al-Dīn Rāwandī, *Dīwān*: p. 52.

towns and villages that had been attacked and despoiled by the besiegers but praises Majd al-Dīn for paying forty thousand dinars to have them lift the siege of Kashan itself. (Later, in 595/1198-9, the Khwarazmshah laid siege to Kashan and sacked it savagely after a resistance lasting four months.)[35]

The list of those praised in this *Dīwān* also shows the connections of these learned men of Kashan with other regions. We have a poem in praise of the vizier Qivām al-Dīn Abu'l-Qāsim al-Nāṣir, delivered in his vizieral palace in Isfahan in 524/1129-30.[36] Ḍiyā' al- Dīn also wrote a poem to Qivām al-Dīn, said to be a crypto-Nizārī Isma'ili, on the occasion of this vizier's visit to Kashan. The same Ḍiyā' al-Dīn also wrote an elegy on the vizier's death and dedicated a further poem to his son.[37]

6 Kashan as a Center of Ceramic Production

The Saljuq period saw the rise of Kashan as a major center of ceramic production. Writing around 380/990, al-Muqaddasī twice mentions Qāšān and praises its "long-necked bottles".[38] By the time of Yāqūt (d. 626/1229), Kashan was famed for its pottery dishes or platters (*ġażā'ir*).[39] Some 50 years later, Zakarīyā' al-Qazvīnī (d. 682/1283) writes:

> From [Kashan] come shiny pottery utensils (*ālāt*). They [the Kashanis] have an outstanding gift in this which other lands do not have. The

35 On the sack of 595, see Rāwandī, *Rāḥat al-ṣudūr*: pp. 493-94. Amanat, Mehrdad, "Kashan iii. History to the Pahlavi Period", *EIr* online (http://www.iranicaonline.org/articles/kashan-iii-history).

36 Ḍiyā' al-Din Rāwandī, *Dīwān*: p. 1.

37 Ibid.: pp. 74-90.

38 See Muqaddasī, *Aḥsan al-taqāsīm fī ma'rifat al-aqālīm*, ed. M.J. de Goeje (Leiden: Brill, 1906): pp. 390 and 396. The Arabic word, *qamāqim* (sing. *qumqum* or *qumquma*), is hard to interpret. As in later Persian, it could mean "brass vessels". In later Arabic it often refers to a flask of glass or pottery. The great lexicographer Ibn Manẓūr (d. 711/1311), Ibn Manẓūr, *Lisān al-'arab* (Beirut: Dār Ṣādir, 1990): XII, p. 495, offers "jar" and "a kind of vessel" but also mentions the interpretation of "copper/brass" vessels, and says that it can mean a vessel for water "of copper, or otherwise, narrow of neck". Schwarz, Paul, *Iran im Mittelalter nach den arabischen Geographen* (Reprint, Hildesheim: Olms, 1969): p. 568, translates the bottles as "Tonkrüge", clay ewers. On ancient copper mining in the Kashan area, see Harrison, J.H., "Minerals", in Fisher, W.B (ed.), *Cambridge History of Iran* (Cambridge: Cambridge University Press, 1968): I, pp. 487-516: here p. 502.

39 Yāqūt, *Mu'jam al-buldān*, ed. F. Wüstenfeld, 6 vols. (Leipzig: Brockhaus, 1869): IV, p. 15.

> utensils (*ālāt*) and vessels (*ẓurūf*) from Kashan are transported from Kashan to other lands.[40]

Pottery is both heavy and breakable, meaning that exportability required high artistic standards and strong fabric. The pottery of Kashan offered both attributes, from the twelfth to the fifteenth century. The body was of a density achieved only much later in European stone-paste earthenware, though not yet to the standard of porcelain. The use of a mixture of ground quartz and white clay had survived from ancient Egyptian pottery techniques, which art historians often call 'faience'.[41] The earliest appearances of enamel overglaze painting (*mīnā'ī* or, in earlier usage, *haft rang*) probably coincided with the invention of this technique in Kashan.[42] In both lusterware and *mīnā'ī* forms (Figs. 2 and 3, the decorative patterns and illustrative painting is of such high quality that these works are widely regarded as related to the art of miniature illustration in books.[43]

It could have been in the twelfth and thirteenth centuries that the adjective *qāšī*, which means "things belonging to Kashan", passed into other languages such as Arabic, with the meaning of "tiles".[44] Some of the ingredients of Kashan pottery could be found nearby, such as quartz pebbles which are "white, clear and shiny, not as clear (*ṣafā*) as rock crystal but more clear than white marble",[45] and finely-divided white stone from the villages near Fīn. From nearby Qamṣar there was a stone "like white sugar" and the famous cobalt called *lājvard*. Some white clays also come from the Kashan district.[46]

---

40 Qazvīnī, Zakarīyā', *Āṯār al-bilād wa-aḫbār al-ʿibād* (Beirut: Dār Ṣādir, 1988): p. 433.

41 Watson, Oliver, *Persian Lusterware* (London: Faber & Faber, 1985): p. 23.

42 Ibid.: p. 24.

43 See Blair, Sheila, "A Brief Biography of Abū Zayd", *Muqarnas*, XXV (2008): pp. 155-76, here p. 157, who states that the second style of Kashan lusterware "was clearly inspired by book painting". For ample literature on the relation of Kashan pottery to miniatures, see Hillenbrand, Robert H., "The Relationship between Book Painting and Luxury Ceramics of 13th Century Iran", in Id. (ed.), *The Art of the Saljuqs in Iran and Anatolia* (Costa Mesa, CA: Mazda, 1994): pp. 134-41.

44 Dozy, R., *Supplément aux dictionnaires arabes*, 2 vols. (Paris: Maisonneuve, 1927): II, pp. 295-6.

45 Kāšānī, Abu'l-Qāsim ʿAbd-Allāh, *ʿArāyis al-javāhir wa-nafāyis al-aṭāyib*, ed. Īraj Afšār (Tehran: Almā'ī, 1386š./2007): pp. 339-48, transl. in Allan, J.W., "Abū 'l-Qāsim's Treatise on Ceramics", *Iran*, XI (1973): pp. 111-20.

46 The line between porcelain and pottery is fairly clear: porcelain usually contains a great deal of kaolin and is fired at temperatures above 2500 centigrade. However, there is some confusion between various kinds of pottery, including fritware (often containing ground glass), stone paste (often containing crushed quartz), faience and soft paste porcelain (terms that are even more loosely defined). Both Kashan lusterware and *mīnā'ī* are best

FIGURE 2 Luster plate signed by Abū Zayd, Kashan, 607/1210
FREER GALLERY OF ART AND ARTHUR M. SACKLER GALLERY, SMITHSONIAN INSTITUTION, WASHINGTON, D.C.: PURCHASE – CHARLES LANG FREER ENDOWMENT, F1941.11

A relatively neglected text written in 592/1195-6 by a jeweler from Nishapur, Muḥammad b. Abi'l-Barakāt Jawharī, contains some of the earliest references to lusterware. He discusses a formula for putting "golden Marcasite and other ingredients on pottery and glass" so that when these vessels emerge from the kiln (*dāš*), they have "a color like gold", and he also reports on enameling techniques on pottery, but the procedures described may fall short of the more elaborate techniques of the Kashan potters. The text offers numerous formulas that require analysis by chemists, but among these is one which Jawharī says is used for "Syrian glass" and "Kashan pottery".[47]

called stone paste. Cf. Watson, Oliver, *Ceramics from Islamic Lands* (London: Thames & Hudson, 2004): pp. 54-5 for a somewhat different view.

47 Jawharī Nīšāpūrī, Muḥammad, *Javāhir-nāma-yi niẓāmī*, ed. Īraj Afšār (Tehran: Mīrāṯ-i Maktūb, 2004): p. 265 (like gold), p. 350 (enamel), p. 354 (Syrian glass). Mīr-Šāfīʿī, Sayyid

FIGURE 3 *Mīnā'ī* bowl with Bahrām Gūr and Āzāda, Kashan, late sixth/twelfth – early seventh/thirteenth century
METROPOLITAN MUSEUM OF ART: PURCHASE – ROGERS FUND, AND GIFT OF THE SCHIFF FOUNDATION, 1957, 57.36.2

Signed works by four generations of Kashani lusterware potters, all of the Abū Ṭāhir family, testify to the development and preservation of the many in-house skills and trade secrets that went into the production of these wares.[48] In the very early seventh/thirteenth century a master ceramicist at the head of

Muḥammad and Bāqir-zāda Kaṯīrī, Mas'ūd, "Muṭāla'a-yi sāḫt-i lu'āb-i zarrīn-fām", *Pažūhiš-hā-yi 'ulūm va-fannāvarī rang*, (2016): pp. 71-8 report using one of Jawharī's formulas for lusterware pottery with success.

48 Given that the techniques used were so complicated and expensive, and were only handed down in a very small number of families, the Kashani potters in some ways resemble the great Italian Della Robbia family. See Karimi, Fatima, "Abū Ṭāhir (Family)", trans. Rahim Gholami and Farzin Negahban, in Madelung, W. and Daftary, F. (eds.), *Encyclopaedia Islamica* (2008): www.referenceworks.brillonline.com/entries/encyclopedia-islamica/abu-tahir-family, viewed 4/24/2017, and Watson, *Persian Lusterware*: p. 177 and Watson, Oliver, s.v. "Abū Ṭāher", in *EIr*, 1/4 (2011): pp. 385-7. The vast literature on Kashan lusterware includes Watson, *Persian Lusterware* and *Ceramics*, Caiger-Smith,

another family of Kashani potters, Abū Zayd, collaborated with the head of the Abū Ṭāhir family on the tile-work in shrines in Qum and Mashhad.[49] Abū Zayd worked in both *mīnā'ī* and luster, and left more signed works than any other potter, the earliest of which is a *mīnā'ī* bowl of 582/1186.[50] He identifies himself as a Ḥasanī *sayyid* on a fine luster plate of 607/1210 (Fig. 2).[51] On other works he not only says that he made the piece but also that he wrote the inscription, perhaps a reflection of Kashani pride in handwriting.[52] Manijeh Bayani, a great expert on medieval Persian inscriptions, says that "the inscriptions of Persian luster-painted, *mīnā'ī* and underglaze-painted wares of the late twelfth to the thirteenth century have no precedent or parallel in other medieval wares".[53] Bayani discusses the various styles of handwriting on Kashan pottery of this period, noting that the cursive style used for the popular poetry on luster-painted wares is closest to *ta'līq*, a style in which hanging and joining of letters are used. Bayani does not regard the cursive style on lusterware as crude, rather as competent and consistent.[54] Certainly it is tempting to connect this type of handwriting to that of the Kashani clerks trained in the *madrasas* of Kashan.

We have a treatise on the methods of fabrication for both lusterware and *mīnā'ī* pottery, written by a later member of the Abū Ṭāhir family, Abu'l-Qāsim 'Abd Allāh Kāšānī, in 700/1300-1.[55] Although the treatise is in part drawn from earlier works on stones and gems by Jawharī and Naṣīr al-Dīn Ṭūsī, who were in turn inspired by Bīrūnī (and ultimately by Aristotle), it contains an original

---

Alan, *Lustre Pottery*, (London: Faber & Faber, 1985), and Soustiel, Jean, *La céramique islamique* (Fribourg, Suisse: Office du livre, 1985).

49 Blair, Sheila and Bloom, Jonathan, "Signatures on Works of Islamic Art and Architecture", *Damaszener Mitteilungen*, XI (1999): pp. 49-66, 54-5. Blair, "A Brief Biography". Interestingly, Abū Zayd is the only Kashani potter known to have produced both vessels and tiles: Blair, "A Brief Biography": p. 161.

50 Blair, "A Brief Biography": p. 156. The survival of dated pieces and signatures of potters such as Abū Zayd, from the late sixth/twelfth century, together with the group of pieces displaying both *mīnā'ī* and luster techniques, proves that these different products were made in the same workshops: Watson, *Persian Lusterware*: p. 84 n. 11.

51 Blair and Bloom, "Signatures": pp. 54-55. Blair, "A Brief Biography".

52 Blair and Bloom, "Signatures": p. 55. See also Blair, "A Brief Biography" and Karimi, Fatima, "Abū Zayd Kāshānī", trans. Farzin Negahban, in Madelung, W. and Daftary, F. (eds.), *Encyclopaedia Islamica* (2008): www.referenceworks.brillonline.com/entries/encyclopedia-islamica/abu-zayd-kashani, viewed 2/21/2017.

53 Pancaroğlu, Oya, *Perpetual Glory: Medieval Islamic Ceramics from the Harvey B. Plotnick Collection*, with transcriptions and translations of inscriptions by Manijeh Bayani (New Haven: Yale University Press, 2007): p. 155.

54 See Pancaroğlu and Bayani, *Perpetual Glory*: pp. 154-5 for Bayani's excellent account of the inscriptions, entitled "A Note on the Content and Style of Inscriptions".

55 For a good discussion of Abu'l-Qāsim himself, see Caiger-Smith, *Lustre Pottery*: p. 73-76. See also Soucek, Priscilla P., "Abu 'l-Qāsem 'Abdallāh Kāšānī", in *EIr*, I/4 (2011): pp. 362-3.

section about the manufacture of Kashan luster and *mīnā'i* pottery. Abu 'l Qāsim also wrote a history of an Ilkhanid ruler, and his brother wrote a treatise on Sufism, all of which testifies to the high literary achievements of these later members of the Abū Ṭāhir family.[56] The family had clearly benefitted and prospered, thanks to the excellent education available in Kashan, and had risen to become authors and scholars.

In the nearby town of Quhrūd we have the last recorded member of the Abū Ṭāhir potter family, who signs: "the work of Yūsuf [...] 'Alī b. Muḥammad [...] the Kāšānī potter (al-Ġażā'irī al-Qāšānī)". In this case the work is not in luster but in blue and white ware, a type of production that would soon dominate.[57] Thanks to this and other individual signatures we can distinguish four different families of potters, and at least seven other potters.[58]

It has been suggested that some of the inspiration for Kashan pottery of this period was due to the arrival of artisans migrating from Fatimid Egypt, or Raqqa in eastern Syria.[59] In any case, given that they were firm Shi'is, the potters of Kashan (including several families of *sayyids*) travelled to the shrine cities such as Qum, Mashhad and Najaf. There is also evidence of their activity in centers not connected with Shi'ism, such as Warāmīn, Damghan and Baku.[60] We know of the specific case of a potter in the seventh/thirteenth century who migrated to Kashan, as seen in the signature to his work: Muḥammad ibn Muḥammad Nīšāpūrī "residing in (*muqīm*) in Kashan".[61]

Some of the themes illustrated on Kashan lusterware are specifically Shi'i, such as the circular tile with a 'horseshoe-shaped' imprint on a foundation

56 Soucek, "Abū l-Qāsem", Graves, Margaret S., s.v. "Kashan vii, Kashanware", in *EIr*: www.iranicaonline.org/articles/kashan-vii-kashan-ware, viewed 4/27/17.

57 Watson, Oliver, "The Masjid-i 'Ali, Quhrūd: An Architectural and Epigraphic Survey", *Iran*, XIII (1975): pp. 59-74, here p. 63 and Watson, *Persian Lusterware*: p. 179. Blair, "A Brief Biography": p. 160. Although the date has been differently read, all readings put the piece in the early eighth/fourteenth century.

58 Watson, *Persian Lusterware*: p. 178.

59 See Lane, Arthur, *Early Islamic Pottery. Mesopotamia, Egypt and Persia* (London: Faber & Faber, 1947): pp. 37-8. See also Pradell, T. et al., "The Invention of Lustre: Iraq 9th and 10th Centuries AD", *Journal of Archaeological Science*, XXXV/5 (2008): pp. 1201-15.

60 We have families of Ḥasanī *sayyids* (Abū Zayd) and Ḥusayni *sayyids* (Abū Ṭāhir family) and some other individual *sayyids* signing Kashan ceramics. Examples of collaboration between individuals and families are well documented. Cf. Watson, *Persian Lusterware*: pp. 176-82 and Blair, "A Brief Biography": pp. 159-60.

61 Bahrami, M., *Gurgan Faiences* (Cairo: Le Scribe Egyptien, 1949): p. 92 (Victoria and Albert Museum, C. 162-1977). Cf. Blair, Sheila, *Islamic Inscriptions* (Edinburgh: Edinburgh University Press, 1998): pp. 148-59, whose chapter on ceramics of the early and middle periods provides a useful summary of the inscriptions and signed and dated pieces, in their historical contexts.

plaque bearing the date 711/1312, intended to commemorate a dream visit by Imam ʿAlī on his famous mount, which left a hoof-print.[62] It also appears that many buyers of Kashan *mīnāʾi* and lusterware pottery preferred themes from the *Šāh-nāma*, such as that of Bahrām Gūr and Āzāda (Fig 3). Complementary themes are the enthroned monarch, musicians, female figures and male-female couples. Most large installations of Kashani lusterware tiles are at mausolea, such as the tomb of ʿAlī in Najaf, yet even in these sacred settings, some 'secular themed' tiles have been used to fill in spaces.[63]

Some tiles spell Kashan in a more Persian style using a 'K' rather than 'Q', as in a star-shaped tile dated 738/1338, for which the full wording is "in the place (*maqām*) of Kashan, may God the Exalted protect it from the accidents of time".[64] The Kashan pride in calligraphy is reflected in the numerous inscriptions in which the potter says that the writing is "in his [own] handwriting (*bi- ḫaṭṭihi*)".[65]

The learned ʿAbd Allāh Qūčānī shows that the majority of verses shown on the ceramics of Kashan are quatrains, some of which are attributed to Kashan's Sufi poet, Afḍal al-Dīn. Qūčānī estimates that the works of some one hundred different poets are recorded on ceramics, although some quatrains are used repeatedly. Some of the poetry is quite difficult, which would be evidence suggesting the production of some pieces on commission and for a fairly literate audience.[66]

The earliest dated Kashan lusterware is from 575/1179[67] and the last dated piece is from 739/1339.[68] One reason for the cessation of production could be that the tastes for this elaborate ware waned, although as noted above, the less expensive blue and white Kashan tiles continued to be popular. The shift of interest from expensive tiles for building interiors to less expensive production for coverage of exteriors, as seen in Timurid architecture, may be another reason for the decline of traditional Kashan pottery. It seems that prior to the waning of lusterware, the production of *haft rang* or *mīnāʾī* had already declined. In this case it could be that the flourishing of painting in book illustration would

62 Watson, *Persian Lusterware*: pp. 146-49, fig. 124.

63 Ibid.: p. 154.

64 Ettinghausen, Richard, "Evidence for the Identification of Kashan Pottery", *Ars Islamica*, III/1 (1936): pp. 44-75, here p. 59.

65 See Graves, "Kashanware".

66 Qūčānī, ʿAbd-Allāh, "Sufālgarān-i Kāšān va-šiʿr-i fārsī", *Našr-i Dāniš*, LXXXIV (1373š.): pp. 31-40.

67 See Graves, "Kashanware", and Porter, Yves, "Lustres fatimides et potiers de Kâshân: Historiographie et obscurs transferts", *Keramik-Freunde der Schweiz*, CXXIX (2015): pp. 12-4.

68 See Graves, "Kashanware".

account for part of the change in taste. Another important reason for cessation of this production could have been the difficulty and cost of obtaining the necessary ingredients for the many colors used in *mīnā'ī* ware.[69]

We have a tile that reads:

> It was written in the place Kashan on the 10th of Rabīʿ II 738 [1337] in the workshop (*kārḫāna*) of the *sayyid* of *sayyids*, Sayyid Rukn al-Dīn Muḥammad, son of the late Sayyid Zayn al-Dīn ʿAlī, the potter [...] The work of the revered and respected Master (*ustād*) Jamāl the designer (*naqqāš*).[70]

The mention of the workshop indicates the significant scale of labor involved. The manufacturers needed not only to produce the pottery but also to build the kilns and fire the pieces.[71] The separate and very respectful mention of the 'designer' tells us that the manufacture of high quality Kashan pottery likely included the presence of such a personage, which in this case probably meant a painter.

7 Agriculture in the Kashan District

As we would expect in any largely agrarian society, the prosperity of Kashan during and subsequent to the Saljuq period would have been based more on agricultural production than on artisanal or scribal patronage. The geographers mention tarragon, plums, peaches, apricots, apricot leather and vegetables.[72] However the range of crops and food products was surely far greater. A government "order" (*farmān*) dated 869/1464-5 lists the production of Kashan as including wheat, barley, cotton, cotton-pods (*kulūza*), beans

69 Abu'l-Qāsim, writing when lusterware was still in production, comments (Allan, "Abū 'l-Qāsim's Treatise on Ceramics": p. 115) that *mīnā'ī* or *haft rang* pottery has "passed into oblivion in this time". Cf. Soustiel, *La céramique*: p. 193.

70 Watson, *Persian Lusterware*: p. 142. Watson also refers here to another inscription in which the artists called himself "the builder" (*al-bannā'*).

71 Some of the best descriptions of the techniques and labor involved in Kashan pottery come from scholars who are themselves potters or who work closely with experienced potters. See Caiger-Smith, *Luster Pottery* : pp. 73-75, who discusses the man hours involved in producing Kashan pottery and Soustiel, *La céramique*: p. 193, who states "la fabrication des céramiques de prestige en Orient ne relevait pas de l'*artisanat*, mais bien d'une entreprise rigoureuse faisant appel à des chimistes expérimentés et dotée d'une infrastructure stable destinée à promouvoir son exploitation".

72 Schwarz, *Iran im Mittelalter*: pp. 568-9.

(*bāqalā*), sorghum (*ḏurra*), millet (*kāvurs*), cumin seed, sesame (*kunjid*), chickpea (*nuḫwud*), almonds, raisins (*mavīz*), grape syrup (*dūšāb*) and other vegetables (*ḫużrāvīyāt*, literally "greens").[73] This document confirms that thanks to its extensive *qanāt* irrigation, the Kashan district was engaged in a wide range of agricultural production.

8 Monuments of the Saljuq Period

Several buildings of the Saljuq period still survive in Kashan and its surroundings. A minaret of the Old Friday mosque (generally known as Masjid-i Maydān, because of its closeness to the central square) bears an inscription – unfortunately somewhat repaired – that reads: "The glorious *qāżī* Majd ... ordered it to be built in 466 [1073-4]" (see Fig. 4).[74] Could this be an indication of Majd al-Dīn Rāwandī, who is known for founding the Madrasa Majdiyya and so many other benefactions? A different member of the Rāwandī family, the abovementioned Ḍiyā' al-Dīn Abū Riḍā, who was known for his poetry, celebrates Majd al-Dīn's construction of the famous shrine of "Sulṭān ʿAlī", son of the Imam Muḥammad al-Bāqir, in nearby Ardahāl.[75] Another Saljuq building is the fortress to the west of Kashan called Qalʿa-yi Jalālī, ascribed to Malik-Šāh (reg. 465/1072-485/1092), a surprisingly large structure for a small city like Kashan.[76]

---

73 Narāqī, Ḥasan, *Āṣār-i tārīḫī-yi šahristānhā-yi Kāšān va-Naṭanz* (Tehran: Anjuman-i āṣār-i millī, 1348š./1969): p. 220, quoting the *farmān* of Jahānšāh, written on stone and installed in the Masjid-i Maydān in Kashan. It is unclear as to why a secondary product, "grape syrup", enters in this list. A slightly different reading of the list is given in Miškātī, Nuṣrat Allāh, "Nāẓarī bi-tārīḫ-i bāstān-šināsī-yi Kāšān", *Hunar va-mardum*, series 5, LIII-LIV (1345š./1966): pp. 7-14 and LV (1346š./1967): pp. 8-13, here pp. 8-9.

74 Narāqī, *Āṣār-i tārīḫī*: p. 115.

75 Ibid.: p. 126. There was a great deal of repair and new building work at this shrine in a further period of prosperity in the ninth/fifteenth century; see Ibid.: pp. 136-7, 172, 184-5. "Sulṭān" is, of course, a purely fictive epithetic added to the name of this *sayyid*.

76 Ibid.: pp. 120ff. Ḍarrābī, ʿAbd al-Raḥīm Kalāntar, *Tārīḫ-i Kāšān*, ed. Īraj Afšār (Tehran: Ibn Sīnā, 1962): p. 469 discusses the dimensions of this castle and the remnant of a town wall attached to it, but given that this is an account, completed in 1288/1871, it may not represent the size in the Saljuq period. See also Haeri, Mohammad-Reza, s.v. "Kashan v. Architecture (2) Historical Monument", in *EIr*, XVI/1 (2012): pp. 12-21.

FIGURE 4 Location of the Old Friday Mosque in Kashan; its minaret with an inscription dated 466/1073-4
PHOTO WIKISOURCE

9 Kashan under the Ilkhanids

The extent of Kashan's suffering under the Mongol invasion is still a matter of debate.[77] However it is clear that Kashan thrived in the subsequent Ilkhanid period. Perhaps the Mongol sack of larger cities such as Rayy and Isfahan permitted a temporary advantage for smaller cities like Kashan. The trade in lusterware platters continued, but artisanal production at Kashan increasingly centered on other wares. Kashan was a major supplier of star and cross tiles for architectural ornamentation, beginning in the heyday of lusterware production and continuing with 'blue and white' technologies.

In the late Mongol and post-Mongol period the potters of Kashan also specialized in blue and white tiles showing decorative patterns with dedicatory or pious inscriptions, used in mosques and tombs. There were also multi-colored tiles forming scenes of hunting, feasting and fighting, for display in private

77 Ibn al-Aṯīr, *al-Kāmil fi'l-ta'rīḫ*, ed. Muḥammad Yūsuf al-Daqqāq, 13 vols. (Beirut: Dār Ṣādir, 2003): sub anno 621. There seems to have been a Mongol raid in 620-21/1223-25. Two manuscripts of Juwaynī say: "And part of the army laid siege to Kashan, which they took [...] with much slaughter and pillage", but the translator, Boyle, takes this to be a later interpolation. See Boyle, John A., *The History of the World Conqueror*, 2 vols. (Manchester: Manchester University Press, 1958): II, p. 437, note 46, and Juwaynī, ʿAṭā Malik, *Tārīḫ-i Jahān-Gušā*, ed. M.M. Qazwīnī, 3 vols. (Leiden: Brill, 1912-37): II, p. 169 note 3.

settings.[78] The excellent cobalt blue mined at the nearby village of Qamṣar gave the coloring of these tiles a unique character. The cobalt itself was exported as far as China, where it played a role in the development of the famous blue and white porcelain of that country.[79]

The *pax mongolica* favored trade, and Kashan benefited due to its location on several major routes. Until the nineteenth century it was an important stop on the road from Rayy/Tehran to Isfahan and beyond to Shiraz and the Persian Gulf, as well as on the route from Yazd to Rayy. Through Simnan, Kashan communicated with Khurasan and Central Asia. The *pax mongolica* attracted European merchant travelers such as Marco Polo, who seems to have stayed at the famous Zoroastrian town of Gabrābād, near Kashan. Marco Polo reported that one of the three Wise Men came from Kashan itself,[80] and his account is partly confirmed by the travel narrative of Friar Odoric of Pordenone, who in approximately 1320 "halted at the city of the three Magi, which is called Cassan, a royal city and of great repute."[81] Another traveler of the late Ilkhanid period was the Moroccan Ibn Baṭṭūṭa, who reported seeing "Qāšānī" tiles on *madrasas* and [Sufi] convents, as well as on the shrine of ʿAlī in Najaf and the bath in Isfahan. He compares these tiles to the faience tiles of North Africa, the *zalīj* or *zulayj*.[82]

The reforming Ilkhanid ruler, Ġāzān Ḫān (reg. 694/1295-703/1304) established "alms houses for *sayyids*" (sing. *dār al-siyāda*), and it is a mark of the importance of Kashan that the city is mentioned in this respect. The historian Ḥāfiẓ-i Abrū writes "A number of 'alms houses for *sayyids*' were constructed in large cities (*šahr-hā-yi buzurg*) like Kashan, Isfahan and Sivas [in Anatolia]".[83] (It is unusual for Kashan to be called a large city.) These "houses" seem to have

78 The shrine of Sulaymān b. Mūsā offers evidence of continuation of the local tile industry: the tile *miḥrāb* bears the inscription "work of Sayyid Quṭb al-Dīn, a descendant of Ḥusayn and potter (*ġażāʾirī*) in the year 902 [1496-7]". Narāqī, *Āṯār-i tārīḫī*: p. 174.

79 Kerr, Rose and Wood, Nigel, *Ceramic Technology. Science and Civilisation in China*, V/12 (Cambridge: Cambridge University Press, 2004): pp. 628-92.

80 Jackson, A.V. Williams, "The Magi in Marco Polo", *JAOS*, XXVI (1905): pp. 79-83. See also Marco Polo, *The Travels of Marco Polo*, trans. Ronald Latham (Harmondsworth: Penguin, 1958): pp. 27-9, which identifies Gabrābād as "Kala Atashparastan", meaning "Castle of the Fire-Worshippers". Jackson is surely correct in identifying this as reference to a town in the Kashan district.

81 Odoric of Pordenone, *Cathay and the Way Thither*, trans. Henry Yule, 2 vols. (London: Hakluyt Society, 1866): I, pp. 50-1. We do not know why Odoric says "royal" city.

82 Ibn Baṭṭūṭa, *The Travels of Ibn Battuta, 1325-1354*, trans. H.A.R. Gibb, 5 vols. (reprint, New Delhi: Munshiram Manoharlal, 1993): I, p. 256 and II, p. 296. *Zalīj* is the term from which Spanish "azulejo" is derived.

83 Narāqī, *Tārīḫ-i ijtimāʿī*: p. 90.

been charitable foundations, since they typically included buildings for a mosque, a school, and a place for travelers to rest. Their roles included fostering orphans and housing the indigent. The expenses were to be paid from income on the endowments left in the nearby towns of Bīdgul, Muḫtaṣṣābād, and Harāskān. Money was also given for shrouds and burial expenses. Supervision of the establishment and the endowment at Kashan was given to Jalāl al-Dīn Murtaḍā, a grandson of Ḍiyā' al-Dīn Abū Riḍā al-Rāwandī. This grandson was called the "model" (*qudwa*) for great men, and supreme marshal (*naqīb al-nuqabā'*) of the descendants of the Prophet.[84]

The Shiʿi Rāwāndī family was still in strong form at the beginning of the eighth/fourteenth century,[85] as demonstrated by the marriage of a daughter into the ruling family of the Muzaffarids.[86] In his history of this period, Nāṣir al-Dīn Kirmānī (fl. 725/1324) mentions that Abū Ṭāhir Ṣafī al-Dīn Ismāʿīl, who held Kashan as a kind of fief (*suyūrġāl*), showed kindness to "the members of ancient houses" (*buyūtāt-i qadīm*).[87] As in neighboring Qum, these "ancient houses" very likely included both descendants of long-term Iranian landowners and Arab elites, especially *sayyids*.

With the breakup of the Ilkhanid empire, Kashan passed briefly under the Injuid ruler of Shiraz, Abū Isḥāq Jamāl al-Dīn, who was in turn defeated and executed by the Muzaffarids in 758/1357. (Both dynasties were patrons of the poet Ḥāfiẓ.) The post-Mongol period continued as a time of prosperity for Kashan. We know that there was a fair amount of building under Muzaffarid patronage.[88] Muzaffarid rule was dispersed among several members of the family, some of whom seem to have resided in Kashan.

10 Tomb Shrines

Just as the Zoroastrians had sacralized the landscape before the coming of Islam (a *čahār-tāq* fire temple still survives in the area[89]) the city of Kashan and surrounding lands were re-sacralized through Shiʿism and Sufism, particularly in the twelfth through fifteenth centuries. The prime Sufi example is the

84 This endowment deed of 703/1304 was published by Afšār, Īraj, "Vaqf-nāma-yi sih dīh dar Kāšān", *Farhang-i Īrānzamīn*, IV (1335š.): pp. 122-38.

85 Narāqī, *Tārīḫ-i ijtimāʿī*: pp. 90-1 quoting *Ḏayl-i Jāmiʿ al-Tavārīḫ-i Rašīdī* and the *vaqf-nāma*.

86 Rajabi, Parvīz, *Kāšān: Nigīn-i anguštarī-yi tārīḫ-i Īrān* (Tehran: Nashr-i Pižvāk-i Kayrān, 2011): p. 131.

87 Narāqī, *Tārīḫ-i ijtimāʿī*: p. 56 quoting Kirmānī.

88 Ibid.: *passim*.

89 Ibid.: pp. 46-7.

tomb of Bābā Afḍal Kāšānī with its handsome conical dome, situated in nearby Maraq.[90] Bābā Afḍal lived in the second half of the sixth/twelfth and first half of the seventh/thirteenth centuries.[91] He was a gifted poet (some of his quatrains were later ascribed to Omar Khayyam) and an outstanding figure in the Sufi philosophical tradition.

The hostility between Shiʿism and Sufism, so strongly developed in later Twelver thought, is not evident in pre-Safavid Kashan.[92] Given the city's loyalty to Shiʿism, tomb shrines of descendants of the Prophet, in Persian called *imāmzādas*, have always been important in the area. The most prominent of these is the tomb of ʿAlī, son of the Imam Muḥammad al-Bāqir,[93] which as we have noted is situated in the nearby town of Ardahāl. ʿAlī is said to have migrated to this town in 113/731, at the request of the early Shiʿis of Fīn (or the Čihil Ḥiṣārān), but he was then martyred by the hostile "governor" of the Umayyads in 116/734. The circumstance that ʿAlī is not buried in the city itself reflects the assumption of Kashan to the role of regional center only through gradual growth.

The tomb had become important at least as early as the Saljuq period, and Ḍiyāʾ al-Dīn Rāwandī mentions ʿAlī in several poems. Very surprisingly, the shrine is now the setting for a broadly diffused pre-Islamic custom. The ceremony of 'carpet washing' in a 'healing spring', associated with the story of the 'Sulṭān' ʿAlī, said to be buried at this location, takes place on the second Friday of the Persian month of Mihr (September-October). The calculation by non-Islamic date suggests that this widely celebrated event is associated with the Zoroastrian festival of Tīrgān, in which water plays a central role.[94]

---

90 Ibid.: pp. 351-5.

91 His *terminus ante quem* is 644/1246. Gurgānī, Murtaḍā Qarāʾī, s.v. "Bābā Afżal Kāšānī", in *Dāyirat al-Maʿārif-i Buzurg-i Islāmī* (*The Great Islamic Encyclopaedia*): x (2001): pp. 735-9, here p. 735.

92 A Sufi treatise by Kamāl al-Dīn ʿAbd al-Razzāq b. Jalāl al-Dīn Kāšānī shows that the tradition was thriving in the eighth/fourteenth century: Kāšānī, Kamāl al-Dīn ʿAbd al-Razzāq b. Jalāl al-Dīn, *Tuḥfat al-iḫwān fī ḫaṣāʾiṣ al-fityān*, ed. Muḥammad Dāmādī (Tehran: Širkat-i intišārāt-i ʿilmī va-farhangī, 1991).

93 Narāqī, *Tārīḫ-i ijtimāʿī*: pp. 125-45.

94 See Borjian, Habib, s.v. "Mašhad-e Ardahāl", in *EIr* online, www.iranicaonline.org/articles/mashad-e-ardahal, viewed 12/11/2016. See Muqaddasī's description (*Aḥsan al-taqāsīm*: p. 396-97) of the festival of Tīr, reporting that the people of the Kashan area carry receptacles and go to a mountain exuding water where they say, "Water us with thy water because of such and such [reason to be supplied] and they carry away water". A similar practice, Ābrīzān or Ābpāšān, observed on the same day of the year was customary among Kashan's coppersmiths until recently: Rajabī, *Kāšān*: p. 98 note 3. For a vigorous defense of the historicity of Sulṭān ʿAlī, see Sajjādī, Aḥmad, "Qālī Šūyān", *Dāyirat al-maʿārif-i tašayyuʿ*, XII (Tehran: Ḥikmat, 1386š./2007): pp. 531-3.

The violent irruption of Tīmūr (Tamerlane) brought about dramatic change in the Iranian dynastic landscape. Kashan in fact submitted to Tīmūr in about 787/1387, and so escaped the horrendous slaughter that befell the inhabitants of nearby Isfahan. After Tīmūr's death (807/1405), his enlightened successors drew on Kashani traditions for the blue and white tile work that became such an important feature of Timurid architecture.

11 Mathematicians and Astronomers

Tīmūr's grandson Uluġ Beg (811-850/1409-1449) invited the mathematician-astronomers of Kashan to his capital, Samarqand. The development of such talents among Kashanis was at least in part due to the curriculum of the *madrasa* founded by ʿAzīz al-Dīn Kāšī (d. 588/1192), as noted above. "The most 'judgely' of judges" ʿImād al-Dīn Yaḥyā b. Aḥmad al-Kāšī (d. 766/1364-5) was a highly respected author who wrote several treatises on mathematics, a tradition that continued among his descendants.[95]

The practical benefits of mathematical knowledge would have been evident to medieval Kashanis, who, like their neighbors in central Iran, faced challenging tasks in areas such as land surveying, water distribution and tile design.[96] Ġiyāṯ al-Dīn Jamšīd Kāšānī (d. 832/1429) was the most brilliant mathematician of the period, having studied in the Madrasa Ḫān, which was founded in Kashan by Sayyid Laṭīf ibn Rukn al-Dīn, a fifth-generation descendant of the poet Ḍiyāʾ al-Dīn al-Rāwandī. Here, the rational sciences, including mathematics and astronomy, were an important part of the curriculum.[97] Early in his career Jamšīd was invited to the Timurid capital of Samarqand, under the patronage of Timur's grandson, Uluġ Beg, one of most mathematically gifted rulers of the medieval era. Jamšīd's enormous talent had supposedly been immediately recognized by another astronomer, on discovering him in an astrolabe shop in Kashan.[98] An entire chapter of Jamšīd's treatise on calculation (*ḥisāb*) is devoted to architectural problems, such as the design of *muqarnas*

95 Narāqī, *Tārīḫ-i ijtimāʿī*: p. 82; Rajabī, *Kāšān*: p. 193.

96 The potential connections between irrigation works and mathematics are shown by Muḥammad Karajī, an Iranian mathematician of the early fifth/eleventh century, in his treatise on math and hydrology; see Nadji, Mehdi and Voigt, Rudolf, " 'Exploration for Hidden Water', by Mohammad Karaji – the Oldest Textbook on Hydrology", *Ground Water* X/5 (1972): pp. 43-6. Nevertheless, the relations between artisans and scientists remain problematic; see Saliba, George, "Book review: Artisans and Mathematicians in Islamic Architecture by Gulru Necipoğlu", *JAOS*, CXIX (1999): pp. 637-45.

97 Narāqī, *Tārīḫ-i ijtimāʿī*: p. 94. On Sayyid Laṭīf, see Marāġī, "Madāris": pp. 73-4.

98 See, among others, Taheri, Jafar, "Mathematical Knowledge of Architecture in the Works of Kâshânî", *Nexus Network Journal*, XI (2009): pp. 77-88, here p. 82.

patterns.[99] He seems to have been in Samarqand at the same time as his nephew, Muʿīn al-Dīn. The latter was another Kashani mathematician-astronomer who worked for the Timurids, as did his astronomer son, Manṣūr al-Kāšī.[100]

We have a letter from Jamšīd to his father, who remained in Kashan, showing that the latter also understood mathematics and astronomy. Applying the characteristic mix of flattery and self-promotion, in correct manner for the times, Jamšīd writes:

> Thanks to divine guidance, to the solicitude of His Majesty who uttered the words 'You can solve it', and by the blessing of the good wishes of the master [my father], which are always with me, the whole content of that book including [that] difficult problem ... was solved.[101]

We know that Jamšīd travelled back and forth between Kashan and Samarqand and that he completed a book in Kashan on 10 Ḏu'l-Ḥijja 818/10 February 1416, under the patronage of Sultan Iskandar, a different Timurid ruler.[102]

12 Turcoman Period

After Timurid rule, Kashan entered under control of the Turcoman dynasties, first the Qara-Qoyunlu and subsequently the Aq-Qoyunlu. The Qara-Qoyunlu ruled Isfahan on and off until their defeat in 874/1469. The wife of Jahān-Šāh Qara-Qoyunlu seems to have been particularly generous to Kashan, both in constructing buildings and endowing charities.[103] Documents and inscriptions from the ninth/fifteenth century provide excellent insight into the city in this period. The foundation deed (*vaqf-nāma*) of the Mīr ʿImād Mosque (or Masjid-i Maydān) is the longest of these documents. The surviving copy is written on 23 Rajab 877/January 1472, following the death of the mosque founder. ʿImād

---

99 Kāšānī, Jamšīd, *Miftāḥ al-ḥisāb*, ed. Nādir al-Nābulsī (Damascus: Wizārat al-taʿlīm al-ʿālī, 1977): pp. 353-91.

100 Barthold, V.V., *Four Studies in the History of Central Asia* (Leiden: Brill, 1956); Rajabī, *Kāšān*: p. 193.

101 Bagheri, Mohammad, "A Newly Found Letter of al-Kashi on Scientific Life in Samarqand", *Historia Mathematica*, XXIV/3 (1997): pp. 241-56, here p. 250.

102 Kennedy, Edward S., "A Letter of Jamshid al-Kāshi to His Father: Scientific Research and Personalities at a Fifteenth Century Court", *Orientalia*, XXIX/2 (1960): pp. 191-213, here p. 206. The same author also completed the further work *Sullam al-samā'* in Kashan, in March of 1407, before travelling to Samarqand; see Schmidl, Petra G., s.v. "Kāshī" in Thomas Hockey et al., (eds.), *The Biographical Encyclopedia of Astronomers* (New York: Springer, 2007): I, pp. 613-5.

103 Narāqī, *Tārīḫ-i ijtimāʿī*: p. 74.

al-Dīn Šīrvānī (d. 872/1467), a minister of Jahān-Šāh, endows the mosque, a Sufi convent (*ḫāniqāh*) and a bath house situated next to a school (Madrasa-i Ḫān), with two of the eighty-eight portions of the *qanāt* named Isfīdābād, of the village of Fīn; the endowment includes further properties, among which are an oil-press, blacksmith shop and bakery. This document details a fair amount of agricultural property (one hundred and eleven *jarībs*), including farms (sing. *mazraʿa*) with accompanying water rights in Muhaḏḏabābād and Naṣrābād.[104] The salaries of the ulema and staff are specified, down to the sweeper, as well as the stipends of the teacher and students in the school to be held at the mosque. The students are to be examined every six months to see if they are worthy of their stipends.[105]

Another deed dated 877/1473 establishes an endowment for a *ḫāniqāh* situated opposite and to the right side of the mosque, facing "a building for time and hour", which would appear to a type of clock or astronomical tower. The new building is to be a residence for the ulema, "the poor" [Sufis], the pious, and "the pure". The endower awards both the *ḫāniqāh* and the clock tower with both water rights and properties "in the neighborhood of Kashan", and some on the road to Bīdgul. There is mention of a hospital (*dār al-šifāʾ*) and salary and equipment for the "astronomer", Faḫr al-Dīn ʿĀlī, who worked in the tower.[106] Many positions, including those for upkeep of the tower, are to be continued by inheritance.

By the time Giosafat Barbaro, Venetian ambassador to Uzun Ḥasan (reg.871-883/1466-1478), visited Kashan, the major exports were in textiles. This would remain true until the early twentieth century. Barbaro wrote:

> We found a well inhabited city called Cassan, where for the most part they make silks and cotton in so great quantity that he who would spend ten thousand ducats in a day may find enough of that merchandise to spend it on.[107]

Other important textile exports included rugs and felt.

---

104 A "small" *jarīb* was sixty cubits by fifty cubits. The Iranian *jarīb* varied locally from 400 to 1450 square meters. See Bosworth, C.E., s.v., "Misāḥa (measurement of plane surfaces), 1. In the central Islamic lands", *EI* 2, VII (1993): pp. 137-8.

105 Ḍarrābī, *Tārīḫ-i Kāšān*: pp. 507-21; Šahšahānī, Ḥusayn, "Ḫulāṣa-ī az vaqfnāma-yi masjid-i Mīr ʿImād", *Farhang-i Īrānzamīn*, V (1336š.): pp. 23-50.

106 Šahšahānī, "Ḫulāṣa": pp. 39-50. Ḍarrābī, *Tārīḫ-i Kāšān*: pp. 523-9. Fifteenth century Italian clocks are common in North Africa; it is unclear what kind of clock is meant here.

107 Barbaro, Giosafat, *Travels to Tana and Persia by Josafa (Giosafat) Barbaro and Ambrogio Contarini*, trans. William Thomas and S. A. Roy (London: Franklin, 1873 [reprint 1963]): p. 72 (adapted to contemporary English).

13 Mint City

Kashan's nearly continuous status as a mint for silver dirhams and gold dinars is testimony to the robustness of its economy throughout the medieval era. The written sources record large issues of gold coins in the fifth/eleventh century under the Buyids.[108] Shortly after that came an issue of gold coins under the Great Saljuq, Alp Arslān, in 457/1064-5.[109] There are also Kashan coins from the times of the Ilkhanids, Jalayirids, Muzaffarids, Timurids, the Qara-Qoyunlu and Aq-Qoyunlu.[110] Coins of the Ilkhanid Abū Saʿīd (reg. 716-36/1316-35), survive in abundance,[111] among them a double dirham of 726. The Abū Saʿīd coins could all be from a single trove, perhaps discovered in the nineteenth century. Nevertheless, the variety of dates testify to a city that thrived under several of the later Ilkhanids. The presence of a mint continued through the Safavid period.

14 Tax Income

The sources offer scarce but valuable information on the tax income from Kashan. However, it is difficult to know the true buying power of the incomes reported, the fineness of the coinage metals, or the underlying principles of tax collection.[112] The *History of Qum* (378/988-89) reports two different surveys, describing the estimated taxes for Kashan as 158,192 dirhams or 181,092 dirhams.[113] Ibn Ḫurradāḏbih gives the combined average annual tax revenues from Qum and Kashan as three million dirhams (compared to the four and

108 Treadwell, Luke, *Buyid Coinage: a die corpus* (Oxford: Ashmolean Museum, 2001), does not list any Buyid coins from Kashan/Qashan. However, Ibn al-Aṯīr, *al-Kāmil*: VIII, pp. 153 and 254, under the years 416 and 428 (1025-6 and 1036-7), mentions Qāšānī dinars, both in his accounts of a wheat price of 200 dinars (in Baghdad, considered extremely high) and a (royal) dowry of 50,000 dinars.

109 Lowick, N.M., "Seljuk Coins", *The Numismatic Chronicle*, 7th series, X (1970): pp. 241-51.

110 Zambaur, Eduard von, *Die Münzprägungen des Islams*: I (Wiesbaden: Steiner, 1968). This very dated but valuable book lists Kashan coins for the dynasties mentioned.

111 See Kolbas, Judith, *The Mongols in Iran* (London: Routledge, 2006): p. 276, note 47. Many of the Abū Saʿīd coins were struck in 729 AH, which suggests that they came from a hoard. See Zambaur, *Münzprägungen*: I, p. 202. There are also numerous Muzaffarid coins from Kashan. On a hoard of Kashan coins minted by Abū Isḥāq, see Rajabī, *Kāšān*: p. 130, note 3.

112 Ibn Ḥawqal, *al-Masālik wa ʾl-mamālik*, ed. M.J. de Goeje (Leiden: Brill, 1872): p. 259, who flourished 331-78/943-88, says that Kashan is a city that "brings in high tax income".

113 Qumī, *Tārīḫ-i Qum*: p. 349. The repetition of ninety-two in both estimates is suspicious.

a half million dirhams from Azerbaijan).[114] Al-Muqaddasī (who is somewhat careless with figures) gives the land tax of Kashan as one million dirhams.[115] In 740/1346 the learned polymath Ḥamd-Allāh Mustawfī Qazvīnī says that the revenue of Kashan and its district was 117,000 dinars (compared to 275,000 dinars for Tabriz), which at the rate of ten dirhams to a dinar would make 1,170,000 dirhams: an income considerably greater than that indicated in the citations from the fourth/tenth century *History of Qum*.[116]

15 Conclusion

Kashan emerged as a small city in part because of the shift in wealth brought about by the Arab conquests. Arabs acquired capital during these events and created new cities or enlarged existing cities, thereby expanding urban life in early Islamic Iran. The long period of stability between the Arab conquest in the seventh century and the invasion by the Ghuzz Turks in the late Saljuq period allowed cities to flourish. This was the case for Kashan, which in some senses was a new city. When there was strong central rule such as under the Ilkhanids, Kashan and other cities with primary resources and artisanal skills could participate advantageously in intercity trade.

Historians of Iran usually consider mountainous territories as areas of refuge. Although not in a mountainous area, Kashan served as a refuge by virtue of its small size and distance from major cities. Military occupations of smaller cities would have been hard to sustain, due to limitations of both food and plunder. It was thanks to its relative remoteness that Kashan was able to have Shiʿi *madrasas* and judges. The Shiʿi families who largely controlled Kashan benefitted the resident population by promoting extensive public works, such as the construction of city walls and maintenance of irrigation systems. Kashan sometimes fell under the sway of the rulers of Rayy, but more continuously under those of Isfahan. The Shiʿi elites of Kashan strove for good relations with the Sunni elites who dominated in these capital cities.

Kashan's comparative prosperity depended on its *qanāt*-based agriculture, a location on important routes, the continual nurturing of artisanal traditions,

114 Ibn Ḫurradāḏbih, *Masālik*: p. 244.

115 Muqaddasī, *Aḥsan al-taqāsīm*: p. 400.

116 Qazvīnī, Ḥamd-Allāh Mustawfī, *The Geographical Part of the Nuzhat al-Qulūb*, trans. Guy LeStrange (London: Luzac/ Leiden: Brill, 1919): p. 72. The non-Islamic *tamġa* tax imposed on commerce in the Ilkhanid period remained in place until the early Safavid period and may be one reason that later tax income is higher. We are assuming that *dinar* here means a gold coin and *dirham* means a silver coin.

and an emphasis on education. It was fortunate to have leaders, and sometimes rulers, who invested in trusts for the support of institutions such as schools and charities. Education enabled Kashanis to acquire patronage as bureaucrats and astronomers. These individuals then often patronized their home city. Non-educational charitable institutions sustained the social fabric of a society with highly disparate levels of income. The continuity of elites, reinforced in Kashan by the presence of *sayyid*s, fostered stability (and surely stifled some talents) across turbulent centuries of changes of regime. Kashan was considerably smaller than Isfahan and somewhat smaller than its Shiʿite neighbor, Qum. However, diverse sources indicate that Kashan maintained a relatively robust civic and regional life from the fourth/tenth century through the ninth/fifteenth century.

Bibliographical References

Afšār, Īraj, "Vaqf-nāma-yi sih dīh dar Kāšān", *Farhang-i Īrānzamīn*, IV (1335š.): pp. 122-38.

Allan, J. W., "ʿAbū'l-Qāsim's Treatise on Ceramics", *Iran*, XI (1973): pp. 111-20.

Amanat, Mehrdad, "Kashan iii. History to the Pahlavi Period", *EIr* online (http://www.iranicaonline.org/articles/kashan-iii-history).

Bagheri, Mohammad, "A Newly Found Letter of al-Kashi on Scientific Life in Samarqand", *Historia Mathematica*, XXIV/3 (1997): pp. 241-56.

Bahrami, M., *Gurgan Faiences* (Cairo: Le Scribe Egyptien, 1949).

Balāḏurī, *Futūḥ al-buldān*, ed. M.J. de Goeje (Leiden: Brill, 1866).

Barbaro, Giosafat, *Travels to Tana and Persia by Josafa (Giosafat) Barbaro and Ambrogio Contarini*, trans. William Thomas and S. A. Roy (London: Franklin, 1873 [reprint 1963]).

Barthold, V.V., *Four Studies in the History of Central Asia*, trans. Svat Soucek (Leiden: Brill, 1956).

Bīdgulī, Maḥmūd Sādāt and Sārūḫānī, Zahrā, "Bar-rasī-yi ʿilal-i tārīḫ-iqlīmī-yi ījād-i ū'ī-hā dar manṭaqa-yi Kāšān", *Ganjīna-yi asnād*, LXXI (Fall 1387š./1967): pp. 25-36.

Blair, Sheila, "A Brief Biography of Abū Zayd", *Muqarnas*, XXV (2008): pp. 155-76.

Blair, Sheila, *Islamic Inscriptions* (Edinburgh: Edinburgh University Press, 1998).

Blair, Sheila and Bloom, Jonathan, "Signatures on Works of Islamic Art and Architecture", *Damaszener Mitteilungen*, XI (1999): pp. 49-66.

Borjian, Habib, s.v. "Mašhad-e Ardahāl", *EIr* online: www.iranicaonline.org/articles/mashad-e-ardahal, viewed 12/11/2016.

Bosworth, C.E., s.v. "Anūšervān Kāšānī", in *EIr*, II/2 (1987): p. 139.

Bosworth, C.E., s.v. "Misāḥa (measurement of plane surfaces), 1. In the central Islamic lands", in *EI* ² VII (1993): pp. 137-8.

Boyle, John A., *The History of the World Conqueror*, 2 vols. (Manchester: Manchester University Press, 1958).

Caiger-Smith, Alan, *Lustre Pottery* (London: Faber & Faber, 1985).

Ḍarrābī, ʿAbd al-Raḥīm Kalāntar, *Tārīḫ-i Kāšān*, ed. Īraj Afšār (Tehran: Ibn Sīnā, 1962).

Dozy, R., *Supplément aux dictionnaires arabes*, 2 vols. (Paris: Maisonneuve, 1927).

Durand-Guédy, David, *Iranian Elites and Turkish Rulers. A History of Iṣfahān in the Saljūq Period*, (London: Routledge, 2010).

Ettinghausen, Richard, "Evidence for the Identification of Kashan Pottery", *Ars Islamica*, III/1 (1936): pp. 44-75.

Graves, Margaret S., s.v. "Kashan vii, Kashanware", in *EIr* (2014): www.iranicaonline.org/articles/kashan-vii-kashan-ware, viewed 4/27/17.

Gurgānī, Murtaḍā Qarāʾī, s.v. "Bābā Afżal Kāšānī", in *Dāyirat al-Maʿārif-i Buzurg-i Islāmī* (*The Great Islamic Encyclopaedia*): X (2001): pp. 735-9.

Haeri, Mohammad-Reza, s.v. "Kashan v. Architecture (2) Historical Monument", in *EIr*, XVI/1 (2012): pp. 12-21.

Harrison, J.H., "Minerals", in Fisher, W.B. (ed.), *Cambridge History of Iran* (Cambridge: Cambridge University Press, 1968): I, pp. 487-516.

Hillenbrand, Robert H., "The Relationship between Book Painting and Luxury Ceramics of 13th Century Iran", in Hillenbrand, Robert H. (ed.), *The Art of the Saljuqs in Iran and Anatolia* (Costa Mesa, CA: Mazda, 1994): pp. 134-41.

Ḥudūd al-ʿālam min al-mašriq ilā al-maġrib, ed. Manūčihr Sutūda (Tehran: Dānišgāh-i Tihrān, 1962).

Ibn al-Aṯīr, *al-Kāmil fiʾl-taʾrīḫ*, ed. Muḥammad Yūsuf al-Daqqāq, 13 vols. (Beirut: Dār Ṣādir, 2003).

Ibn Baṭṭūṭa, *The Travels of Ibn Battuta, 1325-1354*, trans. H.A.R. Gibb, 5 vols. (reprint, New Delhi: Munshiram Manoharlal, 1993).

Ibn Ḥawqal, *al-Masālik waʾl-mamālik*, ed. M.J. de Goeje (Leiden: Brill, 1872).

Ibn Ḫurradāḏbih, *Kitāb al-masālik waʾl-mamālik*, ed. M.J. de Goeje (Leiden: Brill, 1889).

Ibn Manẓūr, *Lisān al-ʿarab* (Beirut: Dār Ṣādir, 1990).

Ibn al-Nadīm, *The Fihrist of al-Nadim*, trans. Bayard Dodge (New York: Columbia University Press, 1970).

Ibn Rusta, *Kitāb al-aʿlāq al-nafīsa*, ed. M.J de Goeje (Leiden: Brill, 1892).

Iqbāl, ʿAbbās, *Vizārat dar ʿahd-i salāṭīn-i buzurg-i Saljūqī* (Tehran: Čāpḫāna-yi dānišgāh, 1959).

Iqbāl, ʿAbbās, *Tāriḫ-i Muġūl* (Tehran: Našr-i Nāmik, 1986).

Iṣṭaḫrī, *Kitāb al-masālik waʾl-mamālik*, ed. M. J. de Goeje (Leiden: Brill, 1870).

Iṣṭaḫrī, *al-Masālik waʾl-mamālik*, ed. Īraj Afšār (Tehran: Bungāh-i tarjama va-našr-i kitāb, 1340š/1961).

Jackson, A.V. Williams, "The Magi in Marco Polo", *JAOS*, XXVI (1905): pp. 79-83.

Jawharī Nīšāpūrī, Muḥammad, *Javāhir-nāma-yi niẓāmī*, ed. Īraj Afšār (Tehran: Mīrāṣ-i Maktūb, 2004).

Jomehpour, Mahmoud, "Qanat Irrigation Systems as Important and Ingenious Agricultural Heritage: Case Study of the *qanats* of Kashan, Iran", *International Journal of Environmental Studies*, LXVI/3 (June 2009): pp. 297-315.

Juwaynī, ʿAṭā Malik, *Tārīḫ-i Jahān-Gušā*, ed. M.M. Qazwīnī, 3 vols. (Leiden: Brill, 1912-37).

Karajī, Muḥammad, *Kitāb inbāṭ al-miyāh al-ḫafīya*, original text and trans. Aly Mazaheri, *La civilisation des eaux cachées* (Nice: Institut d'Études et de Recherches interethniques et interculturelles, 1973).

Karīmī, Fāṭima, "Kāshān", in *Shahr-hā-yi Irān*, III, ed. Muḥammad Yusūf Kīyānī (Tehran: Vizārat-i Farhang, 1989).

Karimi, Fatima, "Abū Ṭāhir (Family)", trans. Rahim Gholami and Farzin Negahban, in Madelung, W. and Daftary F. (eds.), *Encyclopaedia Islamica* (2008): www .referenceworks.brillonline.com/entries/encyclopedia-islamica/abu-tahir-family, viewed 4/24/2017.

Karimi, Fatima, "Abū Zayd Kāshānī", trans. Farzin Negahban, in Madelung W. and Daftary F. (eds.), *Encyclopaedia Islamica* (2008): www.referenceworks.brillonline. com/entries/encyclopedia-islamica/abu-zayd-kashani, viewed 2/21/2017.

Kāšānī, Abu'l-Qāsim ʿAbd-Allāh, *ʿArāyis al-javāhir wa-nafāyis al-aṭāyib*, ed. Īraj Afšār (Tehran: Almāʾī, 1386š./2007).

Kāšānī, Jamšīd, *Miftāḥ al-ḥisāb*, ed. Nādir al-Nābulsī (Damascus: Wizārat al-taʿlīm al-ʿālī, 1977).

Kāšānī, Kamāl al-Dīn ʿAbd al-Razzāq b. Jalāl al-Dīn, *Tuḥfat al-iḫwān fī ḫaṣāʾiṣ al-fityān*, ed. Muḥammad Dāmādī (Tehran: Širkat-i intišārāt-i ʿilmī va-farhangī, 1991).

Kennedy, Edward S., "A Letter of Jamshid al-Kāshi to His Father: Scientific Research and Personalities at a Fifteenth Century Court", *Orientalia*, XXIX/2 (1960): pp. 191-213.

Kerr, Rose and Wood, Nigel, *Ceramic Technology. Science and Civilisation in China*, V/12 (Cambridge: Cambridge University Press, 2004).

Kirmānī, Nāṣir al-Dīn Kirmānī, *Nasāʾim al-asḥār min laṭāʾim al-aḫbār dar tārīḫ al-vuzarāʾ*, ed. Mīr Jalāl ad-Dīn Urmawī (Tehran: Čāpḫāna-yi Dānišgāh, 1364š/1985).

Klausner, Carla L., *The Seljuk Vezirate: A Study of Civil Administration, 1055-1194* (Cambridge, MA: Harvard University Press, 1973).

Kolbas, Judith, *The Mongols in Iran* (London: Routledge, 2006).

Lane, Arthur, *Early Islamic Pottery. Mesopotamia, Egypt and Persia* (London: Faber & Faber, 1947).

Le Strange, Guy, *Lands of the Eastern Caliphate*, (London: Cass, 1966).

Lowick, N.M., "Seljuk Coins", *The Numismatic Chronicle*, 7th series, X (1970): pp. 241-51.

Mahdavī, M. and Anderson, E. W., "The Water-Supply System in the Margin of Dasht-e-Kavir", *Bulletin: British Society for Middle Eastern Studies*, X/2 (1983): pp. 131-47.

Marāġī, Maḥmūd Ṭayyār, "Madāris-i ʿilmīya-yi Kāšān", *Mīrāṯ-i jāvīdān*, VI/2 (1377š./1999): pp. 68-79.

Marco Polo, *The Travels of Marco Polo*, trans. Ronald Latham (Harmondsworth: Penguin, 1958).

Meisami, Julie Scott, "Rāwandī's Rāḥat aṣ-Ṣudūr", *Edebiyat*, V (1994): pp. 183-215.

Mīr-Šāfiʿī, Sayyid Muḥammad and Bāġir-zāda Kaṯīrī, Masʿūd, "Muṭālaʿa-yi sāḫt-i luʿāb-i zarrīn-fām", *Pažūhiš-hā-yi ʿulūm va-fannāvarī rang*, (2016): pp. 71-8.

Miškātī, Nuṣrat Allāh, "Nāẓarī bi-tārīḫ-i bāstān-šināsī-yi Kāšān", *Hunar va-mardum*, series 5, LIII-LIV (1345š/1966): pp. 7-14 and V/55 (1346š./1967): pp. 8-13.

Mudarrisī Ṭabāṭabāʾī, Ḥusayn, *Turbat-i pākān: āṯār wa banāhā-yi qadīmī*, 2 vols. (Qom: Čāpḫāna-yi Mihr, 1976).

Muqaddasī, *Aḥsan al-taqāsīm fī maʿrifat al-aqālīm*, ed. M.J. de Goeje (Leiden: Brill, 1906).

Mustawfī, see Qazvīnī.

Nadji-Esfahani, Mehdi, *Geologie und Hydrogeologie des Gebietes von Kashan* (Aachen: Technische Hochschule, Fakultät für Bergbau und Hüttenwesen, 1971).

Nadji, Mehdi, "Luftbild: Kanate in der Ebene von Kashan, Iran", *Die Erde*, CIII (1972): pp. 209-15.

Nadji, Mehdi and Voigt, Rudolf, "'Exploration for Hidden Water', by Mohammad Karaji – the Oldest Textbook on Hydrology", *Ground Water*, X/5 (1972): pp. 43-6.

Narāqī, Ḥasan, *Tārīḫ-i ijtimāʿī-yi Kāšān*, (Tehran: Institut d'Études et de Recherches sociales, 1345š/1967).

Narāqī, Ḥasan, *Āṯār-i tārīḫī-yi šahristānhā-yi Kāšān va-Naṭanz* (Tehran: Anjuman-i āṯār-i millī, 1348š./1969).

Niẓām al-Mulk, *Siyar al-mulūk*, ed. Hubert Darke (Tehran: Bungāh-i tarjamah va-našr-i kitāb, 1962), trans. Hubert Darke, *Siyāsat-nāma* (*The Book of Government or Rules for Kings*) (London: Routledge & Kegan Paul, 1960).

Odoric of Pordenone in *Cathay and the Way Thither*, trans. Henry Yule, 2 vols. (London: Hakluyt Society, 1866): I, pp. 43-162.

Pancaroğlu, Oya, *Perpetual Glory: Medieval Islamic Ceramics from the Harvey B. Plotnick Collection*, with transcriptions and translations of inscriptions by Manijeh Bayani (New Haven: Yale University Press, 2007).

Peacock, Andrew C.S., *The Great Seljuk Empire* (Edinburgh: Edinburgh University Press, 2015).

Planhol, Xavier de, s.v. "Kashan ii. Historical Geography", in *EIr*, XVI/1 (2012): pp. 2-5.

Planhol, Xavier de, s.v. "Kārīz i. Terminology", in *EIr*, XV (2012): pp. 564-5.

Porter, Yves, "Lustres fatimides et potiers de Kâshân: Historiographie et obscurs transferts", *Keramik-Freunde der Schweiz*, CXXIX (2015): pp. 12-4.

Pradell, T. et al., "The Invention of Lustre: Iraq 9th and 10th Centuries AD", *Journal of Archaeological Science*, XXXV/5 (2008): pp. 1201-15.

Qazwīnī, Ḥamd-Allāh Mustawfī, *The Geographical Part of the Nuzhat al-Qulūb*, trans. Guy Le Strange (London: Luzac/Leiden: Brill, 1919).

Qazwīnī, Zakarīyā', *Āṯār al-bilād wa-aḫbār al-ʿibād* (Beirut: Dār Ṣādir, 1988).

Qazwīnī Rāzī, *Kitāb-i naqż*, ed. Jalāl al-Dīn Muḥaddis (Tehran: Anjuman-i Āṯār-e Millī, 1358š/1980).

Qūčānī, ʿAbd-Allāh, "Sufālgarān-i Kāšān va-šiʿr-i fārsī", *Našr-i Dāniš*, LXXXIV (1373š.): pp. 31-40.

Qudāma b. Jaʿfar, *al-Ḫarāj wa-ṣināʿat al-kitāba*, ed. Muḥammad Ḥusayn Zabīdī (Baghdad: Dār al-Rašīd li l-našr, 1981).

Qumī, Ḥasan ibn Muḥammad, *Tārīḫ-i Qum*, ed. Muḥammad Riḍā Anṣārī Qumī (Qum: Markaz-i Qumšināsī, 2006).

Rajabi, Parwīz, *Kāšān: Nigīn-i anguštarī-yi tārīḫ-i Īrān* (Tehran: Nashr-i Pižvāk-i Kayrān, 2011).

Rāwandī, Muḥammad b. ʿAlī, *Rāḥat al-Ṣudūr*, ed. Muḥammad Iqbāl (Leiden: Brill, 1921).

Rāwandī, Ḍiyā' al-Dīn Abū Riḍā Faḍl Allāh, *Dīwān*, ed. Jalāl al-Dīn Muḥaddiṯ Urmawī (Tehran: Maṭbaʿat al-Majlis, 1334š./1955).

Ṣādiqī, Muṣṭafā, "Dīdgāh-i tārīḫī-yi Rāwandīyān", *Nāma-yi tārīḫ-pažūhišān*, X (1386š./2008): pp. 115-38.

Šahšahānī, Ḥusayn, "Ḫulāṣa-ī az vaqfnāma-yi masjid-i Mīr ʿImād", *Farhang-i Īrānzamīn*, V (1336š.): pp. 23-50.

Sajjādī, Aḥmad, "Qālī Šūyān", *Dāyirat al-maʿārif-i tašayyuʿ*, XII (Tehran: Ḥikmat, 1386š./2007): pp. 531-33.

Saliba, George, "Book review: Artisans and Mathematicians in Islamic Architecture by Gulru Necipoğlu", *JAOS*, CXIX (1999): pp. 637-45.

Šahrestān-i Ērānshahr or *A Catalogue of the Provincial Capitals of Eranshahr*, ed. and trans. J.A. Markwart (Rome: Pontificio Istituto Biblico, 1931).

Schmidl, Petra G., s.v. "Kāshī", in Hockey, Thomas et al., (eds.), *The Biographical Encyclopedia of Astronomers* (New York: Springer, 2007): I, pp. 613-5.

Schwarz, Paul, *Iran im Mittelalter nach den arabischen Geographen* (reprint, Hildesheim: Olms, 1969).

Soucek, Priscilla P., s.v. "Abu 'l-Qāsem ʿAbdallāh Kāšānī", *EIr*, I/4 (2011): pp. 362-3.

Soustiel, Jean, *La céramique islamique* (Fribourg, Suisse: Office du livre, 1985).

Ṭabarī, *Ta'rīḫ al-umam wa'l-mulūk*, ed. M.J. de Goeje, 15 vols. (Leiden: Brill, 1879ff).

Taheri, Jafar, "Mathematical Knowledge of Architecture in the Works of Kâshânî", *Nexus Network Journal*, XI (2009): pp. 77-88.

Treadwell, Luke, *Buyid Coinage: a die corpus* (Oxford: Ashmolean Museum, 2001).

Virani, Shafique, *The Ismailis in the Middle Ages* (Oxford: Oxford University Press, 2007).

Wāṯiqī, Ḥusayn, "Dar-barā-yi tārīḫ-i Kāšān", *Mīrāṯ-i Šahāb*, LXXXIV-LXXXV (1395š.): pp. 147-51.

Watson, Oliver, "The Masjid-i ʿAli, Quhrūd: An Architectural and Epigraphic Survey", *Iran*, XIII (1975): pp. 59-74.
Watson, Oliver, *Persian Lusterware* (London: Faber & Faber, 1985).
Watson, Oliver, *Ceramics from Islamic Lands* (London: Thames & Hudson, 2004).
Watson, Oliver, s.v. "Abū Ṭāher", in *EIr*, I/4 (2011): pp. 385-7.
Yaʿqūbī, *Kitāb al-buldān*, ed. M.J. de Goeje (Leiden: Brill, 1892).
Yāqūt, *Muʿjam al-buldān*, ed. F. Wüstenfeld, 6 vols. (Leipzig: Brockhaus, 1869).
Zambaur, Eduard von, *Die Münzprägungen des Islams*, I (Wiesbaden: Steiner, 1968).
Zanjani, Habibollah, s.v. "Kashan i, Geography", in *EIr*, XVI/1 (2012): pp. 1-2.

Biographical Note

Roy Parviz Mottahedeh is the Gurney Research Professor of History at Harvard University. He has written extensively on the history of the Middle East in the 10th and 11th centuries CE. His publications include *Loyalty and Leadership in an Early Islamic Society* (1980), *The Mantle of the Prophet* (1985) and *Lessons in Islamic Jurisprudence* (2003). He has written numerous articles on the social, intellectual and political history of the Middle East from the 7th century to the present.

Mehrdad Amanat is a Research Fellow at University of California, Santa Barbara. He is the author of *Jewish Identities in Iran, Resistance and Conversion to Islam and the Baha'i Faith* (London I.B.Tauris 2011) and is a regular contributor to *Encyclopedia Iranica.*

The History of Iranian Cities through Their Books: What Ms. Köprülü 01589 Tells Us about 8th/14th Century Shiraz

Sarah Bowen Savant
Aga Khan University
sarah.savant@aku.edu

Majid Montazer Mahdi
University of Exeter
mm755@exeter.ac.uk

Abstract

What can the history of books tell us about Iranian cities and their histories? This article introduces the manuscript of a multi-text compilation (*majmūʿa*) for the purpose of illustrating its potential usefulness as a source for studying the social and cultural history of Shiraz in the turbulent period that followed the collapse of Mongol rule in the area. We specifically seek to show that Köprülü 01589, now housed in Istanbul, helps us to see how books were produced and consumed, and provides insight into the operations of a busy workshop for copying texts. Despite the rarity and historical significance of several of the pieces that it contains, the availability of images of the manuscript for some time in Istanbul and Iran, and attention to it in catalogues, it has not received scholarly attention as a whole.[1] Although this article is only a preliminary study of a single manuscript, we believe it is important for the current volume in showing what manuscripts can reveal of the social world that produced them, the networks of people and ideas that animated city life, and the cultural resources of specific times

1 The manuscript's full reference is Türkiye Yazma Eserler Kurumu Başkanlığı Süleymaniye Yazma Eser Kütüphanesi, Köprülü 01589. The manuscript is catalogued: see Dānišpažūh, Muḥammad Taqī, *Fihrist-i Mīkrufīlm-hā-yi Dānišgāh-i Tihrān* (Tehran: Intišārāt-i Dānišgāh-i Tihrān, 1969): I, pp. 479-87 and also Şeşen, Ramazan; Akpınar, Cemil and İzgi, Cevat, *Catalogue of Manuscripts in the Köprülü Library* (Istanbul: Research Centre for Islamic History, Art and Culture, 1986): II, pp. 254-77. According to Īraj Afšār, former director of Kitābḫāna-yi Markazī Dānišgāh-i Tihrān, Mujtabā Mīnuvī brought a microfilm of the entire Köprülü 01589 to Iran: Afšār, Īraj, "Muqaddama (Introduction)", in *Zangī-Nāma* (Tehran: Intišārat Tūs, 1993): p. VI.

 | DOI:10.1163/9789004434332_016

and places. Furthermore, our approach to Köprülü 01589 can be expanded and applied to other manuscripts originating in Shiraz and other cities.

Keywords

Shiraz – codicology – Köprülü 01589 – *Jāmāsb-nāma* – Ibn Sina

This research was supported by a grant from the Aga Khan University Research Council. We wish to thank Fatemeh Mir Razirenani and Antoine Borrut for their assistance in improving the current publication of our work.

1 Introduction

A few years ago I (Sarah Bowen Savant) was sifting through the digitised catalogue of the Süleymaniye Library in Istanbul, reading entries under the

FIGURE 1 Köprülü 01589, folios 115v and 116r. The interior frame holds a version of a text called the *Jāmāsb-nāma*, which features horoscopes. Here, there are four. In the margin are two texts attributed to Ibn Sīnā, the first entitled *Abṭāl aḥkām al-nujūm* and the second *Risāla fī ḥadd al-jism*, with the latter title illuminated.

category of "İran edebiyatı", when I came across a curious text within a sizable multi-text compilation. The text, *Jāmāsb-nāma*, was written in Persian, and consisted of entries for personages of Iranian and Islamic history (among them, Jamšīd, Noah, Muḥammad, etc.), sequentially arranged with accompanying horoscopes. It seemed to be a late survival of a mostly lost genre treating astrology and history. (I now believe such works continued, though they are scarcely noted in the secondary literature.)[2] I wanted to understand the

2 The bibliography of *Jāmāsp-nāma* (also known as *Jāmāsp-nāmaġ*, or *Jāmāspi*) is complex, but the title refers to at least two sets of texts. There is a Zoroastrian astrological text in Middle Persian, also surviving in part in Pahlavi; a complete version survives in Pāzand. The Zoroastrian text continued in circulation among Parsis. For more details on the Zoroastrian text, see: Agostini, Domenico, *Ayādgār ī Jāmāspīg: Un texte eschatologique zoroastrien* (Rome: Gregorian & Biblical Press, 2013); Agostini, "Rediscovering the Jāmāspi: A Walk in Four Steps", *IrSt*, XLV/2 (2012): pp. 169-80; Sheffield, Daniel, review of Agostini, *Ayādgār ī Jāmāspīg*, *Bulletin of the Asia Institute*, 24 (2014): pp. 207-10 and Modi, J.J. (ed. and trans.), *Jāmāspi, Pahlavi, Pāzend, and Persian Texts* (Bombay: Bombay Education Society Press, 1903). More relevant to the current analysis, the title also refers to a set of Islamic texts copied by both Muslims and Zoroastrians, whose relation to the Middle Persian version is not straightforward. There are at least three different redactions of the text in Persian, and it seems that the piece in Köprülü 01589 might represent one of the earliest prose versions. A Persian manuscript in Bibliothèque nationale de France (BnF), Supplément persan 380, shows very substantial similarities, and has a colophon with date of Šawwāl of 741; see: Blochet, E., *Catalogue des manuscrits persans de la Bibliothèque nationale* (Paris: E. Leroux, 1905-1934): II, pp. 153-4. Also, Richard, Francis, *Catalogue des manuscrits persans. Supplément persan* (Rome: Bibliothèque Nationale de France): I, pp. 517-8. There is, moreover, a rare verse version in Persian, attributed to Naṣīr al-Dīn Ṭūṣī, extant in Aya Sofya 04795. Blochet also provides a description of an Arabic version of the text preserved in the BnF, ms. Arabe 2487 (bearing a colophon with date Ḏu'l-Qa'da 679/ February-March 1281; unlike the Köprülü 01589 text, it lacks *Zayča* (horoscopes). See Blochet, "Études sue le gnosticisme musulman", *Rivista degli studi orientali*, IV/2 (1911): pp. 267-300, and Sezgin, Fuat, *Geschichte des arabischen Schrifttums*, VII: *Astrologie, Meteorologie und Verwandtes bis ca. 430 H* (Leiden: E.J. Brill, 1979): pp. 86-8. Mirzā 'Abd Allāh al-Afandī translated an Arabic version of the text into Persian in the Safavid Period. A Zoroastrian recension of the text was incorporated into the classified *Rivāyat* of Dārāb Hurmazdyār, compiled in the 1670s. This version was lithographed in Unvala Rustamji, Manockji, ed., *Dârâb Hormazyâr's Rivâyat* (Bombay: British India Press, 1922): II, pp. 111-30. The relations between the BnF Arabic version, the Persian text of Köprülü 01589, the Persian translation by al-Afandī, and the Zoroastrian recensions should be investigated thoroughly. It seems that one of the Persian texts, probably al-Afandī's Persian translation, became very popular during the Qajar period. For the manuscripts of the Persian texts extant in Iranian libraries, see: Dirāyatī, Muṣṭafā (ed.), *Fihristigān: nusḫa-hā-yi ḫaṭṭī-yi Īrān (FANḪĀ)* (Tehran: Sāzmān-i Asnād va Kitābḫāna-yi Millī Jumhūrī-yi Islāmī Īrān, 2012): IX, pp. 757-61. A small, dispersed early fourteenth-century manuscript has been studied extensively by art historians; Stefan Kamola has recently discovered its connection to the *Jāmāsp-nāma* and is working now on its text and its scientific, religious, and political contexts: this manuscript appears to also be related to the *Jāmāsp-nāma* text of Köprülü 01589. We thank Fatemeh Mir Razirenani, Daniel Sheffield, and Stefan Kamola for comments on this situation.

historical and literary context of the survival of this unusual text, so as to better understand the community that preserved it.[3]

This began my relationship with what I now understand to be an extremely complex and interesting codex, compiled in the politically chaotic but culturally creative period that followed the demise of the Ilkhanids in the region of Shiraz. To anticipate somewhat, the manuscript contains multiple dated colophons and codicological features suggesting that the bulk of it was written and assembled in Shiraz over a period of less than a year, more specifically between Ḏu'l-Qaʿda 753/December 1352 and Rajab 754/August 1353. Shiraz was at that time ruled by Abū Isḥāq Īnjū, who took undisputed control of Fars in 1342, following the death of the Ilkhanid ruler Abū Saʿīd in 736/1335, and the ensuing years of anarchy and the many-sided contest among regional powers. Abū Isḥāq's family name, Injuid, derived from the Mongolian 'īnjū', royal estates, as the founder of their line, Šaraf al-Dīn Maḥmūd (d. 736/1336), had been sent by the Ilkhanids to administer the state lands of Fars.[4] Abū Isḥāq was widely praised by contemporaries, including the historian Ibn Zarkūb Šīrāzī (d. 789/1387-8) and the poet Ḥāfiẓ (d. 792/1390 or 791/1389), for his intelligence, bravery, chivalry and generosity. Ibn Baṭṭūṭa visited Shiraz during his reign, and the author's flattering comments on the city under Abū Isḥāq would continue to be cited.[5] Shiraz attracted artists, including from Tabriz, which was itself an important cultural centre.[6] But by 1352-3, around the moment when

3 Work on astrological histories is currently being undertaken by Antoine Borrut; see Borrut, Antoine "Court Astrologers and Historical Writing in Early Abbasid Baghdad: An Appraisal", in J. Scheiner and D. Janos (eds), *The Place to Go: Contexts of Learning in Baghdad, 750-1000 CE* (Princeton: The Darwin Press, 2014): pp. 455-501. Among earlier treatments, see, for example, Pingree, David, "Historical Horoscopes", *JAOS*, LXXXII/4 (Oct.-Dec., 1962): pp. 487-502 and Kennedy, E.S. and Pingree, David, *The Astrological History of Māshā'Allāh* (Cambridge: Harvard University Press, 1971).

4 On this short-lived dynasty, see especially Aigle, Denise, *Le Fārs sous la domination mongole: politique et fiscalité* (Paris: Association pour l'avancement des Études Iraniennes, 2005): pp. 158-63; 185-200; Limbert, John, "Inju Dynasty (ca. 725-754/1325-53)", *EIr* (accessed online Feb. 6, 2018); Boyle, J.A., "Īndjū" in *Encyclopaedia of Islam*, 2nd edition, edited by P. Bearman, Th. Bianquis, C.E. Bosworth, E. van Donzel, W.P. Heinrichs (Leiden: E.J. Brill, 1986): III, p. 1208; Bosworth, Clifford Edmund, "141: The Inju'ids, c. 725-54/1325-53, Fars", in Bosworth, Clifford Edmund, *The New Islamic Dynasties: A Chronological and Genealogical Manual*, Rev. ed. (New York: Columbia University Press, 1998): p. 266; and Roemer, H.R., "The Jalayirids, Muzaffarids, and Sarbadārs" in Jackson, Peter and Lockhart, Laurence, *The Cambridge History of Iran* (Cambridge: Cambridge University Press, 1986): VI, pp. 1-41.

5 Gibb, H.A.R. (trans.), *Ibn Battúta: Travels in Asia and Africa, 1325-1354* (London: Routledge & Kegan Paul, 1929; reprint 1953): pp. 92-7; Arberry, Arthur J., *Shiraz: Persian City of Saints and Poets* (Norman: University of Oklahoma Press, 1960): pp. 52-3.

6 Wright, Elaine, "Patronage of the Arts of the Book under the Injuids of Shiraz", in Komaroff, Linda, *Beyond the Legacy of Genghis Khan* (Leiden/Boston: Brill, 2006): pp. 248-68. As Wright

our manuscript was prepared, the regional military ambitions of Abū Isḥāq were proving his undoing, and Shiraz was under siege by a rival, Muḥammad b. Muẓaffar.[7] Perhaps worse, Abū Isḥāq was at odds with important segments of the city population, until a neighbourhood chief opened Shiraz to the Muzaffarid armies. Abū Isḥāq fled to Isfahan, and although his Injuid supporters briefly recovered Shiraz a year later, the Muzaffarids captured Abū Isḥāq at Isfahan in 1357 and sent him as a prisoner to Muḥammad b. Muẓaffar at Shiraz, who had him executed. During the Muzaffarid period, Shiraz was ruled by Shāh Shujāʿ and then his son Zayn al-ʿĀbidīn, until Temür extinguished the independent dynasties of western Iran in 795/1393. It may have been this situation of crisis and threat that motivated the creation of Köprülü 01589, as a way to conserve the city's cultural resources in duplicate, and this situation could also explain why the manuscript left Shiraz early in its existence (on which, see below).[8]

The manuscript is deserving of an exhaustive monographic study covering its codicology and exploring its intellectual contents in detail. Such a study should consider the potential of the work as the sum of something much bigger, revealing the tastes and interests of a reading community at a dramatic juncture in the city's history, in a manner similar to the way Konrad Hirschler has productively studied the Damascene Ashrafiyya library catalogue.[9] There is much to consider, including cultural generation in times of crisis, and the place of Köprülü 01589 in the history of Islamic art.[10] The aim of this current article is limited to introducing the manuscript as an important cultural product of eighth/fourteenth century Shiraz, and as a probable example of how urban centres in the region produced texts in the period. We will suggest that many of its features evidence the working habits of transmission, which are

notes, Abū Isḥāq was the first of his line to claim himself not just *amīr*, but also *sulṭān*, as an assertion of independence.

7 See Aigle's analysis, *Le Fārs*: pp. 186ff.

8 Consider the archival impulse that underlies many large projects, or responses to what Jacques Derrida terms the 'archiviolithic' drive; Derrida, Jacques, *Archive Fever: A Freudian Impression*, translated by Eric Prenowitz (Chicago/London: University of Chicago Press, 1996): pp. 10-1. The many small texts in the margin, for example, might suggest a patron familiar with a library and wishing to preserve his favourite pieces in the face of possible destruction, or given the manuscript's movement, to bring them with him when he moved away from Shiraz.

9 Hirschler, Konrad, *Medieval Damascus: Plurality and Diversity in an Arabic Library; The Ashrafīya Library Catalogue* (Edinburgh: Edinburgh University Press, 2016).

10 Hirschler's description of how competition for political hegemony serves as an impetus for scholarly and artistic activities is likely important for Köprülü 01589 also. See Hirschler, *Medieval Damascus*: pp. 11ff.

more difficult to detect when examining less complicated manuscripts or by reference to narrative sources. Our intention is to contribute to a small but important body of scholarship on cities and book production in the medieval Islamicate world.[11] Manuscripts that contain single texts, or only a small number, rarely reveal so many clues concerning working habits as can be detected in Köprülü 01589. This exemplar might reveal not just how such complex manuscripts were produced, but also provide understanding of the wider practices of manuscript creation that supported transmission of texts in fourteenth-century Shiraz, as well as in other major urban centres of the Middle East in that period. This manuscript represents a moment of full tilt cultural production, with maximum engagement of resources, and so suggests the general system of multi-text compilation. For the moment we put aside the question of how common such compilations were, but note that most studies and resources concerning textual transmission significantly understate this historical form.[12]

Other major issues involve the 'authorial' and other agencies involved in the different stages of the transmission process, and how such a revelatory manuscript can assist us in understanding the choices, decisions, and accidents that have produced what remains of the general textual heritage of the Middle East. It takes 'textual forensics'[13] to gain insight into the many individual decisions that have yielded what is clearly a massively reduced body of material. Since the earliest times, many decisions have depended on pure happenstance, very often with little awareness on the part of the patron or scribe participating in the 'preservation' of the material at any given moment. Examination of the mechanisms of textual transmission – including how, where, and when texts were produced – will offer further insights into the material that has in fact survived.

---

11 The book most deserving of mention would be Wright, Elaine, *The Look of the Book: Manuscript Production in Shīrāz, 1303-1452* (Washington, D.C., Freer Gallery of Art, Smithsonian Institution in association with University of Washington Press, 2013). Wright does not discuss Köprülü 01589, which would place towards the very beginning of her chronology. For Shiraz as a centre in later times, see esp. Uluç, Lâle, "Selling to the Court: Late-Sixteenth-Century Manuscript Production in Shiraz", *Muqarnas*, XVII (2000): pp. 73-96.

12 For a notable exception, preceding our work, see scholarship on the *Safīna-yi Tabrīz*, a multi-text personal compilation consisting of 209 disparate works in Persian and Arabic, collected between 1321 and 1335 by the Tabrizi notable and scholar Tabrīzī, Abu'l-Majd, *Safīna-yi Tabrīz* (Tehran: Markaz-i Našr-i Dānišgāhī, 2003) and Seyed-Gohrab, A.A. and McGlinn, S. eds., *The Treasury of Tabriz: The Great Il-Khanid Compendium* (Amsterdam, West Lafayette, Indiana: Rozenburg Publishers and Purdue University Press, 2007).

13 For the term and its meaning, Greetham, D.C., "Textual forensics", *PMLA*, CXI/1, Special Topic: The Status of Evidence (Jan., 1996): pp. 32-51.

In what follows we touch only briefly on these thoughts, leaving the project of deep investigation of Köprülü 01589 to those future students and researchers seeking a productive monograph project.

2 Working Hypothesis for the Creation of the Manuscript

We found the following features of the manuscript to be the most immediately intriguing, based upon examination of the original in Istanbul and of a pdf provided by the Süleymaniye Library.

2.1 *Size*

The Köprülü 01589 manuscript consists of 433 folios and an antique binding. The folios have been reinforced at least twice in the modern era (resulting in what seem to be air gaps, visible in the pdf). Forty-seven texts are written into the main body of the manuscript, and 69 into the margins.[14] Some folios were shaved slightly, resulting in losses to the margins, perhaps during rebinding(s).

2.2 *Use of Margins*

The margins are used almost in their entirety, and the texts copied into these spaces are often lengthy. In the main body of the manuscript and the margins, the texts are sometimes written by the same copyists, and titles in both the body and the margin are illuminated in the same or similar styles. Seven texts written into the margins contain colophons. In terms of contents, the layout suggests meanings, for example the main body and its margins contain works by the same author (for example Naṣīr al-Dīn Ṭūsī, d. 672/1274) or treat similar themes, and the same occurs within the margins (the texts of Ibn Sīnā[15] represent another example, composing a mini-corpus within the margins). Historians have been accustomed to considering the margins as secondary creations. However, here they seem largely as part of the original design, and important for understanding the manuscript as a book.

2.3 *Bundles*

There seem to be about nine 'bundles' of texts. At least four bundles seem to have originally been bound together, with another three bundles also bound

14 Dānišpažūh, *Fihrist*: I, pp. 479-87.

15 Avicenna in European tradition, d. 428/1037.

together (Appendix 1). We do not know exactly when these were merged into the manuscript, but this was likely done close to the time of the composition of the bundles and the works within them. This is revealed in the fact that in each collection, the main manuscript body has been filled in with poetry written in the same hand. In particular, we observe that 'bundles' 1, 2, 7, and 9 contain poems by Kāšānī (d. ca. 610/1213),[16] copied out to fill extra space in each bundle, by a scribe we are calling FP1; bundles 5, 6, and 9 contain poems by Saʿdī and Ḥāfiẓ, inserted by a scribe we are calling FP2. No text written into the margin continues from one bundle to another. Our identification of these bundles is particularly important to our ideas about the origins of the manuscript.[17]

2.4 *Multiple Hands*

On palaeographic grounds, Majid proposes thirty hands (Appendix 2). The same scribes appear to have worked on different bundles, and on the main texts and in the margins. The handwriting, although not calligraphic, is of fine quality. The manuscript is important to understanding the history of scripts, as it includes the relatively early use of *nasḫ* with features of *taʿlīq* (Fig. 2).[18]

Based on handwriting and watermarks, we can observe that some pieces of the manuscript have been reordered. For example, based on a watermark, the

16 Afḍal al-Dīn Muḥammad b. Ḥasan Maraqī Kāšānī was a Persian gnostic, philosopher and poet, who lived in the second half of the sixth/twelfth and the first half of the seventh/thirteenth centuries. According to Morteza Gharaee Garakani, the first mention of his name seems to be in the works of Naṣīr al-Dīn Ṭūsī (d. 672/1274): who writes that he studied under one of Bābā Afḍal's students. Gharaee Garakani, Morteza and Negahban, Farzin, "Bābā Afḍal", in: *Encyclopaedia Islamica*, Editors-in-Chief: Wilferd Madelung and Farhad Daftary: IV, pp. 1-8.

17 For now we leave aside consideration of the smaller units of quires, which would require further work on the manuscript and its rebindings.

18 These features include, for example, for the second text in the manuscript, the characteristic descent of writing onto the baseline, and specifically, the rightward slope of the letters *alif*, *kāf* and *lām*; the elongated cap of the *kāf*; and the rather weakly defined teeth on the letters *bah*, *pah*, *tah*, *ṯa*, *sīn* and *šīn*, among other features. The script also contains some of the same sense of hurriedness described by Wright. Wright shows that in Shiraz, *nastaʿlīq* is primarily an evolution of *nasḫ*, more specifically a hanging version of the latter (*nasḫ-i taʿlīq*). Whereas *nasḫ* "gives the impression of being the result of a precise and controlled act on the part of the scribe working slowly and with a clear consciousness of the exact form each letter should take", *nastaʿlīq* "has a sense of being written quickly and with great bravado". And "although the impression of great speed is of course false, for even in *nastaʿlīq* the best calligraphers form each letter carefully and slowly, the development of *nastaʿlīq* was nevertheless a process of altering specific *nasḫ* letter forms in ways that allow them to be written more quickly". Wright, "The Calligraphers of Šīrāz and the Development of the *Nastaʿlīq* script", *Manuscripta Orientalia* IX/3 (September 2003): p. 17.

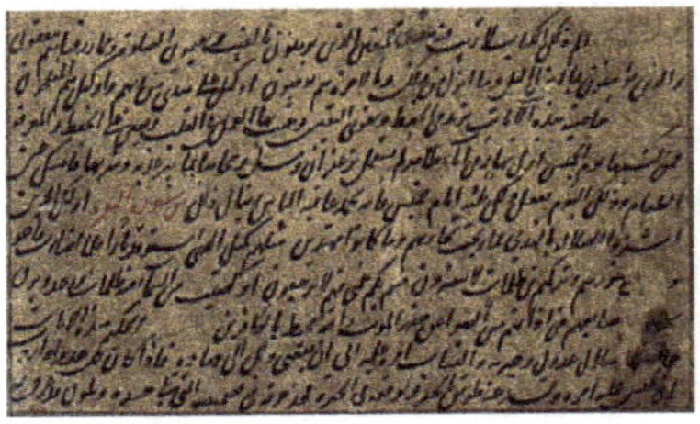

FIGURE 2
Nasḫ with features of ta'līq, from Köprülü 01589, folio 39v, containing the *Ḫawāṣṣ al-Qur'ān* by Abū 'Abd Allāh Muḥammad b. Aḥmad b. Sa'īd.

imprint of which shows on the reverse side of folio 101, this folio onwards to the end of the bundle 3 originally followed folio 411.

2.5 *Dates*

The main text contains 18 colophons, 17 of which have dates (the undated colophon names a scribe; see Appendix 3). Nine of the colophon dates in the main body of the text refer to 753 or 754 (1352 or 1353), as do three further ones in the margins (74v, 286r, 286v). The dates within the manuscript are not in sequential order (e.g., the date of the first text, in Muḥarram 754, is later than a date of the same month, indicated in the sixth text). The colophon of this first text (at 38r) lists Shiraz as the place of copying; the colophon of the twenty-third text (at 410r) indicates Isfahan. Towards the very end of the manuscript, the colophon of a text (running from 431v to 432r) shows a much later date, at 4 Ṣafar 811 (29 June 1408). There is also one marginal text bearing a date of the year 756/1355 (specifically 10 Rabīʿ al-Awwal 756/25 March 1355; written at 174r), and showing the location Dezful (in Khuzistan). Another marginal text has been indicated by the Tehran cataloguer as written in Baghdad (48v), however this information seems illegible from the scanned version.

2.6 *The Works Used by the Creators of the Manuscript*

The question here is what works would have been available for copying at Shiraz in the period of reference. Besides the *Jāmāsb-nāma*, the manuscript includes works by Ṭūsī, many by Ibn Sīnā, a Persian translation of the *Kalīla va Dimna* of Ibn al-Muqaffaʿ (139/756), and among many others, towards the end, several Sufi texts, including from the Kubrawī branch (which had its origins in Transoxania and Khwarazm). Most of the featured authors seem to post-date the early written tradition, falling especially within the twelfth and thirteenth centuries.

2.7 *Arabic-Persian Bilingualism*

The same scribes are observed to write in both Arabic and Persian. The colophons, however, are all written in Arabic.

2.8 *Relationship to Other Manuscripts*

Although we have only begun the investigation of relationships to other manuscripts, we can identify links with two other multi-text compilations from Shiraz, precisely in the mid-750s. Hamidiye 01447, with 419 folios, would appear to have a close relationship with Köprülü 01589, and it is possible that parts of it once formed part of a larger, common manuscript, as hypothesised by ʿAlī Ṣafarī Aq Qalʿa.[19] Among other possibilities, its common features could simply derive from the same workshop or setting; both theories could also be partially correct. The Hamidiye 01447 manuscript shares a number of features with Köprülü 01589: it contains texts with folios of nearly the same dimensions and layout;[20] it shows a similar general style of illuminations (see Images 4 and 5), including for the margins; poetry is copied out to fill space and folios are inserted to gain space in similar manner;[21] the manuscripts present works by some of the same authors (Ṭūsī, Ibn Sīnā, Majd al-Dīn al-Baġdādī); similarly to Köprülü 01589, Hamidye 01447 bears colophons for the *hijrī* years 753-5, one of which mentions Shiraz;[22] and both texts also feature ʿAlid authors. Still, while Hamidiye 01447 is similar to Köprülü in dimensions (i.e., boxed text), there are occasions where it has a different appearance in terms of lining, use of colour, and illuminations.[23]

The other related manuscript is a multi-text compilation in the Ibrāhīm Dihqān library in Arak, Iran, which contains two texts written into the main part of the manuscript and two others in the margins. This is a small collection in comparison to Hamidiye 01447 and Köprülü 01589, containing only 45 folios.[24] The first main-body text is related to *uṣūl al-fiqh* and is titled *Minhāj*

19 Ṣafarī Aq Qalʿa, ʿAlī, "Yādkard-i Tārīḫ-i Bayhaqī dar ʿiqd al-jawāhir fī ansāb al-akābir", *Guzāriš-i mīrāṯ*, no. 58-59 (2013): pp. 54-5. The full reference for the manuscript is Hamidiye 01447 Fatih Collection 00315.

20 According to the catalogue in Istanbul, 330 × 171 mm versus 332 × 165 mm for Köprülü 01589.

21 For example, at 79v and 80r.

22 At 80v.

23 For example f. 231r (for the colour of illumination in the margin) and ff. 258v-264r (for the slanting lines) of Hamidiye 01447. The few names of scribes provided in the colophons do not match between the two manuscripts. There are several other avenues for enquiry, including the relationship of the poetry of Hamidye 01447 to that of Köprülü 01589. Hamidye 01447 shows further similarities with colophons from the years 811 and 812 (like Köprülü) and also seems to have travelled.

24 Ḥaḍratī, Ṣādiq, *Fihrist-i nusḫa-hā-yi ḫaṭṭī-yi kitābḫāna-yi Ustād Ibrāhīm Dihgān* (Qom: Majmaʿ Dhaḫāʾir Islāmī, 2009): II, pp. 335-9. The dimensions for the folios are 340 × 180 mm. It seems that the cataloguer for the Arak library mistakenly recorded 88r as the last folio, whereas the MS has 45 folios, and counted the pages as folios (more correctly written "44v").

al-wuṣūl ilā ʿilm al-uṣūl (1v-28r), by Nāṣir al-Dīn ʿAbd Allāh b. ʿUmar al-Bayḍāwī (d. 719/1319), a Shāfiʿī judge from Shiraz.[25] The next text is a copy of a brief book on *fiqh* by Najm al-Dīn ʿAbd al-Ġaffār b.ʿAbd al-Karīm al-Qazwīnī al-Šāfiʿī (d. 665/1266-7) (29v-44r). In the margin is a treatise called *Kitāb al-Futuwwa*, by the Hanbali jurist ʿAbd Allāh b. Abi'l-Makārim, also known as Ibn al-Miʿmār al-Baġdādī al-Ḥanbalī (d. seventh/thirteenth century). The other text written into the margins is an untitled treatise by Aḥmad b. Ḥusayn b. Aḥmad b. Muḥammad al-Šāfiʿī, answering his friends' question on matters of *fiqh* (31r-42v).[26] This manuscript, despite its thematic divergence from the other two, has similar characteristics in the illuminations, and also has a medallion (1r) similar to ones appearing in Köprülü 01589 (at 1r and 101r), and in Hamidiye 01447 (1r and 154r).

Taking these and other features into account, we present our working hypothesis for Köprülü 01589, as follows.

During the late Injuid era, in or before 753/1352, a bi-lingual Muslim patron or patrons commissioned, in Shiraz, texts to be copied and bound together, perhaps for personal reading. The commission was given to a workshop, where the layout and plans for copying were elaborately drawn up. Some texts were to be in the main body, some in the margins – the latter sometimes featuring very long texts that might offer a different perspective on a topic, a further work by an author, or otherwise fill the space. A coordinator assigned different scribes the units that they were to copy, with tight timelines; folios were passed through the workshop, with one scribe, for example, writing in a text block, and on this unit's completion, two or more scribes dividing his folios amongst themselves as they added texts into the margins. Sometimes they misjudged and had to improvise by adding folios, or where empty space remained, adding text. A copyist working in the margins, for example, misjudged the length of a piece by Ibn Sīnā and inserted extra folios (folios 312:1-4). In at least two other cases, the text in the margins required more pages than the number used for the main body; at that point the main text block was empty, so a scribe added verses to fill the space (Fig. 3).

Most of the manuscript as it exists today was constructed in stages, starting in the mid-750s/1350s. The scribes began with stocks of sheet paper, mostly in what Elaine Wright has described as a "three sided marginal comment format" with "thumb holds". The pages had been variously pre-lined (perhaps through a screening process, during manufacture). Faint boxes were drawn on all sheets.

---

25 Saleh, Walid, "al-Bayḍāwī", in *Encyclopaedia of Islam*, 3rd edition, edited by: Kate Fleet, Gudrun Krämer, Denis Matringe, John Nawas, Everett Rowson. Consulted online on 10 March 2018 <http://dx.doi.org/10.1163/1573-3912_ei3_COM_25267>.

26 Ibid.: II, p. 339.

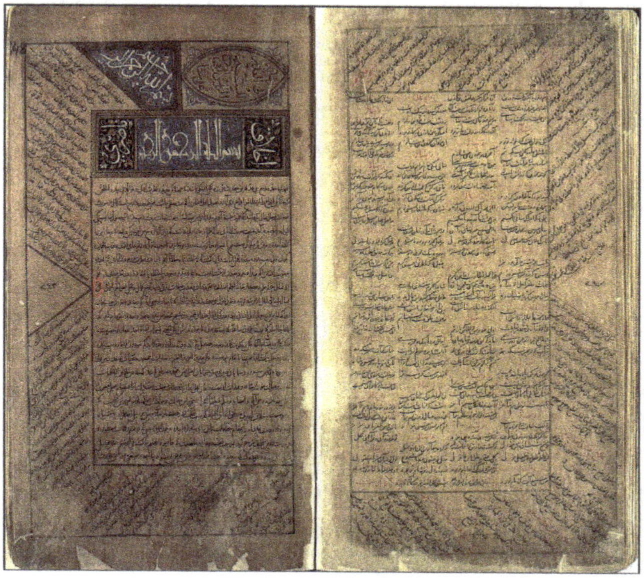

FIGURE 3 Folio 242v of Köprülü 01589, on the right, contains verses by Kāšānī. The text written in the margins (a *Kitāb fī'l furūsiyya wa'l-bayṭara*, 213r-242v) ran longer than expected and required that additional material be written in the main text, to make the page look complete. On 243r, on the left, a new bundle begins with Nāṣir al-Dīn Ṭūṣī's *Aḫlāq-i Nāṣirī*.

The scribes wrote text into their boxes. On completing a particular text the scribe began another, or passed the sheets to another scribe who continued with the work. Bundles of texts were produced by a few scribes; these were then collated. The evidence for the process of building the manuscript out of such bundles lies in observation of where the writing starts and stops, and in the common paper linings within the bundles. The bundles contain the work of multiple scribes. They consist of folios with similar numbers of lines of text, and where they end on a folio (for example, mid-way between the recto side of a folio and the next), the verso side of the folio was blank, but on the recto side of the next folio, text was begun anew. There seem to be nine bundles. The first bundle of the manuscript bears two texts (the writing of both occupying about 37 lines per page). An *Ījāz al-bayān fī tafsīr al-Qur'ān* by Abu'l-Qāsim Maḥmūd b. Abi'l-Ḥusayn al-Nīsābūrī (d. 553/1158),[27] runs 1r-38v, and a *Ḫawāṣṣ*

27 Ḥanīf b. Ḥasan al-Qāsimī consulted Köprülü 01589 as a variant for preparation of his edition, referring to it as "*nusḫa* K"; he used a total of three manuscripts; see pp. 43 and 49-50. He writes that the Köprülü 01589 recension is full of errors. al-Nīsābūrī, Abu'l-Qāsim

al-Qur'ān, by Abū ʿAbd Allāh Muḥammad b. Aḥmad b. Saʿīd (d. 390/999-1000), then runs on to nearly the end of 60r; an empty page follows (60v), then a new bundle begins on 61r, with the third text to appear in the manuscript, a version of the *Maqāmāt* of al-Ḥarīrī (d. 516/1122), running through to the top quarter of 100v (all folios of 37 lines; 101r is then left blank).The *Maqāmāt* is a sort of unit in itself, while the next two texts form another self-contained unit: an *ʿIqd al-jawāhir fī ansāb al-akābir*, by Aḥmad b. Abī Saʿd al-Dāwanī al-Miškānī (alive 560/1164-65), which runs 101v-103v, and the *Jāmāsb-nāma* (mentioned above), which runs 104v-120r. Both of these texts fill 25 lines to the page and are written in the same handwriting. Within the bundle, they are separated by verses of Saʿdī (at 104r), inserted as filler (as described above). We have numbered the bundles in Appendix 1.

The collating of the individual works into bundles was done on at least two occasions, most likely close to their dates of writing, but not necessarily in consecutive order of their creation. Among the colophons, the earliest date, 2 D̲u'l-Qaʿda 753 (10 December 1352), is that of the eighteenth text in the manuscript (within the seventh bundle), the *Aḫlāq-i Nāṣirī*, by Nāṣir al-Dīn Ṭūṣī (d. 672/1274). However, the first text in the manuscript (Abu'l-Qāsim Maḥmūd b. Abi'l-Ḥusayn's *Ījāz al-bayān fī tafsīr al-Qur'ān*) bears a colophon of a date of about three months later (20 Muharram 754/25 February 1353), also mentioning Shiraz. The ninth and last bundle begins with a *Sīrat Al-Šayḫ al-Kabīr Abū ʿAbd Allāh ibn al-Ḫafīf al-Šīrāzī* by Abu'l-Ḥasan ʿAlī b. Muḥammad al-Daylamī, running 383r to 410v and ending with a colophon dated 15 Ṣafar 754 (22 March 1353) in Isfahan. Perhaps this piece was sent from Isfahan to Shiraz.[28] As for the illumination, this seems to have been completed within about the same period as the folios. The artists were developing a new style that would subsequently be recognised as 'Šīrāzī'.

According to our theory, this then was the general situation: a productive workshop, multiple scribes at work, coordination within a relatively narrow time period, in accordance with a plan.[29] But at some point in 753/1354 – owing

---

Najm al-Dīn Maḥmūd, *Ījāz al-bayān fī tafsīr al-Qur'ān*, edited by Ḥanīf b. Ḥasan al-Qāsimī (Beirut: Dār al-Gharb al-Islāmī, 1995).

28 It is also possible that the manuscript travelled to Isfahan and was completed there, however handwriting on the bundles suggests otherwise. On the text, see below, note 36.

29 The workshop most discussed in scholarship for roughly this period is that of Rašīd al-Dīn (d. 718/1318): who established his manuscript production near Tabriz. See, e.g., Ben Azzouna, Nourane and Patricia Roger-Puyo, "The Question of the Formation of Manuscript Production Workshops in Iran According to Rashīd al-Dīn Faḍl Allah al-Hamadhānī's *Majmūʿa Rashīdiyya* in the Bibliothèque nationale de France", *Journal of Islamic Manuscripts*, VII (2016): pp. 152-94.

perhaps to the turbulent situation that led to Shiraz's passing to the Muzaffarids – the work largely ceased. The manuscript was incomplete – the main body texts were nearly finished (not 410ff), as were many or most of the texts written in the margins, or close to finished (e.g. the colophon of 756 at 174r). Illumination of the titles of both the main and marginal texts was partially complete. The bundles had already been formed as units; perhaps the final collation had begun. The manuscript appears to have left Shiraz soon after this halt, judging by the colophon of 756 being inserted in Dezful.

We acknowledge that our hypothesis regarding the process of compilation is just one possibility. In fact we have refined our thinking several times as we continued our work and discovered additional features. The existence of other manuscripts from Shiraz in the same period, for example, raises the possibility that we have only one piece of a large puzzle; the collation phase may have involved removal or addition of bundles from the workshop, for example. But our main argument would remain: that in Köprülü 01589 we can see evidence for the internal workings of a workshop, operating in Shiraz. Moreover, Köprülü 01589, along with Hamidiyye 01447, and indeed the studies for later periods by Wright, suggest the importance of Shiraz as a centre of book production.

Invoking Michel Foucault, the 'book' that resulted from this process represents one mode of what we could call an 'authorial function'.[30] In our case, authorial agency is diffuse, with the patron and workshop dynamics both determining the contents. The process was also guided by the material dimensions of the texts, which for example called for filling out space in the main body of the manuscript and the margins. The cultural meaning of any preserved manuscript fragment lies not just in the 'original' words of an author or authors living centuries prior, or in the exigencies of material textual production, but also in a very large number of relationships, in discourse within specific sets of historical circumstances. Such relationships include those between the patron, the workshop (especially scribes and coordinators) as well as those involving the readers who subsequently inscribed the margins with text, and drew their own associations between the pieces contained within the manuscript (whether we can discern their ideas or not). All those involved in inscription and handling added meanings to the texts. Such manuscripts, given their composition from different pieces and the involvement of collective efforts, call on us to consider the nature of creativity and agency. Relationships lie at the heart of this agency, and indeed at the heart of the written Arabic and

30 Foucault, Michel, "What Is an Author?" in *Language, Counter-Memory, Practice: Selected Essays and Interviews*, edited by Donald F. Bouchard, translated by Donald F. Bouchard and Sherry Simon (Ithaca, NY: Cornell University Press, 1977): pp. 113-38.

Persian traditions themselves, when one considers matters of text 'reuse' and inter-textuality.[31]

Multi-text compilations also challenge us to reconsider the meaning of the term 'book', and to more closely consider the material dimensions of textual transmission. In the field of Islamic history, the study of transmission is almost always based on narrative sources (particularly the authors' comments on what they were doing), a prototype being the *isnād* in Hadith. Also, studies of transmission generally consider a single text and its manuscript witnesses across time. This tends to lead to development – if not of an actual 'critical' edition – of a kind of ideal text that in fact never existed anywhere, at any time. Here, in contrast, we have particularly rich instances of specific titles whose very existence was relational. The manuscript is a corpus, a bearer of meaning and memory, with titles combined to form an artefact of a specific time and place.[32]

3 How the Manuscript was Sourced and Meaning 'Incarnated'

The Sociology of Texts field offers a productive way to consider these relationships. The contents of Köprülü 01589 can be seen as the site, or sites, of many negotiations and productions of meanings. Roger Chartier, specialist in the history of books, writes that the textual units within the manuscript each represent "a unique textual incarnation whose materiality itself crucially shapes meaning". They are valid 'incarnations' of the works themselves.[33] Generally, all works exist in plural forms. This line of thinking offers a way to recognise the roles of multiple agents – including a patron, commissioner, scribes, and readers writing in the margins – as producers of individual titles, but also of

31 The term 'reuse' is employed by scholars of humanities and computer science in reference to the repetition of written words, in whole or in part, including paraphrases. The study of such reuse starts from a position of not knowing why texts share passages, then applies information gained from digital methods, metadata, and relating to historical context and the reading of manuscripts, to form hypotheses. See kitab-project.org, which tracks and visualises reuse in Arabic texts from roughly 700 to 1500, including the networks through which texts passed (Savant is the Principal Investigator).

32 Also, with Foucault generally in mind: "Plainly, we lack a theory to encompass the questions generated by a work and the empirical activity of those who naively undertake the publication of the complete works of an author often suffers from the absence of this framework". "What is an Author?": p. 119.

33 Chartier, Roger, *The Author's Hand and the Printer's Mind*, translated by Lydia G. Cochrane (Cambridge: Polity Press, 2014): p. 159; also, McKenzie, D.F., *Bibliography and the Sociology of Texts* (Cambridge: Cambridge University Press, 1999).

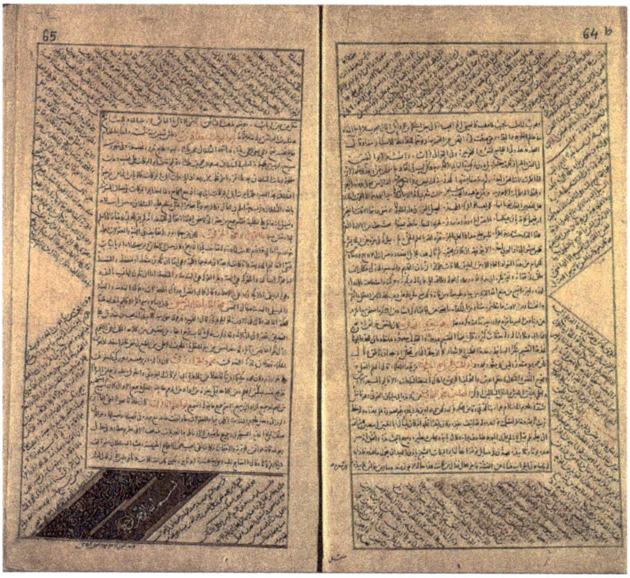

FIGURE 4 Hamidiyye 01447, ff. 64v-65r, has similar features and dates. The illuminated title on folio 65r, bottom left, is for a *Sīrat Ibn Sīna*, and is in a style consistent with Köprülü 01589 and Injuid manuscripts in general, especially as concerns the floral motif. See Wright, *The Look of the Book*, and her discussion of Ilkhanid origins: pp. 19-30.

collected works. This is especially helpful for Köprülü 01589, which is a unique arrangement compared to the example of those manuscripts that witness a single widely reproduced text. Metaphorically, we should see all instances of a text as live performances. Some are monologues – Köprülü 01589 is an opera.

In terms of the contents of Köprülü 01589, four sets of texts stand out. These works in particular deserve more in-depth study for what they can tell us about the intellectual landscape of Shiraz and its environs, eventually generating a detailed picture of cultural materiality, meaning, and memory. What follows is a brief introduction.

1. *Relating to Naṣīr al-Dīn Ṭūsī*

Ṭūsī was one of the most important and influential Šīʿī scholars in the fields of mathematics, geometry, astronomy, philosophy and theology. In later times he was nicknamed 'al-muʿallim al-ṯāliṯ', 'the third teacher' (i.e. after Aristotle and al-Fārābī).

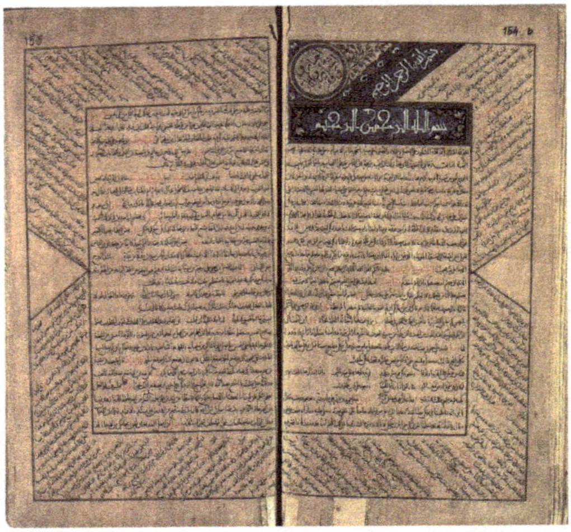

FIGURE 5
Hamidiyye 01447, ff. 154v-155r. The layout and illumination of Hamidiye 01447 resembles that of Köprülü 01589. Compare with Image 3, above.

The works in Köprülü 01589 are among his best known, and these particular instances are important for their relatively early dating. Ṭūsī's *Tajrīd al-ʿaqāʾid* is one of the main theological works for the Twelver Shia, and is still taught in seminaries. Our manuscript contributed to the basis of the modern printed edition, read with the commentary by ʿAllāma al-Ḥillī'.[34] In the margin of the manuscript, opposite the *Tajrīd*, appears text from another work by Ṭūsī, the *Āġāz wa anjām* ("the Beginning and the End"), which S.J. Badakhchani considers his "most important work on eschatology".[35] Köprülü 01589 also contains, in the main body of the manuscript, the *Aḫlāq-i Nāṣirī*, which is one of Ṭūsī's most famous works on ethics, written for the Ismāʿīlī governor of Quhistān, Muḥtašam Nāṣir al-Dīn. Ṭūsī reuses text from Miskawayh's *Tahḏīb al-aḫlāq*, especially in the first discourse,[36] and Ṭūsī's text was itself reused by other Persian authors, including Dawānī (d. 908/1502) in his *Aḫlāq-i Jalālī*. The *Aḫlāq-i Nāṣirī* was revised later in Ṭūsī's life, after he defected to the Mongols. In the margin alongside the *Aḫlāq-i Nāṣirī* is a Persian

34 The *Tajrīd al-ʿaqāʾid*, published as: Nasīr al-Dīn Ṭūsī, *Tajrīd al-Iʿtiqād*, edited by Muḥammad Jawād Ḥusaynī Jalālī (Qom: Maktab al-Iʿlām al-Islāmī, n.d.).

35 Ṭūsī, Nāṣir al-Dīn, *Contemplation and Action: The Spiritual Autobiography of a Shi'i Philosopher*, edited by S. J. Badakhchani (I.B. Tauris in association with the Institute of Ismaili Studies, 1998): p. 7; Ṭūsī, Nāṣir al-Dīn, *Āġāz wa anjām*, edited by Īraj Afšār (Tehran, Intišārāt-i Danišgāh Tihrān, 1956-7).

36 Wickens, G.M. "Introduction" in *The Nasirean Ethics by Naṣīr al-Dīn Ṭūsī* (London: George Allen & Unwin, 1964): p. 16. Köprülü 01589 is not included in Wickens' list of manuscripts, however it is possible that the author used it through one or another of the printed editions cited.

translation of a work attributed, probably falsely, to Ibn al-Muqaffaʿ, known as *al-Adab al-wajīz li' l-walad al-ṣaġīr*, which concerns the correct behaviour of children.[37]

It will be important to read these texts by Ṭūsī in light of others in the manuscript. Ṭūsī is frequently associated with a whole constellation of ideas, repeated rather generally based on citations of works. He was the initiator of an important new phase in the intellectual history of Twelver Shiʿism, and is seen as "the first Twelver scholar to have been at once a theologian and a philosopher, having been particularly influenced by Ibn Sīnā's philosophy".[38] Here, in the artefact of our manuscript, we have a view at a historical moment of how Ṭūsī's work was read, and the ideas that readers associated with his works.[39] The extensive use of Ibn Sīnā's texts in the margins alongside other works requires particular consideration (see Table 1).

2. *Works on astrology*

The *Jāmāsb-nāma*, as noted above, represents a Persian historiographic tradition that is virtually unknown in Europe and North America, surviving in a sizable number of manuscripts, including in Iran. The work provides 55 horoscopes, beginning with Adam and continuing chronologically, to include Jamšīd, Moses, Muḥammad, ʿAlī, Muʿāwiya, as well as other figures of biblical, Iranian, and Arabian history. The sequence ends nearer to the author's own day with Genghis Khan, a flood (the biblical flood receives an earlier horoscope), and a final box which might be dedicated to the end of times (*āḫar ʿālam*). This particular text, as noted above, deserves in-depth comparison with related texts, including the Zoroastrian text of the same name.[40] Other texts in the manuscript also treat astrology, including at 420v-422r.

---

37 Its authenticity is doubtful; see Brockelmann, *S I*, 236 no. 6; Daiber, H. "al-Ṭūsī, Nasīr al-Dīn" in *Encyclopaedia of Islam*, 2nd edition, edited by P. Bearman, Th. Bianquis, C.E. Bosworth, E. van Donzel, W.P. Heinrichs (Leiden: Brill, 2000): x, pp. 746-50.

38 Daftary, Farhad, *A History of Shiʿi Islam* (London/New York: I.B. Tauris/Institute of Ismaili Studies, 2013): pp. 75-7.

39 For example, as Daniel De Smet has written: "Under the influence of Naṣīr al-Dīn Ṭūsī, Twelver Shiʿism was to produce a whole line of thinkers interested both in *falsafa* (especially in the works of Avicenna) in the Muʿtazilī *kalām*, in the 'philosophy of enlightenment' (*ishrāq*) of al-Suhrawardī, in the mysticism of Ibn ʿArabī, in mathematical sciences, in astronomy, but also in Qur'anic commentary, *ḥadīth* and jurisprudence". De Smet, Daniel, "Introduction", translated by Orkhan Mir-Kasimov in Daftary, Farhad and Miskinzoda, Gurdofarid, *The Study of Shiʿi Islam: History, Theology and Law* (London/New York: I.B. Tauris/Institute of Ismaili Studies, 2014): p. 560.

40 See note 2, above. In the manuscript, see 104v-120r.

TABLE 1 Texts attributed to Ibn Sīnā

	Title	ff Location (margins)	Alongside, in the main text …
1	*Abṭāl aḥkām al-nujūm*	108r-116r	*Jāmāsb-nāma*
2	*Risāla fī ḥadd al-jism*	116r-118v	
3	*Tarjuma-yi fārsī-yi Risāla fī daf' al-ġamm 'an al-mawt*[a]	131v-140v	Abu'l-Ma'ālī Naṣr Allāh, *Tarjuma Kalīla va Dimna*
4	*Mi'rāj-nāma*	147r-151v	
5	*Šarḥ va tarjuma-yi Fārsī-i risāla fi'l-qadar*	243r-262r	Ṭūsī, *Aḫlāq-i Nāṣirī* (243r-311r)
6	*'Ahd al-šayḫ Abī 'Alī b. Sīnā*	286v	
7	*Risāla fī daf' al-ġamm 'an al-mawt*	287r-288r	
8	*Risāla fi'l-ḥuzn*	288r-288v	
9	*Risāla fi'l-'ishq*	288v-291r	
10	*Risālat al-Ṭayr*	291v-292v	
11	*Risala al-'arūs*	292v-293v	
12	*Al-Maqāl al-ma'hūda fi'l-aḫlāq*	293v-294v	
13	*Risālat al-birr wa'l-ithm*	294v-298v	
14	*Sabab qiyām al-arḍ fī wasaṭ al-samā'*	298v-301r	
15	*Risālat al-malā'ika*	311v-312r (in the main part of the manuscript)	Ṭūsī, *Risāla fī faḍīlat Amīr al-mu'minīn 'Alī* (311v-312v, in this case in the margins)
16	*Risālat al-ḥukūma fi'l-ḥujaj al-muṯbitīn li'l-māḍī mabda' zamanīya wa bayān taḥlīli-hā*	302v, 312/1-312/4 *4 folios inserted to finish out a marginal text begun from 302v*	Ṭūsī, *Aḫlāq-i Nāṣirī* (243r-311r)
17	*Risāla fī tadāruk al-ḫaṭā' al-wāqi' fi'l-ṭibb*	313v-326v	Abū Bakr al-Muṭahhir b. Muḥammad b. Abī al-Qāsim al-Jamālī al-Yazdī, *Farruḫ-nāma-yi Jamālī* (313v-370r)
18	*Risāla fī masā'il Ḥunayn b. Isḥāq*	327r-333v	
19	*Risāla fi'l-ṭibb*	333v-343v	
20	*Risālat al-fuṣūl*	343	

a Supposedly the Persian translation of the seventh piece in the table. However, the Persian translation seems much lengthier.

3. *Attributed to Ibn Sīnā*

The manuscript contains at least twenty pieces attributed to Ibn Sīnā, most of them rather brief, of which 19 are in the margins. We list them tentatively here, based on the catalogs, with what appears alongside them in the main text (or in the one opposite case, in the margins).

Ibn Sīnā's texts are found in the third, fifth, and seventh bundles, however, texts 5 to 20 belong to the seventh, representing a notable concentration within the margins of a single bundle.

Dmitri Gutas, in several works, has noted the many challenges of Ibn Sīnā's oeuvre, including the author's habit of frequently copying himself in successive works, "which complicates the textual integrity of the repeated sections and passages". There have been bibliographic projects dedicated specifically to unravelling the transmission of his texts, but more and different work can be done.[41] Aside from starting with a title for one of Ibn Sīnā's works, and searching out manuscript witnesses, consideration should be given to how this and other manuscripts provide evidence for the reception of Ibn Sīnā's thought in a specific instance.

One aspect of interest is the continuing arrangement of Ibn Sīnā's works alongside those of Ṭūsī, starting from 243r, in bundle 7. The texts include 11 pieces by Ibn Sīnā which run in the margins alongside Ṭūsī's *Aḫlāq-i Nāsirī*. Also in bundle 7 is Ibn Sīnā's *Risālat al-malāʾika*, covering two folios in the main body of text, while the margin we find a text attributed to Ṭūsī, called *Risāla fī faḍīlat Amīr al-muʾminīn ʿAlī* (311v-312v). These observations suggest, among other things, the importance of the bundles as units of meaning, as well as the high degree of coordination involved in the production of Köprülü 01589.

4. *Sufi texts*

The first text listed in the table below is the *Risālat al-Ṭayr* by Najm al-Dīn Rāzī, born in 573/1177 in Ray, and later known as one of the Kubrawī masters.[42]

---

41 Gutas, "The Study of Avicenna. Status Quaestionis atque Agenda" in Gutas, Dimitri, *Orientations of Avicenna's Philosophy: Essays on his Life, Method, Heritage* (Farnham, Surrey: Ashgate, 2014): section XVII, pp. 1, 45-69. Originally published in *Documenti e studi sulla tradizione filosofica medieval*, XXI (2010): p. 48. Gutas's bibliography is useful for the state of the art in studies of Ibn Sīnā.

42 He travelled around the Muslim world for study of hadith, ultimately arriving in Ḫʷārazm, where he studied under both Najm al-Dīn Kubrā and Majd al-Dīn al-Baġdādī. However, in his magnum opus *Mirṣād al-ʿIbād min al-Mabdaʾ ilā al-Maʿād*, he refers in several instances to Majd al-Dīn as his master. Rāzī, Najm al-Dīn, *The Path of God's Bondsmen: From Origin to Return*. Translated by Hamid Algar. Persian Heritage Series 35 (Delmar; New York: Caravan Books, 1982): pp. 95, 154, 241, 384.

TABLE 2 Sufi Texts

	Title	ff Location	Scribe
1	*Risālat al-Ṭayr*	233r-235v	S12
2	*Sīrat Al-Šayḫ al-Kabīr Abū ʿAbd Allāh ibn al-Ḫafīf al-Šīrāzī*	383r-410v	S25
3	*Risāla fi'l-taṣawwuf*	412r-413r	S12
4	*Asʿila va ajwiba*	413r-413v	S12
5	*Risālat Šaraf al-Dīn al-Balḫī ilā Majd al-Dīn al-Baġdādī*	414r-414v	S12
6	*Risālat al-Jawāb [Majd al-Dīn's answer to the previous letter]*	414v-416r	S12
7	*Risālat al-Safar*	416v-417v	S12
8	*Risālat al-Baġdādī wa jawāb al-Balḫī ʿanhā*	417v	S12
9	*Risālat al-Sayr wa al-ṭayr*	418r-418v	S12
10	*Risāla fi'l-ʿuzla wa al-ḫalwa*	418v-419r	S12
11	*Risāla fi'l-taṣawwuf*	420r-420v	S12

Risālat al-Ṭayr is an allegorical story of birds who take their complaints to the author. Their outcry was the result of social unrest in Ray in that period. The author accepts the task of delivering their request for justice to Sultān Kibrīyā (the Master of Grandeur), asking him to order ʿAnqā Muġrib (a legendary giant bird similar to Sīmurġ in Iranian mythology) to bring back justice and prosperity for the people. Unlike other treatises with similar names and motifs, containing mystic didactic themes, the theme of this work is social unrest.[43] This has led the editor of the text, published on the basis of Köprülü 01589 under the title *Risālat al-Ṭuyūr*,[44] to believe that the work represents the early stages of Najm al-Dīn Rāzī's intellectual life, prior to Sufi recruitment.[45]

The next Sufi text in the manuscript is a biography of a Shirazi saint, Ibn al-Ḫafīf (c. 268-371/882-982), entitled *Sīrat Al-Šayḫ al-Kabīr Abū ʿAbd Allāh ibn*

43 Riyāḥī, Muḥammad Amīn, "Muqaddama-i Muṣaḥḥiḥ", in Hamadānī, Ḫʷāja Yūsuf, and Najm al-Dīn Rāzī, *Rutbat al-ḥayāṭ bi ḍamīma-yi risalat al-ṭuyūr*, edited by Muḥammad Amīn Riyāḥī (Tehran: Intišārat Tūs, 1983): pp. 84-7.

44 Hamadānī, Ḫʷāja Yūsuf, and Najm al-Dīn Rāzī, *Rutbat al-ḥayāṭ bi ḍamīma-i risalat al-ṭuyūr*, edited by Muḥammad Amīn Riyāḥī (Tehran: Intišārat Tūs, 1983). The Tehran cataloguer read the title as *Risālat al-Ṭayr*, but Riyāḥī read it as *Risālat al-Ṭuyūr*. The current condition of the manuscript title box prevents identification of which might be the correct version.

45 *Ibid*, p. 90.

al-Ḫafīf al-Šīrāzī (383r-410v).[46] Ibn al-Ḫafīf's was a master of Daylamī origin, who settled in Fars. According to Ibn Baṭṭūṭa, his shrine was a venerated site of pilgrimage in Fars.[47] He was so famous in this region that some coins minted there were referred to as *'ḍarb-i šayḫ kabīr'* (mint of šayḫ kabīr).[48] Ibn al-Ḫafīf also studied theology under Abu'l-Ḥasan al-Aš'arī (d. 324/935-6) and wrote works on *fiqh* and theology.[49] Al-Sam'ānī considered him to be a master of hadith and enumerated Ibn al-Ḫafīf's masters and pupils in hadith.[50] His biography was written by Abu'l-Ḥasan 'Alī al-Daylamī (fl. 392/1001-2), one of Ibn al-Ḫafīf's disciples, in Arabic. However the original text is not extant, and only the Persian translation by Rukn al-Dīn Yaḥyā ibn Junayd Šīrāzī (eighth/fourteenth century) survives in our manuscript.

The next part of the block of Sufi texts contains 11 short pieces (letters and treatises), including by the Kubrawī shaykh Majd al-Dīn al-Baġdādī (d. c. 607/1210; see Table 2). This shaykh was the second master in the line of the Kubrawī, after the great master and eponym of the order, Najm al-Dīn Kubrā (d. c. 618/1221). Majd al-Dīn was born in 544/1149 in a village called Baġdādak in Khwarazm, to a family related to the ruling dynasty of the Khwarazmshahs. He joined Najm al-Dīn Kubrā to embark on the Sufi path and went on to become one of his successors (*ḫalīfa*). Although Majd al-Dīn was the second master among Kubrā's followers, he was killed earlier than his master. The former's murder is variously narrated in the hagiographical sources. The accounts

46 Annemarie Schimmel published this text during her tenure as a professor at Ankara University (as Annemarie Schimmel-Tari). She used this manuscript and another version of the Persian translation, found in Berlin: Yaḥyā ibn Junayd al-Šīrāzī, Rukn al-Dīn, *Sīrat Al-Šayḫ al-Kabīr Abū 'Abd Allāh ibn al-Ḫafīf al-Šīrāzī*, edited by A. Schimmel-Tari' (Ankara: Ankara Üniversitesi İlâhiyat Fakültesi, 1955). Tawfīq H. Subḥānī republished the same edition in offset format and translated the introduction into Persian: Yaḥyā ibn Junayd al-Šīrāzī, Rukn al-Dīn, *Sīrat al-šayḫ al-kabīr Abū 'Abd Allāh ibn al-Ḫafīf al-Šīrāzī*, edited by A. Schimmel-Tari and translated by Tawfīq H. Subḥānī (Tehran: Intišārāt Bābak, 1984) (references are made to the Tehran reprint).

47 Ibn Baṭṭūṭa, *Ibn Battúta: Travels in Asia and Africa (1325-1354)*, translated by H.A.R. Gibb (London: Routledge & Kegan Paul, 1929; reprint 1953). On Ibn al-Ḫafīf, see also the article by Denise Aigle in this volume.

48 Jiddī, Muḥammad Jawād, "Nigāhī bi ḫuṭūṭ-i muhr-hā wa sikka-hā-yi ḍarb-i Šīrāz" in *Majmū'a maqālāt-i ḫušniwīsī-yi maktab-i Šīrāz* (Tehran: Mu'assisa-yi Ta'līf, Našr wa Tarjuma Āṯār Hunarī-yi Matn, 2001): p. 16.

49 Sobieroj, Florian, "Ibn Khafīf", in Encyclopaedia of Islam, 3rd edition, edited by Kate Fleet, Gudrun Krämer, Denis Matringe, John Nawas, Everett Rowson, consulted online on 03 February 2018 <http://dx.doi.org/10.1163/1573-3912_ei3_COM_30858>; Schimmel-Tari, A. (ed.) "Abū 'Abdullāh b. Ḫafīf Shīrāzī", *Sīrat al-Šayḫ al-Kabīr Abū 'Abd Allāh ibn al-Ḫafīf al-Šīrāzī* (Tehran: Intišārāt Bābak, 1984): p. 221.

50 Al-Sam'ānī, Abū Sa'd 'Abd al-Karīm b. Muḥammad b. Manṣūr al-Tamīmī, *al-Ansāb*, (Hyderabad: Majlis Dā'irat al-Ma'ārif al-Uṯmāniyya, 1977): VIII, p. 221.

include a version in which he was killed by Muḥammad Khwarazmshah because the latter suspected Majd al-Dīn of secretly marrying this king's widowed mother, and another in which he was murdered by the king, but because one of his amirs believed that Majd al-Dīn had an affair with the amir's wife (the common motif being the accusation of an affair and the resulting murder).[51] From Majd al-Dīn's works, two letters, an *ijāza* (permission to practice), an epistle and a book have thus far been published.[52] Of these only the *Risāla-yi dar safar*, an epistle on journeying, is published based on our manuscript (416v-417v), with the aid of two other manuscripts.[53]

4 Conclusion

Köprülü 01589 represents an instance of manuscript production indicating how a specific workshop coordinated its efforts to gather, copy, collate, illuminate, and annotate a large manuscript within a relatively short period of time. Copies of the *Jāmāsb-nāma* and texts by Ṭūsī and Ibn Sīnā were created within this environment and given new meanings as scribes produced their versions across the folios and within the margins. Creative agency was shared, and contributed to a discourse within a specific set of historical circumstances. The collection of texts in a single manuscript, made up of subunits which we call 'bundles', also helps us think anew about the meaning of books as works of particular times and places. The meaning of the work is in no way singular – tied to an original author, in an original place (as scholars attempt to recover in critical editions), but rather something much more interesting, born of relationships. The manuscript opens onto a social world, the networks of people

51 Zaryāb Ḫū'ī, 'Abbās, "Dāstān-i kušta šudan-i Majd al-Dīn Baġdādī manqūl az yik nusḫa-yi ḫaṭṭī", *Yaġmā*, 130 (Urdībihišt 1338š.): pp. 88-94. While Majd al-Dīn was considered among the great shaykhs of the Kubrawī order, his contribution was limited in comparison to that of Raḍī al-Dīn 'Alī Lālā (d. 643/1244), in terms of spiritual lineage, and to that of 'Alā' al-Dawla al-Simnānī (d. 736/1336) in terms of intellectual contribution to Kubrawī theoretical articulation.

52 Two letters are published in: Riyāḥī, Muḥammad Amīn, "Matn-i Nāma-yi Majd al-Dīn Baġdādī", *Yaġmā*, 80 (Isfand 1333š.): pp. 544-8. The *ijāza* is published in: Rajā'ī, Aḥmad 'Alī, "Mu'arrifī-yi Nusḫa-yi Ḫaṭṭī", *Yaġmā*, 210 (Day 1344š.): pp. 531-3; the book, in Arabic, is published with Persian translation in: Baġdādī, Majd al-Dīn, *Tuḥfat al-Barara fī Masā'il al-'Ašara*, edited by Husayn Haydarḫānī Mushtāq-'Alī, translated by Muḥammad Bāqir Sā'idī Ḫurāsānī (Tehran: Marvī, 1368š.).

53 Ra'nā Ḥusaynī, Kirāmat, "Risāla-yi dar safar ta'līf-i Majd al-Dīn Baġdādī", *Jāvidān-i Ḫirad*, 92 (pā'īz 1356š.): pp. 74-82.

and ideas that animated city life, and the cultural resources of a specific time and place.

In the years since its production the manuscript has witnessed some damage, but has largely fared quite well. The Ottomans, in general, had a great interest in the works of Ṭūsī, and the inclusion of several of his pieces may have contributed to its survival. The preservation of the book may also have been assisted by the inclusion of other pieces, or by the presence of illuminations and artistry, which would have appealed to other audiences in different times, or by the subsequent reputation of Shiraz as a noted centre for book production, illumination, and philosophical speculation. The manuscript contains seals which deserve further detailed study, in part concerning the matter of preservation. In fact the survival of the book raises as many questions as does its creation, pertaining to matters such as the enormous role of the Ottomans in passing on the heritage of the Middle East, the chances of preservation of different types of texts, the role of accident and contingency in conservation, and the meanings and significance that we might be permitted to assign to the texts and their fragments as they continued through time.

This article has therefore sought not just to demonstrate the usefulness of Köprülü 01589 as a source for studying the cultural history of Shiraz, but to show that book history offers a view into the cultural horizons of Shiraz and its educated elites, and can serve similarly for other cities. The research climate of the Middle East has been difficult in the twenty-first century, with threats to cultural heritage resulting from insecurity, and access to libraries challenged. Still, sometimes in response to these challenges, we do have increasing access to manuscripts through digitisation efforts (such as in Istanbul at the Süleymaniye), and greater adoption of open access policies. As digitisation progresses we will be better able to mine individual manuscripts for their rare fragments, but we should view this as an opportunity to also investigate much broader patterns in textual production, and probe the roles that particular cities played within them. The digital medium permits advantageous overviews, as well as the detection and visualisation of relationships which can then be subjected to analysis.

Appendices

How was Köprülü 01589 formed? We considered many possible ways of answering this question. The study of the formation of multi-text compilations generally requires attention to the material evidence, such as bindings and paper: ultimately, we arrived at a method in which we first considered the bundles, and what hypothetical situations these could allow or rule out. Next, given these hypotheses, we evaluated them in light of the handwriting, including aspects of its distribution through the manuscript. Thirdly, we considered the colophons, their dates and locations. The order of consideration and priority can thus be summarised as follows:

- First, the codicological features, including bundles;
- Second, the palaeography, including number of hands;
- Third, the colophons and other evidence for dating.

Taking this approach, we progress from physical evidence to narrative, prioritising what the physical object of the text can tell us.

Appendix 1: Bundles

The identification of the bundles was firstly based on consideration of the spacing between the texts and the patterns of folio lining. For example, an empty folio at 60v marks the end of bundle 1.[54] The lining of the folios within each bundle is characteristically similar, although still with some variation.

Text	Author	Title	Ff	Lines/f	Bundle
1	Abu'l-Qāsim Maḥmūd b. Abi'l-Ḥusayn al-Nīsābūrī	*Ījāz al-bayān fī tafsīr al-Qur'ān*	1v-38r	37	1
2	Abū ʿAbd Allāh Muḥammad b. Aḥmad b. Saʿīd	*Ḫawāṣṣ al-Qur'ān*	38v-60r	35/37	1
3	al-Ḥarīrī	*Maqāmāt al-Ḥarīrī*	61r-100v	37	2
4	Aḥmad b. Abī Sa'd al-Dāwanī al-Miškānī	*ʿIqd al-jawāhir fī ansāb al-akābir*	101v-103v	25	3
5	Saʿdī	*Az dīwān-i šʿir-i fārsī-yi Saʿdī* (verses as filler)	104r	32	3
6	Jāmāsb al-Ḥakīm	*Jāmāsb-nāma*	104v-120r	25	3
7	Naṣīr al-Dīn Ṭūṣī	*Tajrīd al-ʿaqā'id*	121v-126v	35	4
8	Abu'l-Maʿālī Naṣr Allāh	*Tarjuma Kalīla va Dimna*	127v-210v	33	5
9	Kāšānī	*Tatimmat Dīwān Kāšānī* (verses as filler)	210v-212v	33	5
10	Majd al-Dīn Ismā'īl al-Faʿalī	*Kitāb al-Tuḥfa*	213r-232v	37 (also 34, 35, 36, 38)	6

54 Author names and titles are entirely tentative and based on Dānišpažūh, *Fihrist*: I, pp. 479-87, and Şeşen, Akpınar, İzgi, *Catalogue*: II, pp. 254-77.

TABLE (*cont.*)

Text	Author	Title	Ff	Lines/f	Bundle
11	Najm al-Dīn Rāzī	*Risālat al-Ṭayr*	233r-235v	37	6
12	al-Ḥusayn b. al-Manṣūr al-Ḥallāj	*Risālat al-Ṭawāsīn*	235v-236r	n/a[a]	6
13	Qusṭā b. Lūqā	*As'ila Abūzār Jumihr az ustāḍ-i ḥūd az Qusṭā b. Lūqā*	236r-236v	n/a	6
14	n/a	prophetic Hadiths in Arabic and Persian	236v-237r	n/a	
15	n/a	*Risāla fī Ḫuṣūṣiyyat nubuwwat Rasūl Allāh*	237r-240r	37	6
16	Niẓām al-Mulk	*Waṣiyat Niẓām al-Mulk*			
17	Kāšānī	*Az dīwān-i š'ir – fārsī-yi Kamāl al-Dīn Kāšānī* (verses as filler)	240v-242v		6
18	Naṣīr al-Dīn Ṭūṣī	*Aḫlāq-i Nāṣirī*	243r-311r	25	7
19	Ibn Sīnā	*Risālat al-malā'ika*	311v-312r	25	7
20	n/a	*Tarjuma Waṣāyā-yi al-ḏahabiyya li Fayṯāġūrus*	312v	n/a	7
21	Ibn Sīnā	4 folios inserted to finish out a marginal text begun from 302v: *Risālat al-Ḥukūma fi'l-ḥujaj al-muṯbitīn li'l-māḍī mabda' zamanīyan wa bayān taḥlīli-hā*	312/1-312/4	30-32	7
22		verses as filler	313r		7
23	Abū Bakr al-Muṭahhir b. Muḥammad b. Abī al-Qāsim al-Jamālī al-Yazdī	*Farruḫ-nāma-yi Jamālī*	313v-370r	25	7
24	Naṣīr al-Dīn Ṭūṣī	*Muḫtaṣar fī ma'rifa al-taqwīm*	372r-375r	37	8
25	n/a	*al-Manṭiq fī waḍ' siyāqa*	375v-382v	unusual layout	8

TABLE (*cont.*)

Text	Author	Title	Ff	Lines/f	Bundle
26	Abu'l-Ḥasan ʿAlī b. Muḥammad al-Daylamī (Arabic original); Rukn al-Dīn Yaḥyā ibn Junayd (translator into Persian)	*Sīrat al-Šayḫ al-Kabīr Abū ʿAbd Allāh ibn al-Ḫafīf al-Šīrāzī*	383r-410v	35 (and 33, 34)	9
27	Saʿdī	verses as filler	410v-411v	39, 36	9
28	Majd al-Dīn al-Baġdādī	*Risāla fi'l-taṣawwuf* (five lines, top of the page – not sure where the rest of the text is)	412r	37	9
29	Majd al-Dīn al-Baġdādī	*Risāla uḫrā*	412r-413r		
30	Majd al-Dīn al-Baġdādī	*As'ila va ajwiba bā Imām Faḫr al-Rāzī*	413r-413v	n/a	9
31	Šaraf al-Dīn al-Balḫī	*Risālat Šaraf al-Dīn al-Balḫī ilā Majd al-Dīn al-Baġdādī*	414r-414v	37	9
32	Majd al-Dīn al-Baġdādī	*Risālat al-jawāb*	414v-416r	37	9
33	Majd al-Dīn al-Baġdādī	*Risālat al-safar*	416v-417v	37	9
34	Majd al-Dīn al-Baġdādī	*Risālat Majd al-Dīn al-Baġdādī dar jawāb Šaraf al-Dīn*	417v	n/a	9
35	Majd al-Dīn al-Baġdādī	*Risālat al-sayr wa'l-ṭayr*	418r-418v	n/a	9
36	Majd al-Dīn al-Baġdādī	*Risāla fi'l-ʿuzla wa'l-ḫalwa*	418v-419r	n/a	9
37	Majd al-Dīn al-Baġdādī (Probably)	*Waṣiyat al-Šayḫ al-Sālik li-l-murīd al-Mustaʿīd al-Nāsik*	419r-419v		
38	Ḥāfiẓ[b]	*Dīwān Ḥāfiẓ*	419v	38	9
39	Majd al-Dīn al-Baġdādī	*Risāla fi'l-taṣawwuf*	420r-420v	n/a	9
40	n/a	*Risāla fi'l-nujūm*	420v-422r	n/a	9
41	Šihāb al-Dīn al-Suhrawardī	*Risālat mu'nis al-ʿuššāq*	422r-423v	35	9
42	n/a	*al-Su'ālāt wa'l-jawābāt*	424r	35	9
43	Kāšānī	*Tatimmat Dīwān Kamāl al-Dīn Kāšānī*	424r-425r	33	9
44	Abu'l-Faraj b. Saʿd know as Ibn al-Ḫallāj	*Šifā' al-ṣudūr wa rāḥat al-masrūr*	425v		9

TABLE (*cont.*)

Text	Author	Title	Ff	Lines/f	Bundle
45	n/a	*Risāla fi'l-kalām al-šiʿī*	426r-428v		9
46	al-Ḥasan b. Abi'l-Ḥasan al-Baṣrī	*Risālat al-Ḥasan b. Abi'l-Ḥasan al-Baṣrī ilā ʿAbd al-Malik b. Marwān*	428v-431r		9
47	Kāšānī	*Tatimmat dīwān Kamāl al-Dīn Kāšānī*	431v-432r		9

a For pages with titles, we do not count lines: all "n/a" in the table refer to such situations.
b The Köprülü catalogue attributes the verses to Kāšānī.

Appendix 2: Palaeography

Scribe (S)	Location in ms (dates from colophons in parentheses)	Bundle	Place
S1	1v-38r (20.01.754),	1	Shiraz
	1mv-25mv[a]	1	
S2	26mr-48mv (15.05.756),	1	Baghdad
	38v-60r (7.754)	1	
FP1[b]	49mr-52mr,	1	
	104r, 104mr	2	
	312/1mr-312/4mv	7	
	313r, 312mr	7	
	410v-411v	9	
	419mr-423mv		
S3	61r-100v	2	
	61mr-74mv (02.754)	2	
S4	74mv-99mr	2	
S5	101v-103v,	3	
	104v-120r,	3	
	104mv-118mv	3	
	313v-370v (754)	7	
S6	121v-126v (21.01.754)	4	

TABLE (*cont.*)

Scribe (S)	Location in ms (dates from colophons in parentheses)	Bundle	Place
S7	121mv-126mv,	4	
	127mv-167mr	4	
S8	127v-210v (05.754)	5	
S9	167mr- 174mr (10.03.756),	5	Dezful
	177mr-211mv	5	
	343mv-367mr (20.03.756)	8	
	367mr-371mv	8	
FP2	210v-212v,	5	
	212rm-212vm	5	
	240v-242v	6	
	424r-425r	9	
	424mr-424mv	9	
	425mr-431mv (04.02.811)	9	
	431v-432r	9	
		9	
S10	213r-232v	6	
S11	213mr-242mv	6	
	432mr-433mv	9	
S12	233r-240r	6	
	412r-419v	9	
	420r-423v	9	
	412mr-419mr	9	
S13	243r-311r (2.11.753)	7	
	243mr-293mv (06.12.753)/ (22.12.753	7	
S14	293mv-298mv	7	
S15	298mv-302mv	7	
	312/1r-312/4v (07.07.754)	-	
S16	303mr-311mr	7	
S17	312mv-329mv	7	
S18	313mv-329mv	7	
S19	330mr-343mv	7	
S20	372r-375r	8	
S21	372mr-375mr	8	

TABLE (*cont.*)

Scribe (S)	Location in ms (dates from colophons in parentheses)	Bundle	Place
S22	375mv-382mv	9	
S23	383mr-385mr	9	
S24	375v-382v	9	
S25	383r-410v (15.02.754)	9	Isfahan
S26	386mv-403mr (764)	9	
S27	403mv-411mv	9	
S28	420v-431v	9	

a mv/r = margin verso/recto.

b In the main part and on the margin of the manuscript, the scribes used poetry pieces to fill the blank pages and spaces caused by miscalculation of the space needed for the different pieces. These poetry pieces, of which there are two different sets, continue throughout the manuscript: we refer to these as Filling Poetry 1 and 2 (FP1, FP2).

Appendix 3: Colophons

date AH	date CE	Text	Place	Page	Main/ margin
20 Muḥarram 754	25/02/1353	*Īǧāz al-bayān fī l-tafsīr al-Qur'ān*	Shiraz	38r	main
Rajab 754	08/1353	*Ḫawāṣṣ al-Qur'ān*	n/a	60r	main
15[a] Jūmādā al-Awwal 754	18/06/1353	*Maqāmāt al-Ḥarīrī*	n/a	100v	main
n/a	n/a	*Jāmāsb-nāma*	n/a	120r	main
21 Muḥarram 754	26/02/1353	*Tajrīd al-ʿaqā'id*	n/a	126v	main
15[b] Jūmādā al-Awwal 754	19/06/1353	*Tarjuma Kalīla va Dimna*	n/a	210v	main
2 Ḏu'l-qaʿda 753	10/12/1352	*Aḫlāq-i Nāṣirī*	n/a	311r	main
7 Rajab 754	08/08/1353	*Risālat al-Ḥukūma fī'l-ḥujaj al-muṯbitīn li'l-māḍī mabda' zamanīyan wa bayān taḥlīli-hā*		312/4v	main
754	1353	*Farruḫ-nāma-yi Jamālī*	n/a	370r	main

TABLE (*cont.*)

date AH	date CE	Text	Place	Page	Main/ margin
15 Ṣafar 754	22/03/1353	*Sīrat al-Šayḫ al-Kabīr Abū ʿAbd Allāh ibn al-Ḫafīf al-Šīrāzī*	Isfahan	410v	main
4 Ṣafar 811	29/06/1408	*Tatimmat dīwān Kamāl al-Dīn Kāšānī*	n/a	432r	main
15[c] Jūmādā al-Awwal 756	28/05/1355	*Mašāriq al-anwār al-nabawiyya*	Baghdad	48v	margin
Ṣafar 754	03/1353	*Aṭwāq al-ḏahab*	n/a	74v	margin
10 Rabīʿ al-awwal 756	25/03/1355	*Risāla Qalamiyya*	Dezful	174r	margin
6 Ḏu'l-Ḥijja 753	13/01/1353	*Ḫilāfat-nāma-yi Ilāhī*	n/a	286r	margin
23 Ḏu'l-Ḥijja 753	30/01/1353	*ʿAhd al-Šayḫ Abī ʿAlī Ibn Sīnā*	n/a	286v	margin
20 Rabīʿ al-Awwal 756	04/04/1355	*al-Ḫawāṣṣ*	n/a	367r	margin
764	1362-1363	*Ḫawāṣṣ al-Aškāl*	n/a	403r	margin

a *Awāsiṭ*.
b *Muntaṣaf*.
c *Muntaṣaf*.

References

Afšār, Īraj, "Muqaddama (Introduction)", in *Zangī-Nāma* (Tehran: Intišārat Tūs, 1993).

Agostini, Domenico, *Ayādgār ī Jāmāspīg: Un texte eschatologique zoroastrien* (Rome: Gregorian & Biblical Press, 2013).

Agostini, Domenico, "Rediscovering the *Jāmāspi*: A Walk in Four Steps", *IrSt*, XLV/2 (2012): pp.169-80.

Aigle, Denise, *Le Fārs sous la domination mongole: politique et fiscalité* (Paris: Association pour l'avancement des Études Iraniennes, 2005).

Arberry, Arthur J., *Shiraz: Persian City of Saints and Poets* (Norman: University of Oklahoma Press, 1960).

Baġdādī, Majd al-Dīn, *Tuḥfat al-Barara fī Masā'il al-ʿAšara*, edited by Husayn Haydarḫānī Mushtāq-ʿAlī, translated by Muḥammad Bāqir Sāʿidī Ḫurāsānī (Tehran: Marvī, 1368š.).

Ben Azzouna, Nourane and Roger-Puyo, Patricia, "The Question of the Formation of Manuscript Production Workshops in Iran According to Rashīd al-Dīn Faḍl Allah al-Hamadhānī's *Majmūʿa Rashīdiyya* in the Bibliothèque nationale de France", *Journal of Islamic Manuscripts*, 7(2016): pp. 152-94.

Blochet, E., *Catalogue des manuscrits persans de la Bibliothèque nationale* (Paris: E. Leroux, 1905-1934).

Blochet, E., "Études sue le gnosticisme musulman", *Rivista degli studi orientali*, IV, Fasc. 2 (1911): pp. 267-300.

Borrut, Antoine, "Court Astrologers and Historical Writing in Early Abbasid Baghdad: An Appraisal", in J. Scheiner and D. Janos (eds), *The Place to Go: Contexts of Learning in Baghdad, 750-1000 CE* (Princeton: The Darwin Press, 2014): pp. 455-501.

Bosworth, Clifford Edmund, *The New Islamic Dynasties: A Chronological and Genealogical Manual*, Rev. ed. (New York: Columbia University Press, 1998).

Boyle, J.A., "Īndjū" in *Encyclopaedia of Islam*, 2nd edition, edited by P. Bearman, Th. Bianquis, C.E. Bosworth, E. van Donzel, W.P. Heinrichs (Leiden: E.J. Brill, 1986): III, p. 1208.

Chartier, Roger, *The Author's Hand and the Printer's Mind*, translated by Lydia G. Cochrane (Cambridge: Polity Press, 2014).

Daiber, H. "al-Ṭūsī, Nasīr al-Dīn" in *Encyclopaedia of Islam*, 2nd edition, edited by P. Bearman, Th. Bianquis, C.E. Bosworth, E. van Donzel, W.P. Heinrichs (Leiden: Brill, 2000): X, pp. 746-50.

Daftary, Farhad, *A History of Shiʿi Islam* (London/New York: I.B. Tauris/Institute of Ismaili Studies, 2013).

Dānišpažūh, Muḥammad Taqī, *Fihrist-i Mīkrufīlm-hā-yi Dānišgāh-i Tihrān* (Tehran: Intišārāt-i Dānišgāh-i Tihrān, 1969)

De Smet, Daniel, "Introduction", translated by Orkhan Mir-Kasimov in Daftary, Farhad and Miskinzoda, Gurdofarid, *The Study of Shiʿi Islam: History, Theology and Law* (London/New York: I.B. Tauris/Institute of Ismaili Studies, 2014): pp. 545-62.

Derrida, Jacques, *Archive Fever: A Freudian Impression*, translated by Eric Prenowitz (Chicago/London: University of Chicago Press, 1996).

Dirāyatī, Muṣṭafā (ed.), *Fihristigān: nusḫa-hā-yi ḫaṭṭī-yi Īrān* (*FANḪĀ*) (Tehran: Sāzmān-i Asnād va Kitābḫāna-yi Millī Jumhūrī-yi Islāmī Īrān, 2012)

Foucault, Michel, "What Is an Author?" in *Language, Counter-Memory, Practice: Selected Essays and Interviews*, edited by Donald F. Bouchard, translated by Donald F. Bouchard and Sherry Simon (Ithaca, NY: Cornell University Press, 1977): pp. 113-38.

Gharaee Garakani, Morteza and Negahban, Farzin, "Bābā Afḍal", in *Encyclopaedia Islamica*, Editors-in-Chief: Madelung, Wilferd and Farhad Daftary: IV, pp. 1-8.

Greetham, D.C., "Textual forensics", PMLA, CXI, no. 1, Special Topic: The Status of Evidence (Jan., 1996): pp. 32-51.

Gutas, Dimitri, "The Study of Avicenna. Status Quaestionis atque Agenda" in Gutas, Dimitri, *Orientations of Avicenna's Philosophy: Essays on his Life, Method, Heritage*,

(Farnham, Surrey: Ashgate, 2014): section XVII, pp. 1, 45-69. Originally published in *Documenti e studi sulla tradizione filosofica medieval*, XXI (2010): pp. 45-69.

Ḥaḍratī, Ṣādiq, *Fihrist-i nusḫa-hā-yi ḫaṭṭī-yi kitābḫāna-yi Ustād Ibrāhīm Dihgān* (Qom: Majmaʿ Dhaḫā'ir Islāmī, 2009).

Hamadānī, Ḫʷāja Yūsuf, and Rāzī, Najm al-Dīn, *Rutbat al-ḥayāṭ bi ḍamīma-yi risalat al-ṭuyūr*, edited by Muḥammad Amīn Riyāḥī (Tehran: Intišārat Tūs, 1983).

Hirschler, Konrad, *Medieval Damascus: Plurality and Diversity in an Arabic Library; The Ashrafīya Library Catalogue* (Edinburgh: Edinburgh University Press, 2016).

Ibn Baṭṭūṭa, *Ibn Battúta: Travels in Asia and Africa (1325-1354)*, translated by H.A.R. Gibb (London: Routledge & Kegan Paul, 1929; reprint 1953).

Jiddī, Muḥammad Jawād, "Nigāhī bi ḫuṭūṭ-i muhr-hā va sikka-hāyi ḍarb-i Šīrāz" in *Majmūʿa maqālāt-i ḫušniwīsī-yi maktab-i Šīrāz* (Tehran: Mu'assisa-yi Ta'līf, Našr va Tarjuma-yi Āṯār Hunarī-yi Matn, 2001): pp. 9-36.

Kennedy, E.S. and Pingree, David, *The Astrological History of Māshā'Allāh* (Cambridge: Harvard University Press, 1971).

Limbert, John, "Inju Dynasty", *EIr*, consulted online Feb. 6, 2018 <http://www.iranicaonline.org/articles/inju-dynasty>.

McKenzie, D.F., *Bibliography and the Sociology of Texts* (Cambridge: Cambridge University Press, 1999).

Modi, J.J. (ed. and trans.), *Jāmāspi, Pahlavi, Pāzend, and Persian Texts* (Bombay: Bombay Education Society Press, 1903).

Al-Nīsābūrī, Abū'l-Qāsim Najm al-Dīn Maḥmūd, *Ījāz al-bayān fī tafsīr al-Qur'ān*, edited by Ḥanīf b. Ḥasan al-Qāsimī (Beirut: Dār al-Gharb al-Islāmī, 1995).

Rajā'ī, Aḥmad ʿAlī, "Muʿarrifī-yi Nusḫa-yi Ḫaṭṭī", *Yaġmā*, 210 (Day 1344š.): pp. 531-3.

Raʿnā Ḥusaynī, Kirāmat, "Risāla-yi dar Safar ta'līf-i Majd al-Dīn Baġdādī", *Jāvidān-i Ḫirad*, 92 (pā'īz 1356š.): pp. 74-82.

Richard, Francis, *Catalogue des manuscrits persans, le Supplément persan* (Rome: Bibliothèque Nationale de France): I, pp. 517-18.

Riyāḥī, Muḥammad Amīn, "Matn-i Nāma-yi Majd al-Dīn Baġdādī", *Yaġmā*, 80 (Isfand 1333š.): pp. 544-8.

Riyāḥī, Muḥammad Amīn, "Muqaddama-yi Muṣaḥḥiḥ" in Hamadānī, Ḫʷāja Yūsuf, and Rāzī, Najm al-Dīn, *Rutbat al-ḥayāṭ bi ḍamīma-yi risalat al-ṭuyūr*, edited by Muḥammad Amīn Riyāḥī (Tehran: Intišārat Tūs, 1983): pp. 84-7.

Roemer, H.R., "The Jalayirids, Muzaffarids, and Sarbadārs" in Jackson, Peter and Lockhart, Laurence, *The Cambridge History of Iran* (Cambridge: Cambridge University Press, 1986): VI, pp. 1-41.

Rustamji, Manockji (ed.), *Dârâb Hormazyâr's Rivâyat* (Bombay: British India Press, 1922).

Ṣafarī Aq Qalʿa, ʿAlī, "Yādkard-i Tārīḫ-i Bayhaqī dar ʿiqd al-jawāhir fī ansāb al-akābir", *Guzāriš-i mīrāṯ*, no. 58-59 (2013): pp. 54-5.

Saleh, Walid, "Al-Bayḍāwī", *Encyclopaedia of Islam*, 3rd edition, Consulted online on 10 March 2018 <http://dx.doi.org/10.1163/1573-3912_ei3_COM_25267>.

Al-Samʿānī, Abū Saʿd ʿAbd al-Karīm b. Muḥammad b. Manṣūr al-Tamīmī, *al-Ansāb* (Hyderabad: Majlis Dā'irat al-Maʿārif al-Uṯmāniyya, 1977), VIII, p. 221.

Schimmel-Tari, A., "Abū ʿAbd Allāh b. Ḫafīf Šīrāzī" in *Sīrat al-Šayḫ al-Kabīr Abū ʿAbd Allāh ibn al-Ḫafīf al-Šīrazī* (Tehran: Intišārāt Bābak, 1984): pp. 5-46.

Şeşen, Ramazan; Akpınar, Cemil and İzgi, Cevat, *Catalogue of manuscripts in the Köprülü Library* (Istanbul: Research Centre for Islamic History, Art, and Culture, 1986).

Seyed-Gohrab, A.A. and McGlinn, S. eds., *The Treasury of Tabriz: The Great Il-Khanid Compendium* (Amsterdam, West Lafayette, Indiana: Rozenburg Publishers and Purdue University Press, 2007).

Sezgin, Fuat, *Geschichte des arabischen Schrifttums*, VII: *Astrologie, Meteorologie und Verwandtes bis ca. 430 H.* (Leiden: E.J. Brill, 1979).

al-Šīrāzī, Rukn al-Dīn Yaḥyā ibn Junayd, *Sīrat al-Šayḫ al-Kabīr Abū ʿAbd Allāh ibn al-Ḫafīf al-Šīrāzī*, edited by A. Schimmel-Tari (Ankara: Ankara Üniversitesi İlâhiyat Fakültesi, 1955).

al-Šīrāzī, Rukn al-Dīn Yaḥyā ibn Junayd, *Sīrat al-Šayḫ al-Kabīr Abū ʿAbd Allāh ibn al-Ḫafīf al-Šīrāzī*, edited by A. Schimmel-Tari and translated by Tawfīq H. Subḥānī (Tehran: Intišārāt-i Bābak, 1984).

Sobieroj, Florian, "Ibn Khafīf", in Encyclopaedia of Islam, 3rd edition, edited by Kate Fleet, Gudrun Kramer, Denis Matringe, John Nawas, Everett Rowson, c Consulted online on 03 February 2018 <http://dx.doi.org/10.1163/1573-3912_ei3_COM_30858>.

Tabrīzī, Abū'l-Majd, *Safīnah-yi Tabrīz* (Tehran: Markaz-i Našr-i Dānišgāhī, 2003).

Ṭūsī, Nāṣir al-Dīn, *Āġāz wa anjām*, edited by Īraj Afšār (Tehran: Intišārāt-i Danišgāh Tihrān, 1956-7).

Ṭūsī, Nāṣir al-Dīn, *Aḫlāq-i Nāṣirī*, ed. Mujtabā Mīnuvī and ʿAlī Riḍā Ḥaydarī (Tehran: Ḫwārazmī, 1978).

Ṭūsī, Nāṣir al-Dīn, *Tajrīd al-Iʿtiqād*, edited by Muḥammad Jawād Ḥusaynī Jalālī (Qom: Maktab al-Iʿlām al-Islāmī, n.d.).

Ṭūsī, Nāṣir al-Dīn, Contemplation and Action: *The Spiritual Autobiography of a Shi'i Philosopher*, Edited by S. J. Badakhchani (New York/London: I.B.Tauris in association with the Institute of Ismaili Studies, 1998).

Uluç, Lâle, "Selling to the Court: Late-Sixteenth-Century Manuscript Production in Shiraz" *Muqarnas*, XVII (2000): pp. 73-96.

Wickens, G.M. "Introduction" in *The Nasirean Ethics by Naṣīr al-Dīn Ṭūsī* (London: George Allen & Unwin, 1964).

Wright, Elaine, *The Look of the Book: Manuscript Production in Shīrāz, 1303-1452* (Washington, D.C.: Freer Gallery of Art, Smithsonian Institution in association with University of Washington Press, 2013).

Wright, Elaine, "The Calligraphers of Šīrāz and the Development of the Nastaʿlīq script", *Manuscripta Orientalia*, IX/3 (September 2003): pp. 16-26.

Wright, Elaine, "Patronage of the Arts of the Book under the Injuids of Shiraz" in Komaroff, Linda, *Beyond the Legacy of Genghis Khan* (Leiden/Boston: Brill, 2006): pp. 248-68.

Zaryāb Ḫū'ī, ʿAbbās, "Dāstān-i kušta šudan-i Majd al-Dīn Baġdādī manqūl az yik nusḫa-yi ḫaṭṭī", *Yaġmā*, 130 (Urdībihišt 1338š.): pp. 88-94

Biographical Notes

Sarah Bowen Savant is an Associate Professor at the Aga Khan University-ISMC in London. She is the author of *The New Muslims of Post-Conquest Iran: Tradition, Memory, and Conversion* (Cambridge: Cambridge University Press, 2013). Her other publications include *The Excellence of the Arabs: A Translation of Ibn Qutaybah's* Faḍl al-ʿArab wa l-tanbīh ʿalā ʿulūmihā (with Peter Webb; The Library of Arabic Literature; Abu Dhabi: New York University Press, 2016), as well as articles and edited volumes dealing with ethnic identity, cultural memory, genealogy, and history writing. Her current book project focuses on the history of books in the Middle East. With a team, she is developing digital methods to study the origins and development of the Arabic and Persian textual traditions (kitab-project.org).

Majid Montazer Mahdi is a PhD student at the University of Exeter. His current research is on biographical dictionaries of Safavid Ulama written in the 16th and 17th centuries. He contributes to the KITAB project by editing the Arabic and Persian manuscripts. His research interests go beyond these two projects. He also is involved in a research project dealing with Muslim bio-medical ethics at the Institute of Ismaili Studies.

Medieval Lexicography on Arabic and Persian Terms for City and Countryside

Roy P. Mottahedeh
Harvard University
mottahed@fas.harvard.edu

Abstract

Medieval Arabic to Persian dictionaries are a relatively untapped source for the conceptual world in the time of their authors. This essay closely examines four such dictionaries from the late fifth/eleventh century to the seventh/thirteenth century written in eastern Iran. These dictionaries are quite rich in terminology for cities, towns, farmland, pasture and desert. They also describe architectural features of buildings. They offer scant but valuable information on markets and social structure. The information from these dictionaries combined with the rich detail available in the Islamic geographers of the third/ninth and fourth/tenth century allows us to form a more perfect picture of medieval Iranian society.

Keywords

city – countryside – suburb – town – market

The Arabic-Persian dictionaries of the period from the tenth to the thirteenth century CE offer a rich and relatively unused source for the cultural history of that period. For this essay, I consider the terms for city and countryside and the characteristic elements of each of these divisions as found in Arabic-Persian dictionaries supplemented by lexicographical evidence found elsewhere.[1]

---

1 Baevskij, Solomon I., *Persian Lexicography: Farhangs of the Eleventh to the Fifteenth Centuries* (Folkestone, U.K.: Global Oriental, 2007), translated by N. Killian, revised and updated by John R. Perry, offers a good survey of the pre-modern Persian dictionaries. However, the author omits several of the Arabic to Persian dictionaries which I discuss here, including Karamīnī, Naṭanzī, and Zamaḫšarī. A very useful list of such dictionaries can be found in Charles A. Storey, *Persian Literature: A Bio-bibliographical Survey*, III/1 (Leiden: Brill, 1984): pp. 78-110. For an important Persian discussion of such material, see Monzavī, ʿAlī -Naqī, *Farhangnāma-hā-yi ʿArabī ba-Fārsī* (Tehran: Tehran University Press, 1959).

 | DOI:10.1163/9789004434332_017

One of the early Arabic to Persian dictionaries of this nature is the *Takmilat al-Aṣnāf* by ʿAlī b. Muḥammad Adīb al-Karamīnī (also vocalized Karmaynī), dated to the eleventh or twelfth century CE.[2] Another lexicographer, securely dated to the late eleventh century is Ḥusayn b. Ibrāhīm Naṭanzī (d. 497/1103), whose *Mirqāt* is one of the earliest extant Arabic to Persian dictionaries.[3] In the same Saljuq era, in 474/1082, Yaʿqūb Kurdī Nīšāpūrī wrote *Kitāb al-Bulġa*, which like many early dictionaries is organized by topic.[4] In this period the very prominent philologist Zamaḫšarī (d. 538/1144) wrote the *Pīšraw-i Adab*, or in Arabic *Muqaddima* (or *Muqaddama*) *al-Adab*.[5] Additional clarifying materials can be found in the earliest extant Persian dictionary by Asadī-Ṭūsī (d. 465/1072-3).[6]

The outstanding geographical literature in Arabic and Persian from the third/ninth century onward contains many references to the terms discussed in this essay. This material is so large that it would require a separate book to evaluate the use of various geographical words. The limited geographical analogs I offer here are derived in part from the magisterial four-volume work on Islamic geography by André Miquel.[7]

The selection of dictionaries I have used are mostly of the middle to late Saljuq era. This closeness in period gives them some coherence in their focus. With the exception of Naṭanzī, these lexicographers come from eastern Iran

2 Perry, John, s.v. "Lexicography: ii Alphabetical dictionaries", in *EIr online*, www.iranicaonline.org/articles/lexicography viewed 9/27/2017, dates Karamīnī to the eleventh century; ʿAlī Navīdī Malāṭi, "Āgāhī-hā-yi tāza dar bāra-yi Karamīnī", *Farhang-nivīsī*, XCVIII (1393š): pp. 69-76, dates this author to the sixth/twelfth century.

3 Naṭanzī, *al-Mirqāt*, ed. Jaʿfar Sajjādī (Tehran: Bunyād-i Farhang, 1346š): p. "w" for death date. This dictionary is organized by subject matter, such as "food", "clothing", "categories of people" and "discussion of professions", etc.

4 Nīšāpūrī, *Kitāb al-Bulġa*, ed. Mujtabā Mīnuvī and Fīrūz Ḥarīrčī (Tehran: Bunyād-i Farhang-i Īrān, 2535). Some of the topics are "artisans and craftsmen" and "springs and canals". In contrast, Karamīnī's *Takmilat* is primarily alphabetical. Both the topical and alphabetical schemes are used in Arabic to Arabic dictionaries.

5 Zamaḫšarī, *Pīšraw-i Adab*, ed. Muḥammad Kāẓim Imām (Tehran: Intišārāt-i Dānišgāh-i Tihrān, 1342š/1963).

6 Asadī -Ṭūsī, *Luġat-i Furs*, ed. Fatḥ Allāh Mojtabāʾī (Tehran: Ḫwārazmī, 1365š). This essay does not consider dictionaries of the fourteenth and fifteenth century in any detail.

7 Miquel, André, *La géographie humaine du monde musulman* (Paris: École des Hautes Études en Sciences Sociales, 1973-1988). I have been unable to consult Miri, Negin, *Administrative and historical geography of Fars province during the Sasanian and early Islamic period (3rd-12th century)* (PhD thesis, University of Sydney, 2007), which deals with administrative divisions of Fars province.

where New Persian first appeared. The Saljuq empire was one of the last political entities to have real presence in both the Arab speaking and Persian speaking worlds. Arabic words entered Persian rapidly, although their survival was not assured. The highly Arabicized style of Naṣīr al-Dīn Ṭūsī (d. 672/1274), who wrote toward the end of this period, is barely comprehensible to a contemporary Persian speaker. Similarly, some Persianized versions of Arabic words, like *mizgīt* for mosque, are no longer used in modern Persian. When reading these dictionaries, it is important to understand that Persian explanations of Arabic terms are offered as well as exact equivalents. Considerations as to how much these dictionaries copy each other are not discussed here.

The greatest modern Persian dictionary, the fourteen-volume *Luġat-Nāma* compiled under the supervision of ʿAlī-Akbar Dihḫudā, is of great assistance in interpreting these early dictionaries, in large part because of its many citations from early poetry and prose.[8] Some of the pre-modern purely Arabic dictionaries, such as the *Lisān al-ʿArab*, written by Ibn Manẓūr (d. 711/1311-2), are also cited for their enlargement on the meanings of words, often tersely defined in the Arabic to Persian dictionaries.[9] There are very few extant Persian to Persian dictionaries from the Saljuq period, rendering these Arabic to Persian dictionaries all the more valuable.

Within any language there are words for reading the land that are very strongly defined by culture, period and location – English is no exception. "Landscape" is originally an old Frisian term for "shoveled land" (particularly between land and sea). From this it has evolved to mean anything from a "view from a particular place" to a "category of painting". It has inspired "seascape" and "cityscape" and continues to evolve. This essay surveys four centuries of reading the land, both the city and the countryside, from the tenth to the thirteenth century CE. However, due to the paucity of secondary literature on these dictionaries, I can only cursorily describe the terms used. Also of enormous interest, but beyond the purview of this essay, are the different attitudes toward these terms, as demonstrated by the European 'Romantic' attachment to landscape that emerged in the later years of the eighteenth century.

8 ʿAlī-Akbar Dihḫudā, *Luġat-Nāma*, 14 vols. (Tehran: Tehran University Publications, 1372š.). The numbering of pages in this dictionary is continuous throughout the volumes and therefore only page numbers will be given.

9 Ibn Manẓūr, *Lisān al-ʿArab* (Beirut: Dār Ṣādir, 1410/1990).

1 The City

The basic Arabic word for city, *al-madīna*, is rendered by the basic Persian word, *šahr*.[10] However, *balad* (town or district) is also translated as *šahr*.[11] Surprisingly, *baladī* (belonging to the city) is translated as *hamšahrī* (fellow city dweller).[12] Why the lexicographers offer the Persian word *hamšahrī*, which is narrower in meaning than the Arabic *baladī*, is unclear to me. *Kūra*, usually understood as "province", is translated as *šahr-i buzurg* or "large city", in which the earlier larger sense of *šahr* as an area, as in Iranshahr, is preserved. However, in contrast, *kūra* is understood by Zamaḫšarī as "little city, city and *šahristān*".[13] In Naṭanzī *miṣr, madīna*, and *fusṭāṭ* are translated as "large city" or *šahr-i buzurg*. *Fusṭāṭ* is also translated by Naṭanzī as *šahr-i jāmiʿ*, a city with a Friday mosque (or city where it is appropriate to congregate on Friday).[14]

2 Streets

One of the various words for a smaller street is the Arabic *darb*, translated as *darvāza* or "gate" in Persian, an understanding supported by other texts, (as is the translation *darband* meaning "barrier" or "bar of a door").[15] *Šāriʿ* and *šāriʿa*, the basic modern Arabic words for street, are rendered in Persian as *šāhrāh* or "principal thoroughfare", while for *fūha al-ṭarīq* (literally "mouth of the road") we have the Persian *sar-i rāh* ("roadside opening").[16] *Zuqāq*, "alley", is appropriately translated as *kūča* (or in the alternate spelling *kūža*).[17] According to Nīšāpūrī, *al-bayḍa wal-qaṣaba*, "the heart and township" is "the center of

10 E.g. Naṭanzī, *Mirqāt*: p. 153.

11 Yet in Zamaḫšarī, *Pīšraw-i Adab*: p. 111, *balda* is translated "small city".

12 See Dihḫudā, *Luġat-Nāma*: p. 20820 on *hamšahrī*, which is only understood as "fellow citizen". See also Nīšāpūrī, *Bulġa*: p. 61.

13 Zamaḫšarī, *Pīšraw-i Adab*: p. 111, who adds: "large road". Yāqūt, *Muʿjam al-Buldān*, 6 vols., ed. de Goeje (Leiden: Brill, 1866-9): I, p. 39 presents the classical geographers' understanding that *kūra* is the subdivision of a province consisting of villages with either a city or a *qaṣaba*. Ibn al-Faqīh *K. al-Buldān*, ed. de Goeje (Leiden: Brill, 1885), p. 105) explains *kūra* as an *iqlīm* (clime), which over time gains currency as a word for "province"; see Miquel, *Géographie*: IV, p. 126. Although this essay hints at the analogs to the words discussed in the early Islamic geographical literature, a thorough treatment of this literature is far beyond the scope of this essay.

14 Naṭanzī, *Mirqāt*: p. 153.

15 Ibid.: p. 154.

16 Nīšāpūrī, *Bulġa*: p. 282; Zamaḫšarī, *Pīšraw-i Adab*: p. 57.

17 Nīšāpūrī, *Bulġa*: p. 318.

the city", *miyān-i šahr*.[18] *Sikka*, meaning "lane" or "road", is translated *kūy* by Zamaḫšarī.[19]

3 Suburbs

The suburbs of the city, *rabaḍ* (an Arabic word which Persian adopted early) is translated idiosyncratically by Karamīnī as *dīvār-i pīrāsta*, "a wall embellished" [by outside gardens?].[20] Naṭanzī explains it more understandably as *girdā-gird-i šahr*, "the surroundings of the city", which Zamaḫšarī repeats and adds "a wall surrounding the city".[21]

The importance of the *rabaḍ* is shown by the great Central Asianist, V.V. Barthold, who states that the cities of the Samanid realm in greater Khurasan consisted of a citadel, a *šahristān* or city proper and the suburb or *rabaḍ*.[22]

Al-Muqaddasī (writing ca. 375/985) says of Kabul, "it has a prosperous *rabaḍ* in which merchants gather as well as a well-fortified citadel (*quhandiz*)."[23] In addition, speaking of Bust, in the province of Sīstān next to contemporary Lashkar-Gah in Afghanistan, he says that the city has a *rabaḍ* in which there are markets (*aswāq*), a theme echoed in descriptions by many of the contemporary and later geographers.[24] Ibn Ḥawqal describes proto-Tashkent as a city (*madīna*) outside which is a citadel (*quhandiz*), and adds that the wall encompasses both citadel and city. It also has a suburb (*rabaḍ*) with its own wall, outside which is another suburb with gardens and dwellings surrounded by yet another wall.[25] The pattern of extension of the city from one suburb to

18 Ibid.: p. 317.

19 Zamaḫšarī, *Pīšraw-i Adab*: p. 118.

20 Karamīnī, ʿAlī b. Muḥammad al-Adīb, *Takmilat al-Aṣnāf*, 2 vols., ed. ʿAlī Ravāqī (Tehran: Anjumān-i Āṣār-u Mafāḫir-i Farhangī, 1385š/2006): p. 243.

21 Naṭanzī, *Mirqāt*: p. 153; Zamaḫšarī, *Pīšraw-i Adab*: p. 112.

22 Barthold, V.V., *An Historical Geography of Iran*, translated by Svat Soucek (Princeton: Princeton University Press, 1984): pp. 13-14, 97. Barthold, who had an outstanding mastery of the primary sources, says of Balkh that it is divided into three elements "like all large towns", and adds that the Persian term for *rabaḍ*, *bīrūn*, is not used in the historians and geographers: see Barthold, V.V., *Turkestan down to the Mongol Invasion* (London: Luzac, 1968 [E.J.W. Gibb Memorial Series]) p. 78. See also Miquel, *Géographie*: IV, p. 218 on this theme.

23 Muqaddasī, *Aḥsan al-Taqāsim fī Maʿrifat al-Aqālīm*, ed. de Goeje (Leiden: Brill, 1906): p. 304.

24 Muqaddasī, *Aḥsan al-Taqāsīm*: p. 304. See also Ibn al-Faqīh, *Al-Buldān*, p. 326 on Samarqand.

25 Ibn Ḥawqal, *Ṣūrat al-Arḍ*, 2 vols., ed. de Goeje (Leiden: Brill, 1873): II, p. 508.

another can be seen elsewhere, as can the presence of agriculture (particularly orchards) within the walls. Ibn al-Faqīh, writing in the late third/ninth century, says the *rabaḍ* of Samarqand extended for thousands of *jarībs* and included fields, gardens and farms.[26] In some cases the suburb and outer wall seem to be identical. A local history tells us that the people of Bukhara asked the Tahirid Muḥammad b. ʿAbd Allāh to build a *rabaḍ*, "so that we can close the gates at night and be safe from thieves and highwaymen. He ordered that a *rabaḍ* be built, which was excellent and strong with towers."[27]

4 Fortress and Palace

The Arabic *quhandiz* (citadel or fortress) is a loan-word from the Persian word for citadel (*kuhan-diz* or "old fort"), encountered during the Arab conquests. *Quhandiz* seems to have been used more in greater Khurasan than elsewhere.[28] Yaʿqūbī says that after the conquest of Samarqand, the Muslims had to capture the *quhandiz* separately.[29] Naršaḫī in his *History of Bukhara*, reports that Qutaiba b. Muslim took the city of Paykand but had a far greater struggle to take the citadel/castle keep (*burj*).[30]

The Persian word *kūšk*, meaning "pavilion" and "palace" in modern Persian, is the translation of a number of different words in medieval Arabic, such as *qaṣr* (fortress), *burj* (tower), and *jawsaq* (an Arabization of *kūšk*).[31] Durand-Guédy

26 Ibn al-Faqīh, *Buldān*: p. 326. Several readings of the text disagree as to how many thousands of *jarībs* there were. The Iranian *jarīb* varied locally from 400 to 1450 square meters; see Bosworth, C.E., s.v. "Misāḥa", in *EI*², viewed 12/21/2017.

27 Naršaḫī, *Tārīḫ-i Buḫārā*: p. 48.

28 According to Dihḫudā it was used for Bukhara, Samarqand, Balkh, Nishapur and Marv. The entry on *arg*/*ark* (p. 1591) in this dictionary particularly associates this word with Sīstān. Yāqūt, *Muʿjam*: IV, p. 210 says that *quhandiz* is a word used in Khurasan and Transoxiana. However, the geographers use it for other parts of Iran. *Ḥudūd al-ʿĀlam* uses it for Qā'in in Central Iran, and Iṣṭaḫrī in his discussion of Fārs in southwest Iran says that many large and small cities such as Shiraz and Fasā have *quhandiz*-s. *Ḥudūd al-ʿĀlam min al-Mašriq ilā al-Maġrib*, ed. Manūčihr Sutūda (Tehran: Intišārāt-i Dānišgāh-i Tihrān, 1340š): p. 90; Iṣṭaḫrī, *Kitāb al-Masālik wal-mamālik*, ed. de Goeje (Leiden: Brill, 1870): p. 116.

29 Yaʿqūbī, *Taʾrīḫ*, 2 vols., ed. M.T. Houtsma (Leiden: Brill, 1883): II, p. 282.

30 Naršaḫī, *Tārīḫ-i Buḫārā*, ed. Modarres Rażavī (Tehran: Bunyād-i Farhang, 1366š): p. 61.

31 Naṭanzī, *Mirqāt*: p. 155 for *burj* and *qaṣr*. Zamaḫšarī, *Pīšraw-i Adab*: p. 114. Karamīnī, *Takmilat*: p. 530. Lane, Edward, *Arabic-English Lexicon*, 2 vols. (Cambridge, U.K.: Islamic Texts Society Trust, 1984): II, p. 486, gives *jawsaq* an extended definition, emphasizing the building's position above others as a belvedere with views. Nīšāpūrī, *Bulġa*: p. 323.

has established that *kūšk* in some instances means "tent".[32] *Ḥiṣn*, another Arabic word for fortress is translated as "a secure refuge" (*panāh* and *diž*).[33] Palace, in Arabic *īwān* (literally "high arch" or "portico"), used to create settings for enthronement, is translated as *dargāh* and *kāšāna* (which in modern Persian is a humble dwelling).[34] For the well-known Persian word for palace *kāḫ* we have *kūšak*, an Arabization of *kūšk*.[35] *Kūšk* in the meaning of "palace pavilion" is used quite frequently. Nāṣir-i Ḫusraw (d. circa 470/1077) in his Persian account of his travels explains that the palace (*qaṣr*) of the Fatimid caliphs in Cairo consists of twelve *kūšk*s.[36] *Kūšk* is also well-attested in the meaning of "garden pavilion". It is uncertain which meaning Nāṣir-i Ḫusraw intends here, "tent" or "garden pavilion", but a castle in Cairo is more likely to consist of pavilions.

5 Mosque

For the Friday mosque, *al-masjid al-jāmiʿ*, one finds *mazgit-i ādīna*.[37] *Mazgit* (also spelled *mazkid, mizkid*, and *mazkit*) is found as early as *Ḥudūd al-ʿĀlam* and Balʿamī.[38] The courtyard (*raḥba*) of the mosque is explained in Persian as *farāḫnā-yi mazgit*.[39]

6 Market

The market (Ar. *sūq*, Pers. *bāzār*) receives more thorough coverage than the mosque. *Kārvānsarāy*, caravanserai, is translated as *tīm* in Persian, for which

---

32 Durand-Guédy, David, "The Tents of the Saljuqs", in Durand-Guédy, David (ed.), *Turko-Mongol Rulers, Cities and City Life* (Leiden: Brill, 2013): pp. 149-89, esp. 172-74.

33 Zamaḫšarī, *Pīšraw-i Adab*: p. 112. Zamaḫšarī, *Pīšraw-i Adab*: p. 113 gives a similar definition for *qalʿa*, but adds "road between two mountains", perhaps meaning "palisade". Miquel, *Géographie*: IV, p. 25 suggests that *ḥiṣn* is more than a wall (*sūr*) because it has defensive features in it.

34 Karamīnī, *Takmilat*: p. 16.

35 Asadī-Ṭūsī, *Luġat-i Furs:* p. 64, has *kāḫ* as equivalent to *kūšk* but adds *qaṣr*.

36 Nāṣir-i Ḫusraw, *Safar-nāma*, ed. Muḥammad Dabīr Siyāqī (Tehran: Zavvār, 1381š): p. 77.

37 Karamīnī, *Takmilat*: p. 101.

38 Balʿami, *Tārīḫ-i Balʿamī* (adapted translation of Ṭabarī's *History*), ed. M.T. Bahār and M.P. Gonābādī (Tehran: Zavvār, 1353š/1974), which was begun in 352/963, is the oldest extensive prose work in New Persian. See Dihḫudā, *Luġat-Nāma*: p. 18333.

39 Nīšāpūrī, *Bulġa*: p. 75. Naṭanzī, *Mirqāt*: p. 160. Dihḫudā suggests that *mazgit* has an Aramaic origin. Prof. Andras Hamori of Princeton University informs me that "masgeda" means mosque in Syriac.

we also have the Arabic *ḫān*.[40] In a more expansive explanation, Zamaḫšarī says: "the place where merchants (*bāzārgānān*) and traders (*sawdāgārān*) come together for exchange".[41] The Arabic word for "drain", *bālūʿa*, is translated as "well/hole in the midst of a caravanserai".[42] Shop (*ḥānūt* in Arabic) is translated as Persian *dukkān*.[43] Arabic *ḥujra*, (still fully usable in modern Persian), literally "room" or assigned space in the market, is translated *sarāča*, small caravanserai.[44] One dictionary explains the Arabic *maʿriḍ* (showroom) as a place for selling slaves (*barda* in Persian).[45] A "moveable shop", *dikka* (literally "bench") in Arabic is defined in Persian as "a shop at the gate of the caravanserai wherever that shop might be".[46]

7 Within the City

The center of the city is *bayḍa* in Arabic, *miyān-i šahr* in Persian.[47] The crossroads (often at the center of the city) in Persian is *čahār-sū-ī gāh*, or *murabbaʿ* (literally "quadrangular meeting of roads") in Arabic.[48] The town or city dump is referred to in several dictionaries as the place "where they throw *ḫāšāk* (refuse) or *ḫāk* (dust)".[49]

A fair number of other institutions in the city are mentioned. Somewhat surprising is the mention of the *nādī*, the Qur'anic and the modern Arabic word for "club", as meaning a "society (*anjuman*) for talking (*jā-yi ḥadīṯ guftan*)".[50] Less surprising is the presence of the *maktab*, the Arabic word for a Qur'an school, translated as *dabistān* and *dabīristān*, the modern words for secular schools.[51] Another somewhat institutionalized space is suggested by the trans-

---

40 Karamīnī, *Takmilat*: p. 189. See also Asadī-Ṭūsī, *Luġat-i Furs*: p. 186. Muqaddasī, *Aḥsan al-taqāsīm*: p. 272, in his description of Isbījāb (on the middle course of the Syr Darya) says that it has *tīmāt*.

41 Zamaḫšarī, *Pīšraw-i Adab*: p. 115 and for the word *tīm*.

42 Nīšāpūrī, *Bulġa*: p. 325.

43 Naṭanzī, *Mirqāt*: p. 154; Zamaḫšarī *Pīšraw-i Adab*: p. 115, spells this word *dūkkān*.

44 Naṭanzī, *Mirqāt*: p. 155. The Arabic word *ḥujra* is also used for rooms or cells for students in a *madrasa*.

45 Zamaḫšarī, *Pīšraw-i Adab*: p. 118.

46 Nīšāpūrī, *Bulġa*: p. 326.

47 Naṭanzī, *Mirqāt*: p. 153.

48 An equivalent already explained by Jāḥiẓ in the third/ninth century. See Dihḫudā, *Luġat-Nāma*: p. 7340.

49 Zamaḫšarī, *Pīšraw-i Adab*: p. 122.

50 Nīšāpūrī, *Bulġa*: p. 319.

51 Naṭanzī, *Mirqāt*: p. 162. Witness to the long use of these words is found in Dihḫudā, *Luġat-Nāma*: pp. 9189 and 9199-9200. In this last entry one finds a quote from Balʿamī that "I memorized this book for children in *dabīristān*".

lation of the Arabic word for a long and wide bench (*maṣṭaba*) which is translated as "place for strangers, place for beggars, place for the impoverished".[52] Some words suggest the layout of the city. Neighborhood, in Arabic *nāḥiya*, is translated as "one side of the city", *yak-sū-yi šahr*.[53] The well-known Arabic word for quarter (*ḥāra*) is translated as "the market place (*sarāygāh*), the old market (*sarāy-i kuhan*)".[54] Arabic *janāb* is translated as "around the caravanserai", *girdā gird-i sarāy*.[55] The Arabic word for "plaza", *finā'*, is translated as "opposite the market", *barābar-i sarāy*.[56] *Maydān*, the word for "square" used in both Persian and in Arabic, is mostly absent from these dictionaries, perhaps because it was so well-known in both languages.[57]

8 Local Hierarchy

A range of words point to the hierarchy of inhabitants in the city. The headman (or mayor), *ra'īs* in Arabic and as a loan word in Persian, is explained as the "superior of the city", *mihtar-i šahr*, and the Arabic term for leader, *za'īm*, is explained in the same way in Persian.[58] *Ra'īs* (with the office called *riyāsat*) has eclipsed *mihtar* in Persian usage but there are echoes of *mihtar* in early Persian literature, such as the *Garšāsp-Nāma* by Asadī-Ṭūsī, written in the fifth/eleventh century. Asadī-Ṭūsī says: *hamī mihtar-i šahr gūyad ki man tu-rā banda-am va-īn buzurg anjuman …*, or "The mayor of the city (*mihtar-i šahr*) kept on saying 'I am your servant. This great assembly …'"[59]

---

52 Zamaḫšarī, *Pīšraw-i Adab*: p. 116. This Arabic word preserves the meaning "refuge for the poor" in later Persian.

53 Naṭanzī, *Mirqāt*: p. 153. Nīšāpūrī, *Bulġa*: p. 317, gives the Arabic *ḥāfiq* as "one side (*kināra*) of the city".

54 Zamaḫšarī, *Pīšraw-i Adab*: p. 122.

55 Naṭanzī, *Mirqāt*: p. 155.

56 Ibid.

57 Nīšāpūrī, *Bulġa*: p. 318 for *maydān* gives "s.p.r.y.s" which I understand as a rendering of *siprīs* or *asbrīs* or *asprīs* meaning "place for running horses".

58 Naṭanzī, *Mirqāt*: p. 27.

59 In the chapter: "The emperor of China (*faġfūr*) becomes aware of the killing of his son", viewed online 27 June 2018: www.ganjoor.net/asadi/garshaspname/sh/120. According to Dihḫudā (*Luġat-Nāma*: p. 19325) s.v. "mihtar", *Ḥudūd* says that every tribe (*qabīla*) or group (*gurūh*) has a *mihtar*. Asadi de Tous, *Le livre de Gerchâsp*, translated by Henri Massé: II, p. 203 (Paris: Imprimerie Nationale, 1951), where the mayoral title is translated "le chef de la cité".

ʿArīf, an "expert", "master craftsman", "head person", or "master of a guild", is translated as a "chosen" or "excellent person" (*guzīda*).[60] The Arabic term *aʿyān*, notables, is translated as "respected or celebrated people", *rūy-šināsān* (literally, "those whose faces are known"). This word, more often in the form *rū-šinās(-ān)*, is well attested in Persian literature.[61] A vast literature on the internal organization of the city exists in the secondary literature: I confine myself to mentioning the appearance of such words in these medieval dictionaries.[62]

9 Town and Village

No generally agreed term for "town" seems to have emerged in pre-modern times, but the modern Persian term *qaṣaba* (a loan word from Arabic) corresponds to this meaning. In medieval usage *qaṣaba* is used to designate a provincial capital.[63] The Arabic word for village, *qarya*, is *dih* (often spelled *dīh*) in Persian. *Qarawī*, Arabic for villager, is translated into Persian as *rūstāʾī*, (countryman) or *ham-dihī* (fellow-villager).[64] One sub-category of village is *daskara*, attested in both Persian and Arabic, and sometimes translated as Persian *kalāt*, all of which are explained as habitation on a high place. Naṭanzī glosses it as "a small village which has been built in a high place".[65] Karamīnī adds to his definition of *daskara/kalāt*, "it is like a castle (*qaṣr*) around which are houses (*buyūt*) belonging to the kings".[66]

André Miquel shows that tenth and eleventh century Muslim geographers had some idea as to what was the smallest size that could be called a "city" and the largest size of "village". Ibn Ḥawqal describes Sūq Ibrāhīm in the Maghrib as "a city but small with a bath and marketplace".[67] He describes some cities

60 Naṭanzī, *Mirqāt*: p. 17. In this edition it is vocalized *ʿirrīf*, but *ʿarīf* in this meaning is well-known.

61 Dihḫudā, *Luġat-Nāma*: p. 10908; Naṭanzī, *Mirqāt*: p. 27.

62 My first article on a city deals with the negotiation in the Buyid period between the ruler and the local community about the appointment of the "head" or mayor of the city; see "Administration in Buyid Qazwin", in *Islamic Civilisation 950-1150* (Oxford: Bruno Cassirer, 1973): pp. 33-45.

63 Dozy, Reinhart, *Supplément aux dictionnaires arabes*, 2 vols. (Paris: Maisonneuve Frères, 1927): II, p. 354, *kursī al-kūra*. It is also understood as "the center of the city" (*mīyān-i šahr*) in Zamaḫšarī, *Pīšraw-i Adab*: p. 114.

64 Nīšāpūrī, *Bulġa*: p. 61.

65 Naṭanzī, *Mirqāt*: p. 154. Nīšāpūrī, *Bulġa*: p. 317.

66 Karamīnī, *Takmilat*: p. 219. This reference may be to royal fortifications in southwestern Iran.

67 Miquel, *Géographie*: IV, p. 207.

in the Middle Euphrates as "the smallest [possible] cities" (*aṣāġir al-mudun*).[68] This issue becomes confused through the differing interpretations of entitlement to a Friday mosque (*jāmiʿ*) in the different schools of law. The Hanafi attitude is that only the capital of a province is entitled to such a mosque, and Hanafism was dominant in many eastern Iranian cities such as Balkh, Bukhara and Samarqand.[69]

10 Urban and Rural

The distinction between urban and rural is fairly clear, *rūstā* being the common Persian word for cultivated countryside, which was adopted into Arabic as *rūstāq* or *rustāq*. Yāqūt gives a very clear explanation of the standard use of *rustāq* by geographers. Fars, the large province of southwest Iran, is divided into five *ustāns*. These are divided into *rustāqs*, which in turn are divided into *ṭassūj-s*, in turn composed of villages. Thus, the *ustān* of Iṣṭaḫr contains the *rustāq* of Nāʾīn, which contains the village of Niyāstāna. Yāqūt adds that in his time among the Iranians (*Furs*) *rustāq* means a place in which there are farms (*mazāriʿ*) and villages, but the term is not applied to cities like Basra and is similar in meaning to the Iraqi word *sawād* (cultivated land).[70] The Arabic word for farm (*mazraʿa*), which in the plural *mazāriʿ* is the modern Persian term for cultivated countryside, is defined in Persian as *kišt-zār*, "a cultivated area".[71]

Very general terms for "urban" exist, such as *ḥaḍarī*, translated into Persian as *šahrī*.[72] *ʿUmrān*, the "culture" of habitation, a well-known term in Ibn Ḫaldūn, is appropriately rendered into Persian as *ābādānī*.[73] Some definitions suggest that the later Persian distinction between winter house (*qišlāq*) and summer house (*yaylāq*) existed before these words were adopted from Turkish. For example, Asadī-Ṭūsī defines *kāšāna* explicitly as a "winter house".[74]

The words for "country" do not have the broad general sense of this word as we know it in modern English but instead show distinctions. *Rīf* in Arabic is

68 Ibid.: p. 207 and note.

69 Baber Johansen, in his excellent essay, "The All-Embracing Town and its Mosques", deals with Ḥanafī law on city and Friday worship, reprinted in his *Contingency in a Sacred Law* (Leiden: Brill, 1998): esp. pp. 89-106.

70 Yāqūt, *Muʿjam*: I, pp. 20-21.

71 Naṭanzī, *Mirqāt*: p. 140.

72 Karamīnī, *Takmilat*: p. 166. Dozy, *Supplément*: II, p. 298) translates *ḥaḍarī* as "bourgeois, citoyen d'une ville".

73 Ibid.: p. 476.

74 Asadī-Ṭūsī, *Luġat-i Furs*: p. 217.

explained in Persian as "the embellished green area on the edge of the plain/ desert (*dašt*)", *sabzī bar-kināra-yi dašt va-pīrāsta*.[75] (*Pīrāsta* can mean a pleasure ground outside a town.) Zamaḫšarī explains *rīf* as "cultivated land" (*zamīn-i ābādān*) and "green and watered land" (*zamīn-i sabz va āb-gāh*).[76] *Barr*, meaning "open country" in Arabic but also in some instances meaning "outside the city", is translated as *dašt* or "plain", "steppe" and "desert" in Persian.[77] *Sawād*, the Arabic word for "black" is explained as "cultivated area outside the city or surrounding the city".[78] *Mazlafa* is defined as "country between green open country (*rīf*) and the outback (*barr*)".[79]

11 Irrigation

Irrigation receives some attention in all these dictionaries. The equivalence of the general Arabic word for water channels, *qanāt*, with the Persian *kārīz* is mentioned. Curiously, Nīšāpūrī translates Arabic *al-ʿidd* (fountain, well) as "that which gives it (water) assistance like the water of a spring and irrigation canal (*kārīz*)".[80] The Arabic word for cistern, *ṣahrīj*, is translated as *ḥawż-e buzurg*,[81] a large reservoir. *Al-faqīr*, Arabic for a watercourse, is explained in Persian as "a place where water comes out of the canal (*kārīz*)".[82] Arabic *jisr* is correctly translated as "small bridge", while *qanṭara* (which often means aqueduct) is translated as "large bridge".[83]

These dictionaries offer insights into the cities and countryside of medieval Iran. They make clear that the land of both city and town could be used for commerce, habitation and even agriculture. Walls (many of which may have appeared with Abbasid decline) gained an important secondary use as points of toll collections useful to rulers who had very limited means to tax trade.

Some of the major features of most cities were the Friday (or congregational) mosque, the central market place with its caravanserais (probably

75 Karamīnī, *Takmilat*: p. 248.
76 Zamaḫšarī, *Pīšraw-i Adab*: p. 38. Dozy, *Supplément*: I, pp. 575-6, who reflects Spanish and Maghrebi Arabic, translates *rīfī* as "villageois, rustique".
77 Karamīnī, *Takmilat*: p. 35.
78 Zamaḫšarī, *Pīšraw-i Adab*: p. 112.
79 Karamīnī, *Takmilat*: p. 44.
80 Nīšāpūrī, *Bulġa*: p. 285. This is probably to be understood as an analogy: "just as spring is to an irrigation canal".
81 Naṭanzī, *Mirqāt*: p. 154.
82 Nīšāpūrī, *Bulġa*: p. 289.
83 Zamaḫšarī, *Pīšraw-i Adab*: p. 62. See also on *qanṭara*, Karamīnī, *Takmilat*: p. 549.

including places both for display and storage of goods), a large plaza often filled with small traders, a citadel for the garrison in or next to the walls, and a town dump.

There is no universal word for countryside. Deserts compose a category in themselves, to be distinguished from pasture and cultivable land, and cultivable land is further divided into several categories. It seems that generally people do not to live on farms scattered over the land: by and large they are together in villages from which they go out to the land. The comparative absence of words for meadow and pasture could be related to the urban origins of the lexicographers.

The limited but very significant evidence of hierarchy in the city tells us that while there was appointment to position from forces outside the community there were also external forces creating hierarchy within the community. Evidence from elsewhere seems to show that at least in some periods, the head of the city, the "mayor", was chosen by negotiation between city and ruler.

Bibliographical References

Asadī -Ṭūsī, *Garšāsp-nāma*, viewed online at www.ganjoor.net/asadi/garshaspname.

Asadī-Ṭūsī, *Luġat-i Furs*, ed. Fatḥ Allāh Mojtabā'ī (Tehran: Khwārazmī, 1365š).

Baevskij, Solomon I., *Persian Lexicography: Farhangs of the Eleventh to the Fifteenth Centuries* (Folkestone, U.K.: Global Oriental, 2007), translated by N. Killian, revised and updated by John R. Perry.

Balʿamī, *Tārīḫ-i Balʿamī* [adapted translation of Ṭabarī's *History*], ed. M.T. Bahār and M.P. Gonābādī (Tehran: Zavvār, 1353š/1974).

Barthold, V.V., *Turkestan down to the Mongol Invasion* (London: Luzac, 1968 [E.J.W. Gibb Memorial Series]).

Barthold, V.V., *An Historical Geography of Iran*, translated by Svat Soucek (Princeton: Princeton University Press, 1984).

Bosworth, C.E., s.v. "Misāḥa (measurement of plane surfaces), 1. In the central Islamic lands", in *EI*² VII (1993): pp. 137-8.

Dihḫudā, ʿAlī-Akbar, *Luġat-Nāma*, 14 vols. (Tehran: Tehran University Publications, 1372š).

Dozy, Reinhart, *Supplément aux dictionnaires arabes*, 2 vols. (Paris: Maisonneuve Frères, 1927).

Durand-Guédy, David, "The Tents of the Saljuqs", in Durand-Guédy, David (ed.), *Turko-Mongol Rulers, Cities and City Life* (Leiden: Brill, 2013): pp. 149-89.

Ḥudūd al-ʿĀlam min al-Mašriq ilā al-Maġrib, ed. Manūčihr Sutūda (Tehran: Intišārāt-i Dānišgāh-i Tihrān, 1340š).

Ibn al-Faqīh, *Al-Buldān*, ed. de Goeje (Leiden: Brill, 1885).

Ibn Ḥawqal, *Ṣūrat al-Arḍ*, 2 vols., ed. de Goeje (Leiden: Brill, 1873).

Ibn Manẓūr, *Lisān al-ʿArab* (Beirut: Dār Ṣādir, 1410/1990).

Iṣṭaḫrī, *Kitāb al-Masālik wal-mamālik*, ed. de Goeje (Leiden: Brill, 1870).

Johansen, Baber, *Contingency in a Sacred Law* (Leiden: Brill, 1998).

Karamīnī, ʿAlī b. Muḥammad al-Adīb, *Takmilat al-Aṣnāf*, 2 vols., ed. ʿAlī Ravāqī (Tehran: Anjumān-i Āṯār-u Mafāḫir-i Farhangī, 1385š/2006).

Lane, Edward, *Arabic-English Lexicon*, 2 vols. (Cambridge, U.K.: Islamic Texts Society Trust, 1984).

Malāti, ʿAlī Navīdī, "Āgāhi-hā-yi tāza dar bāra-yi Adīb-i Karamīnī," *Farhang-nivīsī*, XCVIII (1393š): pp. 69-76.

Miquel, André, *La géographie humaine du monde musulman*, 4 vols. (Paris: École des Hautes Études en Sciences Sociales, 1973-1988).

Mottahedeh, Roy, "Administration in Buyid Qazwin," in Richards, Donald S. (ed.), *Islamic Civilisation 950-1150* (Oxford: Bruno Cassirer, 1973): pp. 33-45.

Munzawī, ʿAlī-Naqī, *Farhangnāma-hā-yi ʿArabī ba-Fārsī* (Tehran: Tehran University Press, 1959).

Muqaddasī, *Aḥsan al-Taqāsim fī Maʿrifat al-Aqālīm*, ed. de Goeje (Leiden: Brill, 1906).

Naršaḫī, *Tārīḫ-i Buḫārā*, ed. Modarres Rażavī (Tehran: Bunyād -i Farhang, 1366š).

Nāṣir-i Ḫusraw, *Safar-nāme*, ed. Muḥammad Dabīr Siyāqī (Tehran: Zavvār, 1381š).

Naṭanzī, *al-Mirqāt*, ed. Jaʿfar Sajjādī (Tehran: Bunyād-i Farhang, 1346š).

Nīšāpūrī, Yaʿqūb Kurdī, *Kitāb al-Bulġa*, ed. Mujtabā Mīnuvī and Fīrūz Ḥarīrčī (Tehran: Bunyād-i Farhang-i Īrān, 2535).

Perry, John, s.v. "Lexicography: ii Alphabetical dictionaries", in *EIr* online, www.iranica online.org/articles/lexicography, viewed 9/27/2017.

Storey, Charles A., *Persian Literature: A Bio-bibliographical Survey*, III/1 (Leiden: Brill, 1984).

Yaʿqūbī, *Taʾrīḫ*, 2 vols., ed. M.T. Houtsma (Leiden: Brill, 1883).

Yāqūt, *Muʿjam al-Buldān*, 6 vols., ed. de Goeje (Leiden: Brill, 1866-9).

Zamaḫšarī, *Pīšraw-i Adab*, or in Arabic *Muqaddimat* (or *Muqaddamat*) *al-Adab*, ed. Muḥammad Kāẓim Imām, 2 vols. (Tehran: Intišārāt-i Dānišgāh-i Tihrān, 1342š /1963).

Biographical Note

Roy Parviz Mottahedeh is the Gurney Research Professor of History at Harvard University. He has written extensively on the history of the Middle East in the 10th and 11th centuries CE. His publications include *Loyalty and Leadership in an Early Islamic Society, The Mantle of the Prophet, and Lessons in Islamic Jurisprudence*. He has written numerous articles on the social, intellectual and political history of the Middle East from the 7th century to the present.

Index of Persons and Groups

Index of Places